Natural Killer Cells: Role in Health and Disease

Natural Killer Cells: Role in Health and Disease

Editor: Maverick Davies

MURPHY & MOORE
www.murphy-moorepublishing.com

www.murphy-moorepublishing.com

MURPHY & MOORE

Cataloging-in-Publication Data

Natural killer cells : role in health and disease / edited by Maverick Davies.
p. cm.
Includes bibliographical references and index.
ISBN 978-1-63987-382-1
1. Killer cells. 2. Killer cells--Therapeutic use. 3. Immunocompetent cells. 4. Cell-mediated cytotoxicity.
5. Cytology. I. Davies, Maverick.
QR185.8.K54 N38 2022
616.079--dc23

Murphy & Moore Publishing,
1 Rockefeller Plaza,
New York City, NY 10020, USA

ISBN 978-1-63987-382-1 (Hardback)

Contents

Preface

Natural killer cells are a type of cytotoxic lymphocytes which are a part of the innate immune system. These cells respond actively to virus infected cells, intracellular pathogens and tumor formation. Major histocompatibility complex (MHC) is presented on the infected cells surfaces. These MHCs are detected by the immune cells and cytokine release is triggered causing lysis or apoptosis of the infected cell. Natural killer cells allow a much faster immune reaction as they can recognize and kill stressed cells even in the absence of antibodies and MHC. They differentiate and mature in the lymph nodes, bone marrow, tonsils, spleen and thymus. Then, they enter into the circulation. The objective of this book is to give a general view of the role of natural killer cells in health and disease. It presents researches and studies performed by experts across the globe. The extensive content of this book provides the readers with a thorough understanding of the subject.

The researches compiled throughout the book are authentic and of high quality, combining several disciplines and from very diverse regions from around the world. Drawing on the contributions of many researchers from diverse countries, the book's objective is to provide the readers with the latest achievements in the area of research. This book will surely be a source of knowledge to all interested and researching the field.

In the end, I would like to express my deep sense of gratitude to all the authors for meeting the set deadlines in completing and submitting their research chapters. I would also like to thank the publisher for the support offered to us throughout the course of the book. Finally, I extend my sincere thanks to my family for being a constant source of inspiration and encouragement.

Editor

Kinetics of Cytotoxic Lymphocytes Reconstitution after Induction Chemotherapy in Elderly AML Patients Reveals Progressive Recovery of Normal Phenotypic and Functional Features in NK Cells

Jérôme Rey[1,2], Cyril Fauriat[2,3], Eloïse Kochbati[2,3], Florence Orlanducci[2,3], Aude Charbonnier[1], Evelyne D'Incan[1], Pascale Andre[4], François Romagne[4,5], Bernadette Barbarat[2,3], Norbert Vey[1,3] and Daniel Olive[2,3]**

[1] Département d'Hématologie, Institut Paoli-Calmettes, Marseille, France, [2] Plateforme d'Immunomonitoring en Cancérologie de Marseille, Institut Paoli-Calmettes, Marseille, France, [3] Centre de Recherche en Cancérologie de Marseille, INSERM U1068, Institut Paoli-Calmettes, Aix-Marseille Université, UM105, CNRS, UMR7258, Marseille, France, [4] Innate-Pharma, Marseille, France, [5] Mi-mAbs, Aix-Marseille Université, Marseille, France

Edited by:
Luigi Daniele Notarangelo, Harvard Medical School, USA

Reviewed by:
Vincent Vieillard, Centre National de la Recherche Scientifique (CNRS), France
Emily Mace, Baylor College of Medicine, USA

***Correspondence:**
Cyril Fauriat
cyril.fauriat@inserm.fr;
Daniel Olive
daniel.olive@inserm.fr

NK cells are defective in acute myeloid leukemia (AML) at diagnosis. Here, we studied the kinetic of expression of the major activating and inhibitory receptors of NK, CD8 T, and γδ T cells in patients undergoing chemotherapy (CT) for the treatment of AML ($n = 29$). We showed that NK cells are the main affected population at diagnosis and that expression of activating receptors is partially restored within a few weeks after CT. CD8 T cells and γδ T cells are only weakly affected at diagnosis. Killer cell immunoglobulin-like receptor expression by NK cells, but not NKG2A and CD85j, was downregulated. Interestingly, the development of NK cells appeared altered as the most immature CD56^{bright} NK cells were seriously underrepresented. Finally, we showed that NK cell functions were only partially restored 6 weeks after CT as degranulation capabilities of NK cells recovered, whereas cytokine production remained low. Our data point out NK cells as antitumor effectors peculiarly hampered by leukemic cells. This study may indicate a timeline when NK-mediated therapies or other immunotherapies could be performed, particularly for patients excluded of hematopoietic stem cell transplantation.

Keywords: NK cells, acute myeloid leukemia, chemotherapy, activating receptors, NK functions

INTRODUCTION

Immunity against cancer and in particular hematological malignancies relies on the capacity of effector immune cells to recognize and kill tumor cells and to alert other immune cells. The armed branch of the immune system responsible for tumor clearance encompasses NK cells, CD8$^+$ αβ T cells (later referred to as CD8 T cells) and γδ T cells.

Abbreviations: NK, natural killer; AML, acute myeloid leukemia; CR, complete remission; CT, chemotherapy; HDs, healthy donors; KIRs, killer cell immunoglobulin-like receptors.

Activation of CD8+ αβ T cells requires the peptidic antigen-specific triggering of the TCR, in association with costimulatory molecules of the B7 family. γδ T cells recognize phospho-antigens originating from the metabolism/mevalonate pathway of cells, which is often up-regulated in cancer cells. Activation of mature CD8 T cells and γδ T cells can be improved by several activating coreceptors such as NKG2C, NKG2D, 2B4, and DNAM-1.

NK cells belong to the innate immune system and secrete many cytokines, such as GM-CSF, IL-10, or pro-inflammatory cytokines (IFN-γ and TNF-α), and chemokines. Activation of NK cells usually requires triggering of several cooperative receptors (1, 2). Some of them are specific for NK cells (natural cytotoxicity receptors, NKp30, NKp46, and NKp44), the others can be shared with cytotoxic T cells (3). Of note, NKG2C is the activating counterpart of NKG2A, and both recognize HLA-E. Activation of NK cells is balanced by inhibitory receptors, most of them recognizing classical HLA class I, or the non-classical class I HLA-E. Killer cell immunoglobulin-like receptors (KIRs) (CD158) are specific for class I HLA-A, B, and C, whereas NKG2A, a C-type lectin, recognizes HLA-E, and CD85j (LILRB1/ILT2) has a broad reactivity of HLA-I. Early immature NK cells express high levels of CD56 ($CD56^{bright}$ NK cells), high levels of NKG2A, and do not express CD16. After an intermediate stage characterized by a dim expression of CD16, NK cells lose partially CD56 expression and start to express inhibitory or activating KIRs, while gradually losing NKG2A expression ($CD56^{dim}$ NK cells) (4). These cells express high levels of CD16. Further maturation involves the acquisition of more KIRs, loss of NKG2A, and finally acquisition of CD57 (5, 6).

Reduced NK cell functions have been shown to promote cancer (7). NK cell importance in leukemia control has been evidenced by the pioneering study of Ruggeri and was followed since by many centers (8, 9). Hence, NK cells are found profoundly altered in solid cancers, such as breast cancer, neuroblastoma, and GIST, and in hematological malignancies such as multiple myeloma or acute myeloid leukemia (AML) (10–13). These phenotypic and functional alterations strongly suggest the need for tumor cells to hamper NK cell-mediated recognition (14). Although the mechanisms behind the interaction and destruction of leukemic cells are not yet clearly defined, the high rate of relapse in AML suggests these mechanisms are altered and profit to leukemic cells that can escape from the immune system (15, 16).

The current mainstay of AML treatment is based on conventional chemotherapy (CT) using anthracyclines and cytarabine, and hematopoietic stem cell transplantation (HSCT) for patients with poor prognosis features. With this treatment, about 50–80% of patients attain complete remission (CR). However, the 2-year survival is only 15–50% due to frequent relapses (17). Although a lot of them have poor-risk disease, HSCT is often not feasible because of comorbidities. There is thus an urgent need to develop new immunotherapeutic approaches that represent alternatives to HSCT such as vaccines, monoclonal antibodies, or immunomodulatory drugs (IMIDs). However, these approaches often require a functional immune system to facilitate the clearance of tumor cells. Therefore, it is relevant to monitor the status of immune cells, since so far, little is known about the effects of CT drugs on immune cells, with very few *in vitro* studies performed on cytotoxic T cells or NK cells (18, 19).

In this study, we analyzed the phenotype of peripheral blood NK cells, γδ T cells, and CD8 αβ T cells (CD8 T cells) of elderly patients treated with CT for AML. Blood samples were collected at diagnosis, remission, and various time points following consolidation CT in order to evaluate potential alterations following CT. Our data revealed important phenotypic alterations of NK cells, contrasting with limited phenotype alteration of γδ T cells and CD8 T cells. The most immature NK cell population was absent at diagnosis and recovered slowly after CT. NK cells presented low cytolytic activity at diagnosis that recovered with time, but their capacity to produce pro-inflammatory cytokines was durably impaired. Overall, these data provide the basic knowledge required for the design of clinical trials of immunotherapies for the treatment of AML in the elderly.

PATIENTS AND METHODS

Patients

We enrolled 29 elderly patients (60–80 years old) with non-promyelocytic AML according to WHO criteria in first CR following induction CT (3 + 7 regimen). All patients have received an induction and one consolidation CT before inclusion. All patients received informed consent. The study was approved by a local ethics committee and the national institution [AFSSAPS (Agence Française de Sécurité Sanitaire des Produits de Santé), No DGS 2006/0396]. Patient peripheral NK, γδ T, and CD8 T cells were analyzed at diagnosis, the day before the second consolidation CT (W0), and every other week after treatment for 8 weeks (Figure S1 in Supplementary Material). Patient characteristics are presented in **Table 1**. All patients were in CR

TABLE 1 | Characteristics of patients.

Characteristic	*N*
Number of patients	29
Age, years	
Mean (SD)	70.17 (1.45)
Median [min–max]	70.00 [38.00–81.00]
≤65	4 (13.79)
>65	25 (86.21)
Sex, *n* (%)	
Male	19 (65.52)
Female	10 (34.48)
FAB category, *n* (%)	
M1	4 (13.79)
M2	8 (27.59)
M4	9 (31.03)
M5	4 (13.79)
M6	2 (6.90)
Unclassified	2 (6.90)
Cytogenetics, *n* (%)	
Normal	21 (72.41)
Favorable	1 (3.45)
Complex	7 (24.14)

at W0. Induction CT was as follows: daunorubicin 45 mg/m^2 D1–D3, cytarabine 100 mg/m^2 D1–D7; consolidation CT 1 is as follows: daunorubicin 45 mg/m^2 D1–D2, cytarabine 50 mg/m^2 subcutaneous twice daily D1–D5; consolidation CT 2 is as follows: idarubicin 8 mg/m^2 D1, cytarabine 50 mg/m^2 subcutaneous BID D1–D5.

Fifteen healthy donors (HD), age-matched, were used as controls and were obtained from the Etablissement Français du Sang. Median age of HD was 72.2 years [65.6–76.4] and the ratio F/M was 8/7. No major past clinical history was noticed for these donors.

Phenotypic Studies

Peripheral blood samples from HD and AML patients were processed and cryopreserved until use. After thawing, PBMCs were processed for flow cytometry experiments. The antibodies used for these experiments are listed in Table S1 in Supplementary Material. 7-AAD was used as a live/dead discrimination marker. Protocols and FACS analysis were performed according to published protocols (1).

Proliferation Assays

PBMCs were thawed up, washed twice in PBS, and incubated 20 min with 2.5 μM CellTrace Violet at 37°C. Cells were then washed twice in PBS before resuspension in RPMI containing 10% FCS, 100 UI/mL IL-2, and 10 ng/mL IL-15. After 6 days of culture, cells were harvested and prepared for flow cytometry analysis. The antibodies used for these experiments are listed in Table S1 in Supplementary Material.

Degranulation and Cytokine Production Assays

PBMCs were thawed up and incubated overnight at 37°C with RPMI 10% FCS (complete medium) alone or with complete medium containing IL-2 (100 UI/mL) + IL-15 (10 ng/mL) or IL-12 (5 ng/ml) + IL-18 (10 ng/mL). Cells were then incubated with K562 cells (ratio 1:10) at 37°C for 4 h in the presence of GolgiPlug (Life Technologies). The antibodies used for these experiments are listed in Table S1 in Supplementary Material. Functional tests with NK cells at diagnosis could not be performed because of lack of material and because of the extremely low frequency of NK cells counts at this time point.

Cytotoxicity Assays

NK cells were isolated using magnetic isolation kit (StemCell Technologies). The purity of NK cells was determined by flow cytometry and was >98%. K562 target cells were labeled with ^{51}Cr (Perkin-Elmer). After three washes, NK cell cytotoxicity against the HLA class I-deficient K562 cell line was evaluated with a standard 4-h ^{51}Cr-release assay at various effector/target ratios (10:1 and 2:1). All experiments were performed in triplicate.

NK Cell Functions

Effector functions of NK cells were assessed by flow cytometry. For target cell stimulation, 1×10^6 PBMCs were mixed with K562 (ratio 10:1) for 4 h at 37°C and 5% CO_2 as previously described (20). Unless otherwise specified, PBMCs were kept unstimulated before functional assays. When indicated, PBMCs were primed overnight with recombinant (r)IL-2 (50 UI/mL) and rIL-15 (5 ng/mL) prior to functional assays (Figure S2A in Supplementary Material). For cytokine production assays after cytokine stimulation, PBMCs were thawed, counted, and incubated for 18 h in the presence or absence of rIL-12 (5 ng/mL) and rIL-18 (20 ng/mL). Cells were then incubated with or without target cells with Golgiplug (BD Biosciences) for four additional hours, and then prepared for FACS analysis.

Statistical Analysis

Statistical analyses were performed using GraphPad Prism software. For comparison between multiple matched samples, a Friedman test with a Dunn's posttest was performed. For comparison between two independent groups, a Mann–Whitney test was performed. Statistical significance was indicated as *$P < 0.05$, **$P < 0.01$, and ***$P < 0.001$. In all graphs, data represent mean ± SEM.

RESULTS

Reconstitution of Antitumor Effectors after CT

We analyzed the peripheral blood of 29 patients diagnosed for a primary AML at diagnosis, after induction CT, and during second consolidation therapy cycle (Figure S1 in Supplementary Material and Section "Materials and Methods"). NK, γδ and CD8$^+$ T cell absolute counts were calculated and compared to those of age-matched HDs. At diagnosis, lymphocyte counts were higher than that of HDs and were restored after CR (**Figure 1A**). As expected, 2 weeks after the second cycle of consolidation CT, lymphocyte counts significantly dropped but recovered by W6. CD8$^+$ T cell numbers were higher at diagnosis, whereas NK and γδ cell numbers were comparable to control (**Figure 1B**). γδ T cell counts were back to normal after CR and remained constant. CD8 T cell counts dropped after induction and reached the levels of HD after the second consolidation, but rised again to stay high compared to HD (not significant). Similar to the other populations, NK cell counts dropped during induction and the first consolidation, eventually increased but cell counts never reached the levels of HD.

Expression of Activating Receptors in Cytotoxic Effectors during CT

We next analyzed whether expression of coactivating receptors was affected during the course of AML treatment (**Figure 2A**). NKG2C expression was found higher on NK cells from patients at diagnosis compared to HDs ($P = 0.0029$) and expression remained high during the treatment. DNAM-1 and 2B4 were found downregulated at diagnosis ($P = 0.0024$ and $P < 0.0001$, respectively). Interestingly 2B4 expression remained lower compared to HD during the study, whereas DNAM-1 expression was completely restored. NKG2D expression was unaffected during the study albeit the comparison with AML at diagnosis could not be tested. Finally, we confirmed in this study the downregulation

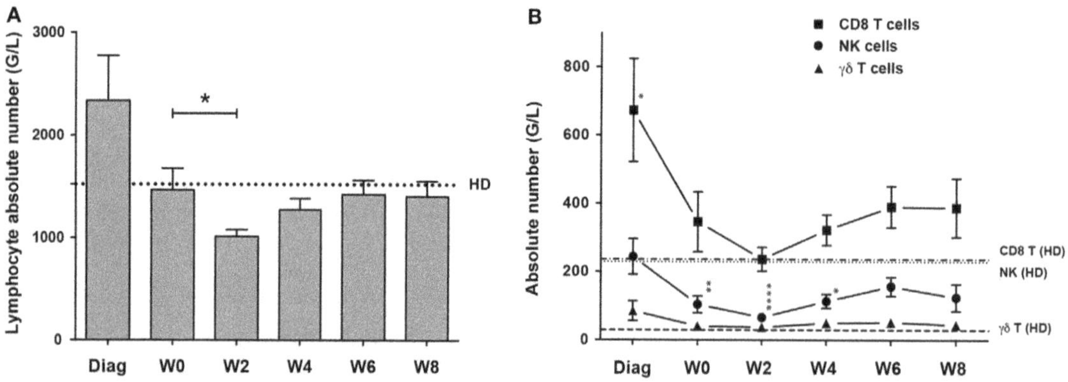

FIGURE 1 | Kinetics of cytotoxic lymphocyte reconstitution. Kinetics of NK cells (round dots), $CD8^+$ T cells (squares) and γδ T cells (triangles) reconstitution at the indicated time points. Cell counts (mean ± SEM) were measured at diagnosis of the disease, before the second consolidation (week 0, W0), at week 2, 4, 6, and 8 (W2, W4, W6, W8). **(A)** Absolute lymphocyte count. Dotted line corresponds to the mean of lymphocyte counts from 15 healthy donors (HD). **(B)** NK, γδ T, and $CD8^+$ αβ T cell absolute counts. Horizontal lines correspond to the mean of NK, γδ T, and $CD8^+$ αβ T cell counts from HD. When indicated a Kruskal–Wallis non-parametric test with Dunn's posttest was performed to compute the *P* value for the comparisons (*$P < 0.05$, **$P < 0.01$, ****$P < 0.0001$).

of NKp30 and NKp46 at the diagnosis ($P < 0.0001$ and $P = 0.0024$, respectively). NKp30 expression was partially restored at W8 (MFI = 953 ± 134 vs HD: MFI = 1,304 ± 108, $P = 0.0365$), whereas that of NKp46 was completely restored, even higher than controls (W6: MFI = 3,340 ± 306 vs HD: MFI = 1,921 ± 212, $P = 0.0017$ and W8: MFI = 3,221 ± 465 vs HD: MFI = 1,921 ± 212, $P = 0.0188$). Noteworthy, neither CD56 nor CD16 expression was altered in patients at diagnosis or during treatment (data not shown). With respect to γδ T cells and CD8 T cells, NKG2C, NKG2D, DNAM-1, and 2B4 expression was also analyzed and revealed a completely different picture. 2B4 expression was downregulated at diagnosis (γδ T cells: MFI: 1,892 ± 237 vs HD: MFI = 2,300 ± 119, $P = 0.0013$; CD8 T cells: MFI: 1,640 ± 146 vs HD: MFI = 1,827 ± 123, $P = 0.05$) and restored at the end of the study. NKG2C, NKG2D, and DNAM-1 were expressed with similar frequencies and MFI compared to controls.

Overall, with respect to activating receptors, NK cells seemed to be the main effector population affected during the treatment of AML patients. We next decided to focus the study on NK cells.

Expression of HLA Receptors in NK Cells during CT

We next analyzed the expression (frequency) of HLA receptors, i.e., NKG2A, CD85j, and KIR/CD158, on NK cells from patients under treatment (**Figure 2B**). The anti-CD158 antibodies used here do not discriminate between inhibitory and activating KIRs. Our analyses revealed a downregulation of CD158a/h and CD158b1/b2/j at diagnosis and after reaching CR (W0) compared to HD. NKG2A expression was not drastically affected during time compared to controls at all time points, although there was some increase at W6 and W8 compared to CR. In contrast, CD85j was more frequent on NK cells from patients at diagnosis and before the second cycle of consolidation CT. Expression tended to decrease over time during treatment, although remaining slightly higher than HD. Of note, frequencies of HLA receptors in γδ T cells and CD8 T cells were slightly altered during the study, with a non-significant increase of CD158b1/b2/j and CD85j (Figure S2 in Supplementary Material).

Kinetics of Early Stages of NK Cell Maturation during CT

We next sought to analyze whether NK cell early maturation was normal in patients before and after AML treatment. Combinations of CD56, CD16, and NKG2A define different steps of NK cell maturation (4, 5). CD56 expression as well as frequencies of CD16 and NKG2A positive cells out of total NK cells were comparable to HD (data not shown and **Figure 2B**). However, it is known that NKG2A is expressed by all $CD56^{bright}$ NK cells and CD16 expression defines two subsets (4). $CD56^{bright}$ NK cells are usually a rare population compared to $CD56^{dim}$ NK cells, and therefore, information about these cells may be masked when looking at the total NK level. Therefore, we analyzed the combinations of these three markers in order to identify the different maturation subsets based on CD56, CD16, and NKG2A markers. We observed a profound loss of total $CD56^{bright}$ and $CD56^{bright}$ $CD16^-$ NK cells at diagnosis [**Figures 3A,B** (left and right panels, respectively); Figure S3 in Supplementary Material]. After induction and reaching CR, total $CD56^{bright}$ were present at low frequency and continued to rise during consolidation (**Figure 3C**).

Functional Properties of NK Cells after CT

NK cells from AML patients have defective cytolytic activities (11, 14, 21). We next sought to verify whether cytotoxicity was restored during treatment and whether other functions of NK cells were affected during treatment by CT. Degranulation (CD107a) and IFN-γ or TNF-α production assays were performed after interaction with K562 cells. Degranulation capacities of total NK cells during recovery could not be compared to NK cells at diagnosis, due to the paucity of NK cells, and was thus compared to HD or total NK cells at CR. Degranulation was impaired in NK cells from patients in CR (W0) but was, at least partially, restored

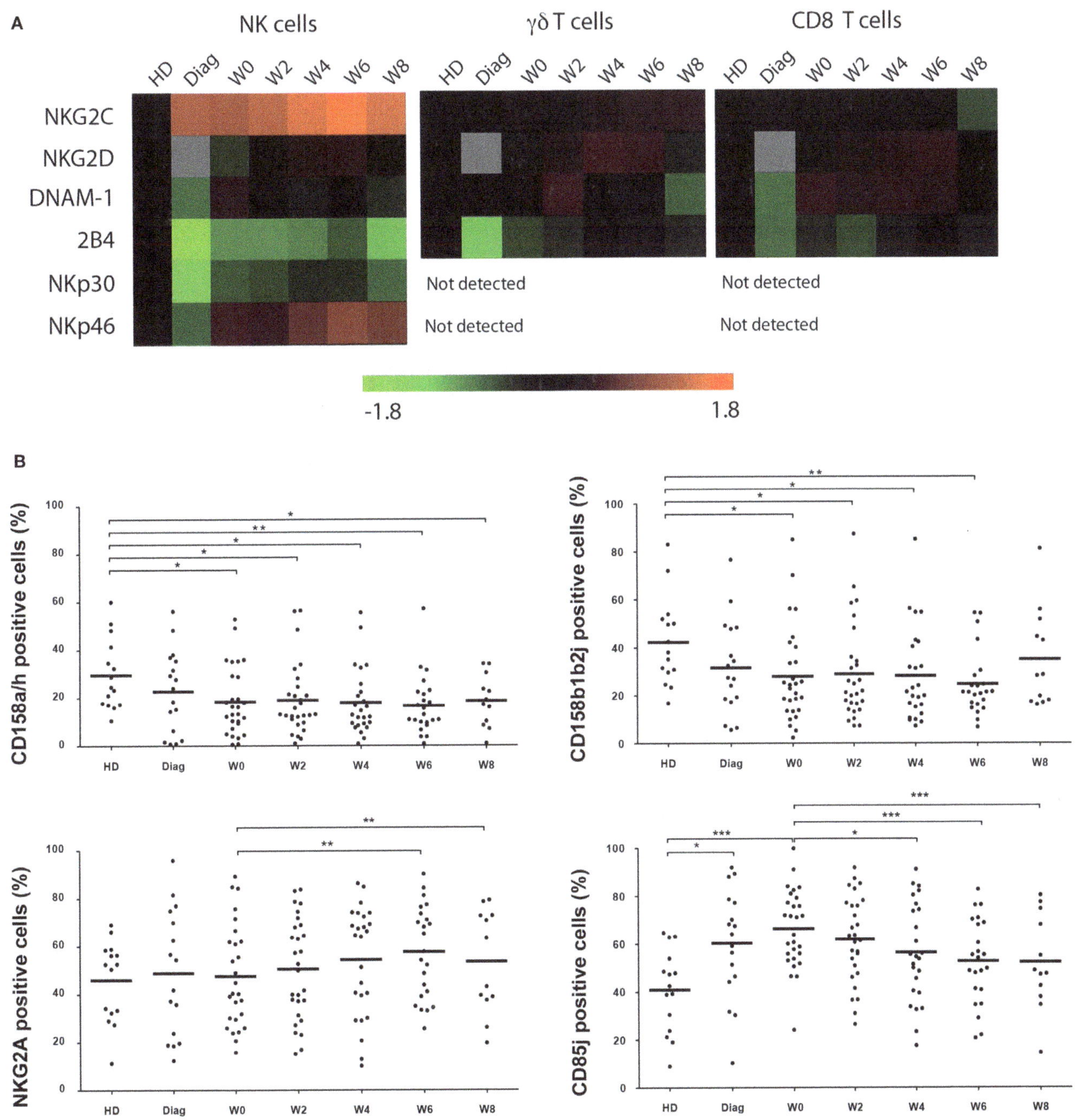

FIGURE 2 | Kinetics of activating and inhibitory receptor expression. (A). Global view of expression of activating receptors was measured by flow cytometry on NK, γδ T, and CD8+ αβ T cells at the indicated time points. MFI (unimodal expression) and percentage of positive cells (bimodal expression) were normalized and hierarchically clusterized with Tmev software. Comparison with healthy donors (HDs) was performed, and data are represented with colors reflecting higher (red) or lower (green) expression compared to HDs. Gray squares correspond to non-determined values. **(B)**. Expression of CD158a/h, CD158b1/b2/j, NKG2A, and CD85j by NK cells from HD and patients at the indicated time points. Bars represent the median expression and when indicated a Kruskal–Wallis non-parametric test with Dunn's posttest was performed to compute the *P* value for the comparisons (**P* < 0.05, ***P* < 0.01, ****P* < 0.001).

during consolidation (**Figure 4A**, black bars). Degranulation never reached the level of HD, and at W8, NK cells displayed a somewhat reduced degranulation capacity compared to W6 (non-significant). Interestingly, despite reduced degranulation at early time points, we found that perforin content in NK cells (and T cells) was higher compared to HD (Figure S4 in Supplementary Material). Similar to degranulation, IFN-γ production was altered in patients in CR, but we did not observe any restoration of the

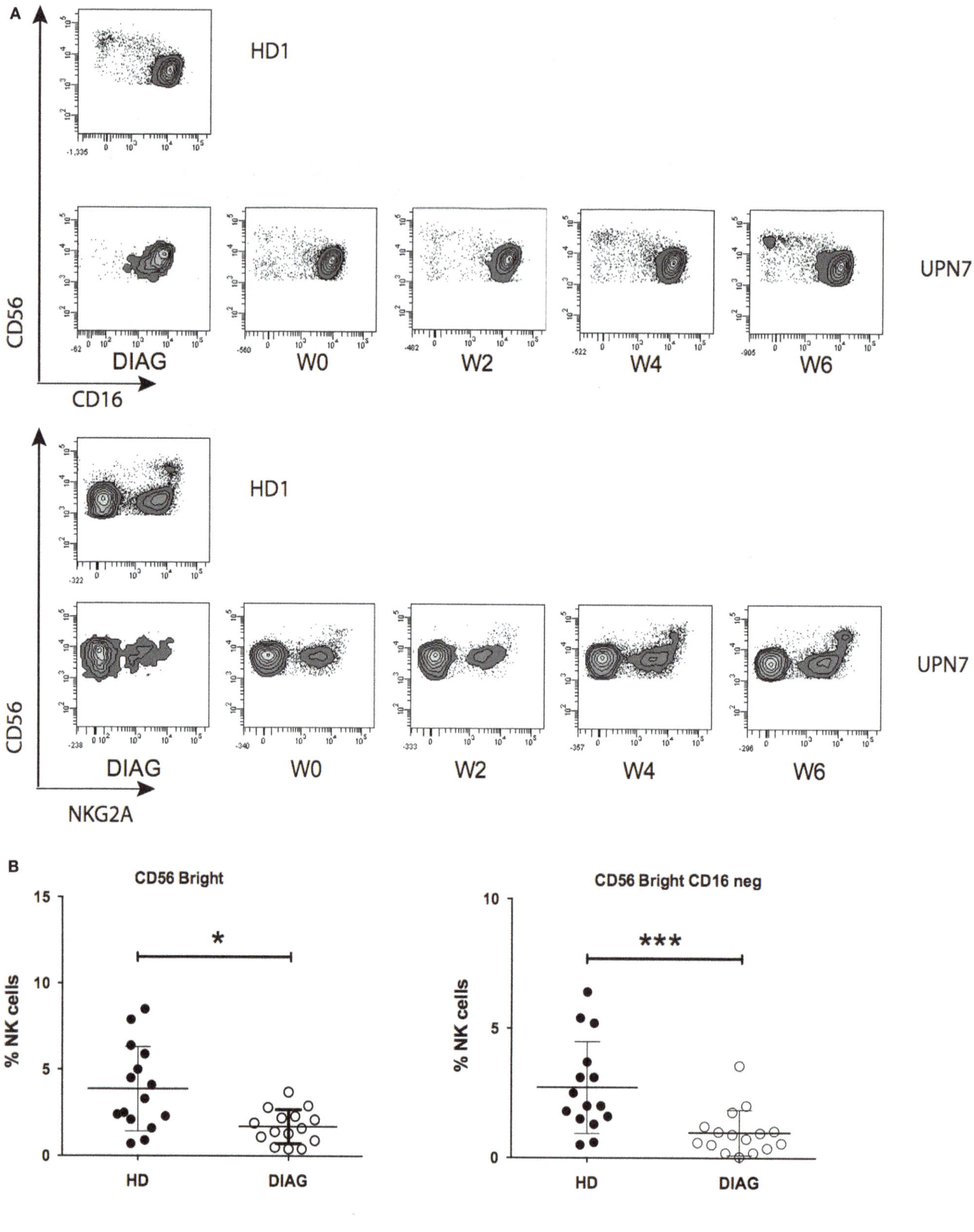

FIGURE 3 | Continued

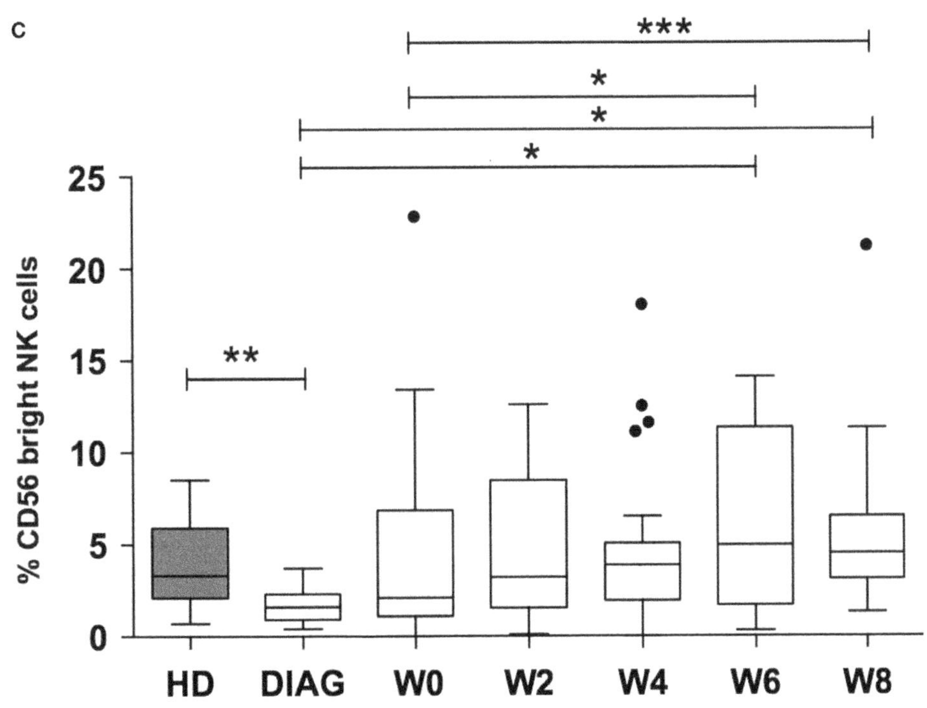

FIGURE 3 | Kinetics of immature NK cells reconstitution. (A) Example of reconstitution of CD56^{bright} NK cells in a patient following treatment. The figure displays the reapparition of CD56^{bright} NK cells expressing or not CD16 (top) or NKG2A (bottom). **(B)** Frequencies of total CD56^{bright} NK cells (left panel) or CD56^{bright} NK cells lacking CD16 expression (right panel). **(C)** Summary figure of reconstitution of the immature CD56^{bright} NK cell subset after induction and consolidation therapies. Data represent Tukey whisker boxes, with outliers represented as round dots. When indicated a Kruskal–Wallis non-parametric test with Dunn's posttest was performed to compute the *P* value for the comparisons (*$P < 0.05$, **$P < 0.01$, ***$P < 0.001$).

capacity of NK cells to produce IFN-γ (**Figure 4A**, gray bars) and TNF-α (not shown) after interaction with K562 compared to HD. In order to verify whether the whole cytokine production machinery was impaired, and not only the capacity to respond to target cells, we incubated overnight NK cells with a cocktail of IL-2/IL-15 or IL-12/IL-18 and then incubated these NK cells with K562 cells (**Figure 4B**). As CD56^{bright} NK cells and CD56^{dim} NK cells were identifiable at W0 (in contrast to diagnosis), these two subsets were gated and analyzed. After interaction with target cells, both CD56^{dim} and CD56^{bright} NK cells from patients at W0 pretreated with IL-2/IL-15 or IL-12/IL-18 displayed a significant increase in degranulation as well as IFN-γ and TNF-α production, similar to controls. Interestingly, functions of cytokine-primed CD56^{dim} NK cells at W4 (4 weeks after the second consolidation cycle) were somewhat decreased compared to functions measured at W0. Regarding CD56^{bright} NK cells, a slight decrease of CD107a and IFN-γ was observed, but generally, the effect was minor compared to CD56^{dim} NK cells.

Finally, we analyzed the capacity of NK cells to proliferate in response to IL-2 and IL-15 stimulation (**Figure 5**). By means of CellTrace Violet dilution assay, we observed that NK cells from patients at W0 and W4 had a high capacity of proliferation, similar to controls. As expected, CD56^{bright} NK cells proliferated more than CD56^{dim} NK cells as revealed by the high frequency of cells with longer history of cell division (>2 generations). Of note, NK cells from patients at diagnosis were not analyzed because of the lack of CD56^{bright} NK cells. Altogether these data revealed that NK cell effector functions that can be triggered by target cells are altered at early steps of AML treatment and that only degranulation is restored over time. By contrast, the sensitivity to exogenous cytokine stimulation was not affected by consolidation CT.

DISCUSSION

The treatment of hematological malignancies, including AML, requires the standardized and recommended use of anthracyclines and aracytine (17). Frequent relapse is the proof that the initial treatment is not sufficient and leaves the patient with a minimal residual disease that eventually overcomes immune surveillance, leading to relapse. Little is known about the status of the immune system during the early steps of CT in AML, in particular with NK cells. Here, we studied effector lymphocytes (NK, CD8, γδ T cells) frequencies during the early course of treatment of elderly AML patients. We analyzed the expression of the main activating and inhibitory receptors in these cells. Focusing on NK cells, we have analyzed their effector functions in response to target cell or cytokine stimulation. Our data revealed that NK cells are the main population affected at diagnosis or during treatment. Although NK cells expand after the induction CT, they remain probably more sensitive to the consolidation CT because the NK cell counts remained lower than that of HD, as described before (22). In contrast, CD8 T cells and γδ T cells were more frequent

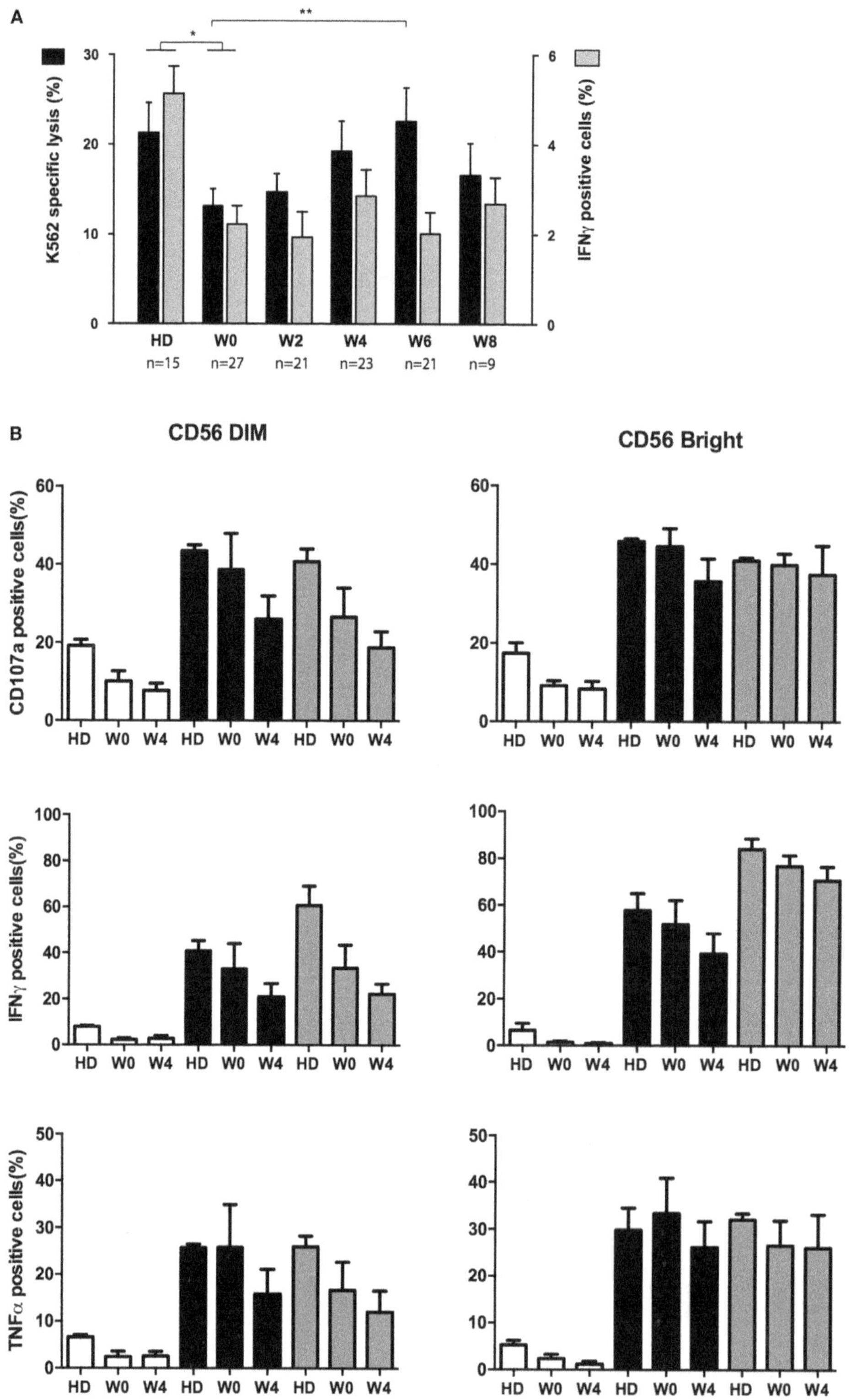

FIGURE 4 | Effector functions of NK cells following chemotherapy. **(A)** Specific lysis of target cells (black bars, left axis) and intracellular production of IFN-γ (gray bars, right axis) after interaction between NK from healthy donors (HDs) or patients at the indicated time points and K562 cells. Histogram represents mean and bars represent SEM. **(B)** Effect of cytokine priming on degranulation and production of IFN-γ and TNF-α upon interaction with K562 cells. NK cells from HDs ($n = 3$) and patients ($n = 5$) at the indicated time points were treated overnight with medium only (white bars), IL-2 + IL-15 (black bars) or IL-12 + IL-18 (gray bars). Cells were then washed and incubated for 4 h with K562 cells. Effector responses were analyzed as indicated in the Section "Materials and Methods." The figure depicts CD56^{dim} (left) and CD56^{bright} (right) identified with the flow cytometry analysis software.

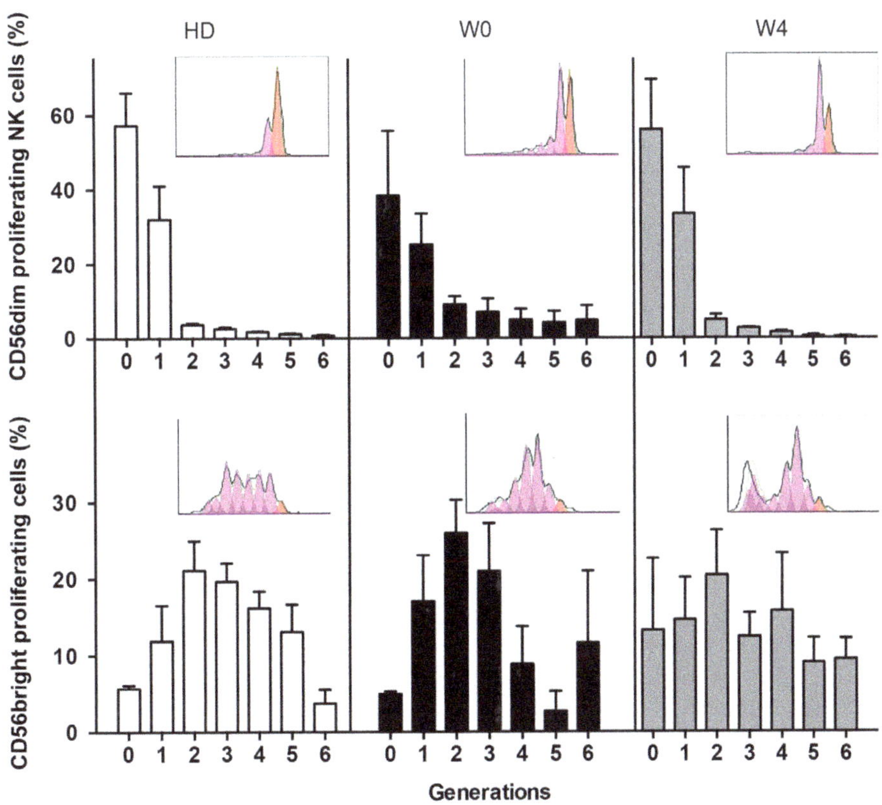

FIGURE 5 | Proliferation capacities of NK cells before and during consolidation. Proliferation capacities of NK cells during chemotherapy (CT). NK cells from healthy donors or patients before (W0) and after the first cycle of consolidation CT were isolated from peripheral blood, stained with CellTrace Violet, and cultured for 6 days in IL-2-containing medium. The figure displays CellTrace Violet dilution reflecting cell proliferation for $CD56^{dim}$ NK cells (top) and $CD56^{bright}$ NK cells (bottom). The histograms indicate the frequency of non-proliferating cells (generation 0) and proliferating cells (generations 1–6). Data represent mean ± SEM. Insets show a representative donor or patient (n = 3 and n = 5, respectively).

than in HDs during the entire study. In addition, NKG2C, DNAM-1, 2B4, and NKG2D were expressed on all T cells at similar levels compared to HD. In contrast, confirming previous studies, DNAM-1, 2B4, NKp30, and NKp46 were downmodulated at diagnosis on NK cells. The impact of treatment resulted in a partial or total restoration of NKp30 and NKp46, respectively, as previously shown (14), and a restoration of DNAM-1 expression. Surprisingly, 2B4 expression remained low during the 8 weeks of observation. The fact that the alterations mostly concerned NK cells likely reflects their involvement in the control of leukemia progression. Alternatively, the general downmodulation of NK-activating receptors shed light on the suppressive pathway that leukemia use to hamper innate immune recognition. Several mechanisms have been suggested explaining how cancer cells impair NK cell functions, such as TGFβ-1 production by tumor cells (10), or the production of histamine and reactive oxygen species by phagocytes (23). A phase III clinical trial using histamine dihydrochloride in addition to IL-2 improved the leukemia-free survival of AML patients (24). Alternatively, it was suggested that AML-induced alteration of NK cells was mediated by IL-10 (21). Our study, along with previous studies, suggests that chemoresistant remaining tumor cells may remain untargeted by defective NK cells. Hence, the weak expression of activating NK receptors is expected to result in a lack of recognition of residual tumor cells. An interesting parameter to analyze would be to compare the level of NK cell receptor expression with the incidence of relapse. In line, a recent studies identified correlations between NK cell receptor profiles with OS and relapse risk (25).

In accordance with defective receptor expression at diagnosis, degranulation, and IFN-γ production upon interaction with target cells were defective at diagnosis. Beside intrinsic insufficiency to degranulate against classical HLA-negative NK target cells, AML-NK cells, and also AML-T cells, display a reduced capacity to form solid conjugates and effective immune synapse with leukemic cells (26, 27). Albeit degranulation properties were restored over time, cytokine production remained impaired along the time course of the study. These data suggest that NK cell recovery after CT is incomplete. Interestingly, at W4 functions of NK cells after exogenous cytokine and target cell stimulation was further decreased compared to W0. Although this finding should be confirmed, it suggests that at this time unknown factors may delay the reconstitution of NK cell function. These data are similar to the reconstitution of NK functions after stem cell transplantation (SCT) where degranulation is functional but not pro-inflammatory cytokine production (1, 28). It seems that degranulation capacities of NK cells during immune reconstitution are a faster process compared to regulatory functions. These data warrant more investigation such as longer time points of observation since in SCT it takes up to 1 year to fully recover IFN-γ or TNF-α production capabilities (1). Furthermore, it would be interesting to determine whether, like for SCT, a fast cytokine production recovery correlates with a lower probability of relapse.

In addition to activating NK receptors, we also analyzed the expression of HLA receptors. Expression of classical HLA-I receptors (i.e., KIRs) was downmodulated in NK cells at the diagnosis and CR, but was restored over time, suggesting a potential break of tolerance, which is unfortunately hampered by the downregulation of activating receptors. NKG2A was not affected at diagnosis nor during treatment. Thus, NK cell activation remains tightly controlled, likely by tumor cells, keeping NK cells tolerant to remaining tumor cells.

Despite the normal expression of CD16 and NKG2A, we could not exclude alterations in the development of NK cells both at diagnosis and during the treatment, since CT induces aplasia. For instance, after HSCT, the population of immature NK cells is prominent during several months after graft (1, 29). Surprisingly, when we measured the ratio between $CD56^{bright}$ NK cells (immature) and $CD56^{dim}$ NK cells, we observed a dramatic reduction in the frequencies of $CD56^{bright}$ NK cells, and particularly in the most immature $CD56^{bright}$ $CD16^{-}$ NK cells. The fact that these cells, as well as $CD56^{bright}$ $CD16^{+}$ NK cells, recover over time suggest that leukemic cells have somehow a deleterious effect on this population. In our cohort, we observed an increase of $CD56^{bright}$ NK cell frequency between the diagnosis and W0 just before the second induction and a continued increase until the end of study. This observation suggests that consolidation CT has little impact on NK cell reconstitution. We cannot exclude that abnormal NK cell differentiation is maintained at these early time point from induction as the cohort is too small. Our observation seems to contradict the study by Dauguet et al. showing that most patients at first CR displayed an unusual high frequency of $CD56^{bright}$ NK cells (30). However, we did not analyze our patients during this very early phase of treatment (15–30 days post induction CT), and we could have missed a critical period of time where NK cells undergo rapid expansion and differentiation as a sign of homeostatic proliferation. This subversion of NK cell maturation by tumor cells has been observed in other cancer settings such as breast cancer (31). Interestingly, a study by Harlin et al. suggested that the microenvironment may alter NK cell biology and notably survival by producing reactive oxygen species, which kill $CD56^{dim}$ NK cells but somehow spare $CD56^{bright}$ NK cells (32). It would be interesting to test whether metabolism, particularly oxygen metabolism, is altered during these early phases of AML treatment. Obviously other mechanisms may be involved. In line, studies in mouse models have also evidenced a direct impact of tumor cells, including AML-like cancer, on NK cell maturation (33, 34). Importantly, the recent study by Mundy-Bosse et al. also showed that $CD56^{bright}$ NK cells are less frequent in AML patients at diagnosis (34). Thus, our study confirms the study of Mundy-Bosse et al. Interestingly, the microRNA mir-29b seems to be involved in the process of maturation blockade both in mice and humans (34). EOMES and T-BET are targets of Mir-29b, and these transcription factors are critical for early NK cell differentiation. Nonetheless, studies are required to elucidate the link between leukemic cells and mir-29b or other potential targets of leukemic cells.

A more thorough analysis with additional differentiation markers, such as combination of KIRs, NKG2A, and CD57, should be performed in order to identify the extent of NK cell maturation recovery during consolidation therapy. Addressing the mechanisms of such a defect would be difficult in humans. A blockade of NK cell differentiation would likely result in a defective production of mature NK cells, which is not the case in our cohort and according to previous studies. Oppositely, an acceleration of maturation of NK cells, while preserving a normal production of NK cells, would potentially result in an accumulation of more mature NK cells. Accordingly, in a different cohort, we have observed an accumulation of late-stage matured $CD56^{dim}$ NK cells expressing the markers CD57 and KIRs in AML patients at diagnosis (35). Interestingly, these two populations (i.e., immature $CD56^{bright}$ and the most mature $CD56^{dim}$ KIR^{+} $CD57^{+}$ NK cells) are distinguished by their differential capacities to respond to cytokines and to regulate other immune cells *via* cytokine production capacities (5, 36). The reduction of $CD56^{bright}$ NK cell pool could then participate to the immune tolerance to leukemic cells. Interestingly, this dichotomy between $CD56^{bright}$ and $CD56^{dim}$ ratio has been recently associated with additional defects in NK cells and correlates with clinical outcome of patients and may have potential consequences on the results of future NK cell-based immunotherapies (25). It seems of importance to note that at W0, $CD56^{bright}$ NK cells that had reappeared behaved in a comparable way to HDs' $CD56^{bright}$ NK cells. Hence, cytokine production, degranulation, and proliferation capacities were similar. These data suggest that these cells may have recovered fully, or simply that the blockade of maturation is paralleled but not linked to alterations of phenotype and functions.

Our data show that NK cells are present in almost normal numbers in patients with AML in CR and NK cells remain present after consolidation CT. Our data provide new insights in the optimal period to introduce NK cell-based immunotherapy, i.e., when NK cells have recovered sufficient effector functions such as cytokine production and cytotoxicity. Many NK cell-based immunotherapies have been developed over the last decades, with allogeneic or autologous NK cells (37). In 2005, haploidentical NK cells were administered in a non-transplantation setting and resulted in a substantial improvement of patient clinical outcome (38). More recently, *in vivo* targeting of NK cells with antibodies was investigated: IPH2101 is a first-in-class anti-KIR mAb that blocks inhibitory KIR–ligand interactions, leading to restoration of NK cell functions (39). A phase II trial in AML elderly patients in first CR1 (NCT01687387) is in progress and several other trials are ongoing in different cancers alone or in combination with other treatments. The future introduction of a first-in-class anti-NKG2A blocking antibody (IPH2201) will also provide a novel strategy to enhance tumor cell recognition (40). Additionally, recognition of leukemic cells by NK cells may be improved by the use of newly engineered antibodies such as CD16xCD33 bispecific antibodies (41). Finally, IMIDs such as lenalidomide represent another immunotherapy enhancing NK and T cell recognition of leukemic cells (26, 42).

Altogether, we present the first longitudinal study allowing determining which time window may be optimal to proceed to the most up-to-date NK cell-based immunotherapies for elderly patients excluded from conventional allogeneic SCT.

ETHICS STATEMENT

All subjects gave written informed consent in accordance with the Declaration of Helsinki.

AUTHOR CONTRIBUTIONS

JR designed the study, performed experiments, analyzed the data, and wrote the manuscript; CF analyzed the data, performed statistical analyses, and wrote the manuscript; EK and FO performed experiments; BB analyzed the data and performed statistical analyses; AC and ED provided samples and clinical expertise; NV, PA, and FR designed the study and wrote the manuscript; NV and DO designed and supervised the study and wrote the manuscript.

ACKNOWLEDGMENTS

The authors thank the Cancer Immunomonitoring platform of the CRCM. The authors thank for their financial support: Agence Nationale pour la Recherche contre le SIDA (ANRS), Groupement des Entreprises Française pour la Lutte contre le Cancer (GEFLUC), Association pour la Recherche contre le Cancer (ARC), Fédération Nationale des Centres de Lutte Contre le Cancer, Fondation Contre la Leucémie, Fondation pour la Recherche Médicale (FRM), Programme Hospitalier de Recherche Clinique, INSERM, and INCa.

REFERENCES

1. Pical-Izard C, Crocchiolo R, Granjeaud S, Kochbati E, Just-Landi S, Chabannon C, et al. Reconstitution of natural killer cells in HLA-matched HSCT after reduced-intensity conditioning: impact on clinical outcome. *Biol Blood Marrow Transplant* (2015) 21:429–39. doi:10.1016/j.bbmt.2014.11.681
2. Bryceson YT, March ME, Ljunggren H-G, Long EO. Activation, coactivation, and costimulation of resting human natural killer cells. *Immunol Rev* (2006) 214:73–91. doi:10.1111/j.1600-065X.2006.00457.x
3. Lanier LL. NK cell recognition. *Annu Rev Immunol* (2005) 23:225–74. doi:10.1146/annurev.immunol.23.021704.115526
4. Beziat V, Duffy D, Quoc SN, Le Garff-Tavernier M, Decocq J, Combadiere B, et al. CD56brightCD16+ NK cells: a functional intermediate stage of NK cell differentiation. *J Immunol* (2011) 186(12):6753–61. doi:10.4049/jimmunol.1100330
5. Björkström NK, Riese P, Heuts F, Andersson S, Fauriat C, Ivarsson MA, et al. Expression patterns of NKG2A, KIR, and CD57 define a process of CD56dim NK-cell differentiation uncoupled from NK-cell education. *Blood* (2010) 116:3853–64. doi:10.1182/blood-2010-04-281675
6. Lopez-Verges S, Milush JM, Pandey S, York VA, Arakawa-Hoyt J, Pircher H, et al. CD57 defines a functionally distinct population of mature NK cells in the human CD56dimCD16+ NK-cell subset. *Blood* (2010) 116:3865–74. doi:10.1182/blood-2010-04-282301
7. Imai K, Matsuyama S, Miyake S, Suga K, Nakachi K. Natural cytotoxic activity of peripheral-blood lymphocytes and cancer incidence: an 11-year follow-up study of a general population. *Lancet* (2000) 356:1795–9. doi:10.1016/S0140-6736(00)03231-1
8. Ruggeri L. Effectiveness of donor natural killer cell alloreactivity in mismatched hematopoietic transplants. *Science* (2002) 295:2097–100. doi:10.1126/science.1068440
9. Tsirigotis PD, Resnick IB, Shapira MY. The role of natural killer cells in hematopoietic stem cell transplantation. *Ann Med* (2012) 44:130–45. doi:10.3109/07853890.2011.554430
10. Mamessier E, Sylvain A, Thibult M-L, Houvenaeghel G, Jacquemier J, Castellano R, et al. Human breast cancer cells enhance self tolerance by promoting evasion from NK cell antitumor immunity. *J Clin Invest* (2011) 121:3609. doi:10.1172/JCI45816
11. Costello RT. Defective expression and function of natural killer cell-triggering receptors in patients with acute myeloid leukemia. *Blood* (2002) 99:3661–7. doi:10.1182/blood.V99.10.3661
12. Delahaye NF, Rusakiewicz S, Martins I, Ménard C, Roux S, Lyonnet L, et al. Alternatively spliced NKp30 isoforms affect the prognosis of gastrointestinal stromal tumors. *Nat Med* (2011) 17(6):700–7. doi:10.1038/nm.2366
13. Costello RT, Fauriat C, Sivori S, Marcenaro E, Olive D. NK cells: innate immunity against hematological malignancies? *Trends Immunol* (2004) 25:328–33. doi:10.1016/j.it.2004.04.005
14. Fauriat C, Just-Landi S, Mallet F, Arnoulet C, Sainty D, Olive D, et al. Deficient expression of NCR in NK cells from acute myeloid leukemia: evolution during leukemia treatment and impact of leukemia cells in NCRdull phenotype induction. *Blood* (2007) 109:323–30. doi:10.1182/blood-2005-08-027979
15. Costello RT, Fauriat C, Rey J, Gastaut J-A, Olive D. Immunobiology of haematological malignant disorders: the basis for novel immunotherapy protocols. *Lancet Oncol* (2004) 5:47–55. doi:10.1016/S1470-2045(03)01323-8
16. Farag SS, Caligiuri MA. Immunologic approaches to acute leukemia in the elderly. *Semin Hematol* (2006) 43:118–25. doi:10.1053/j.seminhematol.2006.01.006
17. Rowe JM, Tallman MS. How I treat acute myeloid leukemia. *Blood* (2010) 116:3147–56. doi:10.1182/blood-2010-05-260117
18. Markasz L, Stuber G, Vanherberghen B, Flaberg E, Olah E, Carbone E, et al. Effect of frequently used chemotherapeutic drugs on the cytotoxic activity of human natural killer cells. *Mol Cancer Ther* (2007) 6:644–54. doi:10.1158/1535-7163.MCT-06-0358
19. Markasz L, Skribek H, Uhlin M, Otvos R, Flaberg E, Eksborg S, et al. Effect of frequently used chemotherapeutic drugs on cytotoxic activity of human cytotoxic T-lymphocytes. *J Immunother* (2008) 31:283–93. doi:10.1097/CJI.0b013e3181628b76
20. Bryceson YT, Fauriat C, Nunes JM, Wood SM, Björkström NK, Long EO, et al. Functional analysis of human NK cells by flow cytometry. *Methods Mol Biol* (2010) 612:335–52. doi:10.1007/978-1-60761-362-6_23
21. Stringaris K, Sekine T, Khoder A, Alsuliman A, Razzaghi B, Sargeant R, et al. Leukemia-induced phenotypic and functional defects in natural killer cells predict failure to achieve remission in acute myeloid leukemia. *Haematologica* (2014) 99:836–47. doi:10.3324/haematol.2013.087536
22. Lichtenegger FS, Lorenz R, Gellhaus K, Hiddemann W, Beck B, Subklewe M. Impaired NK cells and increased T regulatory cell numbers during cytotoxic maintenance therapy in AML. *Leuk Res* (2014) 38:964–9. doi:10.1016/j.leukres.2014.05.014
23. Romero AI, Thorén FB, Brune M, Hellstrand K. NKp46 and NKG2D receptor expression in NK cells with CD56dim and CD56bright phenotype: regulation by histamine and reactive oxygen species. *Br J Haematol* (2005) 132:91–8. doi:10.1111/j.1365-2141.2005.05842.x
24. Brune M, Castaigne S, Catalano J, Gehlsen K, Ho AD, Hofmann W-K, et al. Improved leukemia-free survival after postconsolidation immunotherapy

with histamine dihydrochloride and interleukin-2 in acute myeloid leukemia: results of a randomized phase 3 trial. *Blood* (2006) 108:88–96. doi:10.1182/blood-2005-10-4073

25. Khaznadar Z, Boissel N, Agaugue S, Henry G, Cheok M, Vignon M, et al. Defective NK cells in acute myeloid leukemia patients at diagnosis are associated with blast transcriptional signatures of immune evasion. *J Immunol* (2015) 195:2580–90. doi:10.4049/jimmunol.1500262
26. Khaznadar Z, Henry G, Setterblad N, Agaugue S, Raffoux E, Boissel N, et al. Acute myeloid leukemia impairs natural killer cells through the formation of a deficient cytotoxic immunological synapse. *Eur J Immunol* (2014) 44(10):3068–80. doi:10.1002/eji.201444500
27. Le Dieu R, Taussig DC, Ramsay AG, Mitter R, Miraki-Moud F, Fatah R, et al. Peripheral blood T cells in acute myeloid leukemia (AML) patients at diagnosis have abnormal phenotype and genotype and form defective immune synapses with AML blasts. *Blood* (2009) 114:3909–16. doi:10.1182/blood-2009-02-206946
28. Foley B, Cooley S, Verneris MR, Curtsinger J, Luo X, Waller EK, et al. NK cell education after allogeneic transplantation: dissociation between recovery of cytokine-producing and cytotoxic functions. *Blood* (2011) 118:2784–92. doi:10.1182/blood-2011-04-347070
29. Dulphy N, Haas P, Busson M, Belhadj S. An unusual CD56brightCD16low NK cell subset dominates the early posttransplant period following HLA-matched hematopoietic stem cell transplantation. *J Immunol* (2008) 181:2227–37. doi:10.4049/jimmunol.181.3.2227
30. Dauguet N, Récher C, Demur C, Fournié J-J, Poupot M, Poupot R. Pre-eminence and persistence of immature natural killer cells in acute myeloid leukemia patients in first complete remission. *Am J Hematol* (2010) 86:209–13. doi:10.1002/ajh.21906
31. Mamessier E, Pradel LC, Thibult M-L, Drevet C, Zouine A, Jacquemier J, et al. Peripheral blood NK cells from breast cancer patients are tumor-induced composite subsets. *J Immunol* (2013) 190:2424–36. doi:10.4049/jimmunol.1200140
32. Harlin H, Hanson M, Johansson CC, Sakurai D, Poschke I, Norell H, et al. The CD16- CD56(bright) NK cell subset is resistant to reactive oxygen species produced by activated granulocytes and has higher antioxidative capacity than the CD16+ CD56(dim) subset. *J Immunol* (2007) 179:4513–9. doi:10.4049/jimmunol.179.7.4513
33. Richards JO, Chang X, Blaser BW, Caligiuri MA, Zheng P, Liu Y. Tumor growth impedes natural-killer-cell maturation in the bone marrow. *Blood* (2006) 108:246–52. doi:10.1182/blood-2005-11-4535
34. Mundy-Bosse BL, Scoville SD, Chen L, McConnell K, Mao HC, Ahmed EH, et al. MicroRNA-29b mediates altered innate immune development in acute leukemia. *J Clin Invest* (2016) 126:4404–16. doi:10.1172/JCI85413
35. Chretien A-S, Granjeaud S, Gondois-Rey F, Harbi S, Orlanducci F, Blaise D, et al. Increased NK cell maturation in patients with acute myeloid leukemia. *Front Immunol* (2015) 6:564. doi:10.3389/fimmu.2015.00564
36. Fauriat C, Long EO, Ljunggren H-G, Bryceson YT. Regulation of human NK-cell cytokine and chemokine production by target cell recognition. *Blood* (2010) 115:2167–76. doi:10.1182/blood-2009-08-238469
37. Baier C, Fino A, Sanchez C, Farnault L, Rihet P, Kahn-Perles B, et al. Natural killer cells modulation in hematological malignancies. *Front Immunol* (2013) 4:459. doi:10.3389/fimmu.2013.00459
38. Miller JS, Soignier Y, Panoskaltsis-Mortari A, McNearney SA, Yun GH, Fautsch SK, et al. Successful adoptive transfer and in vivo expansion of human haploidentical NK cells in patients with cancer. *Blood* (2005) 105:3051–7. doi:10.1182/blood-2004-07-2974
39. Vey N, Bourhis J-H, Boissel N, Bordessoule D, Prebet T, Charbonnier A, et al. A phase 1 trial of the anti-inhibitory KIR mAb IPH2101 for AML in complete remission. *Blood* (2012) 120:4317–23. doi:10.1182/blood-2012-06-437558
40. Seymour L, Tinker A, Hirte H, Wagtmann N, Dodion P. O3.2 * Phase I and dose ranging, phase II studies with IPH2201, a humanized monoclonal antibody targeting HLA-E receptor CD94/NKG2A. *Ann Oncol* (2015) 26:ii3. doi:10.1093/annonc/mdv081.2
41. Gleason MK, Ross JA, Warlick ED, Lund TC, Verneris MR, Wiernik A, et al. CD16xCD33 bispecific killer cell engager (BiKE) activates NK cells against primary MDS and MDSC CD33+ targets. *Blood* (2014) 123:3016–26. doi:10.1182/blood-2013-10-533398
42. Chretien A-S, Le Roy A, Vey N, Prebet T, Blaise D, Fauriat C, et al. Cancer-induced alterations of NK-mediated target recognition: current and investigational pharmacological strategies aiming at restoring NK-mediated anti-tumor activity. *Front Immunol* (2014) 5:122. doi:10.3389/fimmu.2014.00122

2

Human Circulating and Tissue-Resident CD56^{bright} Natural Killer Cell Populations

Janine E. Melsen, Gertjan Lugthart, Arjan C. Lankester and Marco W. Schilham*

Department of Pediatrics, Leiden University Medical Center, Leiden, Netherlands

Edited by:
Eric Vivier,
Centre d'Immunologie de Marseille-Luminy, France

Reviewed by:
Francisco Borrego,
Cruces University Hospital, Spain
Francesco Colucci,
University of Cambridge, UK

***Correspondence:**
Janine E. Melsen
j.e.melsen@lumc.nl

Two human natural killer (NK) cell subsets are usually distinguished, displaying the CD56^{dim}CD16^{+} and the CD56^{bright}CD16$^{-/+}$ phenotype. This distinction is based on NK cells present in blood, where the CD56^{dim} NK cells predominate. However, CD56^{bright} NK cells outnumber CD56^{dim} NK cells in the human body due to the fact that they are predominant in peripheral and lymphoid tissues. Interestingly, within the total CD56^{bright} NK cell compartment, a major phenotypical and functional diversity is observed, as demonstrated by the discovery of tissue-resident CD56^{bright} NK cells in the uterus, liver, and lymphoid tissues. Uterus-resident CD56^{bright} NK cells express CD49a while the liver- and lymphoid tissue-resident CD56^{bright} NK cells are characterized by co-expression of CD69 and CXCR6. Tissue-resident CD56^{bright} NK cells have a low natural cytotoxicity and produce little interferon-γ upon monokine stimulation. Their distribution and specific phenotype suggest that the tissue-resident CD56^{bright} NK cells exert tissue-specific functions. In this review, we examine the CD56^{bright} NK cell diversity by discussing the distribution, phenotype, and function of circulating and tissue-resident CD56^{bright} NK cells. In addition, we address the ongoing debate concerning the developmental relationship between circulating CD56^{bright} and CD56^{dim} NK cells and speculate on the position of tissue-resident CD56^{bright} NK cells. We conclude that distinguishing tissue-resident CD56^{bright} NK cells from circulating CD56^{bright} NK cells is a prerequisite for the better understanding of the specific role of CD56^{bright} NK cells in the complex process of human immune regulation.

Keywords: CD56^{bright} NK cell populations, tissue resident, lymphoid tissues, liver, uterus, NK cell development

INTRODUCTION

Since the discovery of natural killer (NK) cells in 1975 (1, 2), major advances were made in deciphering the role of NK cells in health and disease. It is currently accepted that NK cells are not just "killers" that lyse infected or transformed cells but can also play an important role in modulation of immune responses due to the secretion of immunoregulatory cytokines (e.g., IFN-γ and TNF-α) and chemokines (e.g., CCL3 and CCL4). Based on this cytokine secretion profile, NK cells are classified into group 1 of the large family of innate lymphoid cells (ILCs). Developmentally, NK cells are not related to the other (non-cytotoxic) ILCs, and can be distinguished from the remaining

---

Abbreviations: DNAM1, DNAX accessory molecule 1; EOMES, eomesodermin; FGFR1, fibroblast growth factor receptor 1; HEVs, high endothelial venules; HSCT, hematopoietic stem cell transplantation; ILC, innate lymphoid cell; KIR, killer-cell immunoglobulin-like receptor; ltNK cells, lymphoid tissue natural killer cells; MCM4, minichromosome maintenance complex 4; S1PR1, sphingosine-1-phospate receptor 1.

ILCs by the expression of the transcription factor Eomesodermin (EOMES) and the cytolytic protein perforin (3, 4).

In humans, two conventional NK cell subsets have been phenotypically defined based on CD56 and CD16 (FCRγIII) surface expression: $CD56^{bright}CD16^{-/+}$ and $CD56^{dim}CD16^{+}$. While the function of CD56 [neural cell adhesion molecule (NCAM)] on NK cells is not completely understood yet, CD16 can mediate antibody-dependent cellular cytotoxicity (5). Since most research in human NK cell biology is based on peripheral blood, the herein predominant $CD56^{dim}$ NK cell population is most extensively investigated. Based on circulating NK cells, $CD56^{bright}$ and $CD56^{dim}$ NK cells have usually been described as two functionally distinct subsets, cytokine producing and cytolytic, respectively. However, several observations challenge this strict difference, as that both subsets can be cytotoxic or produce cytokines, after appropriate *in vitro* stimulation. Upon target cell recognition, resting $CD56^{dim}$ NK cells are highly cytotoxic, but can produce cytokines as well (6–8). In contrast, $CD56^{bright}$ NK cells require monokine activation (combinations of IL2/IL12/IL15/IL18) to acquire cytolytic activity and produce cytokines (6, 9–11).

Although the $CD56^{dim}$ NK cells predominate in blood, the $CD56^{bright}$ NK cells are far more abundant in the human body due to their enrichment in lymphoid and non-lymphoid tissues (12–18). In addition, $CD56^{bright}$ NK cells comprise the major NK cell population in inflamed and cancer tissues (12, 14, 19). Recently, tissue-resident $CD56^{bright}$ NK cells were identified in liver, uterus, and lymphoid tissues, which points toward a tissue-specific function of $CD56^{bright}$ NK cells (13, 15–17, 20–22). In order to understand the NK cell diversity, it is essential to focus on how $CD56^{bright}$ NK cells develop, distribute, and acquire or alter their phenotype and function specifically in a particular organ. The first four developmental stages (i.e., from hematopoietic stem cell to $CD56^{bright}$ NK cell) were already reviewed extensively elsewhere and will not be discussed here (23, 24). This review attempts to improve the understanding of human circulating and tissue-resident $CD56^{bright}$ NK cells by reappraising their distribution and developmental, functional, and phenotypical characteristics. In addition, we will address to the developmental relationship between $CD56^{bright}$ (stage 4) and $CD56^{dim}$ NK cells (stage 5) and speculate on the position of tissue-resident $CD56^{bright}$ NK cells within the NK cell developmental pathway.

DISTRIBUTION, PHENOTYPE AND FUNCTION

$CD56^{bright}$ NK cells are widely distributed throughout the human body. When compared with blood, $CD56^{bright}$ NK cells are enriched in most human tissues. They represent the majority of NK cells in lymph nodes, tonsil, stomach, gut, liver, uterus, adrenal gland, and visceral adipose tissue (12–18). Although $CD56^{bright}$ NK cells seem to be outnumbered by $CD56^{dim}$ NK cells in lung, kidney, mammillary tissue, bone marrow and spleen, this is probably a reflection of the high blood perfusion of these organs (12, 13, 18, 25). Most knowledge on the phenotype and function of $CD56^{bright}$ NK cells is derived from blood, but it is important to realize that unique subsets of tissue-resident $CD56^{bright}$ NK cells have been described in lymphoid tissues, liver and uterus (13, 15, 22, 26). Conceivably, more organs contain tissue-resident $CD56^{bright}$ NK cell populations. To the best of our knowledge, no tissue-resident $CD56^{dim}$ NK cells have been described to date. Although residency is often used as a term for organ-infiltrating NK cells, it is generally not discussed whether these NK cells are just trafficking through the organ, or truly tissue resident. In this review, we only apply the term "resident" if there is substantial evidence, which allows to distinguish the tissue-resident $CD56^{bright}$ NK cells from circulating $CD56^{bright}$ NK cells. The lack of CD56 expression on murine NK cells hampers the one to one comparison of $CD56^{bright}$ NK cells to their murine counterpart. Due to limitations in obtaining human tissue samples, important findings in mice will be included in this review to cover the lack of human data.

Hallmarks of Tissue-Resident $CD56^{bright}$ NK Cells

In order to be retained within the tissue, tissue-resident $CD56^{bright}$ NK cells should possess characteristics, which prevent egress from the tissue. One of the mechanisms involved in residency is attributed to CD69, which is absent from blood-derived NK cells (13). Originally, CD69 was identified as an early activation marker, but today CD69 is known to be associated with tissue residency by suppressing sphingosine-1-phospate receptor 1 (S1PR1) surface expression (27–29). Although initially identified in the context of T- and B-cell migration, S1PRs have also been proposed to mediate the egress of NK cells from tissues into blood and lymph in mice, driven by a S1P gradient (30–35). It has not been confirmed whether S1PRs are expressed as protein on the cell surface of human NK cell subsets in blood and tissues. At transcriptional level, however, both S1PR1 and S1PR5 are expressed in circulating human NK cells, with the latter being selective for $CD56^{dim}$ NK cells (33, 36, 37). In contrast to S1PR1, S1PR5 is not inhibited by CD69 (32). Another potential mechanism for tissue homing and/or residency is the engagement of chemokine receptors. For instance, CXCR6 and CCR5 are both highly expressed on tissue-resident $CD56^{bright}$ NK cells in lymphoid tissues and liver, but have a low expression on blood-derived $CD56^{bright}$ NK cells, which instead express CCR7 (13, 15, 38). A third mechanism of tissue retention is driven by the expression or absence of adhesion molecules. For instance, the integrin CD49a is highly expressed on uterine $CD56^{bright}$ NK cells, but absent from blood NK cells (20). Furthermore, tissue-resident $CD56^{bright}$ NK cells lack CD62L (L-selectin), which is like CCR7 involved in recruitment of circulating NK cells to lymphoid tissues *via* high endothelial venules (HEVs) (15, 17, 18, 38, 39). Altogether, based on the expression of CD69, chemokine receptors, and adhesion molecules, tissue-resident $CD56^{bright}$ NK cells can be distinguished from circulating $CD56^{bright}$ NK cells. In addition to the phenotypical differences, tissue-resident $CD56^{bright}$ NK cells are functionally different from their circulating counterparts as will be discussed in the next sections.

Lymphoid Tissues

Lymph Node

Lymph nodes contain 40% of the lymphocytes in the human body, of which 2–5% consist of NK cells (**Figure 1A**) (12, 13, 18, 40).

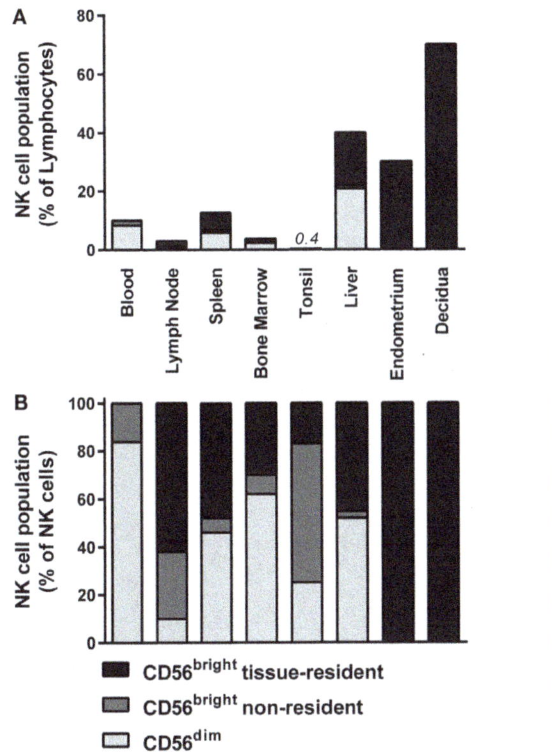

FIGURE 1 | Distribution of NK cell populations in blood and tissues. The distribution of CD56^{dim}, non-resident CD56^{bright}, and tissue-resident CD56^{bright} NK cells is depicted as percentage of **(A)** total lymphocytes and **(B)** total NK cells within blood (13), lymph node (13, 18, 25), spleen (13, 18, 25), bone marrow (13, 18, 25), tonsil (18, 44), liver (15, 45, 46), endometrium (16, 20), and decidua (20, 47). Tissue-resident CD56^{bright} NK cells were defined as CD69$^+$CXCR6$^+$ (lymph node, spleen, and bone marrow), NKp44$^+$CD103$^+$ (tonsil), CD69$^+$CXCR6$^+$ (liver), and CD49a$^+$ (endometrium and decidua). For phenotypical details, see **Figure 2** and **Table 1**.

More than 75% of the NK cells in lymph nodes have a CD56^{bright} phenotype (**Figure 1B**) (13, 18, 25). Accumulating evidence suggests a model in which CD56^{bright} NK cells circulate from the blood to tissues, enter the lymphatic system, and eventually migrate back to the periphery *via* the efferent lymph (41, 42). The mechanisms governing the migration to and infiltration of lymphoid tissues by CD56^{bright} NK cells are mainly deduced from chemokine receptor expression on circulating CD56^{bright} NK cells. As discussed earlier, circulating CD56^{bright} NK cells express CCR7 and CD62L (38, 39). The chemokines engaging CCR7, CCL19, and CCL21 are both highly expressed in lymph nodes (12). HEVs might not be the only route for circulating CD56^{bright} NK cells to enter the lymph node. NK cells in seroma fluid, which represents an accumulation of afferent lymph, resemble circulating CD56^{bright} NK cells regarding low expression of CCR5, killer-cell immunoglobulin-like receptor (KIR), and CD16 and high expression of CCR7 and CD62L (12, 43). This suggests that circulating CD56^{bright} NK cells enter the lymph node both *via* HEVs and afferent lymph vessels.

Recently, we identified a major lymphoid tissue-resident NK cell subset in lymph node, spleen, and bone marrow based on co-expression of CD69 and CXCR6 (**Figure 2**; **Table 1**) (13). In the lymph node, lymphoid tissue NK (ltNK) cells account for 60% of all NK cells and cover the majority of the CD56^{bright} NK cell compartment (**Figure 1B**). LtNK cells display a slightly less intense CD56 and more intense NKp46 expression compared with circulating CD56^{bright} NK cells. In addition, the majority of ltNK cells is CD16$^-$, CD49a$^-$, and CD27$^+$ (13). Interestingly, most ltNK cells do not express DNAX accessory molecule 1 (DNAM1), an activating receptor which is uniformly expressed on circulating CD56^{bright} NK cells (13). The remaining CD56^{bright}CD69$^-$ NK cells closely resemble circulating CD56^{bright} NK cells, suggesting that these cells are blood-derived CD56^{bright} NK cells transiently circulating through the lymph node (13).

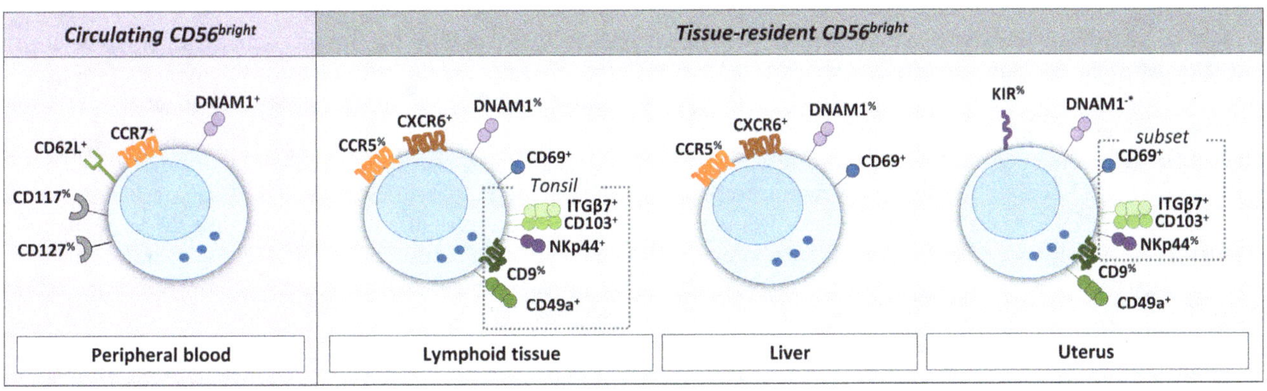

FIGURE 2 | Phenotype of circulating and tissue-resident CD56^{bright} NK cells. The cell surface markers on NK cells that are discriminative between circulating and tissue-resident NK cells in lymphoid tissue, liver, and uterus are shown (see references in text and **Table 1**). Circulating CD56^{bright} NK cells typically express the lymphoid tissue homing makers CD62L (L-selectin) and CCR7. In addition, CD117 (c-kit) and CD127 (IL-7Rα) are expressed by a fraction of circulating CD56^{bright} NK cells. Lymphoid tissue-resident NK cells express CD69 and CXCR6. Tonsil-resident NK cells (defined as NKp44$^+$CD103$^+$) express in addition ITGβ7, CD49a, and partly CD9. The majority of CD69$^+$CXCR6$^+$ liver-resident NK cells express CCR5. In contrast to circulating CD56^{bright} NK cells, only a fraction of lymphoid tissue-, tonsil-, and liver resident NK cells express DNAM1. A subset of CD49a$^+$ uterus-resident NK cells (endometrium and decidua) expresses CD69, ITGβ7, CD103, and NKp44. The reported DNAM1 expression in the uterus is contradicting in the literature and therefore indicated with –*. % indicates that only a fraction of the NK cell population is positive for the marker.

TABLE 1 | Phenotype of circulating and tissue-resident CD56^{bright} NK cells.

Reference	Blood (13, 15, 17, 38)	Lymph node, spleen, marrow (13, 18)	Tonsil (18, 44)	Liver (15, 45, 46)	Uterus (16, 17, 20, 48, 49)	
Definition		CD69$^+$ CXCR6$^+$	NKp44$^+$ CD103$^+$	CD69$^+$ CXCR6$^+$	CD49a$^+$ CD103$^-$	CD49a$^+$ CD103$^+$
CD56	++	+	+/++	+	+++	+++
CD69	–	+	+	+	–	+
Cytokine receptors						
CD117 (c-kit)	%	–	N.A.	N.A.	–	–
CD127 (IL7-Rα)	%	–	–	N.A.	–	–
Chemokine receptors						
CCR7	+	–	–	–	–	–
CCR5	–	%	N.A.	%	–	–
CXCR6	–	+	+	+	N.A.	N.A.
NK cell receptors						
DNAM1	+	%	%	%	–[a]	–[a]
KIR	–	–	%	–	%	%
NKp44	–	–	+	–	–	%
NKp46	+	++	++	++	+	+
Adhesion molecules						
CD9	–	N.A.	%	N.A.	%	+
CD49a (ITGα1)	–	–	+	–	+	+
CD62L (L-selectin)	+	–	–	–	–	–
CD103 (ITGαE)	–	–	+	N.A.	–	+
ITGβ7	N.A.	N.A.	+	N.A.	–	+

The cell surface markers which are discriminative for circulating CD56^{bright} NK cells, lymph node-, spleen-, bone marrow-, tonsil-, liver-, and uterus-resident NK cells are summarized. Uterus-resident NK cells can be subdivided based on CD103 expression.
++ and +++ indicate relatively higher levels of expression.
% indicates that only a fraction of the NK cell population is positive for the marker.
[a]Contradicting literature exist on the DNAM expression on uterine NK cells.
N.A., not assessed.

Lymphoid tissue NK cells were tested in the functional assays classically used for NK cells. The ltNK cells were less potent IFN-γ producers compared with circulating CD56^{bright} NK cells nor did they lyse K562 target cells as efficient as CD56^{dim} NK cells (13). However, the expression of EOMES and perforin distinguishes ltNK cells from the helper-ILC1 group (13). These phenotypical and functional characteristics, combined with their specific location in lymphoid tissues where immune responses are initiated and shaped, point to a distinct yet undefined role of ltNK cells (50).

Notably, without the use of tissue-resident markers, CD69 and CXCR6, ltNK cells could previously not be distinguished from circulating CD56^{bright} NK cells. Therefore, a re-examination of the function of CD56^{bright} NK cells in lymphoid tissue is necessary, in particular in the lymph node where a large population of ltNK cells co-exists next to the non-resident or circulating CD56^{bright} NK cells.

Spleen and Marrow

The spleen and bone marrow contain 14 and 10% of the total lymphocyte pool. NK cells constitute 5–20% and 4% of lymphocytes in spleen and marrow, respectively (**Figure 1A**) (13, 18, 40, 43). The CD56^{bright} and CD56^{dim} NK cells are equally distributed in the spleen, but 90% of CD56^{bright} NK cells consist of ltNK cells (**Figure 1B**) (13, 25, 40). Similar to the spleen, the bone marrow is enriched for CD56^{bright} NK cells, of which the majority consists of ltNK cells (13, 25). The phenotype of ltNK cells in spleen and marrow resembles the ltNK cell population in lymph node. As mentioned before, the non-resident CD56^{bright}CD69$^-$ and CD56^{dim} NK cells in marrow and spleen closely resemble the circulating CD56^{bright} and CD56^{dim} NK cells and are probably circulating NK cells contained in the tissue at time of isolation. Previously, the spleen has been reported to be enriched in CD27$^+$ and NKp46^{bright} NK cells, which could be a reflection of ltNK cells. These findings further illustrate the importance of using tissue-resident markers to distinguish circulating from tissue-resident CD56^{bright} NK cells (51, 52). The manner in which NK cells enter the spleen differs from lymph node entrance, because the spleen does not contain afferent lymphatic vessels or HEVs (53). In mice, NK cells enter the spleen *via* arterioles in the marginal zone, rather than *via* arterioles directly connected to the red pulp, where most NK cells reside (54). Unfortunately, there is a lack of human studies focusing on how NK cells migrate to the spleen and bone marrow.

Tonsil

In tonsil, although the CD56^{bright} subset is predominant, only 0.4% of the total lymphocytes consist of NK cells (**Figure 1A**) (18). Seventeen percent of the total NK cell population in the tonsil co-expresses CD69 and CXCR6 (**Figure 1B**) (44). In contrast to the ltNK cells in lymph node, marrow, and spleen, these tonsil-resident NK cells also express NKp44, CD103, CD49a, Integrinβ7, and partly CD9 (**Figure 2**; **Table 1**) (44). Of note, tonsil-resident NK cells should be distinguished from NKp44$^+$ ILC3s, which are located in the mucosa surrounding the lymphoid follicles and secrete preferentially IL-22 (55, 56). Similar to circulating CD56^{bright} NK cells, the total pool of CD56^{bright} NK cells was shown to produce high levels of IFN-γ and to become cytolytic upon IL-2 and/or IL-12 stimulation (18). Tonsils do not have afferent lymph vessels but HEVs are present, which might support the trafficking of NK cells. Similarly, CCL19 and CCL21 are secreted to attract circulating CD56^{bright} NK cells, which might explain the high content of CD56^{bright} NK cells which lack a tissue-resident phenotype (57).

Liver

Hepatic NK cells comprise 40% of all hepatic lymphocytes (**Figure 1A**) (45). Recently, a major liver-resident EOMES$^+$CD56^{bright} NK cell population has been described, which comprises 45% of the hepatic NK cells and closely resembles ltNK cells phenotypically and functionally (**Figure 1B**) (15, 45, 46). Liver-resident CD56^{bright} NK cells are characterized by a simultaneous expression of CD69 and CXCR6 (46). They have a high expression of CCR5 and NKp46, and low expression of DNAM1, as indirectly concluded from phenotypical analysis on total hepatic CD56^{bright} NK cells (**Figure 2**; **Table 1**) (15, 45).

An independent report demonstrated the presence of a distinct minor liver-resident cell population characterized by CD49a expression (21). Those CD56^{bright}CD49a$^+$ cells make up 2% of the total NK cell compartment in the liver but are not present in every individual (41% of donors) (13, 15, 21). The expression of CXCR6 has not been described. However, these cells do not express EOMES, suggesting that they do not belong to the NK cell lineage (21). Due to the low prevalence of this CD49a$^+$EOMES$^-$ cell population, we can indirectly conclude that the major CD69$^+$CXCR6$^+$ liver-resident NK cell population is negative for CD49a. The IFN-γ production of liver-resident CD56^{bright} NK cells after 4-h stimulation with IL12 and IL18 was lower compared with the non-resident hepatic NK cells (46). Similar to ltNK cells, liver-resident CD56^{bright} NK cells express perforin and granzyme B at a low level, further supporting a non-cytotoxic function (46).

Several studies in mice demonstrated the existence of hapten and virus-specific hepatic NK cell memory, mediated by cells expressing CD49a and CXCR6 (21, 58, 59). In contrast, splenic CXCR6$^+$ NK cells, which potentially resemble the human ltNK cells, were not able to mediate a memory response (58). Thus, although CXCR6 expression is not restricted to the liver, only hepatic NK cells were found to mediate a memory response in mice. Nevertheless, it would be interesting to further study the memory capacities of the highly prevalent CXCR6$^+$ liver- and lymphoid tissue-resident CD56^{bright} NK cells in humans.

Uterus

The uterine mucosa is populated by EOMES$^+$CD56^{bright} NK cells (16, 17, 23). In contrast to blood, there are hardly any CD56^{dim} NK cells detectable in endometrium (no pregnancy) and decidua (pregnancy) (16, 17, 20, 60). Independent of the stage of the menstrual cycle, NK cells make up 30% of the endometrial lymphocytes (**Figure 1A**), although the absolute number of lymphocytes and NK cells increases robustly in the secretory stage (16). During early pregnancy, however, more than 70% of the lymphocytes in the uterine decidua is represented by CD56^{bright} NK cells (**Figure 1A**) (47). Phenotypically, endometrial and decidual CD56^{bright} NK cells closely resemble each other, and will be further referred to as uterine CD56^{bright} NK cells. The CD56 expression of the uterine NK cells is even more intense than their circulating CD56^{bright} counterparts (**Table 1**) (17, 61). All uterine NK cells display CD49a but not CCR5, discriminating them from the circulating, lymphoid tissue, and liver-resident NK cell populations (**Figure 2**) (16, 20). DNAM1 has been reported to be absent on uterine NK cells, although a contradicting report on this observation exists (20, 48, 62). Recently, it was shown that a fraction of uterine NK cells expresses the heterodimer CD103/ITGβ7, NKp44, as well as CD69 (20, 62). Conversely, an earlier study reported that all decidual NK cells express CD69 (17). Despite this discrepancy concerning the CD69 expression, both the CD56^{bright}CD103$^-$ and CD56^{bright}CD103$^+$ NK cells are likely to represent a tissue-resident CD56^{bright} NK cell population, as demonstrated by the expression of KIRs, CD9, and poor IFN-γ production and cytotoxicity (16, 17, 20, 49, 63, 64). Moreover, transcriptome analysis of decidual NK cells and circulating NK cells highlighted the uniqueness of the uterine NK cells (17, 48). To the best of our knowledge, the presence of chemokine receptors, such as CXCR6 and CCR5, has not been reported. Compared with circulating CD56^{bright} NK cells, decidual CD56^{bright} NK cells highly express the activating receptors NKG2C and NKG2E at RNA level; however, $<30\%$ is NKG2C$^+$ on protein level (17, 62, 65).

Initially, a suppressive function of decidual NK cells was thought to be essential to provide maternal–fetal tolerance (17, 64). However, accumulating evidence points toward a more active role of decidual CD56^{bright} NK cells in regulating placentation, as reviewed elsewhere (66). Decidual NK cells are considered to stimulate trophoblast invasion and spinal artery remodeling *via* the production of various chemokines and angiogenic factors (including angiopoietins and GM-CSF) (67–69). Mice lacking decidual NK cells exhibit abnormalities in pregnancy, including abnormal vascular remodeling of decidual arteries (70). Although the process of placentation in humans is different, specific combinations of fetal HLA-C alleles, presented by trophoblasts, and maternal KIR expression were shown to be associated with successful placentation (69, 71). The similarities between endometrial and decidual CD56^{bright} NK cells suggest that decidual CD56^{bright} NK cells are a direct reflection of endometrial CD56^{bright} NK cells in a pregnant tissue microenvironment. Taken together, the phenotypical and functional profile of the uterine CD56^{bright} NK cell compartment supports their unique functional role during pregnancy.

DEVELOPMENTAL RELATIONSHIP BETWEEN CIRCULATING CD56^{dim} AND CD56^{bright} NK CELLS

Thus far, we discussed the tissue-resident and circulating/non-resident CD56^{bright} NK cells within the tissues. Still, the origin of the different CD56^{bright} NK cell populations and their relation to the CD56^{dim} NK cell subset remains unclear. The circulating CD56^{bright} NK cells have been extensively investigated and are generally considered to be the precursors of the CD56^{dim} NK cells. In the last section of this review, we will summarize the current evidence in favor and against the linear relationship between the circulating CD56^{bright} and CD56^{dim} NK cells, and speculate on the position of tissue-resident NK cells in this developmental pathway.

Several studies provided clues about the developmental relationship between CD56^{bright} NK cells and CD56^{dim} NK cells. First, it was shown that CD56^{bright}, but not CD56^{dim} NK cells, constitutively express the high-affinity IL-2Rα (CD25) and display a high proliferative response in the presence of picomolar concentrations of IL-2 (72, 73). Because CD56^{bright} NK cells have significantly longer telomeres compared with CD56^{dim} NK cells, they have been assumed to have a shorter proliferative history (74). A commonly used marker for immaturity, the tyrosine kinase c-kit (receptor for stem cell factor, CD117) is expressed on a fraction of CD56^{bright} NK cells, but is absent on CD56^{dim} NK cells (75, 76). In addition, the recovery of CD56^{bright} NK cells in the first weeks after hematopoietic stem cell transplantation (HSCT) precedes the reconstitution of CD56^{dim} NK, a sequential occurrence potentially pointing toward a developmental relationship (74, 77). Together, these findings resulted in the hypothesis that CD56^{dim} NK cells are derived from CD56^{bright} NK cells.

Differentiation from CD56^{bright} to CD56^{dim} NK Cells *In Vitro*

In efforts to provide evidence for this hypothesis, numerous studies aimed to recapitulate the differentiation from CD56^{bright} to CD56^{dim} NK cells *in vitro*. CD56^{bright} NK cells were shown to acquire a CD56^{dim}-like phenotype upon *in vitro* activation with IL-2, IL-15, and/or co-culture with T cells. This resulted in the upregulation of CD16 and KIRs and the downregulation of IL-7Rα (CD127), CD117, CXCR3, and CCR7 (10, 74). However, the intensity of CD56 expression was not reduced on monokine-activated CD56^{bright} NK cells. The presence of fibroblast growth factor receptor 1 (FGFR1) was demonstrated to be critical for the *in vitro* differentiation of CD56^{bright} NK cells to cytotoxic CD56^{dim} NK cells in a contact-dependent manner (78). FGFR1 is a ligand for CD56 and is constitutively expressed on fibroblasts (79, 80). The high density of CD56 on CD56^{bright} NK cells may thus be of importance in the interaction with fibroblasts and differentiation toward CD56^{dim} NK cells.

Differentiation from CD56^{bright} to CD56^{dim} NK Cells *In Vivo*

The *in vivo* evaluation of the relationship between CD56^{bright} and CD56^{dim} NK cells is hampered by the lack of CD56 expression on murine NK cells. The vast majority of human CD56^{bright} NK cells displayed a reduction of CD56 expression intensity after infusion into immune-deficient mice (78). Whether these *in vivo* differentiated CD56^{dim} NK cells were phenotypically and functionally similar to human blood-derived CD56^{dim} NK cells were not addressed in this study, leaving the possibility that the bright CD56 expression is not sustained in mice lacking human fibroblasts expressing FGFR1. An alternative for murine experiments can be provided by the study of rhesus macaques. Gene tracking data in rhesus macaques transplanted with lentiviral barcoded hematopoietic stem cells demonstrated that the lineage origin of the macaque NK cell homologs of CD56^{bright} (CD56^{+}CD16^{-}) and CD56^{dim} (CD56^{-}CD16^{+}) NK cells is different (81). While the CD56^{bright} homolog was derived from the same progenitors as T-cell, B-cell, and myeloid cells, the CD56^{dim} homolog displayed a unique clonal pattern, suggesting that these cells do not develop from the CD56^{bright} population but may belong to an independent lineage (81).

In addition to mice and macaques studies, human NK cell deficiencies can provide clues about the developmental relationship between CD56^{bright} and CD56^{dim} NK cells. Mutations in the transcription factor gene *GATA2* result in the absence of CD56^{bright} NK cells while CD56^{dim} NK cells are still present (82). This observation argues against the theory that CD56^{dim} NK cells are derived from CD56^{bright} NK cells. On the other hand, humans with a partial minichromosome maintenance complex 4 (MCM4) deficiency, a molecule involved in proliferation, have reduced numbers of circulating CD56^{dim} NK cells but normal numbers of CD56^{bright} NK cells (83). This could indicate that maintenance of the CD56^{dim} NK cell subset requires proliferation, which might be dependent or independent of the CD56^{bright} NK cells. To the best of our knowledge, there are no mutations in transcription factors described, which cause a lack of CD56^{dim} NK cells while the CD56^{bright} NK cells are spared. Recently, Zeb2 was identified as the essential regulator of terminal NK cell maturation in mice and shown to be higher expressed in circulating CD56^{dim} compared with CD56^{bright} NK cells (84). Together, these studies emphasize the need for additional experimental evidence on the transcriptional regulation of human NK cell development.

Intermediate Stages Connecting CD56^{bright} to CD56^{dim} NK Cells

If CD56^{bright} and CD56^{dim} NK cells are successive stages in the NK cell developmental pathway, developmental intermediates should exist. Independent studies reported the existence of phenotypic and functional intermediate stages in the progression from CD56^{bright} to CD56^{dim} NK cells in peripheral blood of healthy donors and patients after HSCT. These studies mainly focused on CD16, CD27, or CD117, for which CD56^{bright} NK cells have a bimodal expression profile (10, 51, 85–88). Both CD16^{+} and CD27^{-} CD56^{bright} NK cells were independently suggested to represent intermediate populations based on phenotype and functional characteristics (10, 51, 88). A relative increase of CD16^{+}, CD27^{-}, and CD117^{-} CD56^{bright} NK cells was observed early after HSCT (85, 86). Notably, the expression of CD16, CD117, and CD27 on CD56^{bright} NK cells can also be modulated by cytokine-activation (10, 51, 74). Because the post-HSCT setting presents a cytokine-rich environment, the "intermediate" CD16^{+}, CD27^{-}, and CD117^{-} CD56^{bright} NK cells may represent cytokine-activated CD56^{bright} NK cells instead of developmental intermediates between CD56^{bright} and CD56^{dim} NK cells (85, 86).

In general, the potential differentiation of circulating CD56^{bright} NK cells to CD56^{dim} NK cells is characterized by loss of CD27, CD117, NKG2A, and CD62L expression and gain of CD16, KIRs, and CD57 expression. Both within the CD56^{bright} compartment (CD117↓, CD27↓, and CD16↑) as well as within the CD56^{dim} compartment (NKG2A↓, CD62L↓, KIRs↑, and CD57↑), the sequential loss and acquisition of these surface markers do not occur in a fixed order (10, 51, 87, 89, 90). Only the extremes of these markers, for instance CD117 and CD57, are mutually exclusively expressed. Together, this illustrates that uniform intermediate stages of differentiation between CD56^{bright} and CD56^{dim} NK cells cannot easily be identified.

DEVELOPMENTAL POSITION OF TISSUE-RESIDENT CD56^{bright} NK CELLS

Studies on the relationship between the NK cell populations have been based on blood-derived CD56^{bright} and CD56^{dim} NK cells. The starting point of most of these studies was a linear developmental relationship between CD56^{bright} and CD56^{dim} NK cells. However, the discovery of distinct tissue-resident CD56^{bright} NK cell populations increases the number of possible relationships between the NK cell populations. Tissue-resident NK cells could be a precursor to circulating NK cells, but the absence of the immature markers CD117 and CD127 argues against this. It also seems unlikely that tissue-resident NK cells represent a transitory population between the circulating CD56^{bright} and CD56^{dim} NK cells. Detailed transcriptome analysis comparing uterine NK cells with both circulating CD56^{bright} and CD56^{dim} NK cells highlighted major differences in gene expression profile between the three NK cell populations (17). Moreover, data from transcription factor-deficient mice suggested that circulating and tissue-resident NK cells are derived from different cell lineages (91). In our opinion, the distinct phenotype and functional signature of the tissue-resident NK cell populations, together with their absence from blood, argues in favor of the hypothesis that tissue-resident CD56^{bright} NK cells develop locally, independently of the circulating NK cells. It seems likely that the organ microenvironment is essential to induce the phenotype and retain tissue localization of tissue-resident cells. Nevertheless, additional studies are needed to shed new light on the developmental relationship between CD56^{bright}, CD56^{dim} and tissue-resident CD56^{bright} NK cell populations.

CONCLUDING REMARKS

The recent identification of tissue-resident CD56^{bright} NK cells in the lymphoid tissues, liver, and uterus led us to reappraise the characteristics of CD56^{bright} NK cell populations in the circulation and tissues. The function of tissue-resident CD56^{bright} NK cells in liver and lymphoid tissues has not been elucidated, although it is very likely that these cells, such as uterine NK cells, exert tissue-specific functions.

The existence of tissue-resident NK cells raises the question whether, and if so how, all the NK cell populations are developmentally related to each other. Based on the available evidence, we conclude that it is still possible that CD56^{dim} NK cells develop independently from the CD56^{bright} NK cells. Tissue-resident NK cells may develop from circulating CD56^{bright} NK cells, or follow their own developmental pathway. Current *in vitro* models do not sufficiently mimic the *in vivo* situation, especially considering the potentially important role of the tissue microenvironment in shaping the features of tissue-resident CD56^{bright} NK cells. As mouse models do not suffice in the evaluation of human NK cell subsets, other animal models might be exploited. Studying patients with aberrations in NK cell development due to genetic mutations could provide novel insights in the origin and development of tissue-resident NK cells. Furthermore, transcriptome analysis of non-resident and resident CD56^{bright} NK cell populations will provide tools to further decipher the role of CD56^{bright} NK cell populations in human immune responses. In conclusion, distinguishing tissue-resident CD56^{bright} NK cells from circulating CD56^{bright} NK cells is a prerequisite for the better understanding of the specific role of CD56^{bright} NK cells in the complex process of human immune regulation.

AUTHOR CONTRIBUTIONS

JM and GL wrote the manuscript. AL and MS critically revised the manuscript and approved it for publication.

REFERENCES

1. Kiessling R, Klein E, Wigzell H. 'Natural' killer cells in the mouse. I. Cytotoxic cells with specificity for mouse Moloney leukemia cells. Specificity and distribution according to genotype. *Eur J Immunol* (1975) 5:112–7. doi:10.1002/eji.1830050208
2. Herberman RB, Nunn ME, Lavrin DH. Natural cytotoxic reactivity of mouse lymphoid cells against syngeneic acid allogeneic tumors. I. Distribution of reactivity and specificity. *Int J Cancer* (1975) 16:216–29. doi:10.1002/ijc.2910160205
3. Serafini N, Vosshenrich CA, Di Santo JP. Transcriptional regulation of innate lymphoid cell fate. *Nat Rev Immunol* (2015) 15:415–28. doi:10.1038/nri3855
4. Artis D, Spits H. The biology of innate lymphoid cells. *Nature* (2015) 517:293–301. doi:10.1038/nature14189
5. Lanier LL, Ruitenberg JJ, Phillips JH. Functional and biochemical analysis of CD16 antigen on natural killer cells and granulocytes. *J Immunol* (1988) 141:3478–85.
6. Fauriat C, Long EO, Ljunggren HG, Bryceson YT. Regulation of human NK-cell cytokine and chemokine production by target cell recognition. *Blood* (2010) 115:2167–76. doi:10.1182/blood-2009-08-238469
7. De Maria A, Bozzano F, Cantoni C, Moretta L. Revisiting human natural killer cell subset function revealed cytolytic CD56dimCD16+ NK cells as rapid producers of abundant IFN-γ on activation. *Proc Natl Acad Sci U S A* (2011) 108:728–32. doi:10.1073/pnas.1012356108
8. Anfossi N, André P, Guia S, Falk CS, Roetynck S, Stewart CA, et al. Human NK cell education by inhibitory receptors for MHC class I. *Immunity* (2006) 25:331–42. doi:10.1016/j.immuni.2006.06.013
9. Cooper M, Fehniger TA, Turner SC, Chen KS, Ghaheri BA, Ghayur T, et al. Human natural killer cells: a unique innate immunoregulatory role

for the CD56bright subset. *Blood* (2001) 97:3146–51. doi:10.1182/blood.V97.10.3146
10. Béziat V, Duffy D, Quoc SN, Le Garff-Tavernier M, Decocq J, Combadière B, et al. CD56brightCD16+ NK cells: a functional intermediate stage of NK cell differentiation. *J Immunol* (2011) 186:6753–61. doi:10.4049/jimmunol.1100330
11. Takahashi E, Kuranaga N, Satoh K, Habu Y, Shinomiya N, Asano T, et al. Induction of CD16+ CD56bright NK cells with antitumour cytotoxicity not only from CD16- CD56bright NK cells but also from CD16- CD56dim NK cells. *Scand J Immunol* (2007) 65:126–38. doi:10.1111/j.1365-3083.2006.01883.x
12. Carrega P, Bonaccorsi I, Di Carlo E, Morandi B, Paul P, Rizzello V, et al. CD56brightPerforinlow noncytotoxic human NK cells are abundant in both healthy and neoplastic solid tissues and recirculate to secondary lymphoid organs via afferent lymph. *J Immunol* (2014) 192:3805–15. doi:10.4049/jimmunol.1301889
13. Lugthart G, Melsen JE, Vervat C, van Ostaijen-Ten Dam MM, Corver WE, Roelen DL, et al. Human lymphoid tissues harbor a distinct CD69+CXCR6+ NK cell population. *J Immunol* (2016) 197:78–84. doi:10.4049/jimmunol.1502603
14. Fehniger T, Cooper MA, Nuovo GJ, Cella M, Facchetti F, Colonna M, et al. CD56bright natural killer cells are present in human lymph nodes and are activated by T cell-derived IL-2: a potential new link between adaptive and innate immunity. *Blood* (2003) 101:3052–7. doi:10.1182/blood-2002-09-2876
15. Hudspeth K, Donadon M, Cimino M, Pontarini E, Tentorio P, Preti M, et al. Human liver-resident CD56bright/CD16neg NK cells are retained within hepatic sinusoids via the engagement of CCR5 and CXCR6 pathways. *J Autoimmun* (2016) 66:40–50. doi:10.1016/j.jaut.2015.08.011
16. Manaster I, Mizrahi S, Goldman-Wohl D, Sela HY, Stern-Ginossar N, Lankry D, et al. Endometrial NK cells are special immature cells that await pregnancy. *J Immunol* (2008) 181:1869–76. doi:10.4049/jimmunol.181.3.1869
17. Koopman LA, Kopcow HD, Rybalov B, Boyson JE, Orange JS, Schatz F, et al. Human decidual natural killer cells are a unique NK cell subset with immunomodulatory potential. *J Exp Med* (2003) 198:1201–12. doi:10.1084/jem.20030305
18. Ferlazzo G, Thomas D, Lin SL, Goodman K, Morandi B, Muller WA, et al. The abundant NK cells in human secondary lymphoid tissues require activation to express killer cell Ig-like receptors and become cytolytic. *J Immunol* (2004) 172:1455–62. doi:10.4049/jimmunol.172.3.1455
19. Dalbeth N, Gundle R, Davies RJ, Lee YC, McMichael AJ, Callan MF. CD56bright NK cells are enriched at inflammatory sites and can engage with monocytes in a reciprocal program of activation. *J Immunol* (2004) 173:6418–26. doi:10.4049/jimmunol.173.10.6418
20. Montaldo E, Vacca P, Chiossone L, Croxatto D, Loiacono F, Martini S, et al. Unique eomes+ NK cell subsets are present in uterus and decidua during early pregnancy. *Front Immunol* (2016) 6:646. doi:10.3389/fimmu.2015.00646
21. Marquardt N, Béziat V, Nyström S, Hengst J, Ivarsson MA, Kekäläinen E, et al. Cutting edge: identification and characterization of human intrahepatic CD49a+ NK cells. *J Immunol* (2015) 194:2467–71. doi:10.4049/jimmunol.1402756
22. Björkström NK, Ljunggren HG, Michaëlsson J. Emerging insights into natural killer cells in human peripheral tissues. *Nat Rev Immunol* (2016) 16:310–20. doi:10.1038/nri.2016.34
23. Montaldo E, Vacca P, Moretta L, Mingari MC. Development of human natural killer cells and other innate lymphoid cells. *Semin Immunol* (2014) 26:107–13. doi:10.1016/j.smim.2014.01.006
24. Yu J, Freud AG, Caligiuri MA. Location and cellular stages of natural killer cell development. *Trends Immunol* (2013) 34:573–82. doi:10.1016/j.it.2013.07.005
25. Eissens DN, Spanholtz J, van der Meer A, van Cranenbroek B, Dolstra H, Kwekkeboom J, et al. Defining early human NK cell developmental stages in primary and secondary lymphoid tissues. *PLoS One* (2012) 7:e30930. doi:10.1371/journal.pone.0030930
26. Manaster I, Mandelboim O. The unique properties of uterine NK cells. *Am J Reprod Immunol* (2010) 63:434–44. doi:10.1111/j.1600-0897.2009.00794.x
27. Bankovich AJ, Shiow LR, Cyster JG. CD69 suppresses sphingosine 1-phosophate receptor-1 (S1P1) function through interaction with membrane helix 4. *J Biol Chem* (2010) 285:22328–37. doi:10.1074/jbc.M110.123299
28. Shiow LR, Rosen DB, Brdicková N, Xu Y, An J, Lanier LL, et al. CD69 acts downstream of interferon-alpha/beta to inhibit S1P1 and lymphocyte egress from lymphoid organs. *Nature* (2006) 440:540–4. doi:10.1038/nature04606
29. Mackay LK, Braun A, Macleod BL, Collins N, Tebartz C, Bedoui S, et al. Cutting edge: CD69 interference with sphingosine-1-phosphate receptor function regulates peripheral T cell retention. *J Immunol* (2015) 194:2059–63. doi:10.4049/jimmunol.1402256
30. Rivera J, Proia RL, Olivera A. The alliance of sphingosine-1-phosphate and its receptors in immunity. *Nat Rev Immunol* (2008) 8:753–63. doi:10.1038/nri2400
31. Mayol K, Biajoux V, Marvel J, Balabanian K, Walzer T. Sequential desensitization of CXCR4 and S1P5 controls natural killer cell trafficking. *Blood* (2011) 118:4863–71. doi:10.1182/blood-2011-06-362574
32. Jenne CN, Enders A, Rivera R, Watson SR, Bankovich AJ, Pereira JP, et al. T-bet-dependent S1P5 expression in NK cells promotes egress from lymph nodes and bone marrow. *J Exp Med* (2009) 206:2469–81. doi:10.1084/jem.20090525
33. Walzer T, Chiossone L, Chaix J, Calver A, Carozzo C, Garrigue-Antar L, et al. Natural killer cell trafficking in vivo requires a dedicated sphingosine 1-phosphate receptor. *Nat Immunol* (2007) 8:1337–44. doi:10.1038/ni1523
34. Pappu R, Schwab SR, Cornelissen I, Pereira JP, Regard JB, Xu Y, et al. Promotion of lymphocyte egress into blood and lymph by distinct sources of sphingosine-1-phosphate. *Science* (2007) 316:295–8. doi:10.1126/science.1139221
35. Schwab SR, Pereira JP, Matloubian M, Xu Y, Huang Y, Cyster JG. Lymphocyte sequestration through S1P lyase inhibition and disruption of S1P gradients. *Science* (2005) 309:1735–9. doi:10.1126/science.1113640
36. Hanna J, Bechtel P, Zhai Y, Youssef F, McLachlan K, Mandelboim O, et al. Novel insights on human NK cells' immunological modalities revealed by gene expression profiling. *J Immunol* (2004) 173:6547–63. doi:10.4049/jimmunol.173.11.6547
37. Wendt K, Wilk E, Buyny S, Buer J, Schmidt RE, Jacobs R. Gene and protein characteristics reflect functional diversity of CD56dim and CD56bright NK cells. *J Leukoc Biol* (2006) 80:1529–41. doi:10.1189/jlb.0306191
38. Campbell JJ, Qin S, Unutmaz D, Soler D, Murphy KE, Hodge MR, et al. Unique subpopulations of CD56+ NK and NK-T peripheral blood lymphocytes identified by chemokine receptor expression repertoire. *J Immunol* (2001) 166:6477–82. doi:10.4049/jimmunol.166.11.6477
39. Frey M, Packianathan NB, Fehniger TA, Ross ME, Wang WC, Stewart CC, et al. Differential expression and function of L-selectin on CD56bright and CD56dim natural killer cell subsets. *J Immunol* (1998) 161:400–8.
40. Westermann J, Pabst R. Distribution of lymphocyte subsets and natural killer cells in the human body. *Clin Investig* (1992) 70:539–44.
41. Caligiuri MA. Human natural killer cells. *Blood* (2008) 112:461–9. doi:10.1182/blood-2007-09-077438
42. Carrega P, Ferlazzo G. Natural killer cell distribution and trafficking in human tissues. *Front Immunol* (2012) 3:347. doi:10.3389/fimmu.2012.00347
43. Montalto E, Mangraviti S, Costa G, Carrega P, Morandi B, Pezzino G, et al. Seroma fluid subsequent to axillary lymph node dissection for breast cancer derives from an accumulation of afferent lymph. *Immunol Lett* (2010) 131:67–72. doi:10.1016/j.imlet.2010.03.002
44. Fuchs A, Vermi W, Lee JS, Lonardi S, Gilfillan S, Newberry RD, et al. Intraepithelial type 1 innate lymphoid cells are a unique subset of IL-12- and IL-15-responsive IFN-γ-producing cells. *Immunity* (2013) 38:769–81. doi:10.1016/j.immuni.2013.02.010
45. Tanimine N, Tanaka Y, Abe T, Piao J, Chayama K, Ohdan H. Functional behavior of NKp46-positive intrahepatic natural killer cells against hepatitis C virus reinfection after liver transplantation. *Transplantation* (2016) 100:355–64. doi:10.1097/TP.0000000000001031
46. Stegmann KA, Robertson F, Hansi N, Gill U, Pallant C, Christophides T, et al. CXCR6 marks a novel subset of T-bet(lo)Eomes(hi) natural killer cells residing in human liver. *Sci Rep* (2016) 6:26157. doi:10.1038/srep26157
47. Bulmer JN, Morrison L, Longfellow M, Ritson A, Pace D. Granulated lymphocytes in human endometrium: histochemical and immunohistochemical studies. *Hum Reprod* (1991) 6:791–8.
48. Wang F, Zhou Y, Fu B, Wu Y, Zhang R, Sun R, et al. Molecular signatures and transcriptional regulatory networks of human immature decidual NK and mature peripheral NK cells. *Eur J Immunol* (2014) 44:2771–84. doi:10.1002/eji.201344183
49. Searle RF, Jones RK, Bulmer JN. Phenotypic analysis and proliferative responses of human endometrial granulated lymphocytes during the menstrual cycle. *Biol Reprod* (1999) 60:871–8. doi:10.1095/biolreprod60.4.871

50. Ferlazzo G, Pack M, Thomas D, Paludan C, Schmid D, Strowig T. Distinct roles of IL-12 and IL-15 in human natural killer cell activation by dendritic cells from secondary lymphoid organs. *Proc Natl Acad Sci U S A* (2004) 101:16606–11. doi:10.1073/pnas.0407522101
51. Vossen MT, Matmati M, Hertoghs KM, Baars PA, Gent MR, Leclercq G, et al. CD27 defines phenotypically and functionally different human NK cell subsets. *J Immunol* (2008) 180:3739–45. doi:10.4049/jimmunol.180.6.3739
52. Tomasello E, Yessaad N, Gregoire E, Hudspeth K, Luci C, Mavilio D, et al. Mapping of NKp46(+) cells in healthy human lymphoid and non-lymphoid tissues. *Front Immunol* (2012) 3:344. doi:10.3389/fimmu.2012.00344
53. Mebius RE, Kraal G. Structure and function of the spleen. *Nat Rev Immunol* (2005) 5:606–16. doi:10.1038/nri1669
54. Grégoire C, Cognet C, Chasson L, Coupet CA, Dalod M, Reboldi A, et al. Intrasplenic trafficking of natural killer cells is redirected by chemokines upon inflammation. *Eur J Immunol* (2008) 38:2076–84. doi:10.1002/eji.200838550
55. Cella M, Fuchs A, Vermi W, Facchetti F, Otero K, Lennerz JK, et al. A human natural killer cell subset provides an innate source of IL-22 for mucosal immunity. *Nature* (2008) 457:722–5. doi:10.1038/nature07537
56. Spits H, Artis D, Colonna M, Diefenbach A, Di Santo JP, Eberl G, et al. Innate lymphoid cells – a proposal for uniform nomenclature. *Nat Rev Immunol* (2013) 13:145–9. doi:10.1038/nri3365
57. Page G, Lebecque S, Miossec P. Anatomic localization of immature and mature dendritic cells in an ectopic lymphoid organ: correlation with selective chemokine expression in rheumatoid synovium. *J Immunol* (2002) 168:5333–41. doi:10.4049/jimmunol.168.10.5333
58. Paust S, Gill HS, Wang BZ, Flynn MP, Moseman EA, Senman B, et al. Critical role for the chemokine receptor CXCR6 in NK cell-mediated antigen-specific memory of haptens and viruses. *Nat Immunol* (2010) 11:1127–35. doi:10.1038/ni.1953
59. Peng H, Jiang X, Chen Y, Sojka DK, Wei H, Gao X, et al. Liver-resident NK cells confer adaptive immunity in skin-contact inflammation. *J Clin Invest* (2013) 123:1444–56. doi:10.1172/JCI66381
60. Kitaya K, Yasuda J, Nakayama T, Fushiki S, Honjo H. Effect of female sex steroids on human endometrial CD16neg CD56bright natural killer cells. *Fertil Steril* (2003) 79:730–4. doi:10.1016/S0015-0282(02)04818-5
61. Lukassen HG, Joosten I, van Cranenbroek B, van Lierop MJ, Bulten J, Braat DD, et al. Hormonal stimulation for IVF treatment positively affects the CD56bright/CD56dim NK cell ratio of the endometrium during the window of implantation. *Mol Hum Reprod* (2004) 10:513–20. doi:10.1093/molehr/gah067
62. Marlin R, Duriez M, Berkane N, de Truchis C, Madec Y, Rey-Cuille MA, et al. Dynamic shift from CD85j/ILT-2 to NKG2D NK receptor expression pattern on human decidual NK during the first trimester of pregnancy. *PLoS One* (2012) 7:e30017. doi:10.1371/journal.pone.0030017
63. Eriksson M, Meadows SK, Wira CR, Sentman CL. Unique phenotype of human uterine NK cells and their regulation by endogenous TGF-beta. *J Leukoc Biol* (2004) 76:667–75. doi:10.1189/jlb.0204090
64. Vacca P, Pietra G, Falco M, Romeo E, Bottino C, Bellora F, et al. Analysis of natural killer cells isolated from human decidua: evidence that 2B4 (CD244) functions as an inhibitory receptor and blocks NK-cell function. *Blood* (2006) 108:4078–85. doi:10.1182/blood-2006-04-017343
65. El Costa H, Casemayou A, Aguerre-Girr M, Rabot M, Berrebi A, Parant O, et al. Critical and differential roles of NKp46- and NKp30-activating receptors expressed by uterine NK cells in early pregnancy. *J Immunol* (2008) 181:3009–17. doi:10.4049/jimmunol.181.5.3009
66. Moffett A, Colucci F. Uterine NK cells: active regulators at the maternal-fetal interface. *J Clin Invest* (2014) 124:1872–9. doi:10.1172/JCI68107
67. Hanna J, Goldman-Wohl D, Hamani Y, Avraham I, Greenfield C, Natanson-Yaron S, et al. Decidual NK cells regulate key developmental processes at the human fetal-maternal interface. *Nat Med* (2006) 12:1065–74. doi:10.1038/nm1452
68. Lash GE, Schiessl B, Kirkley M, Innes BA, Cooper A, Searle RF, et al. Expression of angiogenic growth factors by uterine natural killer cells during early pregnancy. *J Leukoc Biol* (2006) 80:572–80. doi:10.1189/jlb.0406250
69. Xiong S, Sharkey AM, Kennedy PR, Gardner L, Farrell LE, Chazara O, et al. Maternal uterine NK cell-activating receptor KIR2DS1 enhances placentation. *J Clin Invest* (2013) 123:4264–72. doi:10.1172/JCI68991
70. Ashkar AA, Di Santo JP, Croy BA. Interferon gamma contributes to initiation of uterine vascular modification, decidual integrity, and uterine natural killer cell maturation during normal murine pregnancy. *J Exp Med* (2000) 192:259–70. doi:10.1084/jem.192.2.259
71. Hiby SE, Apps R, Sharkey AM, Farrell LE, Gardner L, Mulder A, et al. Maternal activating KIRs protect against human reproductive failure mediated by fetal HLA-C2. *J Clin Invest* (2010) 120:4102–10. doi:10.1172/JCI43998
72. Caligiuri MA, Murray C, Robertson MJ, Wang E, Cochran K, Cameron C, et al. Selective modulation of human natural killer cells in vivo after prolonged infusion of low dose recombinant interleukin 2. *J Clin Invest* (1993) 91:123–32. doi:10.1172/JCI116161
73. Baume DM, Robertson MJ, Levine H, Manley TJ, Schow PW, Ritz J. Differential responses to interleukin 2 define functionally distinct subsets of human natural killer cells. *Eur J Immunol* (1992) 22:1–6. doi:10.1002/eji.1830220102
74. Romagnani C, Juelke K, Falco M, Morandi B, D'Agostino A, Costa R, et al. CD56 bright CD16 – killer Ig-like receptor – NK cells display longer telomeres and acquire features of CD56 dim NK cells upon activation. *J Immunol* (2007) 178:4947–55. doi:10.4049/jimmunol.178.8.4947
75. Nagler A, Lanier LL, Phillips JH. Constitutive expression of high affinity interleukin 2 receptors on human CD16-natural killer cells in vivo. *J Exp Med* (1990) 171:1527–33. doi:10.1084/jem.171.5.1527
76. Matos ME, Schnier GS, Beecher MS, Ashman LK, William DE, Caligiuri MA. Expression of a functional c-kit receptor on a subset of natural killer cells. *J Exp Med* (1993) 178:1079–84. doi:10.1084/jem.178.3.1079
77. Nguyen S, Dhedin N, Vernant JP, Kuentz M, Al Jijakli A, Rouas-Freiss N, et al. NK-cell reconstitution after haploidentical hematopoietic stem-cell transplantations: immaturity of NK cells and inhibitory effect of NKG2A override GvL effect. *Blood* (2005) 105:4135–43. doi:10.1182/blood-2004-10-4113
78. Chan A, Hong DL, Atzberger A, Kollnberger S, Filer AD, Buckley CD, et al. CD56bright human NK cells differentiate into CD56dim cells: role of contact with peripheral fibroblasts. *J Immunol* (2007) 179:89–94. doi:10.4049/jimmunol.179.1.89
79. Root LL, Shipley GD. Normal human fibroblasts produce membrane-bound and soluble isoforms of FGFR-1. *Mol Cell Biol Res Commun* (2000) 3:87–97. doi:10.1006/mcbr.2000.0199
80. Williams EJ, Furness J, Walsh FS, Doherty P. Activation of the FGF receptor underlies neurite outgrowth stimulated by L1, N-CAM, and N-cadherin. *Neuron* (1994) 13:583–94. doi:10.1016/0896-6273(94)90027-2
81. Wu C, Li B, Lu R, Koelle SJ, Yang Y, Jares A, et al. Clonal tracking of rhesus macaque hematopoiesis highlights a distinct lineage origin for natural killer cells. *Cell Stem Cell* (2014) 14:486–99. doi:10.1016/j.stem.2014.01.020
82. Mace EM, Hsu AP, Monaco-Shawver L, Makedonas G, Rosen JB, Dropulic L, et al. Mutations in GATA2 cause human NK cell deficiency with specific loss of the CD56bright subset. *Blood* (2013) 121:2669–77. doi:10.1182/blood-2012-09-453969
83. Gineau L, Cognet C, Kara N, Lach FP, Dunne J, Veturi U, et al. Partial MCM4 deficiency in patients with growth retardation, adrenal insufficiency, and natural killer cell deficiency. *J Clin Invest* (2012) 122:821–32. doi:10.1172/JCI61014
84. van Helden MJ, Goossens S, Daussy C, Mathieu AL, Faure F, Marçais A, et al. Terminal NK cell maturation is controlled by concerted actions of T-bet and Zeb2 and is essential for melanoma rejection. *J Exp Med* (2015) 212:2015–25. doi:10.1084/jem.20150809
85. Vukicevic M, Chalandon Y, Helg C, Matthes T, Dantin C, Huard B, et al. CD56 bright NK cells after hematopoietic stem cell transplantation are activated mature NK cells that expand in patients with low numbers of T cells. *Eur J Immunol* (2010) 40:3246–54. doi:10.1002/eji.200940016
86. Dulphy N, Haas P, Busson M, Belhadj S, Peffault de Latour R, Robin M, et al. An unusual CD56brightCD16low NK cell subset dominates the early posttransplant period following HLA-matched hematopoietic stem cell transplantation. *J Immunol* (2008) 181:2227–37. doi:10.4049/jimmunol.181.3.2227
87. Yu J, Mao HC, Wei M, Hughes T, Zhang J, Park IK, et al. CD94 surface density identifies a functional intermediary between the CD56bright and CD56dim human NK-cell subsets. *Blood* (2010) 115:274–81. doi:10.1182/blood-2009-04-215491
88. Silva A, Andrews DM, Brooks AG, Smyth MJ, Hayakawa Y. Application of CD27 as a marker for distinguishing human NK cell subsets. *Int Immunol* (2008) 20:625–30. doi:10.1093/intimm/dxn022

89. Björkström NK, Riese P, Heuts F, Andersson S, Fauriat C, Ivarsson MA, et al. Expression patterns of NKG2A, KIR, and CD57 define a process of CD56 dim NK-cell differentiation uncoupled from NK-cell education. *Blood* (2010) 116:3853–64. doi:10.1182/blood-2010-04-281675
90. Juelke K, Killig M, Luetke-Eversloh M, Parente E, Gruen J, Morandi B, et al. CD62L expression identifies a unique subset of polyfunctional CD56dim NK cells. *Blood* (2010) 116:1299–307. doi:10.1182/blood-2009-11-253286
91. Sojka DK, Plougastel-Douglas B, Yang L, Pak-Wittel MA, Artyomov MN, Ivanova Y, et al. Tissue-resident natural killer (NK) cells are cell lineages distinct from thymic and conventional splenic NK cells. *Elife* (2014) 2014:1–21. doi:10.7554/eLife.01659

3

NK Cell Subsets in Health and Disease: New Developments

Emanuela Marcenaro[1], Luigi D. Notarangelo[2]*, Jordan S. Orange[3]* and Eric Vivier[4,5]**

[1] Department of Experimental Medicine (DIMES), Center of Excellence for Biomedical Research (CEBR), University of Genova, Genova, Italy, [2] Immune Deficiency Genetics Section, Laboratory of Clinical Immunology and Microbiology, National Institute of Allergy and Infectious Diseases, National Institutes of Health, Bethesda, MD, United States, [3] Department of Pediatrics, Baylor College of Medicine, Center for Human Immunobiology, Texas Children's Hospital, Houston, TX, United States, [4] Centre d'Immunologie de Marseille-Luminy, Aix Marseille Université, INSERM, CNRS, Marseille, France, [5] Service d'Immunologie, Hôpital de la Timone, Assistance Publique-Hôpitaux de Marseille, Marseille, France

Edited and Reviewed by:
Yenan Bryceson,
Karolinska Institute (KI), Sweden

***Correspondence:**
Emanuela Marcenaro
emanuela.marcenaro@unige.it;
Luigi D. Notarangelo
luigi.notarangelo2@nih.gov;
Jordan S. Orange
orange@bcm.edu;
Eric Vivier
vivier@ciml.univ-mrs.fr

Keywords: natural killer cell subsets, immune checkpoints, innate lymphoid cells, anti-tumor responses, anti-viral responses, immunotherapy

Editorial on the Research Topic

NK Cell Subsets in Health and Disease: New Developments

Natural killer (NK) cells were discovered *ca* 1975, as the first group of lymphoid cells that were neither T cells nor B cells. Since then, the dissection of the biology of NK cells has been growing exponentially with many seminal discoveries from the identification of MHC class I-specific inhibitory receptors to the discovery of receptor–ligand pairs involved in NK cell activation and to the manipulation of NK cells in cancer.

In this research topic, we asked a group of thought leaders in NK cell biology to review recent advances in their origins and biology, and their roles in cancer, infection, and inflammation.

Together, these 25 articles provide a timely survey of NK cells as critical immunologic components of health and disease. They will hopefully prompt further dialog and developments in basic and translational immunology.

NK CELL ORIGINS AND BIOLOGY

Nowadays, NK cells are recognized to belong to the family of innate lymphoid cells (ILCs) that include other subsets of lymphoid cells, such as Lymphoid Tissue-inducer cells (LTi), ILC1, ILC2, and ILC3 (Jiao et al.). Human NK cells normally constitute 5–15% of human peripheral blood (PB) lymphocytes. Human PB NK cells can be distinguished in two subsets according to their surface expression of CD56: CD56^{bright} and CD56^{dim}. CD56^{bright} human NK cells express high levels of CD94/NKG2A but low levels of CD16 and lack of KIRs; they predominate in lymph nodes and express low baseline levels of perforin and cytotoxic activity. In addition they produce high levels of cytokines, including IFN-γ, following stimulation by pro-inflammatory cytokines. On the other hand, CD56^{dim} NK cells are CD16^{high} and express KIRs and/or CD94/NKG2A; they predominate in PB (about 90% of PB NK cells); they show high baseline levels of perforin and cytotoxicity against tumor/virus-infected target cells; and they also produce various cytokines in response to direct target cell interactions (Scoville et al.). A third NK cell subset is represented by the CD16^{+}CD56^{neg} NK cells, that are poorly functional, express low levels of natural cytotoxicity receptors (NCRs) and Siglec-7 (an inhibitory lectin-type receptor expressed on the majority of NK cells) and may become more abundant in PB during persistent inflammation, characterizing several human chronic immunological diseases (such as viral infections and autoimmune diseases) (Mikulak et al.).

It has been recently demonstrated that the developmental pathways for NK cell and ILC development are distinct; however, the developmental relationship between the different types of human NK subsets has not been finally clarified. While long held to represent sequential linear stages in maturation, more recent hypotheses propose that CD56^{bright} NK cells may represent NK cells activated *in vivo* and/or that PB NK cell subpopulations derive from different hematopoietic progenitor cells. In this context, a recent study using genetic bar coding of NK cell lineages in rhesus macaques was suggestive of the two subsets having distinct ontologies (Scoville et al.). In addition, the recent characterization of a novel CD34^{+}DNAM-1^{bright}CXCR4^{+} precursor in PB of patients with chronic inflammatory conditions giving rise to apparently licensed and functional maturing NK cells may suggest the possibility for a higher than expected common lymphocyte precursor diversity and a consequently higher peripheral NK cell phenotype variability (Bozzano et al.).

Despite the existence of alternative hypotheses for NK cell development, continually accumulating evidence suggest that a least fraction of CD56^{bright} NK cells represent direct physiologic precursors of CD56^{dim} NK cells. During their process of differentiation, CD56^{dim} NK cells lose expression of CD94/NKG2A and subsequentially acquire inhibitory KIRs and LIR-1, in addition they upregulate T-bet and downregulate Eomes, two key transcription factors regulating NK cell maturation and function during the last steps of their differentiation. In this context, it has been demonstrated that although the number of KIRs correlates with the extent of T-bet/Eomes modulation, "the modulation of these T-box transcription factors during NK cell maturation does not depend on signals conveyed by KIR" (Pradier et al.; Simonetta et al.). The terminally differentiated phenotype of CD56^{dim} cells is marked by the expression of the CD57 molecule that is associated with poor ability to respond to cytokine-mediated stimulation, but preserved cytolytic capacity (Scoville et al.; Bozzano et al.; Della Chiesa et al.).

Notably, within the total CD56^{bright} NK cell subset, a phenotypic and functional heterogeneity can be observed, as shown by the existence of tissue-resident CD56^{bright} NK cells in the uterus/maternal decidual, liver, and lymphoid tissues, where these cells exert tissue-specific functions. Uterine CD56^{bright} NK cells express CD49a while the liver- and lymphoid tissue-resident CD56^{bright} NK cells co-express CD69 and CXCR6. There is increasing evidence that decidual NK (dNK) cells are crucial for pregnancy. Recent data demonstrate that the functions of dNK cells could be suppressed by a reduced HLA-G expression on extravillous trophoblasts suggesting a possible mechanism of recurrent miscarriage (Melsen et al.; Guo et al.). Finally, CD56^{bright} NK cells may also play a major role in controlling T cell responses and maintaining homeostasis. In this context, disorder in the regulation function of NK cells has been recently described in patients with untreated multiple sclerosis (MS), indicating a possible role of this subset in the MS pathogenesis (Gross et al.).

NK CELLS AND VIRUSES

In recent years, it has been appreciated how infection with human cytomegalovirus (HCMV) can shape the NK cell receptor repertoire by inducing an increased frequency of a CD57^{+} NK cell population expressing CD94/NKG2C, the activating counterpart of CD94/NKG2A. This HCMV-induced NK cell subset is characterized by the CD56^{dim}CD16^{bright}LIR-1^{+}KIR^{+}NKG2A^{-} surface phenotype and expresses KIRs interacting with self-HLA class I molecules. In addition, these NK cells can display some hallmarks of adaptive immunity, such as, enhanced effector function, longevity, clonal expansion as well as given epigenetic modifications, including epigenetic remodeling at the *IFNG* locus, decreased expression of certain signaling molecules (i.e., the adaptor protein FcRγ and the tyrosine kinase Syk), and lower expression levels of the transcription factor PLZF. The higher accessibility of the *IFNG* locus can increase IFN-γ production upon target stimulation, while the lack of FcRγ can induce killing *via* antibody-dependent cellular cytotoxicity (ADCC) of opsonized HCMV-infected targets by adaptive NKG2C^{+} NK cells (Della Chiesa et al.). HCMV infection also contributes to age-associated changes in NK cells. In particular, CMV infection and age induce significant changes in the expression of certain markers including CD300a, CD161, T-bet, and Eomes in the CD57^{+} NK cell subset (Lopez-Sejas et al.). Interestingly, HCMV infection may also favor the generation of a recently identified fully mature NK cell subset, characterized by the CD56^{dim}KIR^{+} LIR-1^{+}NKG2A^{-} CD57^{+} phenotype, by a marked downregulation of NCRs and by the unexpected expression of the inhibitory PD-1 immune checkpoint. PD-1^{+} NK cells are also present, and in higher proportions as compared to PB of healthy donors, in both PB and ascitic fluids of ovarian-carcinoma patients suggesting their possible induction/enrichment in the microenvironment associated to the tumor. This phenotype correlates with an impaired NK cell activity toward PD-L^{+} tumor cells that can be partially restored by antibody-mediated disruption of PD-1/PD-L interaction. Importantly, the simultaneous blockade of PD-1 and KIR could amplify the NK-mediated anti-tumor activity. Due to the ineffective anti-tumor functions of this PD-1^{+} NK cell subset, it will be important to carefully evaluate the requirements that can lead to the generation of these cells in health and disease and to understand its role, in particular in patients with advanced cancers (Della Chiesa et al.).

Similar to HCMV infection, Epstein-Barr virus (EBV) infection may change the composition of NK cells. "Early-differentiated NK cells that expand during EBV infection might directly recognize lytically EBV replicating targets, while the terminally differentiated NK cells in HCMV-infected individuals mainly promote ADCC" (Chijioke et al.). Finally, the role of human NK cells in herpesviral control is accentuated by inherent genetic defects of immunity that lead to NK cell aberrations a majority of which have impact upon specific NK cell subsets (Mace and Orange). There is still much to be learned from these rare individuals with regard to specific NK cell populations and their role in human host defense.

NK CELLS AND TUMORS

Natural killer cell-mediated ADCC plays an important role not only in the control of infections but also of tumors and down-regulation of CD16 expression on activated NK cells may limit or regulate this response. In this context, in has been demonstrated that CD56^{dim} CD57^{+} NK cells are particularly prone to losing CD16 after influenza vaccination. This event supports a role for CD16 in early activation of NK cells after vaccination and for

CD16 downregulation as a means to modulate NK cell responses and maintain immune homeostasis of both antibody and T cell-dependent pathways (Goodier et al.).

Several data indicate that the microenvironment, the cytokine milieu, and genetic factors in cancer patients can exert a strong influence on NK cell receptor expression and functional potential (Carrega and Ferlazzo). Thus, the phenotype of the NK cell subpopulation associated with cancer may vary according to the specific kind of tumor and its anatomical location. In this context, chemokines/chemokine receptors play a critical role in the regulation of the distribution of NK cell subpopulations in the various tissues. Remarkably in different tumor types (including lung cancer and melanoma), both migration and homing of NK cells may be altered and even reversed. For example, NK cells present in the tumor microenvironment are often enriched in $CD56^{bright}CD16^{neg/dim}$ NK cells (Carrega and Ferlazzo); in contrast, in melanoma metastatic lymph nodes, it is possible to detect the expansion of an unusual subset characterized by a $CD56^{dim}$ $CD69^{+}CCR7^{+}$ KIR^{+} phenotype (Cristiani et al.; Pesce et al.). A likely explanation of these events is that the release of different types of chemokines by cells of the tumor microenvironment, or the acquisition of different/new chemokine receptors by NK cells, can lead to an altered recruitment of the NK cell subpopulations.

TRANSLATING BASIC NK CELL BIOLOGY INTO THE CLINICS

Natural killer cells may display alloreactive potential in case of mismatch between recipient inhibitory KIRs and graft HLA class I molecules. Several studies have addressed the impact of this variable in the context of hematopoietic stem cell (HSC) and solid organ transplantation (including kidney transplant). In this context, several lines of evidence support that "adaptive" $NKG2C^{+}$ NK cells may contribute to control viral infection in HSC as well as in kidney transplant recipients. On the other hand, increasing evidence supports that alloantibody-mediated NK cell activation *via* FcγRIIIA (CD16) may contribute to rejection in kidney transplant recipients. Further studies integrating phenotypic, functional, and genetic analysis of NK cells should provide valuable insights on the pathogenesis of solid organ transplant complications (Lopez-Botet et al.; Legris et al.). Importantly, in addition to NK cells, ILCs may also represent important players in the early phases after transplantation (Vacca et al.).

A major goal of cancer immunotherapy based on the use of NK cells is to reverse the tumor-induced NK cell impairment observed in cancer patients and to amplify NK cell effector functions. Therapies involving NK cells may either activate endogenous NK cells or involve transfer of exogenous pre-activated NK cells. In order to optimally harness NK cells in the next generation of NK cell-based vaccines and therapeutics single-cell studies of the variety of antigens that can lead to NK cell diversification, the definition of the significance of the human NK repertoire in the context of viral infection and tumors will be extremely important (Strauss-Albee and Blish). Many NK cell-based immunotherapies have been developed over the last decades, with allogeneic or autologous NK cells. In 2005, haploidentical NK cells were administered in a non-transplantation setting and resulted in a substantial improvement of patient clinical outcome and a substantive number of trials have followed. More recently, *in vivo* targeting of NK cells with antibodies was investigated: IPH2101 is a first-in-class anti-KIR antibody that blocks inhibitory KIR–ligand interactions, leading to restoration of NK cell functions. A phase II trial in acute myeloid leukemia elderly patients in first CR1 (NCT01687387) is in progress and several other trials are ongoing in different cancers alone or in combination with other treatments. The future introduction of a first-in-class anti-NKG2A blocking antibody (IPH2201) will also provide a novel strategy to enhance tumor cell recognition (Rey et al.; Chabannon et al.). Remarkably, the recent discovery that a subset of NK cells may express PD-1 open prospects for extending the potential of cancer immunotherapy, by combining drugs blocking PD-1 (or its major ligand PD-L1), with additional antibodies against immune checkpoints (KIR and NKG2A) expressed on this important innate effector cells (Della Chiesa et al.).

Natural killer cells can also be adoptively transferred following solid modifications in order to license them with new or reinforce functions and ensure their controlled persistence and activity in the recipient. In this context, it is crucial to analyze NK cell features in tumor patients in order to indicate a timeline when NK-mediated therapies or other immunotherapies could be performed (Rey et al.). Ongoing developments for innovative cellular therapies suggest that such progress could result in wider clinical applications in the next future (Chabannon et al.). We believe that a deeper investigation into the impact of both conventional (e.g., chemotherapy) and new therapies (e.g., anti-immune checkpoints antibodies) on NK cell homeostasis in cancer patients is now required.

A novel strategy to improve NK cell-mediated immunosurveillance is to enable a proper NK cell migration to target tissues (including tumor sites) by promoting the expression of certain chemoattractant receptors on NK cells to be used for adoptive immunotherapy. In this context, NK cells engineered *ex vivo* to express chemokine receptors by gene transfer or by trogocytosis are under investigation for their better tissue homing and function (Pesce et al.; Bernardini et al.). For example, the acquisition of CCR7 expression by $CD56^{dim}$ KIR^{+} NK cells may play a key role in promoting the migration of this subset to lymph nodes, where $CD56^{dim}$ KIR^{+} NK cells may shape adaptive immune responses by leading to Th1 polarization (through the release of IFNγ and the mechanism of DC "editing") but may also prevent GvHD and HvGD in haplo-HSCT by directly killing patient's DCs and T cells (Pesce et al.). Thus, NK cells can affect DC fate and thereby control T cell responses (Pallmer and Oxenius).

In conclusion, we express our gratitude to all the authors who have contributed to this research topic and to the reviewers for their valuable work.

AUTHOR CONTRIBUTIONS

All authors listed have made a substantial, direct, and intellectual contribution to the work and approved it for publication.

4

Type 1 Innate Lymphoid Cell Biology: Lessons Learnt from Natural Killer Cells

Yuhao Jiao[1,2,3], Nicholas D. Huntington[1,2], Gabrielle T. Belz[1,2]* and Cyril Seillet[1,2]**

[1] Molecular Immunology Division, Walter and Eliza Hall Institute of Medical Research, Melbourne, VIC, Australia, [2] Department of Medical Biology, University of Melbourne, Melbourne, VIC, Australia, [3] School of Medicine, Tsinghua University, Beijing, China

Edited by:
Eric Vivier,
Centre d'Immunologie de Marseille-Luminy, France

Reviewed by:
Hugh Thomson Reyburn,
Spanish National Research Council, Spain
Adelheid Cerwenka,
German Cancer Research Center, Germany
Roland Jacobs,
Hannover Medical School, Germany

***Correspondence:**
Cyril Seillet
seillet@wehi.edu.au;
Gabrielle T. Belz
belz@wehi.edu.au;
Nicholas D. Huntington
huntington@wehi.edu.au

Group 1 innate lymphoid cells (ILCs) comprise the natural killer (NK) cells and ILC1s that reside within peripheral tissues. Several different ILC1 subsets have recently been characterized; however, no unique markers have been identified that uniquely define these subsets. Whether ILC1s and NK cells are in fact distinct lineages, or alternately exhibit transitional molecular programs that allow them to adapt to different tissue niches remains an open question. NK cells are the prototypic member of the Group 1 ILCs and have been historically assigned the functions of what now appears to be a multi-subset family that are distributed throughout the body. This raises the question of whether each of these populations mediate distinct functions during infection and tumor immunosurveillance. Here, we review the diversity of the Group 1 ILC subsets in their transcriptional regulation, localization, mobility, and receptor expression, and highlight the challenges in unraveling the individual functions of these different populations of cells.

Keywords: innate lymphoid cells, immunity, immune protection, lymphocyte subsets, GVHD, tumor rejection

INTRODUCTION

Innate lymphoid cells (ILCs) are members of an expanding family of immune cells that provide the first line of defense against invading pathogens and they also contribute to tissue repair (1). Unlike T and B lymphocytes, ILCs develop independent of the recombinant activating gene (RAG). They have been classified into three main groups (groups 1–3) based on their cytokine and transcription factor expression that aligns them with the subsets used to categorize different $CD4^+$ T helper (Th) cell populations (2). The ILC1 family, such as Th1 cells, is composed of the T-bet-expressing cells that secrete interferon (IFN)-γ. Group 2 ILCs (ILC2s), initially described as "natural helper cells" (3), or nuocytes (4), express GATA-3 similar to Th2 $CD4^+$ T cells, while ILC3s produce IL-17 and/or IL-22 and express the transcription factor RORγt (5). This latter subset is proposed to be equivalent to Th17 and Th22 cells.

Innate lymphoid cells have been the focus of extensive investigation during the last 5 years, in part due to their potential impact on human health and disease. Several studies have uncovered significant heterogeneity and plasticity within the ILC subsets, a feature which highlights that the field may have overlooked delineating the individual contributions of different ILC subsets and identifying their contributions to immunity. This is especially important in the Group 1 ILCs that were originally thought to represent a single homogeneous population composed of natural killer (NK) cells alone. It has now been found that this group is heterogeneous and includes a second subset of T-bet-dependent IFN-γ-producing innate cells that localize principally in tissues (6). Given this diversity within the group 1 ILCs we will refer to the respective subsets as NK cells and

ILC1s. The finding that two IFN-γ-producing subsets exist also raises the possibility that previous analyses of NK cells may have overlooked the contributions of ILC1s as they express similar surface markers, including NKp46, CD122, and NK1.1, to NK cells. However, it is now recognized that NK cells and ILC1s are likely to have quite distinct roles in protective immunity as they differ in multiple aspects, including their transcriptional regulation, localization, mobility, and receptor expression. The challenge in teasing apart the individual functions of these different populations of cells relies largely on the identification of unique markers to characterize each of the populations. In this review, we provide an overview of the different Group 1 ILC populations and highlight the current understanding of their roles in immune protection.

HETEROGENEITY AND PLASTICITY OF THE GROUP 1 ILCs

Group 1 ILCs were initially thought to include only the prototypic member, the NK cell. Even early on though, heterogeneity had already been described among NK cells found in different tissues, such as the liver, the thymus, and the uterus. These NK cells differentially expressed a number of surface molecules, suggesting that much greater diversity existed (7). Classically, NK cells have been defined as $CD3^-NK1.1^+$ with NKp46 expression being identified somewhat later as a more specific marker for mature NK cell cells that are found in secondary lymphoid tissues, such as lymph nodes and the spleen (8). A further subset of ILC1s, initially thought to be NK cells, was found in the liver and expressed the molecule tumor necrosis factor-related apoptosis-inducing ligand (TRAIL) (9). More recently, it has been shown that the intrahepatic ILCs comprise a distinct lineage of ILC1s which develops separately from NK cells (10). ILC1s have also been found in the thymus and express high levels of the IL-7 receptor α (IL-7Rα, CD127) and Gata-3 (11). A further distinction has been elucidated between NK cells and ILC1 as the latter are generally localized within tissues. This has led to their designation as "tissue-resident" cells. This contrasts with the majority of NK cells that at steady-state circulate continuously patrolling for infected or malignant cells, while ILC1 rarely recirculate and appear to be generated by progenitor cells in the tissues or self-renew locally (6, 12). How the behavior of these two subsets of cells might be altered during inflammation is not yet clear. It is possible that in response to infection bone marrow-derived precursors become key contributors to the pool of tissue-resident ILC1s to meet the significantly increased demands of responding to a pathogen infection. Some evidence for such a model is provided by the adoptive transfer of bone marrow ILC progenitors into sublethally irradiated mice that were then able to generate liver and intestinal ILC1s (13, 14).

Conventional NK Cells

Natural killer cells represent 5–15% of circulating lymphocytes in humans and 2–5% in mice. It is well established that tumor growth is in part influenced by the activity of the immune system. However, it has only recently been appreciated that tumor cell avoidance of immune detection also contributes to cancer development in humans. NK cells were identified in the early 1970s due to their ability to spontaneously kill leukemia cells and they have now been strongly implicated as key effectors in cancer immunosurveillance, transplantation rejection, and early viral immunity (15).

The cytokine IL-15 is essential for most facets of NK cell biology. It binds to IL-15Rβ/γ_c heterodimers on the surface of NK cells to drive their survival and proliferation. It also primes NK cell activation resulting in the production of pro-inflammatory cytokines and lytic granules. IL-15 binding results in activation of the JAK1/3 and STAT5 signaling pathways and this induces STAT5-target genes, such as the pro-survival gene *Myeloid cell leukemia 1* (*Mcl1*) and the negative regulator of IL-15 signaling (16), cytokine-inducible SH2-containing protein encoded by *Cish*, which are essential for NK cell survival and homeostasis (17). The NK cells also express multiple activating and inhibitory receptors of the NKG2 and Ly49 (KIR in humans) family (18). The activation of the NK cells relies on the dynamic balance between activating and inhibitory pathways allowing the cells to rapidly sense changes in the environment. NK cells can kill infected or transformed cells through cytolytic mechanisms (granzymes or perforin) or engagement of death receptors.

ILC1s

ILC1s are found at steady state in virtually all tissues, but are enriched in the liver, uterus, skin, salivary glands, and the gut (19). In contrast to NK cells that express integrin α2 (CD49b), ILC1s express high levels of integrin α1 (CD49a) but lack integrin α2 and the T-box transcription factor eomesodermin (Eomes). Conventionally, Eomes expression has been thought to be restricted to NK cells alone and not expressed by ILC1s. However, this does not appear to be a universal rule as ILC1s in some tissues can also express Eomes highlighting that the use of transcription factor expression to demarcate subsets is not necessarily definitive in all settings (**Table 1**; **Figure 1**). Similarly, while CD49a expression is restricted to ILC1 at steady state, following activation, NK cells can also express this marker (20). This incomplete clarity in defining the subsets highlights their complexity and the requirement to dig deeper into understanding the distinct roles of ILC1s and NK cells *in vivo*.

Liver ILC1s

Natural killer cells that differed from conventional NK cells were identified in the liver by the Smyth group who described a unique subset of $NK1.1^+$ cells that expressed TRAIL (9). Initially, they were considered to be immature NK cells as they were enriched for $CD27^+CD11b^-$ cells that lacked expression of most of the Ly49 molecules and CD49b (9). It is now clear that this population is a distinct ILC1 lineage that in contrast to NK cells, which does not require the transcription factor nuclear factor interleukin 3 (Nfil3) for development (10) and is tissue-resident being located almost solely in the liver at steady state (12) (**Figure 1B**). Interestingly, liver ILC1s share with other tissue-resident lymphocytes the dependency on the transcription factor Hobit, while ILC1s in other tissues do not (39). Transcriptomic analysis revealed that liver ILC1 gene expression was quite distinct from NK cells (10). ILC1s display a unique pattern of chemokine receptors and adhesion

TABLE 1 | Phenotype of NK cells and tissue-specific ILC1.

	NK1.1	NKp46	CD49a	CD49b	CXCR6	CD127	TRAIL	CD160	CD226	Ly49E	CD11b	KLRG1	CD62L	EOMES
NK cells	+	+	−	+	−	−	−	−	+	−	+	+	+	+
Liver ILC1	+	+	+	−	+	±	+	+	++	+	−	−	−	−
Thymic ILC1	+	+	n.d.	+	n.d.	+	n.d.	n.d.	n.d.	n.d.	−	−	n.d.	+
IEILC1	+	+	+	−	+	+	+	+	+	−	−	−	−	−
Salivary ILC1	+	+	+	+	+	−	+	+	n.d.	n.d.	−	−	−	+

±, *heterogeneous expression; n.d., not determined.*

molecules distinct from NK cells. These are likely to be important in maintaining their resident positioning in tissues and include the receptors CXCR6, CXCR3, CD103, CD49a, CD69, and CD39 (10, 12, 33). By contrast, NK cells express CX3CR1, CD62L, S1PR1, and S1PR5 that are not found in the ILC1. Liver ILC1s also express their own pattern of cytokine receptors (IL-7R, IL-17RD, IL-21R, and TGF-βR) together with a set of regulatory molecules, such as CD200R, PD1-L, ICOSL, and Lag3 (10, 33).

It is now clear that intrahepatic ILC1s are resident cells; however, how the stability of this population maintained is not clear. Several studies have been undertaken to understand this feature of ILCs but the outcomes have not been consistent (12, 40). One possibility for these discrepancies could be that mouse strains of different backgrounds that have been used as recipients of adoptively transferred ILC1s may impact the development and expansion of these cells. For example, liver ILC1s transferred into sublethally irradiated C57BL/6 mice maintained their original phenotype (12), while ILC1s transferred into immunodeficient $Rag2^{-/-}\gamma_c^{-/-}$ mice differentiated into NK cells that then expressed Eomes and lost surface TRAIL expression (40).

Intrahepatic ILC1s have also been identified in man (41). Interestingly, these cells share many phenotypic traits with mouse liver ILC1s, including the expression of CD49a, NKp46, DNAM-1, and T-bet, but lack Eomes (41). They appear to also share similar functional capacity as they produce high levels of IFN-γ, TNF-α, and GM-CSF, a poor degranulation response and low levels of cytotoxic effector molecules. Despite the conservation of characteristics between man and mouse, the actual role of the ILC1s in the liver is currently unknown. However, it has been shown that these cells do have a unique capacity to generate immune memory responses against hapten and viral antigens (30). Unexpectedly, liver ILC1s could mediate recall responses in an antigen-specific manner. Thus far, however, neither the mechanisms nor the receptors involved in the specific recognition of the antigens are known and will require further investigation.

Thymic ILC1s

NKp46-expressing cells that reside in the thymus were first described as NK cells (11). However, a more recent understanding of their phenotype suggests that they fit more closely with the description for ILC1. Thymic ILC1s differ from conventional NK cells as they express IL-7Rα and require the transcription factor Gata-3 for development (11) similar to other ILC1s (35) and should, therefore, be considered as members of the ILC1 subset (**Figure 1B**). Bone marrow-derived NK cells do not depend on Gata-3 during differentiation although it is important for them to produce IFN-γ during maturation. Thymic ILC1s display a phenotype that is intermediate between classical NK cells and ILC1s again adding to the debate around the classification of these cells. Similar to other ILC1s, thymic ILCs express low levels of Ly49 and high levels of IL-7Rα, CD11b, and CD69. They also express Eomes, a transcription factor normally characteristic of NK cells. Thymic ILC1s can develop *in vitro* and *in vivo* from the $CD4^-CD8^-$ (DN1) subset of immature thymocytes, indicating that they do not arise from committed precursors in the bone marrow (42). To date, relatively little is known about the origins and contributions of thymic ILC1s to immune protection or homeostasis. Further work will be required to characterize these cells and establish their role in the thymus.

Intestinal ILC1s

At least three phenotypically distinct subsets of ILC1s can be found in the gut. In the lamina propria (LP) both NK cells and ILC1s (LP ILC1) have been described, while a third subset of ILC1s reside in the epithelium (ieILC1). LP ILC1s are characterized by high expression of IL-7Rα and the lack of Eomes while NK cells lack IL-7Rα and express Eomes (14). In humans, a subset of $CD127^+$ IFN-γ-producing ILC1s have been identified in the gut and have been shown to be enriched in the intestine of Crohn's disease patients (32). They are also enriched for CD69, but like ILC1s produce high level of IFN-γ and lack of Eomes expression. These cells do not express NKp46 or CD56 but express CD161, a marker commonly expressed by the ILC3 and ILC2 subsets.

An additional ILC1 subset has been reported in the gastrointestinal epithelia in both human and mice (31). These intraepithelial ILC1s (ieILC1s) are characterized by the expression of lymphocyte markers, such as CD103 and CD160 (31). While human ieILC1s express NKp46, CD56, and NKp44, they also express Eomes similar to human LP ILC1s in contrast to murine ILC1 that do not express Eomes.

Salivary Gland ILC1s

ILC1s are found in the salivary gland and like those in some other sites express TRAIL, CD49a, CD103, and CD69 and express Eomes (34) (**Table 1**, **Figure 1**). Interestingly they develop independent of the transcription factor Nfil3 that is similar to liver ILC1s, suggesting that salivary ILC1s may be distinct from NK cells that depend on Nfil3 for development (10). Recently, TGF-β signaling has been implicated in the maintenance of salivary ILC1s (43). The lack of TGF-β signaling was associated with reduced numbers of ILC1 and impaired expression of CD49a, CD103, and CD69 in the salivary glands. By contrast, TGF-βR2 deficiency had minimal impact on the phenotype of ILC1s found

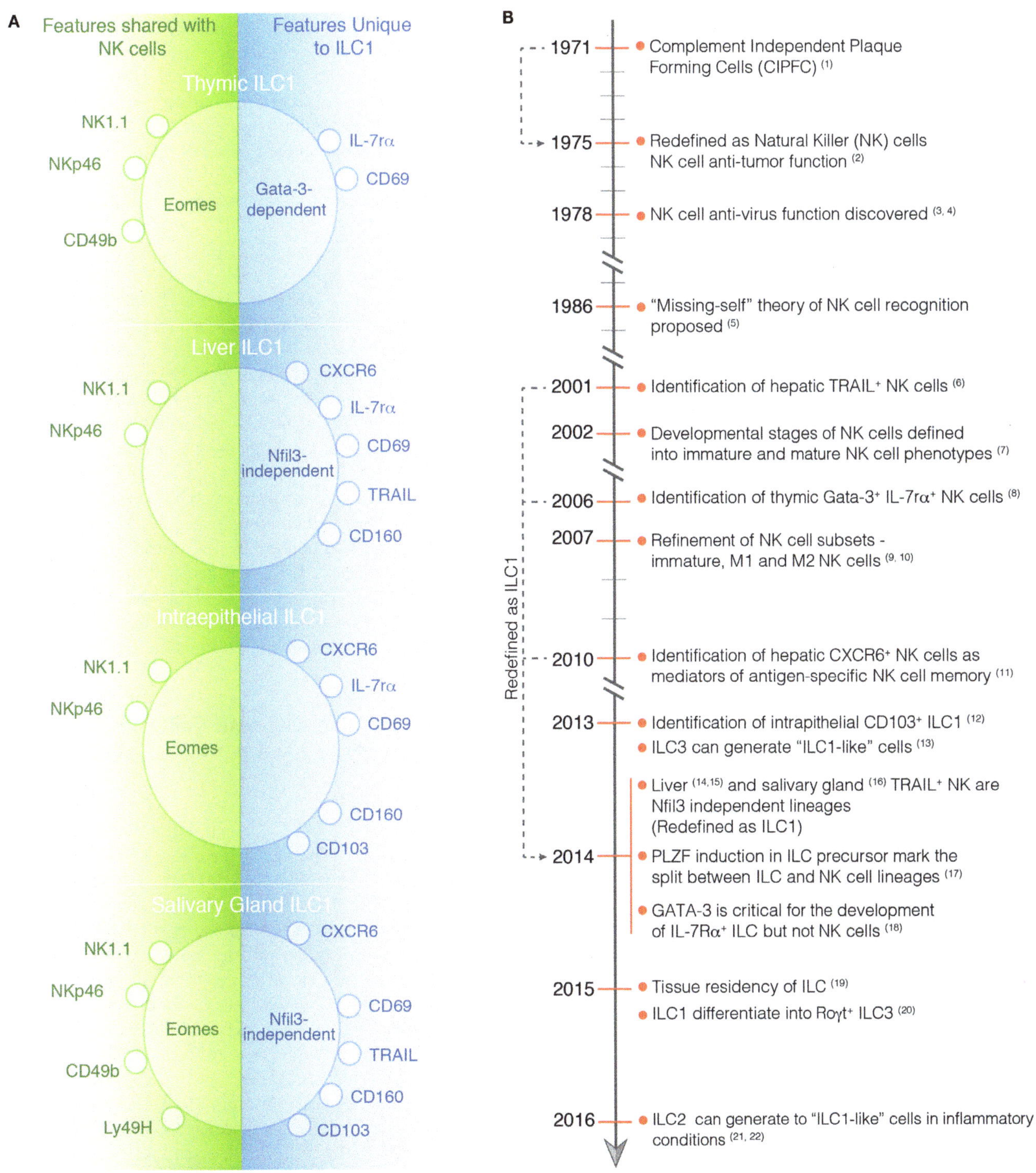

FIGURE 1 | **(A)** Phenotype of tissue-specific ILC1s and NK cells at steady-state. Characteristic markers of ILC1 in the thymus, liver, intestine, and salivary gland are depicted. On the left, the markers that are expressed by both ILC1s and NK cells are shown while on the right the marker that are only expressed by ILC1s are indicated. **(B)** Timeline highlighting important discoveries in the identification and characterization of NK cells and ILC1s. (1) Thornthwaite et al. (21). (2) Kiessling et al. (22). (3) Anderson (23). (4) Santoli et al. (24). (5) Karre et al. (25). (6) Takeda et al. (26). (7) Kim et al. (27). (8) Vosshenrich et al. (11). (9) Hayakawa et al. (28). (10) Huntington et al. (29). (11) Paust et al. (30). (12) Fuchs et al. (31). (13) Bernink et al. (32). (14) Seillet et al. (10). (15) Daussy et al. (33). (16) Cortez et al. (34). (17) Constantinides et al. (13). (18) Yagi et al. (35). (19) Gasteiger et al. (6). (20) Bernink et al. (36). (21) Bal et al. (37). (22) Ohne et al. (38).

in the gut and the liver, suggesting that other factors guide their differentiation (43).

ILC1 PLASTICITY

Tools to investigate the distinct roles of ILC1s and NK cells are currently limited. At steady state, ILC1s and NK cells are distinguished based on the expression of surface markers (e.g., TRAIL, CD49a, CXCR6) and transcription factors (e.g., Eomes, PLZF). However, inflammation can modify these features complicating our ability to track *bona fide* ILC1s and NK cells. Surface markers, such as TRAIL or CD49a can be upregulated during MCMV infection or following exposure to cytokines, such as IL-2, IFN-γ, or IL-15 (20, 44, 45). During inflammation, cytokines can also divert ILC identity as has been shown for ILC2s and ILC3s, which can acquire a phenotype consistent with ILC1s. These so called "ILC1-like" cells that produce IFN-γ can arise from Rorγt$^+$ NKp46-expressing ILC3s in the intestine following stimulation with cytokines, such as IL-12 and IL-18 (32, 46). LP ILC1s are also influenced by the microenvironment differentiating into Rorγt$^+$ ILC3s when exposed to IL-23 and IL-1β. This process was enhanced in the presence of retinoic acid while IL-12 could reverse the transition (36). Similarly in the lung, IL-1β and IL-12 can induce the conversion of ILC2s to ILC1-like cells (37, 38). These findings highlight that ILC are capable of rapid adaption to changes in environmental cues induced by pathogens or inflammation.

CONTRIBUTION OF THE GROUP 1 ILC TO PATHOGEN RESPONSES

Early resistance against intracellular bacteria has been attributed to the activation of phagocytic cells by NK cells through their capacity to produce IFN-γ (47, 48). However, the beneficial role of NK cells in these infections became contentious when more recent studies showed that depletion of NK1.1$^+$ cells led to a reduction of bacterial load, a result that was not be concordant with the proposed positive role for NK cells. It would appear that during infection with *Listeria monocytogenes*, excessive IFN-γ produced by NK cells *via* activation through the costimulatory molecule CD27 impairs innate anti-bacterial defenses (49). This effect was not limited to Listeria infection (49, 50) as NK cell depletion also protected against lethal *Escherichia coli* (51) and *Streptococcus pneumonia* infection (52). An additional explanation for these paradoxical results, however, is that ILC1s and NK cells make differential contributions to defense against infection. These studies were conducted before the discovery of the ILC1s and have inevitably categorized all cells expressing NKp46 and NK1.1 and secrete IFN-γ and TNF-α as putative "NK cells."

Re-investigation of the role of NKp46-expressing cells in the control of parasites, such as *Toxoplasma gondii*, revealed that ILC1s are indeed a major source of IFN-γ and TNF-α following oral infection (14). By contrast, NK cells or NKp46$^+$ ILC3s appeared to have only minor contribution in these settings. However, whether ILC1s are solely responsible for the control of infection remains to be determined (14). In these experiments, the mice carried a germline deletion of T-bet resulting in a deficiency in their entire hematopoietic compartment and loss of IFN-γ expression in ILC1s, NK and NKp46$^+$ ILC3s. These experiments, therefore, do not delineate the role of T-bet in the various different immune populations that might control infection. Although adoptive transfer of ILC1 into Rag2$^{-/-}$γ$_C^{-/-}$ mice was found to reduce the Toxoplasma pathogen load, it only partially restored the recruitment of monocytes to a level similar to those observed in T-bet$^{-/-}$ mice. Similarly, in response to *Clostridium difficile*, ILC1s have been shown to be the major contributor to IFN-γ production during acute infection rather than NK cells (53). Early protection against *C. difficile* was mediated by T-bet-expressing ILC1s in a IFN-γ-dependent manner.

These results collectively imply that the enhanced capacity of ILC1s to produce IFN-γ outstrips that of NK cells during enteric infection such that ILC1s may be the main drivers of immune protection. In addition, the potential plasticity of ILCs allowing the polarization of ILC3s toward an ILC1-like phenotype may further favor efficient IFN-γ production by ILC1s. However, due to the paucity of models that allow targeted deletion of ILC1 alone, it remains complicated to formally demonstrate the specific and non-redundant roles of ILC1. Further characterization of the ILC1 populations and identification of the transcription factors regulating their development may shed light on new approaches for *in vivo* models with which ILC1s functions can be specifically assessed.

CONTRIBUTION OF THE GROUP 1 ILC IN CANCER AND INFLAMMATION

Tumor Rejection

Natural killer cells are able to distinguish transformed cells from normal healthy cells, in part by their reduced expression of major histocompatibility complex class I (MHC-I) molecules, the founding feature of the "missing-self" hypothesis. Both *in vivo* and *in vitro* experiments in humans and mice have demonstrated that MHC-I deficient tumor cells are more sensitive to NK cell-mediated cytolysis (25). However, the mechanisms regulating the activation and inhibition of NK cells are more complex than the original "missing-self" hypothesis (54). Studies have shown that cytotoxicity of NK cells against transformed cells is governed by both cytokines and activating and inhibitory receptors. To suppress tumors, NK cells can cause direct cell target death by granule-mediated lysis, receptor mediated cytotoxicity (Fas/FasL or TRAIL death pathway) (55, 56), or antibody-dependent cell cytotoxicity (ADCC) (57). Besides direct killing of tumor cells, NK cells can secrete TNF-α and IFN-γ to drive the Th1 polarization in the tumor microenvironment to further strengthen adaptive cytotoxic responses to tumor cells (58).

In contrast to NK cell activation, ILC1s have not been reported to exhibit crosstalk with other cells through activating and inhibitory signals. Instead, they are activated by cytokines, such as IL-12, IL-15, and IL-18, which are also able to drive NK cell activation (31). Following activation, ILC1s can also produce TNF-α and

IFN-γ. It has been proposed that ILC1s should be classified as innate "helper-like" cells distinguishing them from NK cells that would represent cytotoxic ILCs (14), however, it has been shown that liver ILC1s are also able to recognize and kill target cells albeit perhaps less efficiently than NK cells (30, 59). This killing capacity would enable tissue-resident ILC1s to have potentially important roles in controlling transformed cells at an early development stage prior to their dissemination throughout the body. Recently, it was reported that early stage cell transformation can induce a local tissue-resident lymphocyte response accompanied by expression of Granzyme B and potentially cytolytic activity toward the transformed cells (60). These tumor-associated ILC1s appear to be transcriptionally distinct from, but functionally related to, conventional NK cells and differ from ILC1s found at steady state. These cells, though, were dependent on IL-15 but did not require *Nfil3* for development similar to ILC1s. These data suggest that tumor-associated ILC1s are developmentally distinct from NK cells as it has been shown for liver (10) and salivary gland ILC1s (34). However, it has been also shown that Nfil3 is not required for mature NK cell homeostasis and that *Nfil3*-deficient mice are able to generate and maintain NK cell memory in response to MCMV infection (61).

Transplantation and Graft-versus-Host Disease

A common clinical situation where both acute and chronic inflammation occurs is transplantation. Either solid organ or hematopoietic cell transplantation (HCT), including bone marrow transplantation (BMT), can lead to both host and donor immune reactions, mainly attributed to mismatched MHC molecules from the host and the donor. These immune reactions involve inflammation of multiple organs and are categorized into host-versus-graft disease (HVGD) and graft-versus-host disease (GVHD) reflecting the origins of target and activated immune cells. When host immune cells react against donor cells, elevation in serum cytokine levels and multi-organ inflammation gives rise to HVGD and typically graft failure results. Conversely, GVHD is the disease associated with systemic inflammation caused by activation of donor immune cells that damage the host tissue, a very common clinical complication after HCT. The alloreactivity from donor NK cells against recipient cells can cause selective cytolysis and damage to host cells and tissues, suggesting that it may be entirely deleterious but paradoxically this effect sets the scene for a potent graft-versus-leukemia (GVL) effect that effectively removes leukemic cells leading to positive outcomes. Alloreactive NK cells reduce GVHD by eliminating recipient antigen-presenting cells that drive immune cell activation (62). In a murine model of MHC-matched allogeneic BMT, the infusion of Ly-49 (a member of the inhibitory receptor family) mismatched NK cells contributed significantly to the anti-tumor effect while preventing GVHD (63). The infusion of pre-activated NK cells with a combination of cytokines, such as IL-12/15/18, strongly suppressed acute GVHD (64). These NK cells were found to maintain their expression of Eomes and T-bet *in vivo* which they exhibited prior to transfer which suggests that the key effectors in this setting are NK cells and not ILC1, which have a different profile (64).

Interestingly, mice reconstituted with cells that lack NKp46 expression had greatly exacerbated GVHD and increased mortality due to increased commensal bacteria infection (65). As both NK cells and ILC1s express NKp46, this clinical outcome may reflect the involvement of both cell types but it is not clear how NK cells and ILC1s individually contribute to the disease progression. To date, only a single study has reported a protective role for ILCs in GVHD (66). In this study, patients that did not develop acute GVHD had increased proportions of skin-homing ILC1s, NCR^- ILC3s and gut-homing ILC2s, suggesting that ILCs can indeed provide protection against the development of GVHD (66). It was observed that following transplantation, patients that developed more severe GVHD had fewer circulating ILC1s when compared with healthy controls. Mobilization of ILC1s following treatment was associated with elevated expression of the early activation marker CD69, and the tissue homing markers CLA and CCR10 which correlated with less severe progression of GVHD. Nevertheless, the recovery of ILCs following conditioning radio-therapy and chemotherapy was slow compared with other immune cell types and the reconstituting ILCs were derived from the donor. It appears that both ILC1s and NK cells from donors can exert protective roles to restrict the progression of GVHD while simultaneously contributing to the GVL effect. This suggests that balancing the ratios of donor and host ILC1s/ NK cells through conditioning prior to BMT may be important for high engraftment and robust GVL effects and substantially negating the risk of GVHD.

CONCLUSION

With the identification of different ILC1 subsets in virtually every tissue in the body, it has become clear that these cells do not express a unique characteristic marker that singularly defines this subset. Some ILC1s, such as liver-resident ILCs, have a relatively well-defined phenotype, location, and origin, and this is clearly distinct from NK cells. In other tissues, such as the thymus or salivary glands, ILC1s appear to share a number of features with NK cells (**Table 1**, **Figure 1**). This includes the expression of Eomes and CD49b. Thus, whether all ILC1 subsets represent one distinct lineage with the same progenitor and blueprint for transcriptional regulation is not clear. It seems unlikely. Liver and salivary gland ILC1s do not require NFIL3 for their development while ILC1s in the gut it is essential (10, 34, 67). More intriguingly, ILC1s exhibit a differential need for IL-15 depending tissue they reside (31). Nevertheless, ILC1s appear to be very adaptable and their phenotype strongly reflects their varied microenvironments in which they exist. They appear to develop from precusors in the fetal liver, bone marrow, or thymus, but can also, somewhat surprisingly, transdifferentiate from other ILC lineages, including ILC2s and ILC3s. Thus, in order to understand the contribution of ILC1 to immune protection, further work will be required to better characterize the origin, transcriptional regulation and perhaps most importantly how microenvironmental factors drive their development and plasticity. This will depend significantly on the generation of novel models in which ILC1s can be specifically deleted without affecting the NK cell compartment and allow detailed understanding of the precise role of ILC1s during infections and inflammatory diseases.

AUTHOR CONTRIBUTIONS

All authors listed, have made substantial, direct and intellectual contribution to the work, and approved it for publication.

REFERENCES

1. Artis D, Spits H. The biology of innate lymphoid cells. *Nature* (2015) 517:293–301. doi:10.1038/nature14189
2. Spits H, Artis D, Colonna M, Diefenbach A, Di Santo JP, Eberl G, et al. Innate lymphoid cells – a proposal for uniform nomenclature. *Nat Rev Immunol* (2013) 13:145–9. doi:10.1038/nri3365
3. Moro K, Yamada T, Tanabe M, Takeuchi T, Ikawa T, Kawamoto H, et al. Innate production of T(H)2 cytokines by adipose tissue-associated c-Kit(+)Sca-1(+) lymphoid cells. *Nature* (2010) 463:540–4. doi:10.1038/nature08636
4. Neill DR, Wong SH, Bellosi A, Flynn RJ, Daly M, Langford TK, et al. Nuocytes represent a new innate effector leukocyte that mediates type-2 immunity. *Nature* (2010) 464:1367–70. doi:10.1038/nature08900
5. Sawa S, Lochner M, Satoh-Takayama N, Dulauroy S, Berard M, Kleinschek M, et al. RORgammat+ innate lymphoid cells regulate intestinal homeostasis by integrating negative signals from the symbiotic microbiota. *Nat Immunol* (2011) 12:320–6. doi:10.1038/ni.2002
6. Gasteiger G, Fan X, Dikiy S, Lee SY, Rudensky AY. Tissue residency of innate lymphoid cells in lymphoid and nonlymphoid organs. *Science* (2015) 350:981–5. doi:10.1126/science.aac9593
7. Shi F-D, Ljunggren H-G, La Cava A, Van Kaer L. Organ-specific features of natural killer cells. *Nat Rev Immunol* (2011) 11(10):658–71. doi:10.1038/nri3065
8. Walzer T, Blery M, Chaix J, Fuseri N, Chasson L, Robbins SH, et al. Identification, activation, and selective in vivo ablation of mouse NK cells via NKp46. *Proc Natl Acad Sci U S A* (2007) 104:3384–9. doi:10.1073/pnas.0609692104
9. Takeda K, Cretney E, Hayakawa Y, Ota T, Akiba H, Ogasawara K, et al. TRAIL identifies immature natural killer cells in newborn mice and adult mouse liver. *Blood* (2005) 105:2082–9. doi:10.1182/blood-2004-08-3262
10. Seillet C, Huntington ND, Gangatirkar P, Axelsson E, Minnich M, Brady HJM, et al. Differential requirement for Nfil3 during NK cell development. *J Immunol* (2014) 192:2667–76. doi:10.4049/jimmunol.1302605
11. Vosshenrich CAJ, García-Ojeda ME, Samson-Villéger SI, Pasqualetto V, Enault L, Richard-Le Goff O, et al. A thymic pathway of mouse natural killer cell development characterized by expression of GATA-3 and CD127. *Nat Immunol* (2006) 7:1217–24. doi:10.1038/ni1395
12. Peng H, Jiang X, Chen Y, Sojka DK, Wei H, Gao X, et al. Liver-resident NK cells confer adaptive immunity in skin-contact inflammation. *J Clin Invest* (2013) 123:1444–56. doi:10.1172/JCI66381
13. Constantinides MG, McDonald BD, Verhoef PA, Bendelac A. A committed precursor to innate lymphoid cells. *Nature* (2014) 508:397–401. doi:10.1038/nature13047
14. Klose CSN, Flach M, Möhle L, Rogell L, Hoyler T, Ebert K, et al. Differentiation of type 1 ILCs from a common progenitor to all helper-like innate lymphoid cell lineages. *Cell* (2014) 157:340–56. doi:10.1016/j.cell.2014.03.030
15. Vivier E, Tomasello E, Baratin M, Walzer T, Ugolini S. Functions of natural killer cells. *Nat Immunol* (2008) 9:503–10. doi:10.1038/ni1582
16. Sathe P, Delconte RB, Souza-Fonseca-Guimaraes F, Seillet C, Chopin M, Vandenberg CJ, et al. Innate immunodeficiency following genetic ablation of Mcl1 in natural killer cells. *Nat Commun* (2014) 5:4539. doi:10.1038/ncomms5539
17. Delconte RB, Kolesnik TB, Dagley LF, Rautela J, Shi W, Putz EM, et al. CIS is a potent checkpoint in NK cell-mediated tumor immunity. *Nat Immunol* (2016) 17:816–24. doi:10.1038/ni.3470
18. Vivier E. What is natural in natural killer cells? *Immunol Lett* (2006) 107:1–7. doi:10.1016/j.imlet.2006.07.004
19. Sojka DK, Plougastel-Douglas B, Yang L, Pak-Wittel MA, Artyomov MN, Ivanova Y, et al. Tissue-resident natural killer (NK) cells are cell lineages distinct from thymic and conventional splenic NK cells. *eLife* (2014) 3:e01659–01659.
20. Schuster IS, Wikstrom ME, Brizard G, Coudert JD, Estcourt MJ, Manzur M, et al. TRAIL+ NK cells control CD4+ T cell responses during chronic viral infection to limit autoimmunity. *Immunity* (2014) 41:646–56. doi:10.1016/j.immuni.2014.09.013
21. Thornthwaite JT, Leif RC. Characterization of immune cells using the plaque cytogram assay. *Reticulendothelial Society* (1972).
22. Kiessling R, Klein E, Wigzell H. "Natural" killer cells in the mouse. I. Cytotoxic cells with specificity for mouse Moloney leukemia cells. Specificity and distribution according to genotype. *Eur J Immunol* (1975) 5:112–7. doi:10.1002/eji.1830050208
23. Anderson MJ. Innate cytotoxicity of CBA mouse spleen cells to Sendai virus-infected L cells. *Infect Immun* (1978) 20:608–12.
24. Santoli D, Trinchieri G, Lief FS. Cell-mediated cytotoxicity against virus-infected target cells in humans. I. Characterization of the effector lymphocyte. *J Immunol* (1978) 121:526–31.
25. Karre K, Ljunggren HG, Piontek G, Kiessling R. Selective rejection of H-2-deficient lymphoma variants suggests alternative immune defence strategy. *Nature* (1986) 319:675–8. doi:10.1038/319675a0
26. Takeda K, Hayakawa Y, Smyth MJ, Kayagaki N, Yamaguchi N, Kakuta S, et al. Involvement of tumor necrosis factor-related apoptosis-inducing ligand in surveillance of tumor metastasis by liver natural killer cells. *Nat Med* (2001) 7:94–100. doi:10.1038/83416
27. Kim S, Iizuka K, Kang HS, Dokun A, French AR, Greco S, et al. In vivo developmental stages in murine natural killer cell maturation. *Nat Immunol* (2002) 3:523–8. doi:10.1038/ni796
28. Hayakawa Y, Smyth MJ. CD27 dissects mature NK cells into two subsets with distinct responsiveness and migratory capacity. *J Immunol* (2006) 176:1517–24. doi:10.4049/jimmunol.176.3.1517
29. Huntington ND, Tabarias H, Fairfax K, Brady J, Hayakawa Y, Degli-Esposti MA, et al. NK cell maturation and peripheral homeostasis is associated with KLRG1 up-regulation. *J Immunol* (2007) 178:4764–70.
30. Paust S, Gill HS, Wang B-Z, Flynn MP, Moseman EA, Senman B, et al. Critical role for the chemokine receptor CXCR6 in NK cell–mediated antigen-specific memory of haptens and viruses. *Nat Immunol* (2010) 11:1127–35. doi:10.1038/ni.1953
31. Fuchs A, Vermi W, Lee JS, Lonardi S, Gilfillan S, Newberry RD, et al. Intraepithelial type 1 innate lymphoid cells are a unique subset of IL-12- and IL-15-responsive IFN-gamma-producing cells. *Immunity* (2013) 38(4):769–81. doi:10.1016/j.immuni.2013.02.010
32. Bernink JH, Peters CP, Munneke M, te Velde AA, Meijer SL, Weijer K, et al. Human type 1 innate lymphoid cells accumulate in inflamed mucosal tissues. *Nat Immunol* (2013) 14:221–9. doi:10.1038/ni.2534
33. Daussy C, Faure F, Mayol K, Viel S, Gasteiger G, Charrier E, et al. T-bet and Eomes instruct the development of two distinct natural killer cell lineages in the liver and in the bone marrow. *J Exp Med* (2014) 11:945. doi:10.1084/jem.20131560
34. Cortez VS, Fuchs A, Cella M, Gilfillan S, Colonna M. Cutting edge: salivary gland NK cells develop independently of Nfil3 in steady-state. *J Immunol* (2014) 192(10):4487–9. doi:10.4049/jimmunol.1303469
35. Yagi R, Zhong C, Northrup DL, Yu F, Bouladoux N, Spencer S, et al. The transcription factor GATA3 is critical for the development of All IL-7R-alpha-expressing innate lymphoid cells. *Immunity* (2014) 40:378–88. doi:10.1016/j.immuni.2014.01.012

36. Bernink JH, Krabbendam L, Germar K, de Jong E, Gronke K, Kofoed-Nielsen M, et al. Interleukin-12 and -23 control plasticity of CD127(+) group 1 and group 3 innate lymphoid cells in the intestinal lamina propria. *Immunity* (2015) 43:146–60. doi:10.1016/j.immuni.2015.06.019
37. Bal SM, Bernink JH, Nagasawa M, Groot J, Shikhagaie MM, Golebski K, et al. IL-1beta, IL-4 and IL-12 control the fate of group 2 innate lymphoid cells in human airway inflammation in the lungs. *Nat Immunol* (2016) 17:636–45. doi:10.1038/ni.3444
38. Ohne Y, Silver JS, Thompson-Snipes L, Collet MA, Blanck JP, Cantarel BL, et al. IL-1 is a critical regulator of group 2 innate lymphoid cell function and plasticity. *Nat Immunol* (2016) 17:646–55. doi:10.1038/ni.3447
39. Mackay LK, Minnich M, Kragten NA, Liao Y, Nota B, Seillet C, et al. Hobit and Blimp1 instruct a universal transcriptional program of tissue residency in lymphocytes. *Science* (2016) 352:459–63. doi:10.1126/science.aad2035
40. Gordon SM, Chaix J, Rupp LJ, Wu J, Madera S, Sun JC, et al. The transcription factors T-bet and eomes control key checkpoints of natural killer cell maturation. *Immunity* (2012) 36:55–67. doi:10.1016/j.immuni.2011.11.016
41. Marquardt N, Beziat V, Nystrom S, Hengst J, Ivarsson MA, Kekalainen E, et al. Cutting edge: identification and characterization of human intrahepatic CD49a+ NK cells. *J Immunol* (2015) 194:2467–71. doi:10.4049/jimmunol.1402756
42. Vargas CL, Poursine-Laurent J, Yang L, Yokoyama WM. Development of thymic NK cells from double negative 1 thymocyte precursors. *Blood* (2011) 118:3570–8. doi:10.1182/blood-2011-06-359679
43. Cortez VS, Cervantes-Barragan L, Robinette ML, Bando JK, Wang Y, Geiger TL, et al. Transforming growth factor-beta signaling guides the differentiation of innate lymphoid cells in salivary glands. *Immunity* (2016) 44(5):1127–39. doi:10.1016/j.immuni.2016.03.007
44. Bezman NA, Kim CC, Sun JC, Min-Oo G, Hendricks DW, Kamimura Y, et al. Molecular definition of the identity and activation of natural killer cells. *Nat Immunol* (2012) 13:1000–9. doi:10.1038/ni.2395
45. Smyth MJ, Takeda K, Hayakawa Y, Peschon JJ, van den Brink MR, Yagita H. Nature's TRAIL – on a path to cancer immunotherapy. *Immunity* (2003) 18:1–6. doi:10.1016/S1074-7613(02)00502-2
46. Klose CS, Kiss EA, Schwierzeck V, Ebert K, Hoyler T, d'Hargues Y, et al. A T-bet gradient controls the fate and function of CCR6-RORgammat+ innate lymphoid cells. *Nature* (2013) 494:261–5. doi:10.1038/nature11813
47. Berg RE, Crossley E, Murray S, Forman J. Relative contributions of NK and CD8 T cells to IFN-gamma mediated innate immune protection against *Listeria monocytogenes*. *J Immunol* (2005) 175:1751–7. doi:10.4049/jimmunol.175.3.1751
48. Dunn PL, North RJ. Early gamma interferon production by natural killer cells is important in defense against murine listeriosis. *Infect Immun* (1991) 59:2892–900.
49. Viegas N, Andzinski L, Wu C-F, Komoll R-M, Gekara N, Dittmar KE, et al. IFN-γ production by CD27 +NK cells exacerbates Listeria monocytogenesinfection in mice by inhibiting granulocyte mobilization. *Eur J Immunol* (2013) 43:2626–37. doi:10.1002/eji.201242937
50. Teixeira HC, Kaufmann SH. Role of NK1.1+ cells in experimental listeriosis. NK1+ cells are early IFN-gamma producers but impair resistance to *Listeria monocytogenes* infection. *J Immunol* (1994) 152:1873–82.
51. Badgwell B, Parihar R, Magro C, Dierksheide J, Russo T, Carson WE III. Natural killer cells contribute to the lethality of a murine model of Escherichia coli infection. *Surgery* (2002) 132:205–12. doi:10.1067/msy.2002.125311
52. Kerr AR, Kirkham LA, Kadioglu A, Andrew PW, Garside P, Thompson H, et al. Identification of a detrimental role for NK cells in pneumococcal pneumonia and sepsis in immunocompromised hosts. *Microbes Infect* (2005) 7:845–52. doi:10.1016/j.micinf.2005.02.011
53. Abt MC, Lewis BB, Caballero S, Xiong H, Carter RA, Susac B, et al. Innate immune defenses mediated by two ILC subsets are critical for protection against acute clostridium difficile infection. *Cell Host Microbe* (2015) 18:27–37. doi:10.1016/j.chom.2015.06.011
54. Ljunggren HG, Karre K. In search of the 'missing self': MHC molecules and NK cell recognition. *Immunol Today* (1990) 11:237–44. doi:10.1016/0167-5699(90)90097-S
55. Arase H, Arase N, Saito T. Fas-mediated cytotoxicity by freshly isolated natural killer cells. *J Exp Med* (1995) 181:1235–8. doi:10.1084/jem.181.3.1235
56. Smyth MJ, Cretney E, Takeda K, Wiltrout RH, Sedger LM, Kayagaki N, et al. Tumor necrosis factor-related apoptosis-inducing ligand (TRAIL) contributes to interferon gamma-dependent natural killer cell protection from tumor metastasis. *J Exp Med* (2001) 193:661–70. doi:10.1084/jem.193.6.661
57. Warren HS, Smyth MJ. NK cells and apoptosis. *Immunol Cell Biol* (1999) 77:64–75. doi:10.1046/j.1440-1711.1999.00790.x
58. Campbell KS, Hasegawa J. Natural killer cell biology: an update and future directions. *J Allergy Clin Immunol* (2013) 132:536–44. doi:10.1016/j.jaci.2013.07.006
59. Ochi M, Ohdan H, Mitsuta H, Onoe T, Tokita D, Hara H, et al. Liver NK cells expressing TRAIL are toxic against self hepatocytes in mice. *Hepatology* (2004) 39:1321–31.
60. Dadi S, Chhangawala S, Whitlock BM, Franklin RA, Luo CT, Oh SA, et al. Cancer immunosurveillance by tissue-resident innate lymphoid cells and innate-like T cells. *Cell* (2016) 164:365–77. doi:10.1016/j.cell.2016.01.002
61. Firth MA, Madera S, Beaulieu AM, Gasteiger G, Castillo EF, Schluns KS, et al. Nfil3-independent lineage maintenance and antiviral response of natural killer cells. *J Exp Med* (2013) 210:2981–90. doi:10.1084/jem.20130417
62. Ruggeri L, Capanni M, Urbani E, Perruccio K, Shlomchik WD, Tosti A, et al. Effectiveness of donor natural killer cell alloreactivity in mismatched hematopoietic transplants. *Science* (2002) 295:2097–100. doi:10.1126/science.1068440
63. Lundqvist A, McCoy JP, Samsel L, Childs R. Reduction of GVHD and enhanced antitumor effects after adoptive infusion of alloreactive Ly49-mismatched NK cells from MHC-matched donors. *Blood* (2007) 109:3603–6. doi:10.1182/blood-2006-05-024315
64. Huber CM, Doisne JM, Colucci F. IL-12/15/18-preactivated NK cells suppress GvHD in a mouse model of mismatched hematopoietic cell transplantation. *Eur J Immunol* (2015) 45:1727–35. doi:10.1002/eji.201445200
65. Ghadially H, Ohana M, Elboim M, Gazit R, Gur C, Nagler A, et al. NK cell receptor NKp46 regulates graft-versus-host disease. *Cell Rep* (2014) 7:1809–14. doi:10.1016/j.celrep.2014.05.011
66. Munneke JM, Bjorklund AT, Mjosberg JM, Garming-Legert K, Bernink JH, Blom B, et al. Activated innate lymphoid cells are associated with a reduced susceptibility to graft-versus-host disease. *Blood* (2014) 124:812–21. doi:10.1182/blood-2013-11-536888
67. Seillet C, Rankin LC, Groom JR, Mielke LA, Tellier J, Chopin M, et al. Nfil3 is required for the development of all innate lymphoid cell subsets. *J Exp Med* (2014) 211:1733–40. doi:10.1084/jem.20140145

5

Human NK Cell Subsets in Pregnancy and Disease: Toward a New Biological Complexity

*Costanza Maria Cristiani[1], Eleonora Palella[1], Rosa Sottile[2], Rossana Tallerico[1], Cinzia Garofalo[1] and Ennio Carbone[1,2]**

[1] Department of Experimental and Clinical Medicine, University Magna Graecia of Catanzaro, Catanzaro, Italy, [2] Department of Microbiology, Cell and Tumor Biology, Karolinska Institutet, Stockholm, Sweden

Edited by:
Emanuela Marcenaro, University of Genova, Italy

Reviewed by:
Francisco Borrego, BioCruces Health Research Institute, Spain
Jacques Zimmer, Centre de Recherche Public de la Santé, Luxembourg

Correspondence:
Ennio Carbone
ennio.carbone@ki.se

In humans, NK cells are mainly identified by the surface expression levels of CD56 and CD16, which differentiate between five functionally different NK cell subsets. However, nowadays NK cells are considered as a more heterogeneous population formed by various subsets differing in function, surface phenotype, and anatomic localization. In human CMV- and hantaviruses-infected subjects, an increased frequency of a $NKG2A^{-}CD57^{+}NKG2C^{+}$ NK cell subset has been observed, while the phenotype of the NK cell subpopulation associated with cancer may vary according to the specific kind of tumor and its anatomical location. The healthy human lymph nodes contain mainly the $CD56^{bright}$ NK cell subset while in melanoma metastatic lymph nodes the $CD56^{dim}CD57^{+}KIR^{+}CCR7^{+}$ NK cell subpopulation prevails. The five NK cell subpopulations are found in breast cancer patients, where they differ for expression pattern of chemokine receptors, maturation stage, functional capabilities. In pregnancy, uterine NK cells show a prevalence of the $CD56^{bright}CD16^{-}$ NK cell compartment, whose activity is influenced by KIRs repertoire. This NK cell subset's super specialization could be explained by (i) the expansion of single mature $CD56^{dim}$ clones, (ii) the recruitment and maturation of $CD56^{bright}$ NK cells through specific stimuli, and (iii) the *in situ* development of tumor-resident NK cells from tissue-resident $CD56^{bright}$ NK cells independently of the circulating NK cell compartment. This new and unexpected biological feature of the NK cell compartment could be an important source of new biomarkers to improve patients' diagnosis.

Keywords: NK cells, chemokines, melanoma, anatomical distribution, diagnosis

INTRODUCTION

Innate lymphoid cells (ILCs) are a family of mononuclear hematopoietic cells participating to tissue immunity (1). Natural killer (NK) cells are prototypic members of the ILCs group 1. They are characterized by the capability to recognize and kill virus-infected or transformed target cells through a balance of activating and inhibitory signals given by different surface receptors, in the absence of antigen-specific priming (2). In humans, NK cells are identified by the surface expression of CD56 and the lack of CD3 (3). One of the first well-defined NK cell activating receptors is CD16, the low-affinity receptor for FcγRIII responsible for Ab-dependent cellular cytotoxicity (4). CD56 and CD16 expression levels discriminate between five NK cell subpopulations. $CD56^{dim}CD16^{+}$

NK cells predominate in blood, at sites of inflammation, and show a high cytotoxic potential and broadly express MHC-I-specific inhibitory receptors. The $CD56^{bright}CD16^{-}$ and $CD56^{bright}CD16^{+}$ NK cell subsets predominate in lymph nodes, produce cytokines upon activation, express only one MHC class I-specific inhibitory receptor (NKG2A), display a low cytotoxic potential, and are considered to be a precursor of the terminally differentiated $CD56^{dim}CD16^{+}$ NK cells (5). $CD56^{bright}CD16^{+}$ and $CD56^{-}CD16^{+}$ NK cells are minor NK cell subpopulations whose role is largely unknown (6).

Today, based on receptor repertoire and expression levels, NK cells are considered as a more heterogeneous population formed by various subsets differing in function, surface phenotype, and anatomic localization. These different NK cell subpopulations are involved both in regulating physiological conditions and in the response against viral and tumor disease. Here, we review the main human NK cell subsets discovered in the context of infection, cancer, and pregnancy.

NATURAL KILLER CELL SUBSETS IN HUMAN CYTOMEGALOVIRUS (CMV) INFECTION

As revealed by murine models, NK cells are essential for an effective defense against the herpes virus family and in particular against human and murine CMV (7). The co-evolution of the CMV and our species' immune system is evident considering the virus escape strategies. Human CMV (HCMV) inhibits the expression of human leukocyte antigen (HLA) class I molecules, interfering with antigen presentation to $CD8^{+}$ T cells (8). HCMV genome encodes for two early genes, US2 and US11, that directly interact with the nascent MHC-I, re-routing the molecules from the endoplasmic reticulum to the proteosomal compartment, thus preventing the correct expression of MHC-I on the infected cells surface (9). On the other side, HLA class I molecules loss in HCMV-infected cells reduces the engagement of inhibitory receptors, prompting NK cell effector functions (10). However, HMCV expresses two late genes, UL40 and UL18, that compensate the loss of classical MHC-I molecules overexpressing non-classical MHC-I (HLA-E) (11) or directly recruiting the NK cell inhibitory receptor ITL-2, respectively (12).

The above described viral escape strategy is counteracted by our immune system due to the expansion of a specific NK cell subset that actively recognizes the molecular signature of the HCMV escape strategy. Indeed, the main surface marker characterizing a specific NK cell subset in the context of HCMV infection is the activating receptor NKG2C, a C-type lectin covalently assembled with CD94 that specifically recognizes HLA-E molecules (13). Studies *in vitro* revealed that the interaction between peripheral blood NK cells and HCMV-infected fibroblasts induces the preferential proliferation of $NKG2C^{+}$ NK cell subset through the direct involvement of the CD94/NKG2C receptor (14). A higher proportion of $NKG2C^{+}$ NK cells after HCMV infection have been further observed in children with symptomatic congenital HCMV infection (15) and in $HCMV^{+}$ healthy adults. In this latter case, $NKG2C^{+}$ NK cells preferentially co-express CD57, a surface marker for highly mature NK cells, while they do not express NKG2A, the inhibitory counterpart of NKG2C. Therefore, these NK cells are a unique population of $NKG2A^{-}CD57^{+}NKG2C^{+}$ NK cell clones that are absent in HCMV-seronegative donors (16).

Analyses performed on solid-organ transplanted (SOT) recipients with acute HCMV infection clarified the development of this subset in several discrete steps marked by the acquisition on the NK cell surface of a specific set of receptors: (a) increase of NKG2C amount, (b) acquisition of CD57 expression, and (c) increase of CD57 expression, resulting in the terminal full mature subset phenotype $CD57^{+}NKG2C^{bright}$ HMCV-associated NK cell subset (17). The mechanism by which this NK cell subset interacts with HCMV-infected fibroblast has been modeled *in vitro* and seems to involve the cell adhesion molecule CD2, a co-activating receptor on NK cells, and its ligand CD58. Indeed, the molecular interference of the CD2–CD58 interaction results in a decreased activation of $CD57^{+}NKG2C^{+}$ NK cells with a reduced secretion of TNFα and IFNγ (18).

A similar increase in $NKG2C^{+}$ NK cells was observed in hematopoietic cell transplantation (HCT) recipients who reactivate HCMV after transplantation. In this context, it has been shown that the $NKG2A^{-}CD57^{+}NKG2C^{+}$ NK cells are also equipped with the killer cell immunoglobulin-like receptors (KIRs), which specifically recognize different HLA class I molecules. This latter immune phenotype feature is associated with a potent IFNγ secretory activity. This indicates that HCMV reactivation after HCT results in the expansion of a more mature and educated NK cell subset: $NKG2A^{-}KIRs^{+}CD57^{+}NKG2C^{+}$ NK cells. In addition, during HCMV reactivation in HCT recipients, $NKG2C^{+}$ expanding NK cells predominantly express KIR2DL3 (19). This NK cell repertoire feature is shared also by $HCMV^{+}$ chronic hepatitis patients, where the KIR expressed on $NKG2C^{+}$ NK cells is in most cases specific for self-HLA class I ligands, making the anti-virus specific NK cell subset able to discriminate between HLA-I self virus-infected and healthy cells (20). Moreover, in heart- and lung-transplanted patients, upon HCMV either reactivation or infection, an increased frequency of the NK cell subset expressing the inhibitory receptor LIR-1 recognizing the MHC class I homolog UL18 has been observed (21).

In $HCMV^{+}$ healthy subjects, the activating KIRs (KIR2DS2, KIR2DS4, and KIR3DS1) also play a role in the adaptation of the NK cell compartment to HCMV infection. This activating receptor clusters mark a highly differentiated NK cell subset present in the periphery of $HCMV^{+}$ healthy subjects regardless of NKG2C expression (22). The appearance and expansion of these NK cell subpopulations seem to be HCMV-specific, since the two phenomena are not induced by other human herpes viruses such as Epstein–Barr virus (22, 23).

A recent study demonstrated that HCMV infection was also related to a distinct subset of NK cells characterized by a deficiency in the expression of FcRγ (also known as FcεRIγ), associated with high amounts of NKG2C and low levels of natural cytotoxicity receptors NKp30 and NKp46. It is conceivable that this finding could be an effect of the HCMV infection. From a functional point of view, this NK cell subset responds poorly to HCMV-infected cells directly, yet it increases its efficiency against infected target cells in the presence of HCMV-specific IgG. FcRγ deficiency and the associated phenotype seemed to be due to a down-modulation of the tyrosine kinase SYK, stably maintained through the hypermethylation of a specific region in the SYK promoter DNA sequences (24).

NATURAL KILLER CELL SUBSETS IN HANTAVIRUSES INFECTION

Pathogenic hantaviruses (HTNV) are zoonotic viruses that cause hemorrhagic fever with renal syndrome or hantavirus pulmonary syndrome in humans (25). Both syndromes are characterized by vascular permeability and increased immune activation, with strong cytotoxic lymphocyte expansion of highly mature NKG2A^{-}CD57^{+}NKG2C^{+}CD56^{dim} NK cells (26). A deeper analysis of this subset showed that it also expresses high levels of the early activation marker CD69, the activating NK cell receptors NKG2D and 2B4 and NCRs. This activated phenotype has been proposed to be dependent on the increased surface expression of IL-15 and IL-15Ra on hantavirus-infected endothelial and epithelial cells. On the other hand, HTNV infection weakens the effector responses of activated NK cells by increasing HLA class I expression on endothelial cells (27).

NATURAL KILLER CELL SUBSETS IN CANCER

Although NK cells recognition and killing of tumor cells has been widely demonstrated *in vitro*, their *in vivo* activity against tumors remains less known. Generally, the presence of tumor-infiltrating NK cells indicates an ongoing immune response against the tumor and a better prognosis; however, some reports suggest that it can not correlate with the prognosis or be associated with advanced disease (28). Therefore, to appreciate the real prognostic value of the tumor NK cell infiltrate more investigations are required.

Nowadays, phenotypically distinct NK cell populations have been identified in different tissues, where they could represent specialized subsets mediating different functions (5). The identification and characterization of various NK cell subsets represents, therefore, an important advance in the understanding of tumors biology and could have deep implications in the monitoring and treatment of the disease.

In non-small cell lung cancer (NSCLC) patients, tumor-infiltrating natural killer (TINK) cells show a very peculiar surface molecular pattern, with high levels of surface CD56 and KIRs but lack of expression of CD16. Based on its perforin content, this CD56^{bright}CD16^{-}KIRs^{+} NK cell subset seems to derive from peripheral CD56^{bright}CD16^{-}KIR^{-} NK cells, suggesting that CD56^{bright}CD16^{-} NK cells can acquire KIRs expression within tumor tissues as a consequence of signals from tumor microenvironment (29). Another study demonstrated that TINK cells from NSCLC patients exhibit a significant reduction in CD11b and CD27 expression, two markers mainly used to distinguish murine NK cell subpopulations. This CD11b^{-}CD27^{-} subset is also characterized by the downregulation of CD57, CD16, DNAM-1, and NKp30, while CD127 and CD117 are clearly expressed, and NKG2A is slightly upregulated, suggesting an immature and/or inactive phenotype of these TINK cells (30).

In breast cancer patients, blood NK cells show strong alterations that became more dramatic with the advance of the disease. Blood NK cell compartment display a reduction in activating receptors (NKp30, DNAM-1, NKG2D, 2B4, CD16) and an upregulation of the inhibitory receptors NKG2A and ILT-2, which are functionally related with defective degranulation and ADCC-mediated killing capabilities. These phenotypic and functional changes are more evident in tumor-infiltrating NK cells, in which a prevalence of the CD56^{bright} compartment has been observed (31). Peripheral NK compartment has been further characterized distinguishing for the five subpopulations, whose proportions varying according to the disease stage. Most CD16^{+} NK cells show different combinations of KIRs except KIR2DL4, that in contrast is expressed on the CD56^{dim}CD16^{-} NK cell subpopulation. The CD56^{bright}CD16^{-} NK cell subset lacks the inhibitory KIRs and display low frequencies of KIR2DL4 and KIR2DS4, possibly indicating a reduced functional ability. CD57 is strongly expressed on CD56^{dim}CD16^{+} and CD56^{bright}CD16^{+} NK cells, while is downregulated in the others three subsets, suggesting that they are less mature and differentiated. CD57 expression also correlates with perforin and granzyme B levels, reflecting the cytotoxic potential of the different NK cell subsets. The five NK cell subsets show also difference in the expression pattern of chemokine receptors. CD56^{dim}CD16^{+} NK cells express CX3CR1 and CXCR1 receptors, in accordance with their ability to reach inflamed tissues. CD56^{dim}CD16^{-} NK cells are characterized by CXCR1, CXCR4, and CD62L expression, suggesting a preferential tropism for both inflamed tissues and lymph nodes. CD56^{bright}CD16^{-} and CD56^{bright}CD16^{+} NK cell subsets display similar patterns, with high levels of CCR7, CD117, and CD62L, underlying a favorite tropism for secondary lymphoid organs. Functionally, CD56^{dim}CD16^{+} NK cells have the strongest cytotoxic with a high degranulation potential. Conversely, CD56^{dim}CD16^{-} and CD56^{bright}CD16^{-} NK cell subsets are poorly cytotoxic, producing low amounts IFNγ and TNFα, while the CD56^{bright}CD16^{+} NK cell subpopulation behaves as an intermediary between CD56^{bright}CD16^{-} and CD56^{dim}CD16^{+} NK cells (32).

An increase of NK cell frequency has been observed within the tumor-invaded lymph nodes (TILNs) of advanced stages melanoma patients, with a prevalence of the CD56^{dim} NK cell subpopulation characterized by a high expression of CD57, KIRs, CD69, and the homing chemokine receptor CCR7, suggesting overall a more differentiated effector phenotype actively migrating in the TILNs. Furthermore, peripheral blood CD56^{dim} NK cell subset of these patients has an increased expression of the chemokine receptor CXCR2 recognizing interleukin-8, one of the chemokines found in TILNs (33). TILNs from melanoma patients also contain a peculiar CD56^{bright} NK cell subset co-expressing CD16. This NK cell subpopulation has high percentages of

TABLE 1 | Markers characterizing NK cell subset in pathological and physiological contexts.

Context	Anatomical localization	NK subpopulation	Upregulated markers	Downregulated markers	Reference
Human cytomegalovirus infection	Peripheral blood	CD56^{dim}	NKG2C CD57 KIRs LIR-1	NKG2A FcRg NCRs	(14, 16, 17, 19–22, 24)
Human hantaviruses infection	Peripheral blood	CD56^{dim}	NKG2C CD57 CD69 NKG2D 2B4 NCRs	NKG2A	(26, 27)
Non-small cell lung cancer	Tumor	CD56^{bright}	NKG2A KIRs CD117 CD127	CD16 CD11b CD27 CD57 DNAM-1 NKp30	(30)
Breast cancer	Peripheral blood	CD56^{bright}CD16^{-}	CCR7 CD117 CD62L	KIRs CD57	(32)
		CD56^{bright}CD16^{+}	KIRs (except for KIR2DL4) CD57 CCR7 CD117 CD62L		
		CD56^{dim}CD16^{+}	KIRs (except for KIR2DL4) CD57 CX3CR1 CXCR1		
		CD56^{dim}CD16^{-}	KIR2DL4 CXCR1 CXCR4 CD62L	CD57	
Melanoma	Tumor-infiltrated lymph node	CD56^{dim}	CD57 KIR2 CD69 CCR7		(33)
		CD56^{bright}	CD16 KIRs NKp46 NKG2D		(34)
Ovarian carcinoma	Peripheral blood	CD56^{dim}	PD-1	NKG2A	(35)
	Ascites		KIRs CD57 LIR-1 CD16	NCRs	
Pregnancy	Uterus	CD56^{bright}	NKG2A KIRs ILT-2	CD16	(39, 41–43, 46).

NKp46^{+}, NKG2D^{+}, and KIRs^{+} NK cells and increased levels of perforin, suggesting a more differentiated CD56^{bright} NK cell subset that may exert antitumor activities (34).

Recently, a PD-1^{+} NK cell subpopulation has been found in peripheral blood and ascites from seropapillary ovarian carcinoma patients. PD-1^{+} NK cells are confined to the CD56^{dim} NK cell subset and are primarily composed by NKG2A^{-}KIRs^{+}LIR-1^{+}CD57^{+} NK cells with low levels of NCRs and high expression of CD16 and perforin/granzyme, indicating that PD-1 expression is confined to terminally differentiated NK cells. However, PD-1^{+} NK cells show a low cytolytic activity against tumor targets cells and are poor cytokines producers (35).

NATURAL KILLER CELL SUBSETS IN PREGNANCY

The immunological features of the placentation and pregnancy is another research field where the expansion of specific NK subsets has been appreciated to play a crucial role.

In the first trimester of pregnancy, NK cells represent the 70% of CD45$^+$ decidual leukocytes together with CD8$^+$, CD4$^+$, $\gamma\delta$ T cells, and dendritic cells. These cells are defined "uterine NK" (uNK, also called decidual or dNK) and play a role both in the success of pregnancy and in the early human immunity response during infections. uNK cells are predominant in non-pregnant endometrium, increasing and changing their morphology during the secretory phase of the menstrual cycle (36). Prolactin and endometrial-derived IL-15 promote uNK cells stimulation and differentiation. During early pregnancy, uNK cells infiltrate the trophoblast supporting the placentation process. uNK cells control the remodeling of uterine spiral arteries, increasing the contact area between maternal blood and trophoblast cells (37, 38). uNK cells are identified by a different phenotype and functional characteristics compared to peripheral blood NK cells (pNK); they show a CD56^{bright}CD16$^-$ phenotype and express a repertoire of activating and inhibitory receptors (NKRs) belonging to early differentiation stages of pNK cells (39). uNK cells secrete vascular endothelial growth factor C, angiopoietin 2 (ANG2), placental growth factor (PIGF), cytokines that are involved in angiogenesis, and that do not belong to pNK cytokine profile (40).

Trophoblast cells express MHC-I molecules HLA-C, HLA-E, and HLA-G. HLA-E has high affinity for inhibitory receptor CD94/NKG2A expressed by all uNK cells (41); HLA-C is recognized by KIRs receptor family whose expression is induced directly *in utero*, while HLA-G is bound by ILT-2, that is present on uNK cells (42), and by the activating KIR2DL4 (43). The role of uNK cells during pregnancy has not been completely understood, but the mouse is used as model due to the presence of uNK cells in the decidua and to the expression of MHC repertoire on trophoblast cells. When the MHC-I is bound to NK cell receptors, the uNK cells are inhibited and this interferes with fetal growth and remodeling of decidual spiral arteries (44). In murine pregnancy, only maternal MHC contributes to educate uNK cells to acquire functional properties, while the presence of paternal-inherited allogeneic MHC allele inhibits uNK cells with complications on both fetal growth and arterial remodeling (45). In human pregnancy, KIR repertoire is specific for each mother, and fetal HLA-C (group C1 or C2) is specific for maternal and paternal HLA-C contribution. The presence of maternal KIR AA haplotype, constituted by inhibitory KIRs genes only, is associated with pregnancy disorders but the risk of pre-eclampsia is increased when the fetus presents more C2 genes than the mother or when it carries a paternal HLA-C2 epitope. Instead, KIR genes that are protective are localized on Tel-B region of B haplotype, that contains the inhibitory KIR2DL5A and the activating KIR3DS1, KIR2DS5, and KIR2DS1. Among them, only KIR2DS1 binds to HLA-C, and its maternal frequency correlates with protection from pre-eclampsia (46).

KIR2DL4 has been proposed to favor reproduction through a different mechanism. This activating KIR is expressed in the endosomal compartment of uNK cells, where it binds and accumulates the soluble form of HLA-G secreted by extravillous trophoblasts. KIR2DL4 engagement has been shown to lead to the activation of DNA damage response signaling pathways, resulting in the reprograming of uNK cells to a senescent state characterized by a very active metabolism and secretion of an array of soluble mediators. This regulated and sustained secretory response would shape and remodel the local environment to increase vascularization required for a growing fetus (43).

It has been shown that uNK cells have the capability to control HCMV infection, and actually they acquire cytotoxic phenotype against HCMV-infected decidual fibroblasts *in vitro*, modulating their receptor repertoire. NKG2D and CD94/NKG2C have an important role in uNK cell response, also demonstrated by cytokine production. Moreover, uNK cells are also able to infiltrate HCMV-infected tissues and to generate an immunological synapse. Further investigations are needed to characterize the innate immune defense mechanisms in controlling viral response during pregnancy *in vivo* (47).

CONCLUSION

The deep and new knowledge on the composition of the human NK cell compartment reveals a more complex landscape than that appreciated before. Here, we gave an overview of the NK cell subsets enter recruited in specific pathological and physiological conditions (**Table 1**). This NK cell subsets' super specialization could be possibly explained at least in the following three ways:

(1) ontogenetically determined mature CD56^{dim} NK cell subsets, arising from the circulating or secondary lymphoid tissues NK cell population in humans and equipped with clonally distributed activating and inhibitory NK cell receptors repertoire, scrutinize our tissues continuously preserving our body homeostasis. When the single NK cell subset equipped with the appropriate receptor/s to recognize the pathological associated molecular pattern is called in action as effector or regulatory cells upon the pathological stimulation, it will expand as classically described in mouse upon MCMV infection (48), in a process reminiscent of the adoptive immune memory.

(2) CD56^{bright} immature NK cells are recruited in the lymph nodes, other secondary lymphoid organs or inflammated tissues where they are exposed to stimuli that induce their activation and maturation, i.e., cytokines, cell-to-cell interactions, and chemokine stimulation. The different combinations of biological stimuli will dictate the maturation toward a given subset.

(3) Tumor-resident NK cells could develop locally from tissue-resident CD56^{bright} NK cells independently of the circulating

NK cell compartment, likely due to the organ and/or tumor microenvironment (49).

Whatever the genesis of the repertoire of NK cell subsets in the health and pathology, we foresee in this new biological feature of the NK cell compartment a gold mine to obtain new biomarkers to improve patients' diagnosis.

AUTHOR CONTRIBUTIONS

All the authors contributed in writing and revising the manuscript.

REFERENCES

1. Hazenberg MD, Spits H. Human innate lymphoid cells. *Blood* (2014) 124(5):700–9. doi:10.1182/blood-2013-11-427781
2. Kiessling R, Klein E, Pross H, Wigzell H. "Natural" killer cells in the mouse. II. Cytotoxic cells with specificity for mouse Moloney leukemia cells. Characteristics of the killer cell. *Eur J Immunol* (1975) 5(2):117–21. doi:10.1002/eji.1830050208
3. Lanier LL, Chang C, Azuma M, Ruitenberg JJ, Hemperly JJ, Phillips JH. Molecular and functional analysis of human natural killer cell-associated neural cell adhesion molecule (N-CAM/CD56). *J Immunol* (1991) 146(12):4421–6.
4. Moretta A, Bottino C, Vitale M, Pende D, Cantoni C, Mingari MC, et al. Activating receptors and coreceptors involved in human natural killer cell-mediated cytolysis. *Annu Rev Immunol* (2001) 19:197–223. doi:10.1146/annurev.immunol.19.1.197
5. Cooper MA, Fehniger TA, Caligiuri MA. The biology of human natural killer-cell subsets. *Trends Immunol* (2001) 22(11):633–40. doi:10.1016/S1471-4906(01)02060-9
6. Poli A, Michel T, Thérésine M, Andrès E, Hentges F, Zimmer J. CD56bright natural killer (NK) cells: an important NK cell subset. *Immunology* (2009) 126(4):458–65. doi:10.1111/j.1365-2567.2008.03027.x
7. Biron CA, Nguyen KB, Pien GC, Cousens LP, Salazar-Mather TP. Natural killer cells in antiviral defense: function and regulation by innate cytokines. *Annu Rev Immunol* (1999) 17:189–220. doi:10.1146/annurev.immunol.17.1.189
8. Hengel H, Brune W, Koszinowski UH. Immune evasion by cytomegalovirus - survival strategies of a highly adapted opportunist. *Trends Microbiol* (1998) 6(5):190–7. doi:10.1016/S0966-842X(98)01255-4
9. Story CM, Furman MH, Ploegh HL. The cytosolic tail of class I MHC heavy chain is required for its dislocation by the human cytomegalovirus US2 and US11 gene products. *Proc Natl Acad Sci U S A* (1999) 96(15):8516–21. doi:10.1073/pnas.96.15.8516
10. López-Botet M, Angulo A, Gumá M. Natural killer cell receptors for major histocompatibility complex class I and related molecules in cytomegalovirus infection. *Tissue Antigens* (2004) 63(3):195–203. doi:10.1111/j.1399-0039.2004.00210.x
11. Cerboni C, Mousavi-Jazi M, Wakiguchi H, Carbone E, Kärre K, Söderström K. Synergistic effect of IFN-gamma and human cytomegalovirus protein UL40 in the HLA-E-dependent protection from NK cell-mediated cytotoxicity. *Eur J Immunol* (2001) 31(10):2926–35. doi:10.1002/1521-4141(2001010)31:10<2926::AID-IMMU2926>3.0.CO;2-2
12. Cerboni C, Achour A, Wärnmark A, Mousavi-Jazi M, Sandalova T, Hsu ML, et al. Spontaneous mutations in the human CMV HLA class I homologue UL18 affect its binding to the inhibitory receptor LIR-1/ILT2/CD85j. *Eur J Immunol* (2006) 36(3):732–41. doi:10.1002/eji.200425220
13. Llano M, Lee N, Navarro F, García P, Albar JP, Geraghty DE, et al. HLA-E-bound peptides influence recognition by inhibitory and triggering CD94/NKG2 receptors: preferential response to an HLA-G-derived nonamer. *Eur J Immunol* (1998) 28(9):2854–63. doi:10.1002/(SICI)1521-4141(199809)28:09<2854::AID-IMMU2854>3.0.CO;2-W
14. Gumá M, Budt M, Sáez A, Brckalo T, Hengel H, Angulo A, et al. Expansion of CD94/NKG2C+ NK cells in response to human cytomegalovirus-infected fibroblasts. *Blood* (2006) 107(9):3624–31. doi:10.1182/blood-2005-09-3682
15. Noyola DE, Fortuny C, Muntasell A, Noguera-Julian A, Muñoz-Almagro C, Alarcón A, et al. Influence of congenital human cytomegalovirus infection and the NKG2C genotype on NK-cell subset distribution in children. *Eur J Immunol* (2012) 42(12):3256–66. doi:10.1002/eji.201242752
16. Muntasell A, López-Montañés M, Vera A, Heredia G, Romo N, Peñafiel J, et al. NKG2C zygosity influences CD94/NKG2C receptor function and the NK-cell compartment redistribution in response to human cytomegalovirus. *Eur J Immunol* (2013) 43(12):3268–78. doi:10.1002/eji.201343773
17. Lopez-Vergès S, Milush JM, Schwartz BS, Pando MJ, Jarjoura J, York VA, et al. Expansion of a unique CD57+ NKG2Chi natural killer cell subset during acute human cytomegalovirus infection. *Proc Natl Acad Sci U S A* (2011) 108(36):14725–32. doi:10.1073/pnas.1110900108
18. Rölle A, Halenius A, Ewen EM, Cerwenka A, Hengel H, Momburg F. CD2-CD58 interactions are pivotal for the activation and function of adaptive natural killer cells in human cytomegalovirus infection. *Eur J Immunol* (2016) 46(10):2420–5. doi:10.1002/eji.201646492
19. Foley B, Cooley S, Verneris MR, Pitt M, Curtsinger J, Luo X, et al. Cytomegalovirus reactivation after allogeneic transplantation promotes a lasting increase in educated NKG2C+ natural killer cells with potent function. *Blood* (2012) 119(11):2665–74. doi:10.1182/blood-2011-10-386995
20. Béziat V, Dalgard O, Asselah T, Halfon P, Bedossa P, Boudifa A, et al. CMV drives clonal expansion of NKG2C+ NK cells expressing self-specific KIRs in chronic hepatitis patients. *Eur J Immunol* (2012) 42(2):447–57. doi:10.1002/eji.201141826
21. Berg L, Riise GC, Cosman D, Bergström T, Olofsson S, Kärre K, et al. LIR-1 expression on lymphocytes, and cytomegalovirus disease in lung-transplant recipients. *Lancet* (2003) 361(9363):1099–101. doi:10.1016/S0140-6736(03)12855-3
22. Béziat V, Liu LL, Malmberg JA, Ivarsson MA, Sohlberg E, Björklund AT, et al. NK cell responses to cytomegalovirus infection lead to stable imprints in the human KIR repertoire and involve activating KIRs. *Blood* (2013) 121(14):2678–88. doi:10.1182/blood-2012-10-459545
23. Hendricks DW, Balfour HH Jr, Dunmire SK, Schmeling DO, Hogquist KA, Lanier LL. Cutting edge: NKG2C(hi)CD57+ NK cells respond specifically to acute infection with cytomegalovirus and not Epstein-Barr virus. *J Immunol* (2014) 192(10):4492–6. doi:10.4049/jimmunol.1303211
24. Lee J, Zhang T, Hwang I, Kim A, Nitschke L, Kim M, et al. Epigenetic modification and antibody-dependent expansion of memory-like NK cells in human cytomegalovirus-infected individuals. *Immunity* (2015) 42(3):431–42. doi:10.1016/j.immuni.2015.02.013
25. Vaheri A, Strandin T, Hepojoki J, Sironen T, Henttonen H, Mäkelä S, et al. Uncovering the mysteries of hantavirus infections. *Nat Rev Microbiol* (2013) 11(8):539–50. doi:10.1038/nrmicro3066
26. Björkström NK, Lindgren T, Stoltz M, Fauriat C, Braun M, Evander M, et al. Rapid expansion and long-term persistence of elevated NK cell numbers in humans infected with hantavirus. *J Exp Med* (2011) 208(1):13–21. doi:10.1084/jem.20100762
27. Braun M, Björkström NK, Gupta S, Sundström K, Ahlm C, Klingström J, et al. NK cell activation in human hantavirus infection explained by virus-induced IL-15/IL15Rα expression. *PLoS Pathog* (2014) 10(11):e1004521. doi:10.1371/journal.ppat.1004521
28. Lefebvre S, Antoine M, Uzan S, McMaster M, Dausset J, Carosella ED, et al. Specific activation of the non-classical class I histocompatibility HLA-G antigen and expression of the ILT2 inhibitory receptor in human breast cancer. *J Pathol* (2002) 196(3):266–74. doi:10.1002/path.1039
29. Carrega P, Morandi B, Costa R, Frumento G, Forte G, Altavilla G, et al. Natural killer cells infiltrating human nonsmall-cell lung cancer are enriched in CD56 bright CD16(-) cells and display an impaired capability to kill tumor cells. *Cancer* (2008) 112(4):863–75. doi:10.1002/cncr.23239
30. Jin J, Fu B, Mei X, Yue T, Sun R, Tian Z, et al. CD11b(-)CD27(-) NK cells are associated with the progression of lung carcinoma. *PLoS One* (2013) 8(4):e61024. doi:10.1371/journal.pone.0061024

31. Mamessier E, Sylvain A, Thibult ML, Houvenaeghel G, Jacquemier J, Castellano R, et al. Human breast cancer cells enhance self tolerance by promoting evasion from NK cell antitumor immunity. *J Clin Invest* (2011) 121(9):3609–22. doi:10.1172/JCI45816
32. Mamessier E, Pradel LC, Thibult ML, Drevet C, Zouine A, Jacquemier J, et al. Peripheral blood NK cells from breast cancer patients are tumor-induced composite subsets. *J Immunol* (2013) 190(5):2424–36. doi:10.4049/jimmunol. 1200140
33. Ali TH, Pisanti S, Ciaglia E, Mortarini R, Anichini A, Garofalo C, et al. Enrichment of CD56(dim)KIR + CD57 + highly cytotoxic NK cells in tumour-infiltrated lymph nodes of melanoma patients. *Nat Commun* (2014) 5:5639. doi:10.1038/ncomms6639
34. Messaoudene M, Fregni G, Fourmentraux-Neves E, Chanal J, Maubec E, Mazouz-Dorval S, et al. Mature cytotoxic CD56(bright)/CD16(+) natural killer cells can infiltrate lymph nodes adjacent to metastatic melanoma. *Cancer Res* (2014) 74(1):81–92. doi:10.1158/0008-5472.CAN-13-1303
35. Pesce S, Greppi M, Tabellini G, Rampinelli F, Parolini S, Olive D, et al. Identification of a subset of human natural killer cells expressing high levels of programmed death 1: a phenotypic and functional characterization. *J Allergy Clin Immunol* (2016). doi:10.1016/j.jaci.2016.04.025
36. Bulmer JN, Morrison L, Longfellow M, Ritson A, Pace D. Granulated lymphocytes in human endometrium: histochemical and immunohistochemical studies. *Hum Reprod* (1991) 6(6):791–8.
37. Gubbay O, Critchley HO, Bowen JM, King A, Jabbour HN. Prolactin induces ERK phosphorylation in epithelial and CD56(+) natural killer cells of the human endometrium. *J Clin Endocrinol Metab* (2002) 87(5):2329–35. doi:10.1210/jcem.87.5.8515
38. Dunn CL, Critchley HO, Kelly RW. IL-15 regulation in human endometrial stromal cells. *J Clin Endocrinol Metab* (2002) 87(4):1898–901. doi:10.1210/ jcem.87.4.8539
39. El Costa H, Casemayou A, Aguerre-Girr M, Rabot M, Berrebi A, Parant O, et al. Critical and differential roles of NKp46- and NKp30-activating receptors expressed by uterine NK cells in early pregnancy. *J Immunol* (2008) 181(5):3009–17. doi:10.4049/jimmunol.181.5.3009
40. Li XF, Charnock-Jones DS, Zhang E, Hiby S, Malik S, Day K, et al. Angiogenic growth factor messenger ribonucleic acids in uterine natural killer cells. *J Clin Endocrinol Metab* (2001) 86(4):1823–34. doi:10.1210/jc.86.4.1823
41. King A, Allan DS, Bowen M, Powis SJ, Joseph S, Verma S, et al. HLA-E is expressed on trophoblast and interacts with CD94/NKG2 receptors on decidual NK cells. *Eur J Immunol* (2000) 30(6):1623–31. doi:10.1002/1521-4141(200006)30:6<1623::AID-IMMU1623>3.0.CO;2-M
42. Sharkey AM, Gardner L, Hiby S, Farrell L, Apps R, Masters L, et al. Killer Ig-like receptor expression in uterine NK cells is biased toward recognition of HLA-C and alters with gestational age. *J Immunol* (2008) 181(1):39–46. doi:10.4049/jimmunol.181.1.39
43. Rajagopalan S. HLA-G-mediated NK cell senescence promotes vascular remodeling: implications for reproduction. *Cell Mol Immunol* (2014) 11(5):460–6. doi:10.1038/cmi.2014.53
44. Long EO, Barber DF, Burshtyn DN, Faure M, Peterson M, Rajagopalan S, et al. Inhibition of natural killer cell activation signals by killer cell immunoglobulin-like receptors (CD158). *Immunol Rev* (2001) 181:223–33. doi:10.1034/j.1600-065X.2001.1810119.x
45. Kieckbusch J, Gaynor LM, Moffett A, Colucci F. MHC-dependent inhibition of uterine NK cells impedes fetal growth and decidual vascular remodelling. *Nat Commun* (2014) 5:3359. doi:10.1038/ncomms4359
46. Hiby SE, Apps R, Sharkey AM, Farrell LE, Gardner L, Mulder A, et al. Maternal activating KIRs protect against human reproductive failure mediated by fetal HLA-C2. *J Clin Invest* (2010) 120(11):4102–10. doi:10.1172/ JCI43998
47. Siewiera J, El Costa H, Tabiasco J, Berrebi A, Cartron G, Le Bouteiller P, et al. Human cytomegalovirus infection elicits new decidual natural killer cell effector functions. *PLoS Pathog* (2013) 9(4):e1003257. doi:10.1371/journal. ppat.1003257
48. Sun JC, Beilke JN, Lanier LL. Adaptive immune features of natural killer cells. *Nature* (2009) 457(7229):557–61. doi:10.1038/nature07665
49. Melsen JE, Lugthart G, Lankester AC, Schilham MW. Human circulating and tissue-resident CD56(bright) natural killer cell populations. *Front Immunol* (2016) 7:262. doi:10.3389/fimmu.2016.00262

6

Natural Killer Cell Development and Maturation Revisited: Possible Implications of a Novel Distinct Lin^{-}CD34^{+}DNAM-1^{bright}CXCR4^{+} Cell Progenitor

Federica Bozzano[1,2], *Francesco Marras*[3] *and Andrea De Maria*[2,4,5]*

[1] Department of Experimental Medicine (DIMES), University of Genova, Genova, Italy, [2] Center of Excellence for Biomedical Research (CEBR), University of Genova, Genova, Italy, [3] Istituto Giannina Gaslini, Genova, Italy, [4] Clinica Malattie Infettive, IRCCS AOU San Martino-IST Genova, Istituto Nazionale per la Ricerca sul Cancro, Genova, Italy, [5] Department of Health Sciences, DISSAL, University of Genova, Genova, Italy

Edited by:
Luigi Daniele Notarangelo, Harvard Medical School, USA

Reviewed by:
Rafael Solana, University of Córdoba, Spain
Emily Mace, Baylor College of Medicine, USA

***Correspondence:**
Andrea De Maria
de-maria@unige.it

Since the first description of natural killer (NK) cells, the view on their role in innate immunity has evolved considerably. In addition to first-line defense against transformed and pathogen-infected autologous cells, NK cells contribute to modulate adaptive immune responses and in some cases acquire specialized functions, including exhausted, adaptive, and decidual NK cells. NK cells derive from CD34^{+} progenitors, *in vivo* and *in vitro*; however, it is unclear whether the high phenotype diversity *in vivo* may be generated from these precursors alone. The recent characterization of a novel CD34^{+}DNAM-1^{bright}CXCR4^{+} precursor giving rise to apparently licensed and functional maturing NK cells may suggest the possibility for a higher than expected common lymphocyte precursor diversity and a consequently higher peripheral NK cell phenotype variability. Here, we review the evidences on NK cell central and peripheral development from CD34^{+} precursors and propose a possible updated reading frame based on the characterization of CD34^{+}DNAM-1^{bright}CXCR4^{+} cell progenies, which favors the possibility of concurrent NK cell maturation from different CD34^{+} precursors.

Keywords: natural killer, NK cell development, CD34, DNAM-1, common lymphoid progenitors

INTRODUCTION

Natural killer (NK) cells, are a central component of the innate immune response (1) and constitute the first line of defense against a variety of tumors and microbial pathogens (2–5).

Over recent years, it has become clear that their role exceeds the boundaries of the original assignment to patrol the tissues as a first line of defense and rather also includes regulatory and editing functions of the innate and adaptive immune response.

Opposite to T or B cells, NK cells do not undergo somatic rearrangement of genes coding for antigen-specific receptors. Their functional characteristics include production and release of IFNγ and TNFα and also G-CSF. In addition they may produce chemokines and IL-8, and under specialized and limited conditions also IL-10, IL-6, and IL-1. Following the identification and characterization of innate lymphoid cells (ILCs) into at least three lineages characterized by different phenotype, homing, and function (6, 7), NK cells have been proposed to be included into group 1 ILC (1) based on their expression of NFIL3, Tbet, and Eomes transcription factors (1, 8).

Human NK cells normally constitute 5–15% of peripheral blood (PB) lymphocytes. The majority of NK cells, are present in relative abundance in bone marrow (BM), liver, uterus, spleen, and lungs, as well as to a lesser extent in secondary lymphoid tissue (SLT), mucosa-associated lymphoid tissue, and the thymus. The classical description of NK cell phenotypes relies on CD56 and CD16 expression with the distinction of three broad phenotypes which include $CD16^{+/-}CD56^{bright}$ cells representing a minority of circulating and the majority of tissue-associated NK cells, $CD16^{+}CD56^{dim}$ cells that constitute the majority of circulating NK cells and are viewed as effectors cells, and $CD16^{+}CD56^{-}$ exhausted NK cells that are poorly functional express low levels of natural cytotoxicity receptors (NCRs) and may become more abundant in PB during chronic infections, such as HIV-1 infection. A peripheral development of NK cells has been shown to take place beyond BM and lymphnode niches. NK cells undergo a progressive development of $CD56^{bright}$ into $CD56^{dim}$ NK cells with a progressive loss of NKG2A and concomitant progressive expression of KIRs, CD57, and NKG2C into terminally differentiated NK cells (9–11).

The original view encompassing a tripartite subset characterization has recently been updated in view of NK cell phenotype diversity. Using mass cytometry, it has become evident that in healthy adult humans a much larger number of distinct NK cell phenotypes are simultaneously present at any time. From 6,000 to 30,000 different NK cell phenotypes have been identified by mass in a single donor, and in any small group of persons up to 100,000 different NK cell phenotypes may be detected (12). In addition, by computer-assisted flow-cytometric analysis, at least 5–8 distinct subsets may be identified using a set of three NK cells specific monoclonal antibodies beyond CD16 and CD56 (11). This abundance of NK cell phenotype diversity is determined by combinatorial expression of the multitude of receptors and co-receptors present on their surface. In this regard, there are both evidences for an inherent intrinsic or genetically determined driver for the persistence of some phenotypes (mostly accounting for KIR variability), and for an extrinsic or environmental influence on the prevalence of other phenotypes supported by foreign antigenic stimulation supporting the diversity for NCR representation and expression (12).

CD34 NK CELL PRECURSORS

Similar to other blood cells, NK cells derive from hematopoietic stem cells (HSCs) and can be grown *in vitro* from lymphoid-restricted multipotent progenitors that may retain B and/or T lymphocyte developmental potential (13–15). The classical model of hematopoiesis postulates that the earliest fate decision toward NK cells downstream of HSCs is represented by the divergence of lymphoid and myeloid lineages. Erythroid and megakaryocyte lineages branch off before the lymphoid–myeloid split. This step is followed by myeloid–lymphoid divergence where common lymphoid progenitors (CLPs), and common myeloid progenitors (6) are generated. Accordingly, the CLP group would not include cell progenitors with myeloid potential. In contrast to mouse hematopoiesis, definitive evidence for a comprehensive model that best describes human hematopoiesis is still to be completely defined (16). Recently, a different pattern of cell maturation has been proposed following *ex vivo* and *in vivo* results in humans. Analysis of human cord blood (CB) and BM using seven distinct markers, including CD45RA, CD135 (Flt3), CD7, CD10, CD38, and CD90, allowed the identification of seven distinct progenitor cell classes (17). In this setting, some cells are described as multi-lymphoid progenitors (MLPs), defined by $CD34^{+}CD38^{-}Thy\text{-}1^{neg\text{-}low}CD45RA^{+}$, belong to the CLP group and are able, in specific culture conditions, to give rise to all lymphoid cells as well as monocytes, macrophages, and dendritic cells (DCs) (18, 19). Among these MLPs included in this last model, NK cells derive from $CD34^{+}$ hematopoietic stem cells (HPC) precursors originally identified in BM (20). However, $CD34^{+}$ cells giving rise to NK cell progeny have been detected also in PB, thymus, lymphnodes, CB, GALT, and decidua (21, 22). In addition, other reports indicate that T and NK cells are generated from non-characterized bipotent T/NK common progenitors, which may circulate in PB of healthy donors (HDs), albeit at very low frequencies (23, 24). While it is agreed that $CD34^{+}$ NK cell progenitors reside in the BM, there is a less clear view on whether seeding of these cells into other organs generates organ-specific NK cell maturation, or whether a predefined CLP or MLP with specific developmental and homing characteristics would exit under certain conditions from the BM and specifically seed into the final sites of maturation.

NK CELL MATURATION

Distinct stages of development of NK cells from HPC have been described with an orderly and staged acquisition of NK cell markers, and distinct maturational stages (1). Five stages of human NK cell development have been described (25). Stage 1–2 $CD34^{+}CD45RA^{-/+}Cd10^{+/-}CD117^{-/+}$ cells have been observed in human SLT and retain non-NK cell lineage potential since under optimal *in vitro* conditions they can develop into T and DC cells. This development potential is lost in the third stage in which may identify committed immature NK (iNK) cells.

The acquisition of the interleukin 15 (IL-15R) receptor beta chain (CD122) marks an important step of NK cell differentiation, since IL15 promotes NK cell differentiation, functional maturation, and survival in both mouse and human (26). Thus, IL-15R expression identifies an NK cell precursor subset defined by developmental potential in response to IL-15, by lack of functional immunophenotype observed in mature NK cells and by lack of other Lineage specific surface antigen as CD3, CD14, and CD19. Two populations of IL-15-responsive $Lin^{-}CD94^{-}$NK differentiating intermediates have been identified ($Lin^{-}CD34^{dim}CD45RA^{+}$ $alpha4beta7^{bright}CD117^{+}CD161^{+/-}CD94^{-}$ stage 2 and $Lin^{-}CD34^{-}$ $alpha4beta7^{-}CD117^{+}CD161^{+}CD94^{-}$ stage 3). They are enriched in the interfollicular T cell-rich areas of secondary lymphoid organs where their putative progeny, $CD56^{bright}CD94^{+}$ NK cells, also resides (25, 27, 28). This anatomical localization has been attributed to specific trafficking of BM derived NK cell precursors to SLT *via* high endothelial venules and would be mediated by high expression of CD62L on circulating $Lin^{-}CD94^{-}NK$ differentiating intermediates (28). NK cell differentiation then progresses by orderly acquisition of CD161, CD56, CD94/NKG2A, NKp46, NKG2D, KIRs and functional receptors CD16 (27, 29, 30). The role of CD56 during NK cell development is yet undefined while acquisition of CD94, which then persists on PB $CD56^{bright}$ NK cells and is needed for surface expression of NKG2A or NKG2C, signals the transition to stage 4 in SLT and NK cell maturation is completed with transition from stage 4 $CD56^{bright}CD16^{+/-}$ to stage 5 in $CD56^{dim}CD16^{+}$ NK cells (9, 31). A source of possible confusion is represented by work showing that cells of myeloid lineage may, under certain specific conditions, generate NK cells *in vitro* (32). This work concentrates only on cord-blood $CD34^{+}$ cells under particular conditions (32). The view that NK cells may be derived *in vitro* together with myelomonocytes without evidence for T cell growth reflects work by other groups as well (27, 33).

Some caution is needed when considering these models, which are nevertheless useful to provide a general scaffolding to understand NK cell peripheral maturation. Opposite to the model of peripheral T cell maturation from which it has been shifted for practical purposes, NK cell maturative migration between stages is not a one-way process. For example, NK cells may revert form terminal differentiation and, under favorable conditions *in vitro* (e.g., IL18 supplementation), may modify surface receptor expression with upregulation of CCR7, CD83, and CD25 and downregulation of CD16 (34). Furthermore, *in vivo*, NK cell education, epistatic interaction with KIR genes and viral infection or other environmental stimuli have a marked bearing on NK cell repertoire phenotype and activating and inhibitory receptor expression (35, 36).

An additional layer of entanglement to a linear model of NK cell development has been represented by the suggestion that NK cells may represent a subset of ILC (37). ILC have been shown *in vitro* derived from $CD34^{+}$ cells isolated preferentially from the CB compared to PB (38). Indeed, the recent demonstration of the possibility of an elective ILC deficiency in humans without NK cell deficiency shows that ILCs might be dispensable in natural conditions and that developmental pathways for NK cell and ILC development are distinct (37).

OUTLIERS TO A LINEAR SINGLE-CELL MODEL OF NK CELL DEVELOPMENT

There are additional outliers to a model of sequential NK cell development that cannot be apparently reconciled with a single-cell maturation scheme for NK cells, so far. These include the observation of adaptive or memory-like NK cell responses, the appearance of $CD56^{-}CD16^{+}$ exhausted NK cells in some clinical conditions, and the origin of NK cells in decidua.

In mice, infection with MCMV determines the expansion of specific NK cell subsets (39–41), which maintain for prolonged periods of time the ability to produce increased amounts of TNFα and IFNγ. This observation is reminiscent of memory T cell function thus suggesting a possible memory-like or adaptive feature of NK cells. This pattern has been observed also in humans and predominantly relates to HCMV previous infection (42–45). Human adaptive NK cell expansions are monomorphic. Indeed, only increased proportions of $NKG2C^{+}$ cells appear in PB, irrespective of the different invading pathogens that have been so far able to induce such NK cell expansions, including HCMV or Hantavirus or Chikungunya (46, 47). $NKG2C^{+}$ NK cells expansions persist (48) after acute HCMV infection into latency, and may be observed also after BM transplantation (48, 49). Similar to virus-induced adaptive NK cells (50, 51), $NKG2C^{+}$ NK cells may be obtained after cytokine induction (52). Active research in this area so far did not reach conclusive evidence that these memory-like NK cell expansions occur as a single terminal event along the previously described pathway of peripheral NK cell development. Additional work and efforts are needed to directly answer some crucial questions in this area. Specific trials and work will need to be designed to understand (a) whether the increase in adaptiveNKG2C^{+} NK cells during Hantavirus or other RNA virus infections represents an HCMV-independent event, or rather reflects a recall response of $NKG2C^{+}$ adaptive NK cells in $HCMV^{+}$ patients with latent infection, (b) whether only viruses or rather other pathogens may associate with $NKG2C^{+}$ NK cell expansions, (c) why only $NKG2C^{+}$ adaptive NK cells represent a recall response to invading viruses with different antigenic and PAMP characteristics (e.g., HCMV is a DNA-virus, Chikungunya is a RNA virus, no shared molecular patterns have been described), and (d) what is the advantage in terms of virus control or host survival provided by this quite specific HCMV-associated adaptive NK cell response.

Another apparent outlier to sequential NK cell subset development is represented by $CD56^{neg}CD16^{+}$ exhausted NK cells, which have been described for the first time over 20 years ago (53). These cells may represent up to 20–40% of all NK cells or 3–6% of all lymphocytes in HIV^{+} patients while they represent a rare population in the PB of HDs. Immunophenotypic analyses revealed that cell surface receptors expressed on $CD56^{neg}CD16^{+}$ cells overlap with that of so-called "stage 3" iNK cells and are able, albeit to a reduced extent, to kill target cells and produce chemokines (54). Thus, it appears unlikely that $CD56^{neg}$ NK cells represent the progeny of iNK cells. Effector molecule expression by $CD56^{neg}$ NK cells further support the possibility that these cells are more closely related to and share characteristics with more highly differentiated $CD56^{dim}$ NK cells. Additional comprehensive studies of this subset are needed, in order to clarify when and under which stimuli aberrant differentiation into $CD56^{neg}CD16^{+}$ NK cells occurs as well as whether it is reversible or not.

Natural killer cells also localize in differentiated tissues including non-pregnant endometrium (55). Following embryo implantation, decidualization of human endometrium is associated with a massive recruitment of NK cells that will build up and may represent as many as 50–90% of lymphoid cells present in decidual tissue. Decidual NK cells (dNK) numbers progressively decrease from mid-gestation onwards (56). dNK cells have unique phenotypic properties and functional profile and are $CD56^{bright}CD16^{+/-}$ KIR^{+} cells.

Analysis of gene expression in dNK revealed relevant differences with both CD56^{bright} and CD56^{dim} peripheral NK cell subset. CD9, galectin, alpha-1 integrin, and other adhesion molecules are over-expressed in dNK (57), express major activating NK receptors, including NKp46, NKp30, NKG2D, and DNAM-1, and contain high levels of perforin and granzymes (comparable to CD56^{dim} peripheral NK cells), but have a poor ability to kill classical NK target cells (22, 58–60). dNK cells are able to release high amounts of cytokines and chemokines (including IL-8, VEGF, SDF-1, and IP-10), which are involved in tissue remodeling, trophoblast migration, and/or neo-angiogenesis and placentation. Thus, dNK cells appear to play an unexpected role in vessel formation and tissue building rather than their classical defensive role and these findings further underline the plasticity of NK cells, possibly induced by the particular microenvironment. Whether dNK cells represent a peripheral differentiation of other NK subsets or whether they represent a specific lineage derivation from a common precursor is so far poorly understood.

A NOVEL CD34$^+$ PRECURSOR AND A NEW PERSPECTIVE

Recently, a novel, previously uncharacterized CLP has been identified and is defined by Lin$^-$CD34$^+$ DNAM-1^{bright}CXCR4$^+$ markers (60). These precursors were found enriched in PB of patients with chronic inflammatory conditions of either infectious or non-infectious origin. Based on available data on CD34$^+$ cell maturation niches (61–64), the phenotype of these cells suggest that they represent recent migrants from BM that still retain CXCR4 and DNAM-1, which are derived from endosteal niches following bone remodeling during chronic inflammation (65, 66). In HSC, BM donor Lin$^-$CD34$^+$DNAM-1^{bright} cells represent only a fraction (15%) of mobilized CD34$^+$ cells. A comparison of chemokine receptor expression provided clues to a different end-organ circulation of these cells compared to classical CD34$^+$DNAM$^-$ progenitors. Lin$^-$CD34$^+$DNAM-1^{bright}CXCR4$^+$ cells appear to have the potential of trafficking not only into lymphnodes or GALT *via* CD62L/L interactions, but also (or rather) to peripheral inflamed tissues along fractalkine or IL-8 in view of their higher expression of CX3CR1$^+$ and CXCR1$^+$ and lower proportion of CD62L.

Analysis of transcription factors of these novel CD34$^+$ cells showed that, they have a different array of transcription factors, including Tbet and FoxP3 in addition to Id2, E4BP4, which are expressed in classical CD34$^+$DNAM$^-$CXCR4$^-$ cells purified from CB. A wide difference in transcriptional signature was further confirmed and expanded by microarray analysis of purified CD34$^+$ cells. Interestingly, their abundant transcription of metalloproteases supports the idea of a direct exit in areas of osteoclast resorption. Further, and in line with previous reports that failed to pinpoint the exact nature of the progenitor cells (29, 67), this novel CLP could give rise to NK and T cells but not to myelomonocytes. On the contrary, CB-derived CD34$^+$ cells give rise *in vitro* only to NK cells and myelomonocytes but not to T cells or NKT cells.

Interestingly, when considering the characteristics of NK cells *in vitro* derived from these precursors, some remarkable differences are evident in comparison to NK cells derived from CD34$^+$DNAM$^-$CXCR4$^-$ progenitors. Lin$^-$CD34$^+$DNAM-1^{bright}CXCR4$^+$ derived maturing NK cells appear to have a much more mature and licensed phenotype, as they express KIRs and perforin, high levels of NCRs, DNAM-1 and NKG2D and also produce IFNγ when triggered. These characteristics are unseen in maturing NK cells derived from CD34$^+$ CB cells, which under the same culture conditions do not produce IFNγ, and are NCR low, KIR$^-$, DNAM-1$^{+/-}$, and NKG2D$^{low/neg}$.

The question remains open on where Lin$^-$CD34$^+$DNAM-1^{bright}CXCR4$^+$ cells fit with the classical known human NK cell progenitor hierarchy (18, 19) and why only very low levels of circulating CLP are detectable in HDs (16), while they may be greatly increased during systemic inflammation. According to the study by Doulatov et al. (19), Lin$^-$CD34$^+$DNAM-1^{bright}CXCR4$^+$ cells would (surprisingly) fit in the group of megakaryocyte/erythroid precursors, characterized by the CD38$^+$CD10$^-$CD7$^-$Flt3 phenotype similar to that of Lin$^-$CD34$^+$DNAM-1^{bright} cells. In the absence of experiments carried out with culture conditions favoring different pathways for precursor differentiation, one cannot exclude that different progenies might be obtained.

Lin$^-$CD34$^+$DNAM-1^{bright}CXCR4$^+$ NK cell progeny includes a full array of the classical phenotypes, including CD56^{bright}, CD56^{dull}, and CD56$^-$CD16$^+$ NK cells subsets in addition to NKT CD3$^+$CD56$^+$ cells and to T cells but no cells of monocyte/myelomonocytic lineage (68). In view of the quite different phenotype of maturing NK cells derived from these precursors *in vitro* as compared to NK cells maturing from CD34$^+$DNAM-1$^-$CXCR4$^-$ CB cells, it is tempting to hypothesize that the so far acknowledged model for NK cell differentiation and maturation from a single progenitor into all the known phenotypes and subsets may need renewed evaluation (**Figures 1A,B**).

Since an until recently uncharacterized CD34 precursor with distinct transcriptional signature and phenotype gives rise to NK cells with different phenotypic and functional characteristics, the hypothesis may be proposed that a parallel development of some NK cell phenotypes may take place *in vivo* from two different CD34$^+$ precursors (i.e., Lin$^-$CD34$^+$DNAM-1^{bright}CXCR4$^+$ and CD34$^+$DNAM-1$^-$CXCR4$^-$). It is, therefore, possible that a good number of the diverse NK cell phenotypes observed by mass cytometry (12) and possibly also some of the special subsets of NK cells observed *in vivo* may derive from different developmental stages of the two CD34 precursors. This view could be in line with the data by Doulatov et al. and Laurenti et al. (18, 19) where different CLP may give rise to different progenies but the same progeny may derive from different CLPs (17, 18). In addition, it should be underscored that recent work by Wu and colleagues (69) in elegant experiments of clonal tracking has identified a quite surprising origin of NK cells in macaques. Barcoding experiments show that in these animals, a progenitor different from B/T/Myeloid lineage stem cells gives rise to CD16$^+$CD56$^-$ NK cell progeny, and more importantly, parallel development of different NK cell phenotypes derive

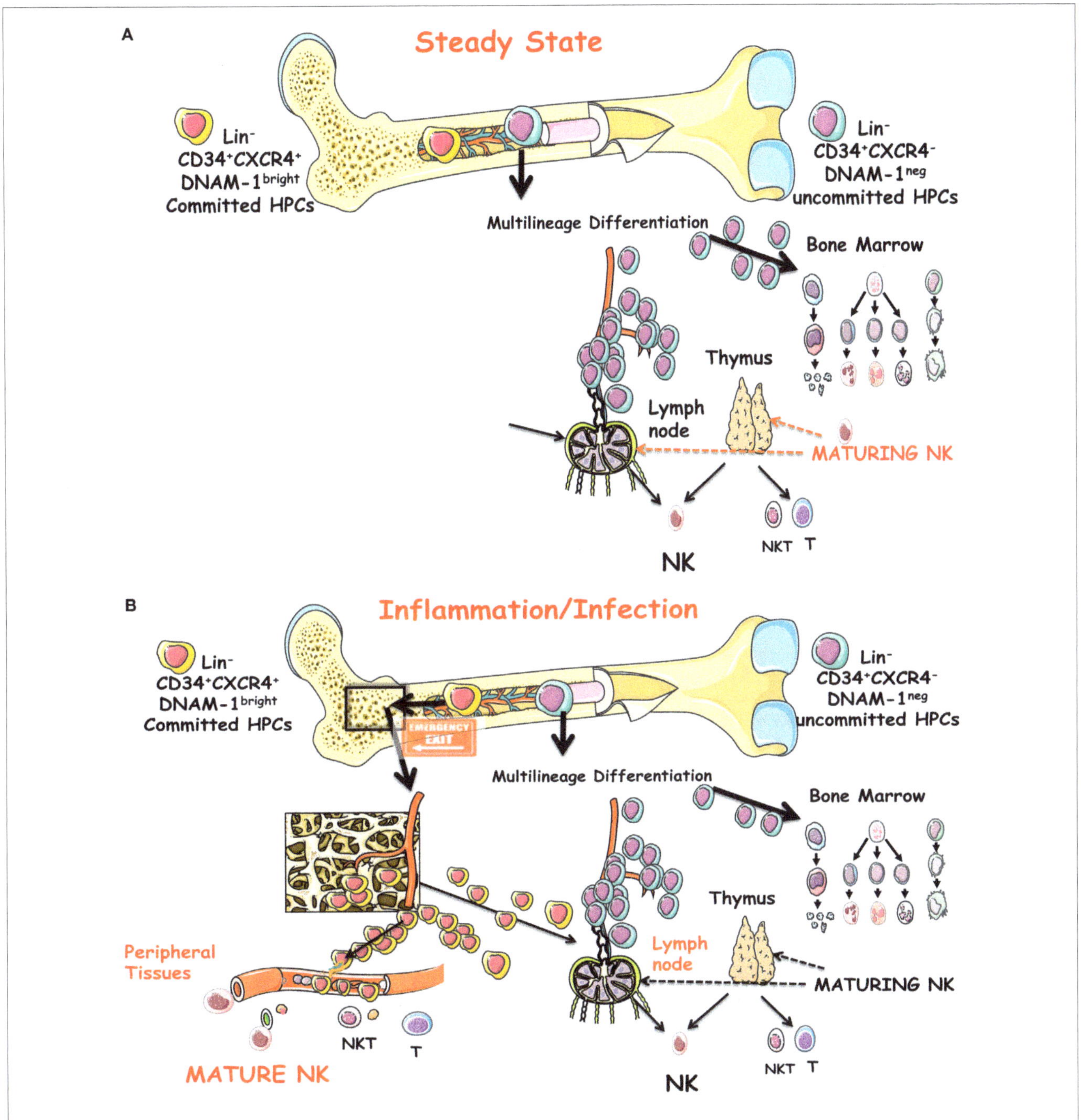

FIGURE 1 | **(A)** Schematic representation of the precursor genesis and maturation process of NK cells *in vivo* under steady-state conditions. $CD34^+DNAM\text{-}1^-CXCR4^-$ cells are indicated as general precursors to subsequent stages of multilineage commitment. A committed lymphocyte precursor to NK cells in the bone marrow (BM) or entering the thymus or lymph nodes gives rise to canonical maturing NK cells. **(B)** Under inflammatory conditions, a $CD34^+DNAM\text{-}1^{bright}CXCR4^+$ cell precursor presumably resident in a BM niche proximal to osteoclasts is released into the bloodstream. These cells may travel to peripheral tissues in addition to secondary lymphoid organs and generate differently mature NK cells in addition to T and NKT cells.

from different progenitors. Given the difficulties in defining $CD56^{bright}$ and $CD56^{dim}$ NK cell subsets in chimpanzees and macaques (70), the possible correlate of these findings on clonal tracking in macaques (69) needs to be evaluated with caution and may deserve evaluation also in other non-human primates.

Overall, the identification of a novel $CD34^+$ cell, giving rise to NK cells with distinct characteristics may represent a parallel and concurrent reading frame for the established model of NK cell development from $CD34^+$ cells to $CD56^{bright}$ to $CD56^{dim}$ NK cells (9, 10).

CONCLUDING REMARKS

The identification of a novel CD34$^+$ cell with distinct transcriptional and phenotypic characteristics to standard CD34$^+$ cells, endowed with the ability to generate NK cells of a special phenotype, opens new possibilities to improve our understanding of NK cell development and maturation, particularly under special or emergency conditions including systemic inflammation.

In this regard, several points need to be further understood to formulate a clear picture.

First, it will be important to understand whether additional different CD34$^+$ precursors exist that may give rise preferentially to specific NK cell (or ILC) progenies or whether the identification of Lin$^-$CD34$^+$DNAM-1^{bright}CXCR4$^+$ cells represents the only other CD34$^+$ precursor, giving rise to NK cells. In other terms, we do not yet know whether these cells were the only ones that needed characterization in a sort of "dark side of the (BM) niche" that eluded identification until recently.

Next, it will be important to understand the relationship of these NK cells that are actively released from osteoclast niches to classical CD34$^+$DNAM-1$^-$ NK cells that are passively released in the sinusoids.

Finally, careful analysis of NK cells developing from Lin$^-$CD34$^+$DNAM-1^{bright}CXCR4$^+$ cells is needed to understand whether these presumably "tissue-bound" precursors follow a predefined program or rather whether signals delivered in peripheral tissues guide and sustain the development of NK cells with specialized function.

AUTHOR CONTRIBUTIONS

FB analyzed, discussed, and interpreted data and literature and wrote the manuscript; FM analyzed, discussed, and interpreted data and literature and contributed to manuscript preparation; AM supervised, discussed, and interpreted data and literature, verified data analysis, and wrote the manuscript.

REFERENCES

1. Yu J, Freud AG, Caligiuri MA. Location and cellular stages of natural killer cell development. *Trends Immunol* (2013) 34(12):573–82. doi:10.1016/j.it.2013.07.005
2. De Maria A, Bozzano F, Cantoni C, Moretta L. Revisiting human natural killer cell subset function revealed cytolytic CD56(dim)CD16+ NK cells as rapid producers of abundant IFN-gamma on activation. *Proc Natl Acad Sci U S A* (2011) 108(2):728–32. doi:10.1073/pnas.1012356108
3. Vankayalapati R, Klucar P, Wizel B, Weis SE, Samten B, Safi H, et al. NK cells regulate CD8+ T cell effector function in response to an intracellular pathogen. *J Immunol* (2004) 172(1):130–7. doi:10.4049/jimmunol.172.1.130
4. Warfield KL, Perkins JG, Swenson DL, Deal EM, Bosio CM, Aman MJ, et al. Role of natural killer cells in innate protection against lethal Ebola virus infection. *J Exp Med* (2004) 200(2):169–79. doi:10.1084/jem.20032141
5. Moretta A. Natural killer cells and dendritic cells: rendezvous in abused tissues. *Nat Rev Immunol* (2002) 2(12):957–65. doi:10.1038/nri956
6. Zook EC, Kee BL. Development of innate lymphoid cells. *Nat Immunol* (2016) 17(7):775–82. doi:10.1038/ni.3481
7. Juelke K, Romagnani C. Differentiation of human innate lymphoid cells (ILCs). *Curr Opin Immunol* (2016) 38:75–85. doi:10.1016/j.coi.2015.11.005
8. Montaldo E, Vacca P, Moretta L, Mingari MC. Development of human natural killer cells and other innate lymphoid cells. *Semin Immunol* (2014) 26(2):107–13. doi:10.1016/j.smim.2014.01.006
9. Romagnani C, Juelke K, Falco M, Morandi B, D'Agostino A, Costa R, et al. CD56brightCD16- killer Ig-like receptor- NK cells display longer telomeres and acquire features of CD56dim NK cells upon activation. *J Immunol* (2007) 178(8):4947–55. doi:10.4049/jimmunol.178.8.4947
10. Moretta L. Dissecting CD56dim human NK cells. *Blood* (2010) 116(19):3689–91. doi:10.1182/blood-2010-09-303057
11. Gondois-Rey F, Granjeaud S, Rouillier P, Rioualen C, Bidaut G, Olive D. Multiparametric cytometry from a complex cellular sample: improvements and limits of manual versus computational-based interactive analyses. *Cytometry A* (2016) 89(5):480–90. doi:10.1002/cyto.a.22850
12. Horowitz A, Strauss-Albee D, Leipold M, Kubo J, Nemat-Gorgani N, Dogan OC, et al. Genetic and environmental determinants of human NK cell diversity revealed by mass cytometry. *Sci Transl Med* (2013) 5(208):208ra145. doi:10.1126/scitranslmed.3006702
13. Bar-Ephraïm YE, Mebius RE. Innate lymphoid cells in secondary lymphoid organs. *Immunol Rev* (2016) 271(1):185–99. doi:10.1111/imr.12407
14. Geiger TL, Sun JC. Development and maturation of natural killer cells. *Curr Opin Immunol* (2016) 39:82–9. doi:10.1016/j.coi.2016.01.007
15. Jensen CT, Strid T, Sigvardsson M. Exploring the multifaceted nature of the common lymphoid progenitor compartment. *Curr Opin Immunol* (2016) 39:121–6. doi:10.1016/j.coi.2016.01.009
16. Kyoizumi S, Kubo YF, Kajimura J, Yoshida K, Imai K, Hayashi T, et al. Age-associated changes in the differentiation potentials of human circulating hematopoietic progenitors to T- or NK-lineage cells. *J Immunol* (2013) 190(12):6164–72. doi:10.4049/jimmunol.1203189
17. Doulatov S, Notta F, Eppert K, Nguyen LT, Ohashi PS, Dick JE. Revised map of the human progenitor hierarchy shows the origin of macrophages and dendritic cells in early lymphoid development. *Nat Immunol* (2010) 11(7):585–93. doi:10.1038/ni.1889
18. Laurenti E, Doulatov S, Zandi S, Plumb I, Chen J, April C, et al. The transcriptional architecture of early human hematopoiesis identifies multilevel control of lymphoid commitment. *Nat Immunol* (2013) 14(7):756–63. doi:10.1038/ni.2615
19. Doulatov S, Eppert K, Nguyen LT, Ohashi PS, Dick JE. Revised map of the human progenitor hierarchy shows the origin of macrophages and dendritic cells in early lymphoid development. *Nat Immunol* (2010) 11(7):585–93. doi:10.1038/ni.1889
20. Yin T, Li L. The stem cell niches in bone. *J Clin Invest* (2006) 116(5):1195–201. doi:10.1172/JCI28568
21. Siewiera J, Gouilly J, Hocine HR, Cartron G, Levy C, Al-Daccak R, et al. Natural cytotoxicity receptor splice variants orchestrate the distinct functions of human natural killer cell subtypes. *Nat Commun* (2015) 6:10183. doi:10.1038/ncomms10183
22. Vacca P, Pietra G, Falco M, Romeo E, Bottino C, Bellora F, et al. Analysis of natural killer cells isolated from human decidua: evidence that 2B4 (CD244) functions as an inhibitory receptor and blocks NK-cell function. *Blood* (2006) 108(13):4078–85. doi:10.1182/blood-2006-04-017343
23. Klein Wolterink RG, Me G-O, Vosshenrich CAJ, Hendriks RW. The intrathymic crossroads of T and NK cell differentiation. *Immunol Rev* (2010) 238(1):126–37. doi:10.1111/j.1600-065X.2010.00960.x
24. Izon DJ. T-cell development: thymus-settling progenitors: settled? *Immunol Cell Biol* (2008) 86(7):552–3. doi:10.1038/icb.2008.50
25. Freud AG, Yu J, Caligiuri MA. Human natural killer cell development in secondary lymphoid tissues. *Semin Immunol* (2014) 26(2):132–7. doi:10.1016/j.smim.2014.02.008

26. Becknell B, Caligiuri MA. Interleukin-2, interleukin-15, and their roles in human natural killer cells. *Adv Immunol* (2005) 86:209–39. doi:10.1016/S0065-2776(04)86006-1
27. Freud AG, Yokohama A, Becknell B, Lee MT, Mao HC, Ferketich AK, et al. Evidence for discrete stages of human natural killer cell differentiation in vivo. *J Exp Med* (2006) 203(4):1033–43. doi:10.1084/jem.20052507
28. Freud AG, Becknell B, Roychowdhury S, Mao HC, Ferketich AK, Nuovo GJ, et al. A human CD34(+) subset resides in lymph nodes and differentiates into CD56bright natural killer cells. *Immunity* (2005) 22(3):295–304. doi:10.1016/j.immuni.2005.01.013
29. Grzywacz B, Kataria N, Sikora M, Oostendorp RA, Dzierzak EA, Blazar BR, et al. Coordinated acquisition of inhibitory and activating receptors and functional properties by developing human natural killer cells. *Blood* (2006) 108(12):3824–33. doi:10.1182/blood-2006-04-020198
30. Perussia B, Chen Y, Loza MJ. Peripheral NK cell phenotypes: multiple changing of faces of an adapting, developing cell. *Mol Immunol* (2005) 42(4):385–95. doi:10.1016/j.molimm.2004.07.017
31. Ouyang Q, Baerlocher G, Vulto I, Lansdorp PM. Telomere length in human natural killer cell subsets. *Ann N Y Acad Sci* (2007) 1106:240–52. doi:10.1196/annals.1392.001
32. Grzywacz B, Kataria N, Kataria N, Blazar BR, Miller JS, Verneris MR. Natural killer-cell differentiation by myeloid progenitors. *Blood* (2011) 117(13):3548. doi:10.1182/blood-2010-04-281394
33. Costa P, Sivori S, Bozzano F, Martini I, Moretta A, Moretta L, et al. IFN-α-mediated increase in cytolytic activity of maturing NK cell upon exposure to HSV-infected myelomonocytes. *Eur J Immunol* (2009) 39(1):147–58. doi:10.1002/eji.200838532
34. Mailliard RB, Alber S, Shen H, Watkins SC, Kirkwood JM, Herberman RB, et al. IL-18-induced CD83+CCR7+ NK helper cells. *J Exp Med* (2005) 202(7):941–53. doi:10.1084/jem.20050128
35. Beziat V, Liu L, Malmberg J-A, Ivarsson MA, Sohlberg E, Bjorklund AT, et al. NK cell responses to cytomegalovirus infection lead to stable imprints in the human KIR repertoire and involve activating KIRs. *Blood* (2013) 121(14):2678–88. doi:10.1182/blood-2012-10-459545
36. Yawata M, Yawata N, Draghi M, Little A-M, Partheniou F, Parham P. Roles for HLA and KIR polymorphisms in natural killer cell repertoire selection and modulation of effector function. *J Exp Med* (2006) 203(3):633–45. doi:10.1084/jem.20051884
37. Spits H, Artis D, Colonna M, Diefenbach A, Di Santo JP, Eberl G, et al. Innate lymphoid cells – a proposal for uniform nomenclature. *Nat Rev Immunol* (2013) 13(2):145–9. doi:10.1038/nri3365
38. Moretta F, Petronelli F, Lucarelli B, Pitisci A, Bertaina A, Locatelli F, et al. The generation of human innate lymphoid cells is influenced by the source of hematopoietic stem cells and by the use of G-CSF. *Eur J Immunol* (2016) 46(5):1271–8. doi:10.1002/eji.201546079
39. Paust S, Gill HS, Wang B-Z, Flynn MP, Moseman EA, Senman B, et al. Critical role for the chemokine receptor CXCR6 in NK cell-mediated antigen-specific memory of haptens and viruses. *Nat Immunol* (2010) 11(12):1127–35. doi:10.1038/ni.1953
40. Sun JC, Lanier LL. Natural killer cells remember: an evolutionary bridge between innate and adaptive immunity? *Eur J Immunol* (2009) 39(8):2059–64. doi:10.1002/eji.200939435
41. Min-Oo G, Kamimura Y, Hendricks DW, Nabekura T, Lanier LL. Natural killer cells: walking three paths down memory lane. *Trends Immunol* (2013) 34(6):251–8. doi:10.1016/j.it.2013.02.005
42. Guma M, Angulo A, Vilches C, Gomez-Lozano N, Malats N, Lopez-Botet M. Imprint of human cytomegalovirus infection on the NK cell receptor repertoire. *Blood* (2004) 104(12):3664–71. doi:10.1182/blood-2004-05-2058
43. Guma M, Budt M, Saez A, Brckalo T, Hengel H, Angulo A, et al. Expansion of CD94/NKG2C+ NK cells in response to human cytomegalovirus-infected fibroblasts. *Blood* (2006) 107(9):3624–31. doi:10.1182/blood-2005-09-3682
44. Romo NMG, Muntasell A, Heredia G, Baía D, Angulo A, Guma M, et al. Natural killer cell-mediated response to human cytomegalovirus-infected macrophages is modulated by their functional polarization. *J Leukoc Biol* (2011) 90(4):717–26. doi:10.1189/jlb.0311171
45. Della Chiesa M, Falco M, Podesà M, Locatelli F, Moretta L, Frassoni F, et al. Phenotypic and functional heterogeneity of human NK cells developing after umbilical cord blood transplantation: a role for human cytomegalovirus? *Blood* (2011) 119(2):399–410. doi:10.1182/blood-2011-08-372003
46. Petitdemange C, Becquart P, Wauquier N, Béziat V, Debré P, Leroy EM, et al. Unconventional repertoire profile is imprinted during acute chikungunya infection for natural killer cells polarization toward cytotoxicity. *PLoS Pathog* (2011) 7(9):e1002268. doi:10.1371/journal.ppat.1002268
47. Bjorkstrom NK, Lindgren T, Stoltz M, Fauriat C, Braun M, Evander M, et al. Rapid expansion and long-term persistence of elevated NK cell numbers in humans infected with hantavirus. *J Exp Med* (2011) 208(1):13–21. doi:10.1084/jem.20100762
48. Foley B, Cooley S, Verneris MR, Pitt M, Curtsinger J, Luo X, et al. Cytomegalovirus reactivation after allogeneic transplantation promotes a lasting increase in educated NKG2C+ natural killer cells with potent function. *Blood* (2012) 119(11):2665–74. doi:10.1182/blood-2011-10-386995
49. Rolle A, Pollmann J, Cerwenka A. Memory of infections: an emerging role for natural killer cells. *PLoS Pathog* (2013) 9(9):e1003548. doi:10.1371/journal.ppat.1003548
50. Romee R, Schneider S, Leong JW, Chase JM, Keppel CR, Sullivan RP, et al. Cytokine activation induces human memory-like NK cells. *Blood* (2012) 120(24):4751–60. doi:10.1182/blood-2012-04-419283
51. Sun JC, Beilke J, Lanier LL. Adaptive immune features of natural killer cells. *Nature* (2009) 457(7229):557–61. doi:10.1038/nature07665
52. Cooper MA, Elliott J, Keyel PA, Yang L, Carrero JA, Yokoyama WM. Cytokine-induced memory-like natural killer cells. *Proc Natl Acad Sci U S A* (2009) 106(6):1915–9. doi:10.1073/pnas.0813192106
53. Hu PF, Hultin L, Hultin P, Hausner MA, Hirji K, Jewett A, et al. Natural killer cell immunodeficiency in HIV disease is manifest by profoundly decreased numbers of CD16+CD56+ cells and expansion of a population of CD16dimCD56- cells with low lytic activity. *J Acquir Immune Defic Syndr Hum Retrovirol* (1995) 10(3):331–40. doi:10.1097/00042560-199511000-00005
54. Björkström NK, Ljunggren H-G, Sandberg JK. CD56 negative NK cells: origin, function, and role in chronic viral disease. *Trends Immunol* (2010) 31(11):401–6. doi:10.1016/j.it.2010.08.003
55. Le Bouteiller P, Siewiera J, Casart Y, Aguerre-Girr M, El Costa H, Berrebi A, et al. The human decidual NK-cell response to virus infection: what can we learn from circulating NK lymphocytes? *J Reprod Immunol* (2011) 88(2):170–5. doi:10.1016/j.jri.2010.12.005
56. Lash GE, Robson S, Bulmer JN. Review: functional role of uterine natural killer (uNK) cells in human early pregnancy decidua. *Placenta* (2010) 31(Suppl):S87–92. doi:10.1016/j.placenta.2009.12.022
57. Koopman LA, Kopcow H, Rybalov B, Boyson JE, Orange JS, Schatz F, et al. Human decidual natural killer cells are a unique NK cell subset with immunomodulatory potential. *J Exp Med* (2003) 198(8):1201–12. doi:10.1084/jem.20030305
58. Moffett-King A. Natural killer cells and pregnancy. *Nat Rev Immunol* (2002) 2(9):656–63. doi:10.1038/nri967
59. Tabiasco J, Rabot MF, Aguerre-Girr M, El Costa H, Berrebi A, Parant O, et al. Human decidual NK cells: unique phenotype and functional properties – a review. *Placenta* (2006) 27(Suppl A):S34–9. doi:10.1016/j.placenta.2006.06.003
60. Bozzano F, Marras F, Ascierto ML, Cantoni C, Cenderello G, Dentone C, et al. 'Emergency exit' of bone-marrow-resident CD34(+)DNAM-1(bright) CXCR4(+)-committed lymphoid precursors during chronic infection and inflammation. *Nat Commun* (2015) 6:8109. doi:10.1038/ncomms9109
61. Kunisaki Y, Bruns I, Scheiermann C, Ahmed J, Pinho S, Zhang D, et al. Arteriolar niches maintain haematopoietic stem cell quiescence. *Nature* (2013) 502(7473):637–43. doi:10.1038/nature12612
62. Greenbaum A, Hsu Y-MS, Day RB, Schuettpelz LG, Christopher MJ, Borgerding JN, et al. CXCL12 in early mesenchymal progenitors is required for haematopoietic stem-cell maintenance. *Nature* (2013) 495(7440):227–30. doi:10.1038/nature11926
63. Ding L, Morrison SJ. Haematopoietic stem cells and early lymphoid progenitors occupy distinct bone marrow niches. *Nature* (2013) 495(7440):231–5. doi:10.1038/nature11885
64. Kollet O, Dar A, Shivtiel S, Kalinkovich A, Lapid K, Sztainberg Y, et al. Osteoclasts degrade endosteal components and promote mobilization of hematopoietic progenitor cells. *Nat Med* (2006) 12(6):657–64. doi:10.1038/nm1417
65. Takayanagi H. Osteoimmunology: shared mechanisms and crosstalk between the immune and bone systems. *Nat Rev Immunol* (2007) 7(4):292–304. doi:10.1038/nri2062

66. Kulak CA, Borba V, Jorgetti V, Dos Reis LM, Liu XS, Kimmel DB, et al. Skeletal microstructural abnormalities in postmenopausal women with chronic obstructive pulmonary disease. *J Bone Miner Res* (2010) 25(9):1931–40. doi:10.1002/jbmr.88
67. Sivori S, Cantoni C, Parolini S, Marcenaro E, Conte R, Moretta L, et al. IL-21 induces both rapid maturation of human CD34+ cell precursors towards NK cells and acquisition of surface killer Ig-like receptors. *Eur J Immunol* (2003) 33(12):3439–47. doi:10.1002/eji.200324533
68. Alter G, Jost S, Rihn S, Reyor LL, Nolan BE, Ghebremichael M, et al. Reduced frequencies of NKp30+NKp46+, CD161+, and NKG2D+ NK cells in acute HCV infection may predict viral clearance. *J Hepatol* (2011) 55(2):278–88. doi:10.1016/j.jhep.2010.11.030
69. Wu C, Li B, Lu R, Koelle Samson J, Yang Y, Jares A, et al. Clonal tracking of rhesus macaque hematopoiesis highlights a distinct lineage origin for natural killer cells. *Cell Stem Cell* (2014) 14(4):486–99. doi:10.1016/j.stem.2014.01.020
70. Rutjens E, Mazza S, Biassoni R, Koopman G, Ugolotti E, Fogli M, et al. CD8+ NK cells are predominant in chimpanzees, characterized by high NCR expression and cytokine production, and preserved in chronic HIV-1 infection. *Eur J Immunol* (2010) 40(5):1440–50. doi:10.1002/eji.200940062

7

Modulation of T-bet and Eomes during Maturation of Peripheral Blood NK Cells does not Depend on Licensing/Educating KIR

*Amandine Pradier[1], Federico Simonetta[1], Sophie Waldvogel[1], Carine Bosshard[1], Jean-Marie Tiercy[2] and Eddy Roosnek[1]**

[1] Division of Hematology, Department of Medical Specialties, Geneva University Hospitals, University of Geneva, Geneva, Switzerland, [2] Transplantation Immunology Unit, National Reference Laboratory for Histocompatibility, Department of Genetics and Laboratory Medicine, University Hospital Geneva, Geneva, Switzerland

Edited by:
Jordan Orange, Baylor College of Medicine, USA

Reviewed by:
Jeroen Van Bergen, Leiden University, Netherlands
Nicholas D. Huntington, The Walter and Eliza Hall Institute of Medical Research, Australia

***Correspondence:**
Eddy Roosnek
eddy.roosnek@unige.ch

Peripheral natural killer (NK) cells upregulate T-bet and downregulate Eomes, the key transcription factors regulating NK cell maturation and function during the last maturation steps toward terminally differentiated effector cells. During this process, NK cells acquire killer immunoglobulin-like receptors (KIR) and effector functions, such as cytotoxicity and target cell-induced cytokine production. Inhibitory KIR are pivotal in the control of effector functions, but whether they also modulate T-bet/Eomes expression is unknown. We have measured T-bet/Eomes levels, KIR expression, and effector functions of maturing CD94^{neg}CD56^{dim}NK cells using CD57 as surface marker for maturation. Our cohort consisted of 23 healthy blood donors (HBD) homozygous for the KIR A haplotype that contains only inhibitory KIR2DL1 (ligand HLA-C2), KIR2DL3 (ligand HLA-C1), and KIR3DL1 (ligand HLA-Bw4). We confirm that during maturation of NK cells, the number of KIR increases, levels of T-bet/Eomes are modulated, and that cells acquire effector functions, such as cytotoxicity (CD107) and target cell-induced cytokine production (TNF-α). Because maturation was associated with the increase of the number of KIR as well as with the modulation of T-bet/Eomes, the number of KIR correlated with the extent of T-bet/Eomes modulation. However, whether the KIR were triggered by their cognate HLA ligands or not had no impact on T-bet and Eomes expression, indicating that modulation of T-box transcription factors during NK cell maturation does not depend on signals conveyed by KIR. We discuss the relevance of this finding in the context of models of NK cell maturation while cautioning that results obtained in a perhaps quite heterogeneous cohort of HBD are not necessarily conclusive.

Keywords: NK cell, maturation, KIR, licensing, T-Bet, Eomes

INTRODUCTION

Natural killer (NK) cells are involved in the early response to pathogens as well as in the recognition of autologous cells under stress induced by infection or transformation (1). Maturation toward an end-stage cytotoxic effector cell is associated with the loss of CD94 and acquisition of killer immunoglobulin-like receptors (KIR), of markers such as CD57 and of fine-tuning of effector functions (2–5). Activation of NK cells depends on an aggregate of signals conveyed by various activating and/or inhibitory receptors. Responsiveness to self is reduced through the triggering of inhibitory KIR by their ligands that consist of allelic variants of MHC molecules that are clustered in three groups, HLA class I ligands (HLA-C1, HLA-C2, and HLA-Bw4). NK cells without inhibitory KIR for self may remain harmless either because acquisition of the most potent effector functions (arming) requires interaction of inhibitory KIR with self-ligands or because NK cells stimulated by activating receptors in the absence of interaction of inhibitory KIR with self reduce responsiveness. These latter two processes are referred to as "licensing" (6, 7) and "education" (8, 9). The ensuing level of responsiveness may still be adapted to their immunological environment in mature NK cells, but the molecular mechanisms have not been established yet [recently reviewed in Ref. (10)].

T-bet and Eomesodermin (Eomes), two T-box transcription factors, are master regulators of T cell effector functions, including cytotoxicity and interferon-gamma (IFN-γ) production (11, 12). Furthermore, recent reports have shown that these two transcription factors also regulate maturation and function of NK cells. Murine (13) and human (14, 15) NK cells express T-bet and Eomes constitutively, and mice lacking both T-bet and Eomes are completely deprived of NK cells (16). Moreover, the two T-box transcription factors are modulated during NK cell differentiation (14, 17, 18) and are necessary for maintenance and differentiation of peripheral NK cells, while their deletion in mature NK cells results in reversion to an immature phenotype (16).

Mature NK cells express more licensing/educating KIR, are more cytotoxic, and proliferate less (2, 4). Whether they produce more cytokines after stimulation with target cells remains under debate (4, 19). Furthermore, they have downregulated Eomes and upregulated T-bet (14, 20). The concurrence of these processes has made it difficult to determine whether licensing is associated with, or the cause of the modulation of effector functions as well as T-bet/Eomes expression. It has been difficult to establish that signals conveyed by inhibitory KIR encountering their ligands induce maturation. Indeed, one could also argue that the increase in effector functions found to be correlated with the expression of such KIR simply reflects a licensing-independent concomitant NK cell maturation (4).

T-bet and Eomes induce NK cell maturation by suppressing CD27 and c-kit and upregulating S1P5 and KLRG1 [recently reviewed in Ref. (21)]. Hence, if signals through licensing receptors induce maturation, one would expect that the same signals would modulate the levels of T-bet and Eomes first. In this report, we measured T-bet and Eomes levels in maturing NK cells in 23 healthy blood donors (HBD) that were homozygous for the KIR A haplotype, which comprises only inhibitory KIR for their respective HLA-C or HLA-Bw4 ligands. We found that T-bet and Eomes were modulated in parallel with the increase of KIR, but whether the ligand for the KIR was present or not had no impact. This finding strengthens the model of a recently put forward (4), licensing-independent concomitant NK cell maturation.

MATERIALS AND METHODS

DNA Extraction and KIR Genotyping

Killer immunoglobulin-like receptors genotyping was performed on DNA extracted from 400 µl of blood using QIAmp Blood Mini kit (Qiagen) with the KIR Genotyping SSP kit (Invitrogen) according to manufacturer's instructions. The AA haplotype was defined as KIR2DL5A^{neg}/KIR2DL5B^{neg}/KIR2DS1^{neg}/KIR2DS2^{neg}/KIR2DS3^{neg}/KIR2DS5^{neg}/KIR2DS1^{neg}.

HLA Typing

HLA-A, B, C typing was performed by PCR-SSO hybridization on microbeads arrays (Luminex technology), using the LabType high definition reagents (One Lambda, Canoga Park, CA, USA). When required for the assignment of the HLA-C1 and -C2 groups, typing ambiguities were resolved by PCR-SSP using Olerup SSP kits (Milan Analytika AG, La Roche, Switzerland).

FACS Analysis, Effector Function Tests

Peripheral blood mononuclear cells (PBMCs) from AA haplotype homozygous donors blood from the Geneva University Hospitals blood transfusion center who gave informed consent by signing a standard form approved by the hospital's ethical commission were isolated from anticoagulated blood by Ficoll density gradient centrifugation. We performed FACS analysis with monoclonal antibodies specific for the following antigens: CD158a (FITC, HP-MA4, Biolegend), CD57 (PECF594, clone NK1, BD Biosciences), CD94 (PerCPCy5.5, clone HP-3D9, BD Biosciences), CD158b (PEVio770, clone DX27, Miltenyi), CD158e (Alexa700, clone DX9, Biolegend), CD3 (APCH7, clone SK7, BD Biosciences), CD56 (BV421, cloneHCD56, Biolegend), CD107a (BV605, clone H4A3, Biolegend), TNF-α (BV605, clone MAb11, Biolegend), IFN-γ (BV605, clone 4S.B3, Biolegend), Eomes (eFluor 660, clone WD1928, e-Bioscience), and T-bet (PE, clone 4B10, e-Bioscience).

Effector functions were measured by stimulating PBMC at 37°C with K562 (cytotoxicity, TNF-α production) or with IL2/12/18 (IFN-γ production) as described before (20). Anti-CD107a was added at the start of the culture set up to measure degranulation. After 1 h, GolgiStop and GolgiPlug (BD Biosciences) were added, and cells were cultured for a further 3 h (CD107 and TNF-α) or 5 h (IFN-γ detection).

Intracellular staining for TNF-α, IFN-γ, Eomes, and T-bet was performed overnight at 4°C on permeabilized cells with FoxP3/transcription factor staining buffer set (e-Bioscience). CD3^{neg}CD56^{neg} lymphocytes in the same blood sample were considered as Eomes and T-bet negative cells.

Data acquired on a Gallios 3 cell analyzer (BD Biosciences) were analyzed with FlowJo software (Tree Star Inc.) with the gates depicted in **Figure 1**. The first two gates (**Figures 1A,B**) were used to demarcate NK cells as CD3^{neg}CD56^{pos} cells with the FSC/SSC of lymphocytes. CD56^{bright} and CD56^{dim} NK cells were discriminated on basis of the level of CD56-expression (**Figure 1C**). Maturation state (CD57), effector functions, and T-bet/Eomes levels of CD94^{neg} (**Figure 1D**), CD56^{dim} NK cells were measured in the respective eight subsets defined by the expression KIR2DL1, KIR2DL3, and KIR3DL1 (**Figures 1E,F**). Levels of T-bet and Eomes are expressed as mean fluorescence intensity ratio (MFIR),

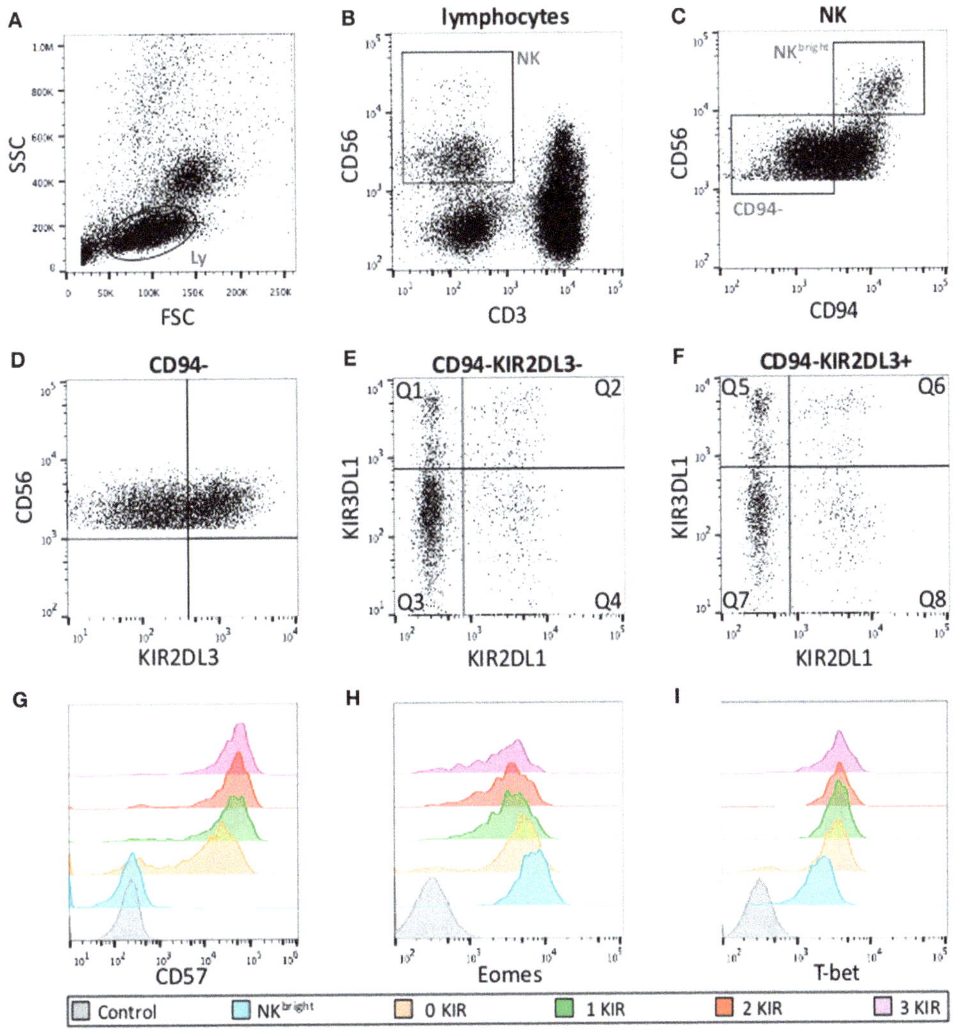

FIGURE 1 | Gating strategies for KIR subset analysis. (A) Lymphocyte gates on FSC and SSC. **(B)** NK cell defined as CD56^{pos}CD3^{neg} lymphocytes. **(C)** CD56^{bright} and CD56^{dim} NK cells according to the level of CD56-expression and gate on CD56^{dim}CD94^{neg} NK cells used for further analysis. **(D–F)** CD56^{dim}CD94^{neg} cells according to KIR expression (eight subsets). Q1: KIR2DL1^{neg}KIR2DL3^{neg}KIR3DL1^{pos}, Q2: KIR2DL1^{pos}KIR2DL3^{neg}KIR3DL1^{pos}, Q3: KIR2DL1^{neg}KIR2DL3^{neg}KIR3DL1^{neg}, Q4: KIR2DL1^{pos}KIR2DL3^{neg}KIR3DL1^{neg}, Q5: KIR2DL1^{neg}KIR2DL3^{pos}KIR3DL1^{pos}, Q6: KIR2DL1^{pos}KIR2DL3^{pos}KIR3DL1^{pos}, Q7: KIR2DL1^{neg}KIR2DL3^{pos}KIR3DL1^{neg}, Q8: KIR2DL1^{pos}KIR2DL3^{pos}KIR3DL1^{neg}. Examples of fluorescence intensity of CD57 **(G)**, Eomes **(H)** and T-bet **(I)** according to the number of KIR expressed with their respective negative controls.

representing the mean fluorescence of T-bet/Eomes divided by the fluorescence of T-bet/Eomes negative CD3^{neg}CD56^{neg} cells in the same sample. The percentage of CD107^{pos} cells is expressed as %CD107^{pos} NK cell after stimulation with K562 – % CD107^{pos} unstimulated NK cell.

Statistical Analysis

Non-parametric Wilcoxon matched-pairs signed rank test or Mann–Whitney test were used to compare groups, p values >0.05 were considered as not statistically significant.

RESULTS

We have studied maturing NK cells in 23 KIR A haplotype homozygous HBD that comprise only the inhibitory variant of KIR2DL1, KIR2DL3, and KIR3DL1. This allowed us to measure the effect of the presence or absence of HLA ligands on inhibitory KIR using monoclonal antibodies that also recognize their activatory variants. **Table 1** shows the KIR and HLA class I molecules expressed by the 23 HBD tested. As expected for KIR A haplotype homozygous individuals, NK cell subpopulations expressing KIR2DL1 and KIR2DL3 were present in all HBD. Two of 23 HBD lacked NK cells expressing KIR3DL1, which is owed to the fact that some KIR3DL1 allelic variants are not expressed at the protein level (22). The last three columns of the table show in which HBD the respective KIR encounter their cognate ligands [KIR2DL1↔C2, KIR2DL3↔C1, and KIR3DL1↔Bw4 epitope on HLA-B antigens as well as the HLA-A antigens (23, 24) marked in bold]. Hence, KIR2DL1 encountered its cognate HLA ligand in 10/23 and KIR2DL3

TABLE 1 | HLA class I and KIR expression in the 23 HBD.

HBD	HLA-A	HLA-A	HLA-B	HLA-B	Epitopes B	HLA-C	HLA-C	Epitopes C	KIR3DL1	KIR2DL3	KIR2DL1
1	02:01	**32:01**[c]	15:01	40:02	Bw6/Bw6	02:02	03:03	C1/C2	a	a	a
2	02:01	30:02	15:01	35:01	Bw6/Bw6	01:02	04:01	C1/C2		a	a
3	24:02	30:01	13:02	44:03	Bw4/Bw4	06:02	16:01	C1/C2	a	a	a
4	02:01	02:01	18:01	51:01	Bw4/Bw6	07:01	14:02	C1/C2	a	a	
5	02:01	11:01	18:03	39:09	Bw6/Bw6	07:01	07:02	C1/C2		a	
6	02:01	25:01	13:02	35:01	Bw4/Bw6	07:01	07:02	C1/C2	a	a	
7	01:02	02:01	07:02	49:01	Bw4/Bw6	07:01	07:02	C1/C2	a	a	
8	02:01	23:01	49:01	51:01	Bw4/Bw4	02:02	07:01	C1/C2	a	a	a
9	01:01	01:01	57:01	57:01	Bw4/Bw4	06:02	07:01	C1/C2	a	a	a
10	01:01	11:01	49:01	52:01	Bw4/Bw4	07:01	12:02	C1/C2	a	a	
11	02:01	32:01	35:01	44:02	Bw4/Bw6	04:01	05:01	C2/C2	a		a
12	01:01	11:01	15:01	35:01	Bw6/Bw6	03:03	04:01	C1/C2		a	a
13	02:01	03:01	56:01	56:01	Bw6/Bw6	01:02	01:02	C1/C2		a	
14	02:01	24:02	15:01	44:02	Bw4/Bw6	03:03	05:01	C1/C2	a	a	a
15	03:01	29:02	07:02	44:03	Bw4/Bw6	07:02	16:01	C1/C2	a	a	
16	02:01	26:01	13:02	58:01	Bw4/Bw4	06:02	07:01	C1/C2	b	a	a
17	24:02	30:02	38:01	40:01	Bw4/Bw6	03:04	12:03	C1/C2	b	a	
18	01:01	11:01	08:01	51:01	Bw4/Bw6	07:01	15:02	C1/C2	a	a	a
19	26:01	32:01	38:01	44:03	Bw4/Bw4	12:03	16:01	C1/C2	a	a	
20	02:01	24:02	35:01	44:02	Bw4/Bw6	01:02	01:02	C1/C2	a	a	
21	26:01	33:01	14:02	44:02	Bw4/Bw6	08:02	12:03	C1/C2	a	a	
22	01:01	68:01	08:01	39:01	Bw6/Bw6	07:01	12:03	C1/C2		a	
23	23:01	31:01	07:02	08:01	Bw6/Bw6	07:01	07:02	C1/C2		a	

[a]*Licensed.*
[b]*KIR3DL1 not expressed (22).*
[c]*HLA-A*3201 with Bw4 epitope (23, 24).*

in 22/23 HBD, while KIR3DL1 was able to license NK cells in 15/23 HBD.

Expression of Eomes/T-bet/CD57 in Relation to the Number of KIR

Several reports have shown that NK cells upregulate CD57 and increase the number of KIR expressed during maturation. In parallel, CD56^{dim} NK cells downregulate Eomes and upregulate T-bet. Using the gating strategy shown in **Figure 1**, we measured the levels of CD57, Eomes, and T-bet in CD56^{bright} NK cells and in the eight NK cell subpopulations of CD56^{dim}CD94^{neg} NK cells expressing different combinations of KIR2DL1, KIR2DL3, and KIR3DL1. **Figure 2** shows the results of our panel of 23 HBD and confirms (**Figures 2A,B**) that indeed CD57 levels on CD94^{neg}KIR^{pos}CD56^{dim} NK cells are significantly higher than on CD94^{neg}KIR^{neg}CD56^{dim} NK cells, and that CD57 levels increase further with the number of KIR expressed (for comparison, CD57, Eomes and T-bet levels on CD56^{bright} NK cells are shown). This was true when all KIR (**Figure 2A**) or only licensing KIR (**Figure 2B**) were considered. Furthermore, levels of Eomes that are significantly lower in CD56^{dim} NK cells than in CD56^{bright} NK cells (**Figures 2C,D**) are further downregulated with the acquisition of KIR. As expected, T-bet levels were the opposite of Eomes levels. CD56^{dim} NK cells expressed more T-bet than their CD56^{bright} counterparts, and these levels increased with the number of KIR expressed (**Figures 2E,F**). Again, whether all KIR were licensing (**Figures 2D,F**) or not (**Figures 2C,E**) did not seem to have a considerable impact.

Eomes/T-bet/CD57 Expression and Effector Function of Licensed and Unlicensed NK Cells

To investigate the effect of licensing on Eomes/T-bet levels more precisely, we gated on CD56^{dim} NK cells expressing a single KIR and tested the impact of the presence or absence of its HLA ligand. Furthermore, we gated on CD94^{neg} NK cells to disregard the weaker inhibitory signals through CD94/NKG2A (25), of which the contribution to licensing is unclear. **Figures 3A,B** show that the expression of Eomes/T-bet in licensed NK cells was identical to the level in unlicensed NK cells. This was true for single KIR2DL1^{pos} NK cells in the presence (HBD 1–3, 8, 9, 11, 12, 14, 16, 18) or in the absence (HBD 4–6, 7, 10, 13, 15, 17, 19–23) of its ligand HLA-C2 (**Figure 3C**) as well as for single KIR3DL1^{pos} NK cells in the presence (HBD 1, 3, 4, 6–11, 14, 15, 18–21) or in the absence (HBD 2, 5, 12, 13, 16, 17, 22, 23) of its ligand HLA-Bw4 (**Figure 3D**). We were not able to compare the effect of the ligand HLA-C1 on single KIR2DL3^{pos} NK cells because our panel comprised only one HLA-C1 negative HBD (HBD 11), but the fact that the Eomes/T-bet levels in licensed single KIR2DL3^{pos} NK cells (HBD 1–10, 12–23) was identical to the levels in (un)licensed single KIR2DL1^{pos} or KIR3DL1^{pos} NK cells did suggest that the effect of licensing in single KIR2DL3^{pos} NK cells may also be negligible.

Licensed and unlicensed NK cells expressed similar levels of CD57 (**Figure 3E**), but effector functions that were higher in cells expressing more KIR (data not shown) were to some extent affected by licensing. Licensed NK cells

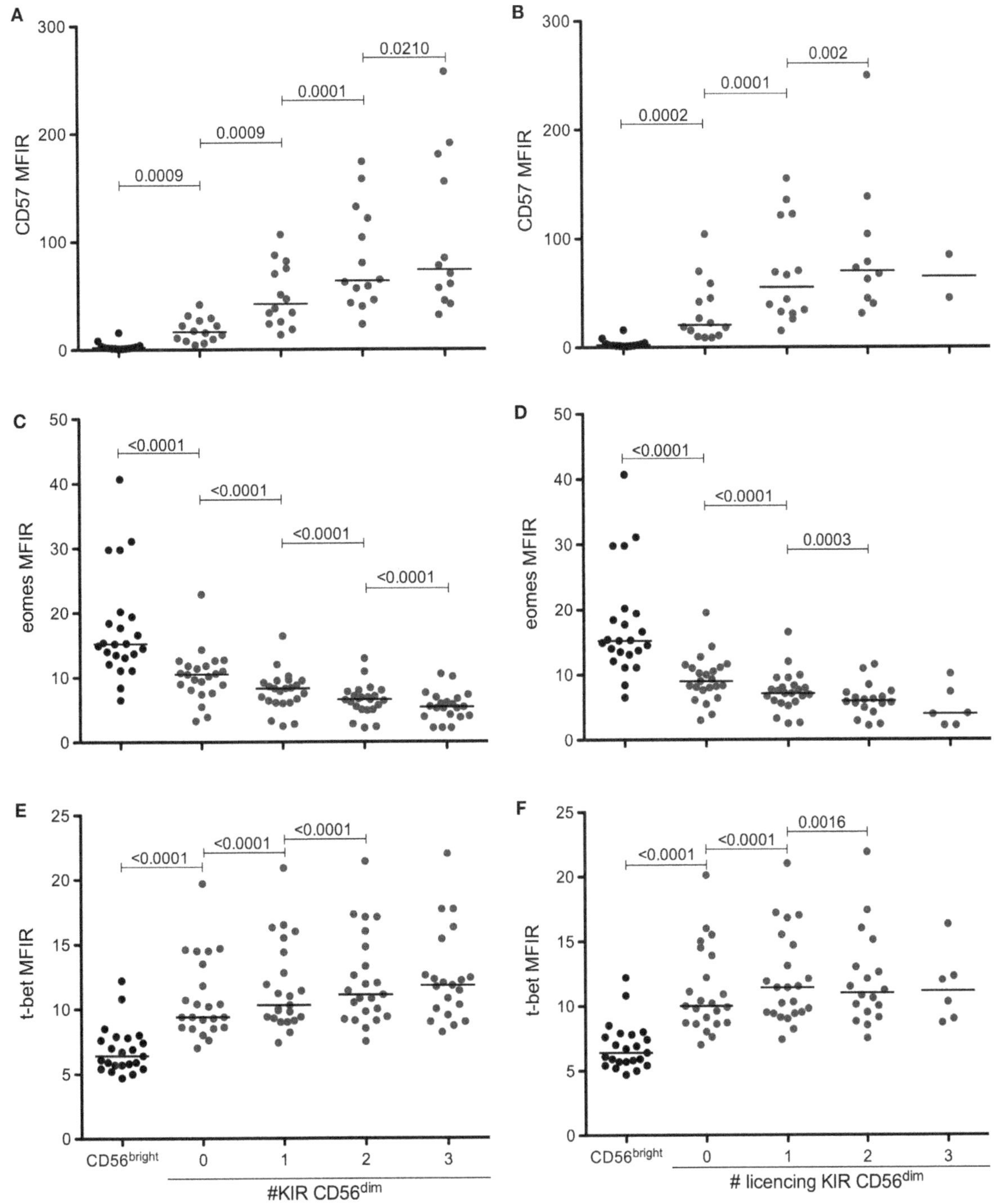

FIGURE 2 | CD57 and T-bet are upregulated, while Eomes is downregulated during NK cell maturation and KIR acquisition. Expression levels of CD57 **(A,B)**, eomes **(C,D)**, and T-bet **(E,F)** in CD56^{bright} or in CD56^{dim} CD94^{neg} cells according to the number of KIR **(A,C,E)** or licensing KIR **(B,D,F)** expressed. Medians are shown, and Wilcoxon matched-pairs signed rank test was used for statistical analysis, *p* values are indicated. As mentioned in the text, CD94^{pos} NK cells have been excluded from analysis because, although the results are very similar (not shown), the inclusion of CD94^{pos} NK cells would render the interpretation of the results more difficult.

produced significantly more TNF-α than unlicensed NK cells (**Figure 4A**), and the most cytotoxic NK cells (**Figure 4B**) were licensed. Hence, we found that the effect of licensing on maturation and effector functions of human NK cells may be of secondary importance only, which in our opinion accords well with the variegated results on the effect of licensing on different effector functions of human NK cells reported by others (2, 4, 19).

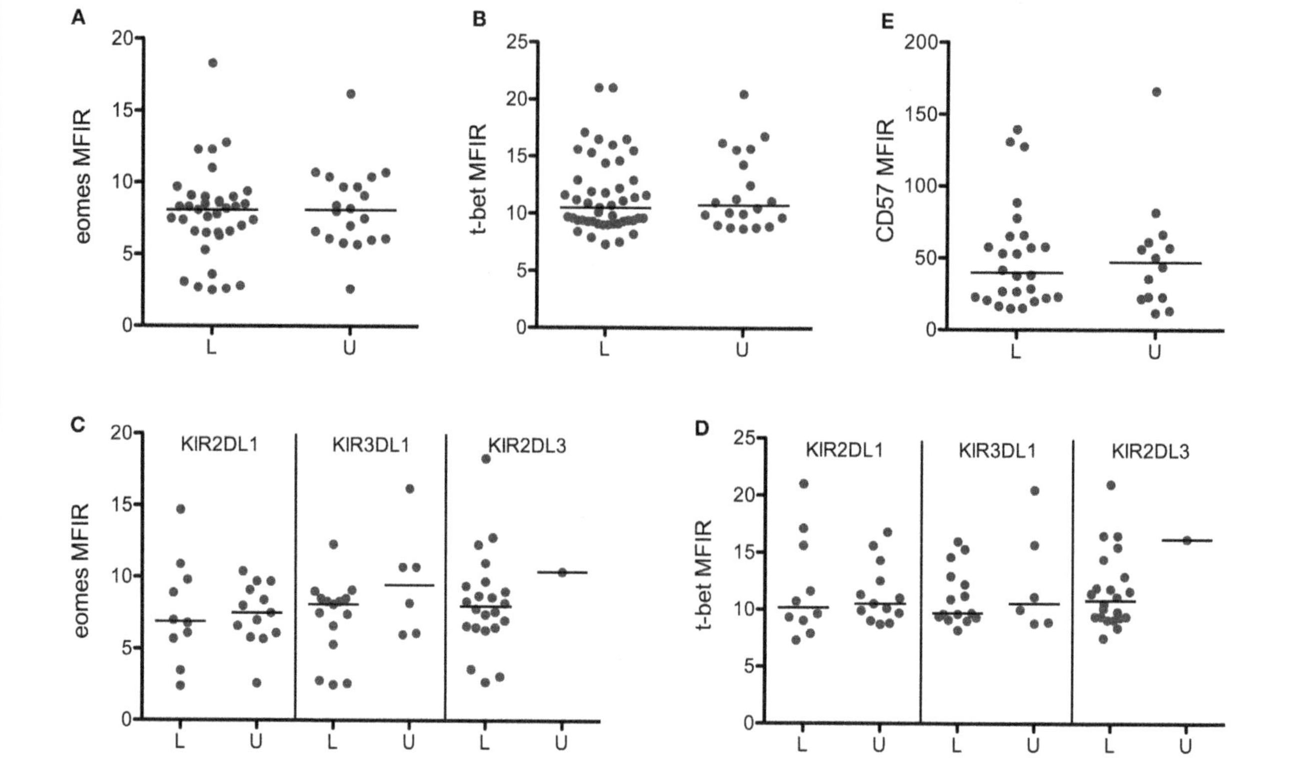

FIGURE 3 | Licensing has no effect on eomes, T-bet, and CD57 expression. Eomes **(A)**, T-bet **(B)**, and CD57 **(E)** MFIR are shown in CD56^{dim}CD94^{neg} cells expressing a single licensed (L) or unlicensed (U) KIR. Eomes **(C)** and T-bet **(D)** MFIR are represented for each KIR separately. Medians are shown, and Mann–Whitney test was used for statistical analysis.

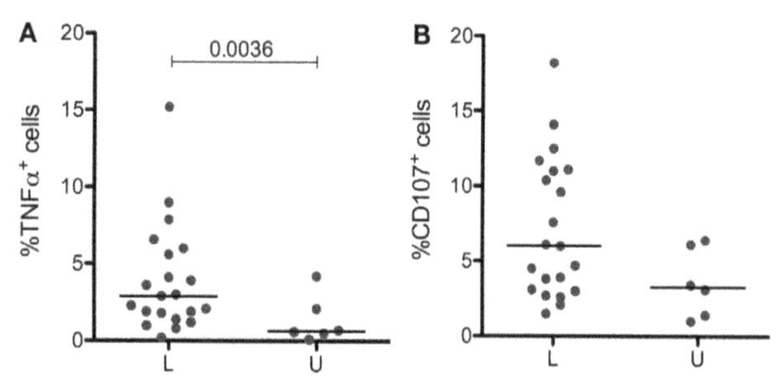

FIGURE 4 | Impact of licensing on TNF-α production and cytoxicity in CD56^{dim}CD94^{neg} cells. TNF-α^{pos} **(A)** and CD107^{pos} **(B)** cells in CD56^{dim}CD94^{neg} cells expressing a single licensed (L) or unlicensed (U) KIR after culture for 3 h in presence of K562 target cells. Medians are shown, and Mann–Whitney test was used for statistical analysis, *p* values are indicated when significant.

DISCUSSION

The expression of the master regulators of NK cell differentiation T-bet and Eomes is modulated during the maturation toward end-stage cytotoxic NK cells. Cytokine-producing CD56^{bright} NK cells express higher levels of Eomes and lower levels of T-bet than cytotoxic CD56^{dim} NK cells, and the T-bet/Eomes ratio increases further during the last maturation steps toward terminally differentiated CD56^{dim}CD57^{pos} NK cells (14, 15, 20). During maturation, the number of KIR increases, markers such as CD94 and CD62L are downregulated, CD57 is upregulated, and effector functions change (2–5).

There are indications that licensing favors maturation, but the extent thereof remains under debate (2, 4, 19). It is conceivable that licensing modulates T-bet and Eomes and that the ensuing upregulation of T-bet is at the origin of the increased cytotoxicity that is characteristic of mature NK cells. However, it may also be that licensing, upregulation of T-bet, and the increase of cytotoxicity are maturation-associated, parallel processes.

To answer this question, we measured T-bet and Eomes levels in maturing NK cells that were licensed or not. We found no differences in T-bet/Eomes levels in licensed or unlicensed NK cells. Although this would indicate that licensing and modulation of T-bet and Eomes are independent processes, we remain somewhat reluctant to draw such a firm conclusion based on the analysis of a panel of perhaps quite heterogeneous HBD. In humans, it might not be that easy to discriminate between cells that have matured in response to signals through licensing KIR or in response to cytokines (6, 7, 26, 27) produced during an inflammatory immune response. Indeed, mouse models have been somewhat more consistent regarding the effect of licensing on effector functions (6, 28, 29) than human studies (2, 4, 8, 19, 30). Likewise, our results regarding the acquisition of effector functions or of a more mature phenotype of CD56^{dim}CD94^{neg} single KIR^{pos} NK cells in the presence or absence of their cognate HLA ligand were not very conclusive. Licensed NK cells produced more TNF-α, seemed to be somewhat more cytotoxic but did not express more CD57 than unlicensed cells. We also measured differences in IFN-γ production, the cytokine that is commonly tested in functional studies of NK cells. We found no differences, but again, we consider these results with caution because in our hands only few NK cells *ex vivo* produce IFN-γ when stimulated with target cells without being cultured in the presence of IL2/IL-12/IL-18 (20). Licensed and unlicensed CD56^{dim} NK cells produced similar quantities of IFN-γ after stimulation with cytokines (data not shown), but this result is not more than had to be expected because the culture with cytokines would simply annul potential differences between licensed and unlicensed NK cells.

CONCLUSION

In conclusion, we found no impact of licensing on the expression of T-bet and Eomes.

Because T-bet and Eomes are the master regulators of NK cell maturation, these results are hard to reconcile with the model in which licensing induces maturation directly. However, because we also found that the effect of licensing was not easily substantiated for classical maturation-associated attributes, such as cytokine production, cytotoxicity, and expression of CD57, we believe that this finding should be interpreted with caution. In fact, human NK cells may not be ideal to study the effect of licensing on maturation because the circumstances *in vivo* that may bypass the need for licensing are hard to define.

AUTHOR CONTRIBUTIONS

AP, FS, and ER designed the study; AP and CB performed the experiments; SW organized the cohort of HBD; J-MT helped with HLA and KIR-typing; AP and ER wrote the manuscript.

REFERENCES

1. Trinchieri G. Biology of natural killer cells. *Adv Immunol* (1989) 47:187–376. doi:10.1016/S0065-2776(08)60664-1
2. Juelke K, Killig M, Luetke-Eversloh M, Parente E, Gruen J, Morandi B, et al. CD62L expression identifies a unique subset of polyfunctional CD56dim NK cells. *Blood* (2010) 116:1299–307. doi:10.1182/blood-2009-11-253286
3. Lopez-Verges S, Milush JM, Pandey S, York VA, Arakawa-Hoyt J, Pircher H, et al. CD57 defines a functionally distinct population of mature NK cells in the human CD56dimCD16+ NK cell subset. *Blood* (2010) 116(19):3865–74. doi:10.1182/blood-2010-04-282301
4. Bjorkstrom NK, Riese P, Heuts F, Andersson S, Fauriat C, Ivarsson MA, et al. Expression patterns of NKG2A, KIR, and CD57 define a process of CD56dim NK cell differentiation uncoupled from NK cell education. *Blood* (2010) 116:3853–64. doi:10.1182/blood-2010-04-281675
5. Yu J, Mao HC, Wei M, Hughes T, Zhang J, Park IK, et al. CD94 surface density identifies a functional intermediary between the CD56bright and CD56dim human NK cell subsets. *Blood* (2010) 115(2):274–81. doi:10.1182/blood-2009-04-215491
6. Kim S, Poursine-Laurent J, Truscott SM, Lybarger L, Song YJ, Yang L, et al. Licensing of natural killer cells by host major histocompatibility complex class I molecules. *Nature* (2005) 436(7051):709–13. doi:10.1038/nature03847
7. Cooley S, Xiao F, Pitt M, Gleason M, McCullar V, Bergemann T, et al. A subpopulation of human peripheral blood NK cells that lacks inhibitory receptors for self MHC is developmentally immature. *Blood* (2007) 110:578–86. doi:10.1182/blood-2006-07-036228
8. Anfossi N, Andre P, Guia S, Falk CS, Roetynck S, Stewart CA, et al. Human NK cell education by inhibitory receptors for MHC class I. *Immunity* (2006) 25:331–42. doi:10.1016/j.immuni.2006.06.013
9. Shifrin N, Raulet DH, Ardolino M. NK cell self tolerance, responsiveness and missing self recognition. *Semin Immunol* (2014) 26:138–44. doi:10.1016/j.smim.2014.02.007
10. Kadri N, Wagner AK, Ganesan S, Karre K, Wickstrom S, Johansson MH, et al. Dynamic regulation of NK cell responsiveness. *Curr Top Microbiol Immunol* (2016) 395:95–114. doi:10.1007/82_2015_485
11. Pearce EL, Mullen AC, Martins GA, Krawczyk CM, Hutchins AS, Zediak VP, et al. Control of effector CD8+ T cell function by the transcription factor eomesodermin. *Science* (2003) 302(5647):1041–3. doi:10.1126/science.1090148
12. Cruz-Guilloty F, Pipkin ME, Djuretic IM, Levanon D, Lotem J, Lichtenheld MG, et al. Runx3 and T-box proteins cooperate to establish the transcriptional program of effector CTLs. *J Exp Med* (2009) 206:51–9. doi:10.1084/jem.20081242
13. Intlekofer AM, Takemoto N, Wherry EJ, Longworth SA, Northrup JT, Palanivel VR, et al. Effector and memory CD8+ T cell fate coupled by T-bet and eomesodermin. *Nat Immunol* (2005) 6:1236–44. doi:10.1038/ni1268
14. Luetke-Eversloh M, Cicek BB, Siracusa F, Thom JT, Hamann A, Frischbutter S, et al. NK cells gain higher IFN-gamma competence during terminal differentiation. *Eur J Immunol* (2014) 44(7):2074–84. doi:10.1002/eji.201344072
15. Knox JJ, Cosma GL, Betts MR, McLane LM. Characterization of T-bet and eomes in peripheral human immune cells. *Front Immunol* (2014) 5:217. doi:10.3389/fimmu.2014.00217
16. Gordon SM, Chaix J, Rupp LJ, Wu J, Madera S, Sun JC, et al. The transcription factors T-bet and Eomes control key checkpoints of natural killer cell maturation. *Immunity* (2012) 36:55–67. doi:10.1016/j.immuni.2011.11.016
17. Daussy C, Faure F, Mayol K, Viel S, Gasteiger G, Charrier E, et al. T-bet and Eomes instruct the development of two distinct natural killer cell lineages in the liver and in the bone marrow. *J Exp Med* (2014) 211(3):563–77. doi:10.1084/jem.20131560
18. Gotthardt D, Prchal-Murphy M, Seillet C, Glasner A, Mandelboim O, Carotta S, et al. NK cell development in bone marrow and liver: site matters. *Genes Immun* (2014) 15:584–7. doi:10.1038/gene.2014.55
19. Lin L, Ma C, Wei B, Aziz N, Rajalingam R, Yusung S, et al. Human NK cells licensed by killer Ig receptor genes have an altered cytokine program that modifies CD4+ T cell function. *J Immunol* (2014) 193(2):940–9. doi:10.4049/jimmunol.1400093
20. Simonetta F, Pradier A, Bosshard C, Masouridi-Levrat S, Chalandon Y, Roosnek ENK. Cell functional impairment after allogeneic hematopoietic stem cell transplantation is associated with reduced levels of T-bet and eomesodermin. *J Immunol* (2015) 195:4712–20. doi:10.4049/jimmunol.1501522
21. Simonetta F, Pradier A, Roosnek E. T-bet and eomesodermin in NK cell development, maturation, and function. *Front Immunol* (2016) 7:241. doi:10.3389/fimmu.2016.00241
22. Thomas R, Yamada E, Alter G, Martin MP, Bashirova AA, Norman PJ, et al. Novel KIR3DL1 alleles and their expression levels on NK cells: convergent evolution of KIR3DL1 phenotype variation? *J Immunol* (2008) 180:6743–50. doi:10.4049/jimmunol.180.10.6743
23. Stern M, Ruggeri L, Capanni M, Mancusi A, Velardi A. Human leukocyte antigens A23, A24 and A32 but not A25 are ligands for KIR3DL1. *Blood* (2008) 112(3):708–10. doi:10.1182/blood-2008-02-137521
24. Foley BA, De Santis D, Van Beelen E, Lathbury LJ, Christiansen FT, Witt CS. The reactivity of Bw4-positive HLA-B and HLA-A alleles with KIR3DL1:

implications for patient and donor suitability for haploidentical stem cell transplants. *Blood* (2008) 112(2):435–43. doi:10.1182/blood-2008-01-132902

25. Yu J, Heller G, Chewning J, Kim S, Yokoyama WM, Hsu KC. Hierarchy of the human natural killer cell response is determined by class and quantity of inhibitory receptors for self-HLA-B and HLA-C ligands. *J Immunol* (2007) 179:5977–89. doi:10.4049/jimmunol.179.9.5977
26. Grau R, Lang KS, Wernet D, Lang P, Niethammer D, Pusch CM, et al. Cytotoxic activity of natural killer cells lacking killer-inhibitory receptors for self-HLA class I molecules against autologous hematopoietic stem cells in healthy individuals. *Exp Mol Pathol* (2004) 76(2):90–8. doi:10.1016/j.yexmp.2003.12.002
27. Orr MT, Murphy WJ, Lanier LL. 'Unlicensed' natural killer cells dominate the response to cytomegalovirus infection. *Nat Immunol* (2010) 11(4):321–7. doi:10.1038/ni.1849
28. Johansson S, Johansson M, Rosmaraki E, Vahlne G, Mehr R, Salmon-Divon M, et al. Natural killer cell education in mice with single or multiple major histocompatibility complex class I molecules. *J Exp Med* (2005) 201(7):1145–55. doi:10.1084/jem.20050167
29. Brodin P, Lakshmikanth T, Johansson S, Karre K, Hoglund P. The strength of inhibitory input during education quantitatively tunes the functional responsiveness of individual natural killer cells. *Blood* (2009) 113(11):2434–41. doi:10.1182/blood-2008-05-156836
30. Morvan M, David G, Sebille V, Perrin A, Gagne K, Willem C, et al. Autologous and allogeneic HLA KIR ligand environments and activating KIR control KIR NK-cell functions. *Eur J Immunol* (2008) 38:3474–86. doi:10.1002/eji.200838407

8

Features of Memory-Like and PD-1$^+$ Human NK Cell Subsets

Mariella Della Chiesa[1], Silvia Pesce[1], Letizia Muccio[1], Simona Carlomagno[1], Simona Sivori[1,2], Alessandro Moretta[1,2] and Emanuela Marcenaro[1,2]*

[1] *Dipartimento di Medicina Sperimentale, Università degli Studi di Genova, Genova, Italy,* [2] *CEBR, Università degli Studi di Genova, Genova, Italy*

Edited by:
Sophie Ugolini, Institut national de la santé et de la recherche médicale (INSERM), France

Reviewed by:
Christine Susanne Falk, Hannover Medical School, Germany
Rafael Solana, University of Córdoba, Spain

***Correspondence:**
Alessandro Moretta
alemoret@unige.it

Human NK cells are distinguished into CD56^{bright}CD16$^-$ cells and CD56^{dim}CD16$^+$ cells. These two subsets are conventionally associated with differential functional outcomes and are heterogeneous with respect to the expression of KIR and CD94/NKG2 heterodimers that represent the two major types of HLA-class I-specific receptors. Recent studies indicated that immature CD56^{bright} NK cells, homogeneously expressing the inhibitory CD94/NKG2A receptor, are precursors of CD56^{dim} NK cells that, in turn, during their process of differentiation, lose expression of CD94/NKG2A and subsequentially acquire inhibitory KIRs and LIR-1. The terminally differentiated phenotype of CD56^{dim} cells is marked by the expression of the CD57 molecule that is associated with poor responsiveness to cytokine stimulation, but retained cytolytic capacity. Remarkably, this NKG2A$^-$KIR$^+$LIR-1$^+$CD57$^+$CD56^{dim} NK cell subset when derived from individuals previously exposed to pathogens, such as human cytomegalovirus (HCMV), may contain "memory-like" NK cells. These cells are generally characterized by an upregulation of the activating receptor CD94/NKG2C and a downregulation of the inhibitory receptor Siglec-7. The "memory-like" NK cells are persistent over time and display some hallmarks of adaptive immunity, i.e., clonal expansion, more effective antitumor and antiviral immune responses, longevity, as well as given epigenetic modifications. Interestingly, unknown cofactors associated with HCMV infection may induce the onset of a recently identified fully mature NK cell subset, characterized by marked downregulation of the activating receptors NKp30 and NKp46 and by the unexpected expression of the inhibitory PD-1 receptor. This phenotype correlates with an impaired antitumor NK cell activity that can be partially restored by antibody-mediated disruption of PD-1/PD-L interaction.

Keywords: human NK cells, NKG2C, PD-1, memory, HCMV, immune checkpoint, CD57

INTRODUCTION

In physiological conditions, human peripheral blood NK cells include different cell subsets corresponding to different stages of NK cell differentiation. These subsets are characterized by the different expression of some receptors and distinct functional capabilities (1, 2).

The two major peripheral blood NK cell subsets are distinguished on the basis of their relative surface expression of CD56 molecule. In particular, CD56^{bright} NK cells (around 10% of peripheral blood NK cells) are CD16 (Fcgamma RIII)$^{dim/negative}$, CD117/c-kit^{positive} and express the high affinity IL-2Rα chain (CD25), whereas CD56^{dim} NK cells (around 90%) are CD16^{bright} and express only the intermediate affinity IL-2Rβ and γ chains (CD122/132) (3). In addition, $CD56^{bright}$ NK cells are characterized by higher IL18Rα surface expression than $CD56^{dim}$ subset (4). The same NK cell subsets are also characterized by distinct homing properties due to the different surface expression of chemokine receptors: $CD56^{dim}$ NK cells, expressing CXCR1, CX3CR1, and ChemR23, preferentially migrate to inflamed peripheral tissues (5, 6), whereas $CD56^{bright}$ NK cells, thanks to their CCR7 and CD62L expression, preferentially migrate to secondary lymphoid organs (SLOs) (6). Interestingly, recent data indicate that, in some cases, the $CD56^{dim}$ subset may also *de novo* express CCR7 and migrate toward SLOs (7–9).

Differently from $CD56^{dim}$ NK cells, $CD56^{bright}$ NK cells are characterized by low expression of lytic granules and by production of high amounts of cytokines, such as IFN-γ, TNF-α, and GM-CSF (10, 11). Thus, $CD56^{bright}$ NK cells have been usually considered as "regulatory NK cells" and $CD56^{dim}$ NK cells as "cytotoxic NK cells" (notably $CD56^{dim}$ NK cells can also release large amounts of cytokines but only upon receptor-mediated triggering) (12).

These two NK cell subsets also differ in terms of surface expression of HLA-I-specific receptors. Indeed, $CD56^{bright}$ NK cells express only CD94/NKG2A, whereas $CD56^{dim}$ NK cells may also express KIRs, and/or LIR-1 (13, 14). Since inhibitory and activating receptors can be distinguished within the KIR family (15), two broad groups of KIR haplotypes have been identified on the basis of gene content. A haplotypes express only one activating KIR whereas B haplotypes up to five (16). Also CD94/NKG2A has an activating counterpart represented by CD94/NKG2C (17).

In this context, several studies indicated that $CD56^{dim}$ KIR^{+} NK cells derive from $CD56^{bright}$ KIR^{-} $NKG2A^{+}$ NK cells and that late stages of NK cells maturation are associated with the expression of CD57. This molecule is expressed on a fraction of $CD56^{dim}$ NK cells and is believed to mark a subpopulation of terminally differentiated NK cells that are mainly characterized by the KIR^{+}, $LIR\text{-}1^{+}$, and $CD94/NKG2A^{-}$ phenotype (18, 19).

In addition to $CD56^{bright}$ and $CD56^{dim}$ NK cell subsets, low frequencies of $CD56^{neg}$ $CD16^{bright}$ NK cells are also detected in healthy donors. In patients with chronic viral infections, this $CD56^{neg}$ NK cell subset expands and a pathological redistribution of the various NK cell subsets occurs. Indeed increments in the percent of $CD56^{neg}$ NK cells have been reported in several pathological conditions, including hepatitis C virus (HCV) (20, 21), human cytomegalovirus (HCMV) (22), hantavirus infections (23), and autoimmune disorders (24–26).

The fact that the $CD56^{dim}$ NK cell subset is often heterogeneous in terms of expression levels of natural cytotoxicity receptors (NCRs: NKp46, NKp30, and NKp44) (27) led to the distinction of two additional NK cell subsets termed NCR^{dull} and NCR^{bright} (28). The demonstration that the NCR surface density correlates with the magnitude of the NK-mediated natural cytotoxicity provided a rational explanation for the clonal heterogeneity of NK cells in killing autologous or allogeneic NK-susceptible targets.

In this context, it is important to consider that, in healthy donors, most $CD56^{dim}$ $KIR^{+}NKG2A^{-}CD57^{+}$ NK cells are characterized by a lower surface expression of NCRs (18, 19). On the other hand, $CD56^{bright}$ NK cells are characterized by higher NKp46 surface expression as compared to $CD56^{dim}$ NK cells.

Finally, despite the fact that NK cells have always been considered members of the innate immune system, new increasing evidences suggest that NK cells can display some features that are usually attributed to adaptive immune cells, such as expansion and contraction of subsets, increased longevity, and a more potent response upon secondary challenge with the same antigen (memory-like properties) (29).

MEMORY-LIKE NK CELL SUBSETS EMERGING UPON HCMV INFECTION

In the last years, it has been observed how HCMV infection can shape the NK cell receptor repertoire inducing the expansion of a specific NK cell population expressing the activating receptor CD94/NKG2C (30, 31) and the marker of terminal differentiation CD57 (32). This HCMV-induced $NKG2C^{+}CD57^{+}$ NK cell subset displays a highly differentiated surface phenotype, $CD56^{dim}CD16^{bright}LIR\text{-}1^{+}KIR^{+}NKG2A^{-}$, and is characterized by the expression of self KIRs (33). More recently, it has been proposed that, upon HCMV infection, NK cells might acquire some hallmarks of adaptive immunity, i.e., clonal expansion, enhanced effector function, longevity, as well as given epigenetic modifications (34–36). Indeed, in HCMV seropositive healthy individuals (HD), the memory-like $NKG2C^{+}CD57^{+}$NK cell subset is characterized by an epigenetic remodeling at the IFN-γ locus similar to that found in memory T cells, which is likely responsible for the enhanced IFN-γ production upon target recognition observed in $NKG2C^{+}$ NK cells (37). Interestingly, the HCMV-induced $NKG2C^{+}$ subset is also characterized by a decreased expression of certain signaling molecules, i.e., the adaptor protein FcεRγ and the tyrosine kinase Syk, and by lower expression levels of the transcription factor PLZF, which is involved in regulating epigenetic modifications (e.g., DNA methylation) (34, 35). While the higher accessibility of the IFN-γ locus can enhance IFN-γ production upon target recognition, the lack of FcεRγ could favor a more efficient killing *via* ADCC of opsonized HCMV-infected targets by memory-like $NKG2C^{+}$ NK cells as compared to conventional NK cells (**Figure 1**). Indeed, in the absence of FcεRγ that only bears one immunoreceptor tyrosine-based activatory motif (ITAM) sequence, CD16 engagement would involve exclusively the adaptor protein CD3ζ, which contains three ITAMs, possibly delivering a stronger signal inside the cell (38, 39). Along this line, a crucial role for CD16 engagement by anti-HCMV antibodies has been proposed not only in promoting ADCC and cytokine release by $NKG2C^{+}$ NK cells but also in favoring their preferential expansion (34, 35, 40). Moreover, the effector function triggered by CD16 engagement in adaptive $NKG2C^{+}$ NK cells (i.e., degranulation and cytokine release) can be enhanced by CD2 costimulation, suggesting a synergy between these receptors in regulating anti-HCMV responses (39).

However, although new insights on memory-like (or adaptive) NK cells are continuously collected, the signals responsible for the described epigenetic modifications and protein expression alterations as well as the exact mechanisms regulating the generation of memory-like $NKG2C^{+}$ NK cells are not completely understood.

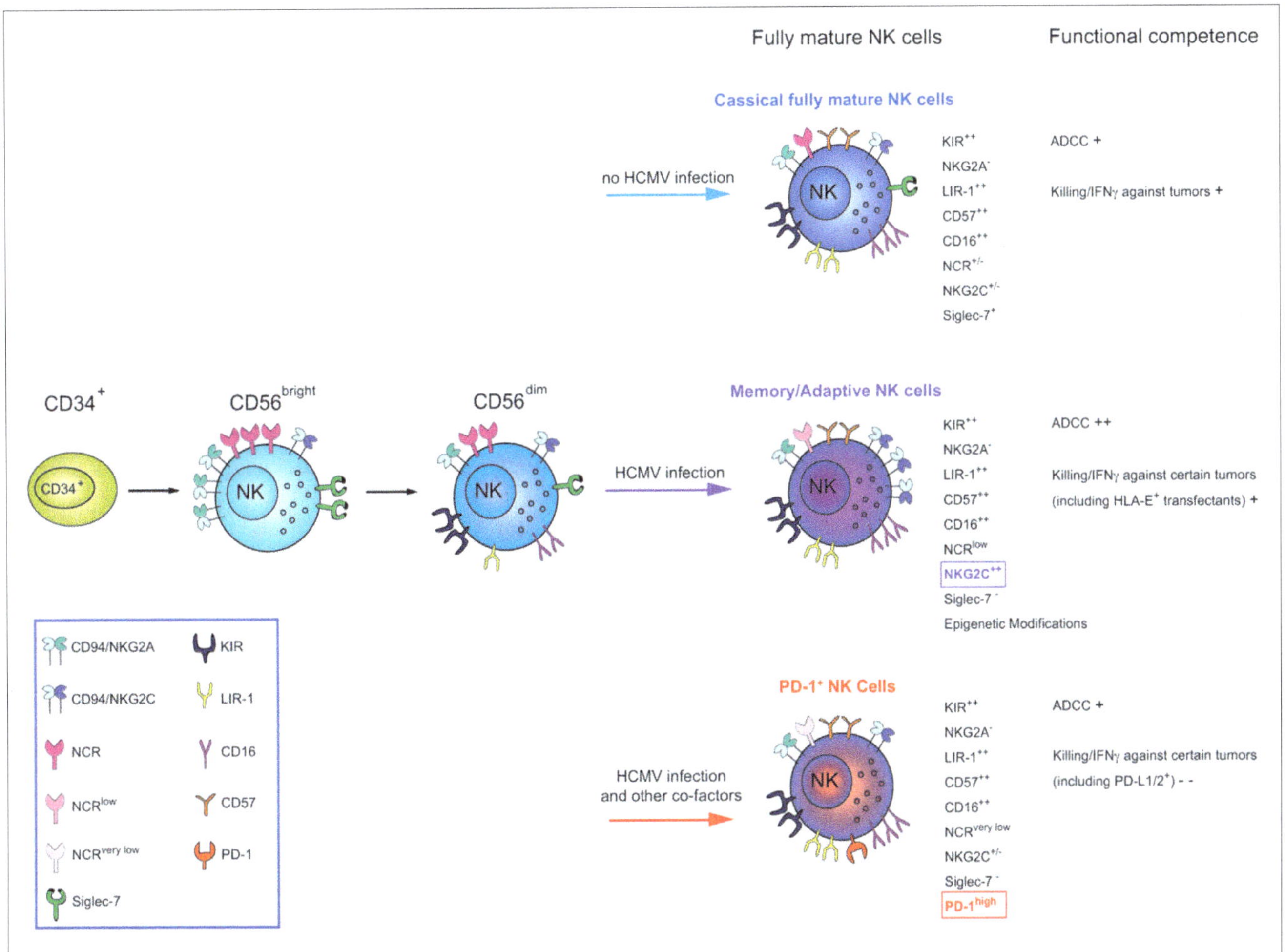

FIGURE 1 | NK cells differentiate starting from CD34$^+$ bone marrow precursors into CD56^{bright} (CD94/NKG2A$^+$KIR$^-$) and then to CD56^{dim} cells. CD56^{dim} NK cells, in turn, continue to differentiate throughout their life-span, acquiring novel functional and phenotypic properties. During this process, they lose expression of CD94/NKG2A, sequentially acquire inhibitory KIRs and, at the final step, CD57. This molecule appear to define a subpopulation of highly differentiated NK cells (classical fully mature NK cells), characterized by the KIR$^+$LIR-1$^+$NKG2A$^-$NCR$^{+/-}$ phenotype. Functionally, these cells display natural cytolytic activity and ADCC against tumor targets, but poor responsiveness to cytokine stimulation. After HCMV infection/reactivation, increased proportions of a subset of terminally differentiated CD57$^+$ NK cells, characterized by high expression of NKG2C and downregulation of Siglec-7 receptors are induced (the so-called memory-like NK cells). These cells display increased functional capability in terms of ADCC and IFN-γ production/killing in response to HLA-E$^+$ and opsonized HCMV-infected targets but decreased function after cytokine stimulation. Following HCMV infections accompanied by other cofactors (infections?), an additional type of CD57$^+$ NK cell subset can be generated. This subset is characterized by the expression of the inhibitory PD-1 receptor (not necessary co-expressed with NKG2C) and a very low expression of the NCRs, NKp30 and NKp46. These cells, called PD-1$^+$ NK cells, are characterized by compromised effector functions against tumor cells expressing ligands for PD-1 as well as against tumors primarily killed on NCRs/NCR-ligands interactions.

The studies carried out in immunocompromised individuals undergoing hematopoietic stem cell transplantation (HSCT), where HCMV infection exerts the strongest effect on NK cells skewing, could help clarifying these aspects. The first observations reported that, in adult UCBT recipients, a rapid accumulation of mature CD56^{dim} KIR$^+$NKG2A$^-$NKG2C$^+$ NK cells occurred following HCMV infection (22, 41). This accelerated maturation was usually accompanied by the downregulation of the inhibitory receptor Siglec-7, which, thus, represents a typical hallmark of the anti-HCMV response in NK cells, along with the expansion of cells expressing NKG2C. Notably, in uninfected UCBT recipients, NK cells were characterized by a more immature phenotype (high frequencies of CD56^{bright} NK cells and NKG2A$^+$CD56^{dim} NK cells) even at late time points after UCBT. Remarkably, in some HCMV-infected UCBT recipients, a subset of hyporesponsive CD56$^-$CD16^{bright} NK cells displaying a mature phenotype was also observed which likely reflected a condition of severely impaired T cell immunity (22, 42). More recently, a remarkable acceleration of NK cell maturation was described also in pediatric patients receiving a type of allograft different from UCBT, i.e., a HLA-haploidentical HSCT, depleted of both α/β$^+$ T cells and B cells and containing variable numbers of donor-derived NK cells and γ/δ$^+$ T cells (43). In most recipients, HCMV reactivation favored the preferential expansion of highly differentiated NKG2C$^+$CD57$^+$selfKIR$^+$NKG2A$^-$Siglec-7$^-$NCR^{low}IL18Rα^{low} NK cells and their persistence over time. These cells could kill certain tumor targets, release IFN-γ, display efficient reverse ADCC, and could recognize HLA-E^+ targets through NKG2C (putative receptor for HCMV). On the other hand, they showed an impaired ability to release IFN-γ upon IL-12+IL-18 exposure. The particular signature shown by these HCMV-induced NK cells may suggest their skewing toward an adaptive condition specialized in controlling HCMV (43). The $NKG2C^+CD57^+$ memory-like NK cell subset could also contribute to protect against leukemia relapse (44). Interestingly, an expansion of memory-like $NKG2C^+CD57^+$ NK cells could be observed also in some recipients who did not reactivate HCMV and who received grafts containing high numbers of mature NK cells derived from a $HCMV^+$ donor. Thus, donor-derived transplanted NK cells, primed by a previous encounter with HCMV in the donor, could have persisted and proliferated in the recipient in response to a subclinical reactivation, favoring antiviral responses (43, 45).

Notably HCMV-induced memory NK cell subsets could be represented not only by the described $NKG2C^+CD57^+$ population but also by $NKG2C^-$ NK cell subsets, expressing activating receptors different from NKG2C, such as activating KIRs, or other still undefined activating receptors (33, 39, 46, 47).

Thus, in the HSCT setting, HCMV clearly reveals as a key driving force regulating the differentiation of functionally and phenotypically skewed NK cell subsets characterized by memory-like properties.

HUMAN RESTING NK CELLS CAN EXPRESS HIGH LEVELS OF PD-1 RECEPTOR

NK cells are believed to play a critical role in the recognition and eradication of tumors by using different killing strategies; however, tumor cells often develop immunosuppressive mechanisms to avoid NK cell-mediated killing, allowing for tumor escape (48, 49).

An improved understanding of the molecular mechanisms involved in tumor recognition and eradication has led to the identification of checkpoint signaling pathways involved in limiting the anticancer immune response.

One of the most critical checkpoint pathways responsible for mediating tumor-induced immune suppression is the programmed death-1 (PD-1) pathway. This receptor, by modulating the duration and the amplitude of physiological immune responses, is capable of promoting tolerance and preventing tissue damage in settings of chronic inflammation, as well as autoimmune pathologies (48, 49). The induction of the PD-1-dependent inhibitory pathway is mediated by the interaction of this receptor with its ligands, PD-L1 and PD-L2 (50).

The constitutive or inducible expression of the PD-1 receptor has been described on both adaptive and innate immune cells, including T, B, and dendritic cells.

In T cells, binding of PD-1 to its ligands inhibits T cell activation, proliferation, and cytokine production and eventually may result in T lymphocyte exhaustion. Thus, tumors and viruses, by expressing PD-1 ligands, have evolved a remarkable mechanism to hijack the PD-1-dependent regulatory mechanism to avoid T cell-mediated surveillance of cancer or infected cells. Remarkably, however, mAb-mediated blockade of PD-1/PD-L interactions, by disrupting the immune checkpoint-based inhibitory pathway, provides an important opportunity to enhance antitumor immunity particularly in the case of tumor antigen-specific T cells (51, 52).

Similar to T cells, NK cells express surface inhibitory receptors that can be targeted in checkpoint blockade strategies, including the HLA-class I-specific KIR family and CD94/NKG2A heterodimer. Blocking KIR/NKG2A–HLA-class I interactions resulted in potent NK cell-mediated antitumor efficacy (53). Phase I/II trials testing human anti-KIR and anti-NKG2A antibody are ongoing (54–57). Regarding PD-1, the antitumor effect of specific antibodies has been always considered to depend mainly on the rescue effect on activated PD-1^+ tumor-specific T cells recruited in the tumor environment (58). On the contrary, very little was known on the activated PD-1^+ NK cells expansions that had been seldom reported in patients with certain tumors or chronic viral infections (59–63). In fact, a precise information on the actual function of these cells was not possible due to the low levels of PD-1 expression and the consequent difficulty in distinguishing positive from negative NK cells.

More insights on the expression and function of PD-1 on NK cells could be recently obtained thanks to the demonstration that the PD-1 receptor is brightly expressed on a discrete cell subset of peripheral blood NK cells from one-fourth of otherwise healthy individuals (64).

PD-1^+ cells are confined to $CD56^{dim}$ NK cells, and (if present) on $CD56^{neg}$ NK cells, whereas the $CD56^{bright}$ cell subset is consistently PD-1^-. In addition, a remarkable difference exists among donors regarding the size of the PD-1^+ NK cell subset. Importantly, the analysis at different time points of the size of PD-1^+ cell subset in given individuals indicated that this population remains substantially stable over time (64).

The fact that only some of the individuals analyzed are characterized by a PD-1^+ NK cell subset may be the result of given acute or chronic infection affecting only part of the population (an increase in PD-1^+ lymphocytes has been associated with HCV, HBV, and HIV) (63, 65–68).

Interestingly, our analysis indicates that a direct correlation between HCMV infection and presence of a PD-1^+ NK cell subset in the healthy donors analyzed could be established. In particular, we found that PD-1^+ individuals are in all instances seropositive for HCMV and display higher frequencies of $NKG2C^+$ and Siglec-7^- NK cells.

By comparing the PD-1^+ and PD-1^- NK cell subsets derived from seropositive PD-1^+ HD, it was possible to show that the PD-1^+ subset is confined to cells displaying the phenotypic features of fully mature NK cells, characterized homogeneously by the $CD56^{dim}$ $KIR^+LIR\text{-}1^+NKG2A^-CD57^+$ phenotype (22, 30). Moreover, only a minor fraction of these cells expressed Siglec-7, whereas, unexpectedly, NKG2C was not necessarily co-expressed with PD-1 receptor. These data could indicate that, in addition to HCMV, additional factors (infections?) may contribute to the induction of PD-1 expression (64).

Further phenotypic characterization indicates that the PD-1^+ NK cell subset, when compared with PD-1^- NK cells, has lower expression of NCRs (NKp46 and NKp30). In addition, the comparison between PD-1^+ and PD-1^- NK cells that are contained within the highly differentiated $KIR^+NKG2A^-CD57^+$ subset showed that the expression of NCRs is maximally reduced in the PD-1^+ subset (**Figure 1**).

Functional analysis of PD-1^+ NK cells indicated that they display a low cytolytic activity and impaired degranulation against tumor targets, even when these cells lack PD-L1/PD-L2 expression (i.e., K562). The impaired degranulation in response to PD-L^{neg} tumor target cells may be a consequence of the defective expression of NCRs, since these target cells may express a series of ligands for activating NK receptors, such as B7-H6 (69, 70). Remarkably, the reduced degranulation of PD-1^+ cells following interaction with tumor targets expressing PD-1 ligands is reflecting not only the poor NCR-mediated cell activation but also the inhibitory signal mediated by PD-1 upon interaction with PD-L1/PD-L2 expressed on tumor targets. In this context, it is of note that the inhibition of NK cell degranulation induced by PD-1/PD-L interaction on tumor cells could be partially reverted by mAbs specific for PD-L1/PD-L2 (64).

PD-1^+ NK cells also display an altered capability of releasing IFN-γ and TNF-α cytokines after stimulation with the same tumor targets used in degranulation assays. Finally, in line with previous data on classical $CD57^+$ NK cells, PD-1^+ NK cells appear to represent a population of poorly proliferating cells, rescued to divide only in the presence of high concentrations of microenvironmental cytokines. This suggests that PD-1^+ NK cells, like $CD57^+CD8^+$ T cells, have a proliferation defect *in vitro* (e.g., lower expression of IL-2Rβ).

Remarkably, PD-1^+ NK cells are present in higher proportions in the ascites of ovarian-carcinoma patients (71), suggesting their possible induction/enrichment in tumor microenvironment. Also, in this case, the PD-1^+ NK cell subset of these patients displayed a functional defect against PD-L1/PD-$L2^+$ tumor targets. However, disruption of PD-1 receptor/ligands interaction by specific anti-PD-L mAbs restored degranulation against these tumor target cells.

In conclusion, these findings support the notion that PD-1 signaling may inhibit/block not only T lymphocytes-mediated adaptive responses but also NK cell-mediated innate responses (58, 72).

Therefore, it cannot be excluded that PD-1 may represent an inhibitory checkpoint expressed on NK cells in various cancers of different histotype and that this inhibitory receptor may be involved in the impaired antitumor NK cell responses by these patients.

In this context, it should be stressed/emphasized that, while, in conventional NK cells, the effector function against tumors is primarily regulated by the interactions between HLA-class I-specific inhibitory receptors (KIR and CD94/NKG2A) and HLA-class I molecules, in the case of PD-1^+ NK cells, the simultaneous expression of PD-1 together with given inhibitory HLA-specific receptors may provide an additional level of suppression of NK cell-mediated antitumor responses. In this case, downregulation of HLA-class I molecules on tumor cells may not be sufficient to induce efficient NK cell responses. These, however, could be restored, at least in part, by mAb-mediated disruption of PD-1/PD-L interaction. On the other hand, in case of PD-L^+ tumors that do not downregulate HLA-class I molecules, it may be necessary the combined blocking of different inhibitory checkpoints by anti-KIRs and anti-PD-1 mAbs.

CONCLUSION

In conclusion, recent studies led to the identifications of novel unexpected properties of NK cells, including the generation of fully mature NK cells displaying some functional characteristics that are reminiscent of cells of adaptive immunity. These cells are generally considered "memory-like" and are characterized by the expression of given set of inhibitory checkpoints mainly represented by different KIRs and LIR-1. Memory-like NK cells that have been originally identified in $HCMV^+$ individuals are characterized by functional enhancement in terms of ADCC and IFN-γ production, features linked to changes in the expression of multiple intracellular proteins and transcription factors. On the other hand, these "adaptive" NK cells respond weakly to certain tumors (due to the reduced expression of NKp46 and NKp30) and cytokine receptor-based activation compared to classical fully mature NK cells. Recently, a further novel NK cell subset has been identified in $HCMV^+$ individuals. This subpopulation, called "PD-1^+ NK cells," is mainly composed by fully mature NK cells and displays strongly reduced capacity to kill PD-L^+ tumor cells, due to the expression on their surface of high levels of the inhibitory checkpoint PD-1 and of very low levels of NCRs. Due to its ineffective antitumor functions, it would be important to better evaluate the conditions that lead to the generation of this subset and to understand its role in health and disease, in particular in patients with advanced cancers.

In this context, drugs blocking PD-1 and its major ligand PD-L1 have shown great promise in treating many different cancer types. However, the focus is currently only on T cell responses. The fact that a fraction of NK cells express PD-1 can open prospects for extending the potential of cancer immunotherapy to this important innate effector cells.

Innovative treatment could be designed to combine innate immune activation with activation of the adaptive immune system (73).

AUTHOR CONTRIBUTIONS

All authors listed have made substantial, direct, and intellectual contribution to the work and approved it for publication.

ACKNOWLEDGMENTS

Supported by grants awarded by Associazione Italiana Ricerca per la Ricerca sul Cancro (AIRC)-Special Project 5 × 1000 no. 9962 (Alessandro Moretta), AIRC IG 2014 Id. 15704 (Alessandro Moretta); Progetto di Ricerca di Ateneo 2014 (Emanuela Marcenaro).

REFERENCES

1. Moretta L, Montaldo E, Vacca P, Del Zotto G, Moretta F, Merli P, et al. Human natural killer cells: origin, receptors, function, and clinical applications. *Int Arch Allergy Immunol* (2014) 164(4):253–64. doi:10.1159/000365632
2. Cichocki F, Sitnicka E, Bryceson YT. NK cell development and function – plasticity and redundancy unleashed. *Semin Immunol* (2014) 26(2):114–26. doi:10.1016/j.smim.2014.02.003
3. Cooper MA, Fehniger TA, Caligiuri MA. The biology of human natural killer-cell subsets. *Trends Immunol* (2001) 22(11):633–40. doi:10.1016/S1471-4906(01)02060-9
4. Pesce S, Carlomagno S, Moretta A, Sivori S, Marcenaro E. Uptake of CCR7 by KIR2DS4(+) NK cells is induced upon recognition of certain HLA-C alleles. *J Immunol Res* (2015) 2015:754373. doi:10.1155/2015/754373
5. Parolini S, Santoro A, Marcenaro E, Luini W, Massardi L, Facchetti F, et al. The role of chemerin in the colocalization of NK and dendritic cell subsets into inflamed tissues. *Blood* (2007) 109(9):3625–32. doi:10.1182/blood-2006-08-038844
6. Robertson MJ. Role of chemokines in the biology of natural killer cells. *J Leukoc Biol* (2002) 71(2):173–83.
7. Mailliard RB, Alber SM, Shen H, Watkins SC, Kirkwood JM, Herberman RB, et al. IL-18-induced CD83+CCR7+ NK helper cells. *J Exp Med* (2005) 202(7):941–53. doi:10.1084/jem.20050128
8. Marcenaro E, Cantoni C, Pesce S, Prato C, Pende D, Agaugue S, et al. Uptake of CCR7 and acquisition of migratory properties by human KIR+ NK cells interacting with monocyte-derived DC or EBV cell lines: regulation by KIR/HLA-class I interaction. *Blood* (2009) 114(19):4108–16. doi:10.1182/blood-2009-05-222265
9. Marcenaro E, Pesce S, Sivori S, Carlomagno S, Moretta L, Moretta A. KIR2DS1-dependent acquisition of CCR7 and migratory properties by human NK cells interacting with allogeneic HLA-C2+ DCs or T-cell blasts. *Blood* (2013) 121(17):3396–401. doi:10.1182/blood-2012-09-458752
10. Moretta A, Marcenaro E, Parolini S, Ferlazzo G, Moretta L. NK cells at the interface between innate and adaptive immunity. *Cell Death Differ* (2008) 15(2):226–33. doi:10.1038/sj.cdd.4402170
11. Fauriat C, Long EO, Ljunggren HG, Bryceson YT. Regulation of human NK-cell cytokine and chemokine production by target cell recognition. *Blood* (2010) 115(11):2167–76. doi:10.1182/blood-2009-08-238469
12. De Maria A, Bozzano F, Cantoni C, Moretta L. Revisiting human natural killer cell subset function revealed cytolytic CD56(dim)CD16+ NK cells as rapid producers of abundant IFN-gamma on activation. *Proc Natl Acad Sci U S A* (2011) 108(2):728–32. doi:10.1073/pnas.1012356108
13. Vitale M, Castriconi R, Parolini S, Pende D, Hsu ML, Moretta L, et al. The leukocyte Ig-like receptor (LIR)-1 for the cytomegalovirus UL18 protein displays a broad specificity for different HLA class I alleles: analysis of LIR-1 + NK cell clones. *Int Immunol* (1999) 11(1):29–35. doi:10.1093/intimm/11.1.29
14. Marcenaro E, Carlomagno S, Pesce S, Della Chiesa M, Parolini S, Moretta A, et al. NK cells and their receptors during viral infections. *Immunotherapy* (2011) 3(9):1075–86. doi:10.2217/imt.11.99
15. Moretta A, Sivori S, Vitale M, Pende D, Morelli L, Augugliaro R, et al. Existence of both inhibitory (p58) and activatory (p50) receptors for HLA-C molecules in human natural killer cells. *J Exp Med* (1995) 182(3):875–84. doi:10.1084/jem.182.3.875
16. Parham P. MHC class I molecules and KIRs in human history, health and survival. *Nat Rev Immunol* (2005) 5(3):201–14. doi:10.1038/nri1570
17. Braud VM, Allan DS, O'Callaghan CA, Soderstrom K, D'Andrea A, Ogg GS, et al. HLA-E binds to natural killer cell receptors CD94/NKG2A, B and C. *Nature* (1998) 391(6669):795–9. doi:10.1038/35869
18. Bjorkstrom NK, Riese P, Heuts F, Andersson S, Fauriat C, Ivarsson MA, et al. Expression patterns of NKG2A, KIR, and CD57 define a process of CD56dim NK-cell differentiation uncoupled from NK-cell education. *Blood* (2010) 116(19):3853–64. doi:10.1182/blood-2010-04-281675
19. Lopez-Verges S, Milush JM, Pandey S, York VA, Arakawa-Hoyt J, Pircher H, et al. CD57 defines a functionally distinct population of mature NK cells in the human CD56dimCD16+ NK-cell subset. *Blood* (2010) 116(19):3865–74. doi:10.1182/blood-2010-04-282301
20. Gonzalez VD, Falconer K, Michaelsson J, Moll M, Reichard O, Alaeus A, et al. Expansion of CD56- NK cells in chronic HCV/HIV-1 co-infection: reversion by antiviral treatment with pegylated IFNalpha and ribavirin. *Clin Immunol* (2008) 128(1):46–56. doi:10.1016/j.clim.2008.03.521
21. Gonzalez VD, Falconer K, Bjorkstrom NK, Blom KG, Weiland O, Ljunggren HG, et al. Expansion of functionally skewed CD56-negative NK cells in chronic hepatitis C virus infection: correlation with outcome of pegylated IFN-alpha and ribavirin treatment. *J Immunol* (2009) 183(10):6612–8. doi:10.4049/jimmunol.0901437
22. Della Chiesa M, Falco M, Podesta M, Locatelli F, Moretta L, Frassoni F, et al. Phenotypic and functional heterogeneity of human NK cells developing after umbilical cord blood transplantation: a role for human cytomegalovirus? *Blood* (2012) 119(2):399–410. doi:10.1182/blood-2011-08-372003
23. Bjorkstrom NK, Lindgren T, Stoltz M, Fauriat C, Braun M, Evander M, et al. Rapid expansion and long-term persistence of elevated NK cell numbers in humans infected with hantavirus. *J Exp Med* (2011) 208(1):13–21. doi:10.1084/jem.20100762
24. Nguyen S, Morel V, Le Garff-Tavernier M, Bolgert F, Leblond V, Debre P, et al. Persistence of CD16+/CD56-/2B4+ natural killer cells: a highly dysfunctional NK subset expanded in ocular myasthenia gravis. *J Neuroimmunol* (2006) 179(1–2):117–25. doi:10.1016/j.jneuroim.2006.05.028
25. Lugli E, Marcenaro E, Mavilio D. NK cell subset redistribution during the course of viral infections. *Front Immunol* (2014) 5:390. doi:10.3389/fimmu.2014.00390
26. Bjorkstrom NK, Ljunggren HG, Sandberg JK. CD56 negative NK cells: origin, function, and role in chronic viral disease. *Trends Immunol* (2010) 31(11):401–6. doi:10.1016/j.it.2010.08.003
27. Moretta A, Bottino C, Vitale M, Pende D, Cantoni C, Mingari MC, et al. Activating receptors and coreceptors involved in human natural killer cell-mediated cytolysis. *Annu Rev Immunol* (2001) 19:197–223. doi:10.1146/annurev.immunol.19.1.197
28. Sivori S, Pende D, Bottino C, Marcenaro E, Pessino A, Biassoni R, et al. NKp46 is the major triggering receptor involved in the natural cytotoxicity of fresh or cultured human NK cells. Correlation between surface density of NKp46 and natural cytotoxicity against autologous, allogeneic or xenogeneic target cells. *Eur J Immunol* (1999) 29(5):1656–66. doi:10.1002/(SICI)1521-4141(199905)29:05<1656::AID-IMMU1656>3.0.CO;2-1
29. Della Chiesa M, Sivori S, Carlomagno S, Moretta L, Moretta A. Activating KIRs and NKG2C in viral infections: toward NK cell memory? *Front Immunol* (2015) 6:573. doi:10.3389/fimmu.2015.00573
30. Guma M, Angulo A, Vilches C, Gomez-Lozano N, Malats N, Lopez-Botet M. Imprint of human cytomegalovirus infection on the NK cell receptor repertoire. *Blood* (2004) 104(12):3664–71. doi:10.1182/blood-2004-05-2058
31. Guma M, Budt M, Saez A, Brckalo T, Hengel H, Angulo A, et al. Expansion of CD94/NKG2C+ NK cells in response to human cytomegalovirus-infected fibroblasts. *Blood* (2006) 107(9):3624–31. doi:10.1182/blood-2005-09-3682
32. Lopez-Verges S, Milush JM, Schwartz BS, Pando MJ, Jarjoura J, York VA, et al. Expansion of a unique CD57+NKG2Chi natural killer cell subset during acute human cytomegalovirus infection. *Proc Natl Acad Sci U S A* (2011) 108(36):14725–32. doi:10.1073/pnas.1110900108
33. Beziat V, Liu LL, Malmberg JA, Ivarsson MA, Sohlberg E, Bjorklund AT, et al. NK cell responses to cytomegalovirus infection lead to stable imprints in the human KIR repertoire and involve activating KIRs. *Blood* (2013) 121(14):2678–88. doi:10.1182/blood-2012-10-459545
34. Schlums H, Cichocki F, Tesi B, Theorell J, Beziat V, Holmes TD, et al. Cytomegalovirus infection drives adaptive epigenetic diversification of NK cells with altered signaling and effector function. *Immunity* (2015) 42(3):443–56. doi:10.1016/j.immuni.2015.02.008
35. Lee J, Zhang T, Hwang I, Kim A, Nitschke L, Kim M, et al. Epigenetic modification and antibody-dependent expansion of memory-like NK cells in human cytomegalovirus-infected individuals. *Immunity* (2015) 42(3):431–42. doi:10.1016/j.immuni.2015.02.013
36. Rolle A, Brodin P. Immune adaptation to environmental influence: the case of NK cells and HCMV. *Trends Immunol* (2016) 37(3):233–43. doi:10.1016/j.it.2016.01.005

37. Luetke-Eversloh M, Hammer Q, Durek P, Nordstrom K, Gasparoni G, Pink M, et al. Human cytomegalovirus drives epigenetic imprinting of the IFNG locus in NKG2Chi natural killer cells. *PLoS Pathog* (2014) 10(10):e1004441. doi:10.1371/journal.ppat.1004441
38. Hwang I, Zhang T, Scott JM, Kim AR, Lee T, Kakarla T, et al. Identification of human NK cells that are deficient for signaling adaptor FcRgamma and specialized for antibody-dependent immune functions. *Int Immunol* (2012) 24(12):793–802. doi:10.1093/intimm/dxs080
39. Liu LL, Landskron J, Ask EH, Enqvist M, Sohlberg E, Traherne JA, et al. Critical role of CD2 co-stimulation in adaptive natural killer cell responses revealed in NKG2C-deficient humans. *Cell Rep* (2016) 15(5):1088–99. doi:10.1016/j.celrep.2016.04.005
40. Costa-Garcia M, Vera A, Moraru M, Vilches C, Lopez-Botet M, Muntasell A. Antibody-mediated response of NKG2Cbright NK cells against human cytomegalovirus. *J Immunol* (2015) 194(6):2715–24. doi:10.4049/jimmunol.1402281
41. Foley B, Cooley S, Verneris MR, Pitt M, Curtsinger J, Luo X, et al. Cytomegalovirus reactivation after allogeneic transplantation promotes a lasting increase in educated NKG2C+ natural killer cells with potent function. *Blood* (2012) 119(11):2665–74. doi:10.1182/blood-2011-10-386995
42. Mavilio D, Lombardo G, Benjamin J, Kim D, Follman D, Marcenaro E, et al. Characterization of CD56-/CD16+ natural killer (NK) cells: a highly dysfunctional NK subset expanded in HIV-infected viremic individuals. *Proc Natl Acad Sci U S A* (2005) 102(8):2886–91. doi:10.1073/pnas.0409872102
43. Muccio L, Bertaina A, Falco M, Pende D, Meazza R, Lopez-Botet M, et al. Analysis of memory-like natural killer cells in human cytomegalovirus-infected children undergoing alphabeta+T and B cell-depleted hematopoietic stem cell transplantation for hematological malignancies. *Haematologica* (2016) 101(3):371–81. doi:10.3324/haematol.2015.134155
44. Cichocki F, Cooley S, Davis Z, DeFor TE, Schlums H, Zhang B, et al. CD56dimCD57+NKG2C+ NK cell expansion is associated with reduced leukemia relapse after reduced intensity HCT. *Leukemia* (2016) 30(2):456–63. doi:10.1038/leu.2015.260
45. Foley B, Cooley S, Verneris MR, Curtsinger J, Luo X, Waller EK, et al. Human cytomegalovirus (CMV)-induced memory-like NKG2C(+) NK cells are transplantable and expand in vivo in response to recipient CMV antigen. *J Immunol* (2012) 189(10):5082–8. doi:10.4049/jimmunol.1201964
46. Muntasell A, Pupuleku A, Cisneros E, Vera A, Moraru M, Vilches C, et al. Relationship of NKG2C copy number with the distribution of distinct cytomegalovirus-induced adaptive NK cell subsets. *J Immunol* (2016) 196(9):3818–27. doi:10.4049/jimmunol.1502438
47. Della Chiesa M, Falco M, Bertaina A, Muccio L, Alicata C, Frassoni F, et al. Human cytomegalovirus infection promotes rapid maturation of NK cells expressing activating killer Ig-like receptor in patients transplanted with NKG2C-/- umbilical cord blood. *J Immunol* (2014) 192(4):1471–9. doi:10.4049/jimmunol.1302053
48. Vivier E, Daeron M. Immunoreceptor tyrosine-based inhibition motifs. *Immunol Today* (1997) 18(6):286–91. doi:10.1016/S0167-5699(97)80025-4
49. Keir ME, Butte MJ, Freeman GJ, Sharpe AH. PD-1 and its ligands in tolerance and immunity. *Annu Rev Immunol* (2008) 26:677–704. doi:10.1146/annurev.immunol.26.021607.090331
50. Ghiotto M, Gauthier L, Serriari N, Pastor S, Truneh A, Nunes JA, et al. PD-L1 and PD-L2 differ in their molecular mechanisms of interaction with PD-1. *Int Immunol* (2010) 22(8):651–60. doi:10.1093/intimm/dxq049
51. Topalian SL, Hodi FS, Brahmer JR, Gettinger SN, Smith DC, McDermott DF, et al. Safety, activity, and immune correlates of anti-PD-1 antibody in cancer. *N Engl J Med* (2012) 366(26):2443–54. doi:10.1056/NEJMoa1200690
52. McDermott DF, Atkins MB. PD-1 as a potential target in cancer therapy. *Cancer Med* (2013) 2(5):662–73. doi:10.1002/cam4.106
53. Moretta A, Bottino C, Vitale M, Pende D, Biassoni R, Mingari MC, et al. Receptors for HLA class-I molecules in human natural killer cells. *Annu Rev Immunol* (1996) 14:619–48. doi:10.1146/annurev.immunol.14.1.619
54. Romagne F, Andre P, Spee P, Zahn S, Anfossi N, Gauthier L, et al. Preclinical characterization of 1-7F9, a novel human anti-KIR receptor therapeutic antibody that augments natural killer-mediated killing of tumor cells. *Blood* (2009) 114(13):2667–77. doi:10.1182/blood-2009-02-206532
55. Kohrt HE, Thielens A, Marabelle A, Sagiv-Barfi I, Sola C, Chanuc F, et al. Anti-KIR antibody enhancement of anti-lymphoma activity of natural killer cells as monotherapy and in combination with anti-CD20 antibodies. *Blood* (2014) 123(5):678–86. doi:10.1182/blood-2013-08-519199
56. Vey N, Bourhis JH, Boissel N, Bordessoule D, Prebet T, Charbonnier A, et al. A phase 1 trial of the anti-inhibitory KIR mAb IPH2101 for AML in complete remission. *Blood* (2012) 120(22):4317–23. doi:10.1182/blood-2012-06-437558
57. Godal R, Bachanova V, Gleason M, McCullar V, Yun GH, Cooley S, et al. Natural killer cell killing of acute myelogenous leukemia and acute lymphoblastic leukemia blasts by killer cell immunoglobulin-like receptor-negative natural killer cells after NKG2A and LIR-1 blockade. *Biol Blood Marrow Transplant* (2010) 16(5):612–21. doi:10.1016/j.bbmt.2010.01.019
58. Okazaki T, Chikuma S, Iwai Y, Fagarasan S, Honjo T. A rheostat for immune responses: the unique properties of PD-1 and their advantages for clinical application. *Nat Immunol* (2013) 14(12):1212–8. doi:10.1038/ni.2762
59. Benson DM Jr, Bakan CE, Mishra A, Hofmeister CC, Efebera Y, Becknell B, et al. The PD-1/PD-L1 axis modulates the natural killer cell versus multiple myeloma effect: a therapeutic target for CT-011, a novel monoclonal anti-PD-1 antibody. *Blood* (2010) 116(13):2286–94. doi:10.1182/blood-2010-02-271874
60. Wiesmayr S, Webber SA, Macedo C, Popescu I, Smith L, Luce J, et al. Decreased NKp46 and NKG2D and elevated PD-1 are associated with altered NK-cell function in pediatric transplant patients with PTLD. *Eur J Immunol* (2012) 42(2):541–50. doi:10.1002/eji.201141832
61. MacFarlane AW IV, Jillab M, Plimack ER, Hudes GR, Uzzo RG, Litwin S, et al. PD-1 expression on peripheral blood cells increases with stage in renal cell carcinoma patients and is rapidly reduced after surgical tumor resection. *Cancer Immunol Res* (2014) 2(4):320–31. doi:10.1158/2326-6066.CIR-13-0133
62. Alvarez IB, Pasquinelli V, Jurado JO, Abbate E, Musella RM, de la Barrera SS, et al. Role played by the programmed death-1-programmed death ligand pathway during innate immunity against *Mycobacterium tuberculosis*. *J Infect Dis* (2010) 202(4):524–32. doi:10.1086/654932
63. Norris S, Coleman A, Kuri-Cervantes L, Bower M, Nelson M, Goodier MR. PD-1 expression on natural killer cells and CD8(+) T cells during chronic HIV-1 infection. *Viral Immunol* (2012) 25(4):329–32. doi:10.1089/vim.2011.0096
64. Pesce S, Greppi M, Tabellini G, Rampinelli F, Parolini S, Olive D, et al. Identification of a subset of human natural killer cells expressing high levels of programmed death 1: a phenotypic and functional characterization. *J Allergy Clin Immunol* (2016). doi:10.1016/j.jaci.2016.04.025
65. Watanabe T, Bertoletti A, Tanoto TA. PD-1/PD-L1 pathway and T-cell exhaustion in chronic hepatitis virus infection. *J Viral Hepat* (2010) 17(7):453–8. doi:10.1111/j.1365-2893.2010.01313.x
66. Day CL, Kaufmann DE, Kiepiela P, Brown JA, Moodley ES, Reddy S, et al. PD-1 expression on HIV-specific T cells is associated with T-cell exhaustion and disease progression. *Nature* (2006) 443(7109):350–4. doi:10.1038/nature05115
67. Barber DL, Wherry EJ, Masopust D, Zhu B, Allison JP, Sharpe AH, et al. Restoring function in exhausted CD8 T cells during chronic viral infection. *Nature* (2006) 439(7077):682–7. doi:10.1038/nature04444
68. Wenjin Z, Chuanhui P, Yunle W, Lateef SA, Shusen Z. Longitudinal fluctuations in PD1 and PD-L1 expression in association with changes in anti-viral immune response in chronic hepatitis B. *BMC Gastroenterol* (2012) 12:109. doi:10.1186/1471-230X-12-109
69. Brandt CS, Baratin M, Yi EC, Kennedy J, Gao Z, Fox B, et al. The B7 family member B7-H6 is a tumor cell ligand for the activating natural killer cell receptor NKp30 in humans. *J Exp Med* (2009) 206(7):1495–503. doi:10.1084/jem.20090681
70. Pesce S, Tabellini G, Cantoni C, Patrizi O, Coltrini D, Rampinelli F, et al. B7-H6-mediated downregulation of NKp30 in NK cells contributes to ovarian carcinoma immune escape. *Oncoimmunology* (2015) 4(4):e1001224. doi:10.1080/2162402X.2014.1001224

71. Tabellini G, Benassi M, Marcenaro E, Coltrini D, Patrizi O, Ricotta D, et al. Primitive neuroectodermal tumor in an ovarian cystic teratoma: natural killer and neuroblastoma cell analysis. *Case Rep Oncol* (2014) 7(1):70–8. doi:10.1159/000357802
72. Hamanishi J, Mandai M, Ikeda T, Minami M, Kawaguchi A, Murayama T, et al. Safety and antitumor activity of anti-PD-1 antibody, nivolumab, in patients with platinum-resistant ovarian cancer. *J Clin Oncol* (2015) 33(34):4015–22. doi:10.1200/JCO.2015.62.3397
73. Topalian SL, Drake CG, Pardoll DM. Immune checkpoint blockade: a common denominator approach to cancer therapy. *Cancer Cell* (2015) 27(4):450–61. doi:10.1016/j.ccell.2015.03.001

9

Recognition and Regulation of T Cells by NK Cells

*Katharina Pallmer and Annette Oxenius**

Institute of Microbiology, ETH Zürich, Zürich, Switzerland

Edited by:
Eric Vivier,
Centre d'Immunologie de Marseille-Luminy, France

Reviewed by:
Sophie Caillat-Zucman,
INSERM, France
Stephen Noel Waggoner,
Cincinnati Children's Hospital Medical Center, USA
Catharina C. Gross,
University Hospital Münster, Germany

***Correspondence:**
Annette Oxenius
oxenius@micro.biol.ethz.ch

Regulation of T cell responses by innate lymphoid cells (ILCs) is increasingly documented and studied. Direct or indirect crosstalk between ILCs and T cells early during and after T cell activation can affect their differentiation, polarization, and survival. Natural killer (NK) cells that belong to the ILC1 group were initially described for their function in recognizing and eliminating "altered self" and as source of early inflammatory cytokines, most notably type II interferon. Using signals conveyed by various germ-line encoded activating and inhibitory receptors, NK cells are geared to sense sudden cellular changes that can be caused by infection events, malignant transformation, or cellular stress responses. T cells, when activated by TCR engagement (signal 1), costimulation (signal 2), and cytokines (signal 3), commit to a number of cellular alterations, including entry into rapid cell cycling, metabolic changes, and acquisition of effector functions. These abrupt changes may alert NK cells, and T cells might thereby expose themselves as NK cell targets. Here, we review how activated T cells can be recognized and regulated by NK cells and what consequences such regulation bears for T cell immunity in the context of vaccination, infection, or autoimmunity. Conversely, we will discuss mechanisms by which activated T cells protect themselves against NK cell attack and outline the significance of this safeguard mechanism.

Keywords: T cell, natural killer cell, immune regulation, virus infection, autoimmune disease

NK CELLS: REGULATION, EFFECTOR FUNCTIONS, AND EDUCATION

Natural killer (NK) cells, presenting at a frequency of around 5% in the blood, belong to the family of group 1 innate lymphocytes (ILC1) and are functionally characterized by their cytotoxicity and their ability to produce cytokines, most prominently interferon γ (IFNγ). NK cells belong to the innate immune system, and they can react to rapid changes in host cells without prior sensitization. As part of the first line defense, they recognize and lyse virally infected cells and tumor cells. NK cells are activated *via* innate cytokines such as IL-2, IL-12, IL-15, IL-18, and type I IFNs as well as *via* the recognition of sudden cellular changes perceived *via* different inhibitory and activating receptors expressed on their surface (1–7). Additionally, direct triggering of toll-like receptors (TLRs) on NK cells can further stimulate their activation (8–10).

Regulation of NK Cell Activity

Compared with T and B cells whose antigen receptors are highly variable and specific for a specific antigen, NK cells express various germ-line encoded activating and inhibitory receptors. Depending on the net balance of signals perceived by activating and inhibitory receptors, NK cells are either activated and exert effector functions or are restrained (11, 12).

Healthy cells constitutively express ligands for inhibitory receptors on NK cells in order to protect themselves against NK-mediated killing. Classical MHC-I molecules are expressed on every nucleated cell in the body and bind to the inhibitory receptor killer immunoglobulin-like receptors (KIRs) in humans and Ly49A C, D in mice, respectively. The non-classical MHC-I molecule HLA-E in humans and Qa-1 in mice binds to the heterodimeric inhibitory receptor CD94/NKG2A and CD48 binds to the inhibitory receptor 2B4, leading to a repressed state of the NK cell (13, 14).

Infected or malignant cells can downregulate MHC-I, also known as "missing-self hypothesis," to become invisible for CD8 T cells; however, the loss of MHC-I ligands for inhibitory receptors on NK cells sensitizes these cells for NK-mediated killing. Conversely, overexpression of ligands engaging NK-activating receptors ("induced self-recognition") also renders these cells NK cell targets (14, 15). Activating ligands are not expressed at steady-state, but tumorigenesis, virus infection, or DNA damage can activate stress pathways, leading to upregulation of various activating ligands that bind to NK cell-activating receptors and thereby promote NK cell activation, resulting in cytotoxicity and cytokine secretion (16). NKG2D is a well-studied NK cell-activating receptor, it has multiple cellular ligands including MHC-I homologs such as MHC class I chain-related proteins A and B (MICA and MICB) and UL16-binding proteins (ULBPs) (17). As a result of the activation of heat-shock transcription elements in the promoters of the genes, MICA and MICB are upregulated on NK target cells. The sensing of type I IFN can also trigger MICA and MICB expression on dendritic cells (DCs) (18, 19). Moreover, HCMV-infected cells upregulate MICA and ULBP3 (20, 21).

The DNAX accessory molecule-1 (DNAM-1 or CD226) is an adhesion molecule, which is expressed on multiple cells including NK cells. DNAM-1 serves as an activating receptor on NK cells, the engagement by its ligands poliovirus receptor (PVR), and nectin-2 leads to increased cytotoxicity in NK cells (22, 23). The cellular ligands of DNAM-1 are induced upon cellular stress (24, 25). Interestingly, regulatory T cells (Tregs) can also use DNAM-1–DNAM-1L interaction to modulate T cell responses, indicating that some receptors shared by innate and adaptive immunity are involved in regulating T cell responses (26).

Another family of NK cell-activating receptor is the natural cytotoxicity receptor family, consisting of NKp30, NKp44, and NKp46 in humans. Of note, NCR1 is the NKp46 ortholog and the only member of the NCR family in rodents (27). NKp30 and NKp46 are expressed on resting NK cells in contrast to NKp44 which is found only on activated NK cells. Cellular ligands for NKp44 are partly known and include proliferating cell nuclear antigen (PCNA) and mixed lineage leukemia 5 (MML5). Ligands which bind to NKp30 comprise HLA-B-associated transcript 3 (BAT 3) and B7-H6 (member of the B7 family of immunoreceptors) (28). However, the cellular ligands for NKp46 remain elusive (29, 30). Activating NK cell receptors may also directly be triggered by microbial constituents, such as Ly49H in C57BL/6 mice, recognizing the murine *Cytomegalovirus* (MCMV) encoded protein m157 and NCR1 recognizing influenza virus hemagglutinin proteins (31, 32) (**Table 1**).

TABLE 1 | Viral-derived ligands for NK cell receptors.

Receptor	Ligand	Source	Effect of ligand/ receptor interaction	Reference
NKp46	Viral HA and NA	Influenza virus, Poxvirus, Sendai virus, Newcastle disease virus	Activating	(31, 145–147)
NKp44	Viral HA and NA	Influenza virus, Sendai virus, Newcastle disease virus	Activating	(146, 148)
NKp30	Viral HA	Poxvirus	Inhibitory	(147)
	pp65	HCMV	Inhibitory	(149)
TLR7/8	Single-stranded RNA	HIV	Activating	(150)
Ly49H	m157	MCMV	Activating	(32)
KIR3DL1	NS-1	DENV	Activating	(151)

HA, hemagglutinin; NA, neuraminidase; NS-1, non-structural protein 1; HCMV, human Cytomegalovirus; HIV, human immunodeficiency virus; MCMV, murine Cytomegalovirus; DENV, dengue virus.

The fact that ligands for activating NK cell receptors are regulated *via* stress pathways and that microorganisms have evolved mechanisms to downregulate ligands for specific activating receptors might explain why there are different ligands for one receptor (33).

Effector Functions

Key effector functions of NK cells comprise cytokine secretion and cytolytic granule-mediated cell apoptosis. The secretion of IFNγ and tumor necrosis factor (TNF) by NK cells promotes APC and phagocyte function, including enhanced phagocytosis, production of antimicrobial peptides, oxidative burst, and upregulation of MHC molecules. The granule exocytosis pathway is activated by the net balance of activation/inhibition signals conveyed by inhibitory and activating receptors and involves the secretion of granules which contain perforin and granzymes (16). Perforin/granzyme A and B trigger caspase-dependent and -independent death pathways. FasL and TRAIL are expressed on the surface of NK cells, which bind to the death receptors Fas (CD95) and TRAILR (DR4 and DR5), respectively. The engagement of FasL–Fas/TRAIL–TRAILR results in the induction of apoptosis of virally infected cells or tumor cells. Antibody-dependent cellular cytotoxicity (ADCC) can be exerted by NK cells when target cells are coated with IgG antibodies that bind the Fcγ receptor CD16 on NK cells, overriding inhibitory signals and triggering cytotoxicity and cytokine secretion (16, 34, 35).

Education

As NK cells recognize mostly cellular ligands *via* their activating receptors, NK cells are potentially able to induce tissue damage. Thus, a tight regulation is required to avoid self-induced damage by NK cells. To this end, NK cells undergo a process called education, or "licensing," describing a phase of NK maturation in which NK cells acquire effector functions and, at the same time, adapt their responsiveness to the steady-state expression levels of NK receptor ligands in host cells. For instance, mature NK cells recognize MHC-I molecules on the cell surface *via* their inhibitory receptors, resulting in a suppressed state of the NK cell toward healthy host cells (36). Surprisingly, the absence of MHC-I during NK cell education does not provoke overt NK cell activity, but NK cells appear to be hyporesponsive upon stimulation and fail to eliminate cells lacking MHC-I. This demonstrates that the environment, in which the NK education occurs, effectively determines the responsiveness of NK cells [(37, 38) and reviewed in Ref. (39)].

Interestingly, NK cells are not only killer cells as part of the first line defense but they also have the capacity to shape adaptive immunity by regulating T cell responses (40, 41). Of note, NK cells are not the only ILCs which are capable of modulating T cell responses. There is emerging evidence that other ILC subsets can contribute to shape T cell immunity, either by enhancing or suppressing the size of T cell immunity or by regulating the differentiation of T cell responses which was reviewed in Ref. (42, 43). In this review, we illustrate how specifically NK cells can regulate T cell immunity from a "T cell centric" point of view, and we will provide further insights into the relevance of NK–T cell interaction in various disease settings. In particular, we elucidate the role of NK–T cell interaction in acute and chronic viral infections. The profound understanding on how NK–T cell interaction occurs in different stages of T cell activation in different disease settings, and how it affects the size and quality of T cell responses might open new perspectives on the development of specific and powerful therapies.

STAGES IN THE LIFE OF A T CELL

T cells need to perceive three signals for proper activation: antigen, costimulation, and cytokines. Antigen is processed and presented in the context of MHC class I and MHC class II molecules by APCs and determines the specificity of the response. Activated APCs provide further costimulation and cytokines, both needed for T cell activation. Among APCs, DCs that have been activated by engagement of pattern recognition receptors (PRRs) potently stimulate naive T cells (44). The signal of costimulation is provided by a number of molecules including CD28–CD80/86, CD27–CD70, OX40L–OX40, 4-1BB–4-1BBL, and RANK–RANKL (45–49). The specific cytokines required for T cell activation (i.e., proliferation, differentiation, and survival) result from the inflammatory milieu triggered by an infection or vaccination. In case of viral infections, most prominent pro-inflammatory cytokines are type I IFNs and IL-12, which have redundant functions as signal 3 cytokines (50). Integration of all three signals leads to fast T cell proliferation, known as clonal burst, reaching peak expansion before contraction and formation of a pool of memory cells.

At the peak of T cell expansion, T cells predominantly exhibit an effector phenotype (51, 52). Effector CD8 T cells can be split into two main subsets, a smaller subset of memory precursor effector cells (MPECs) that have the potential to become long-lived memory cells and short-lived effector cells (SLECs) that lack this ability (53).

Impact of NK Cells on T Cells during Priming Phase

At steady-state conditions, NK cells are mainly excluded from the LN (54, 55). However, early during infections activated NK cells can enter into LN and localize in close proximity to T cells in the LN, enabling them to influence T cells during early stages of activation and thereby shaping the ensuing size and quality of the T cell responses. NK cells circulate in blood and migrate in a CXCR3-dependent, but CCR7-independent manner into activated LNs (56).

Since APCs are essential for cell activation, any changes in APC activation, maturation, and function are tightly associated with the emerging T cell response. Although NK cells were also shown to directly interact with T cells, this direct interaction seems to become more important after initial activation of the T cells. During the initial priming phase, T cell regulation by NK cells is mainly occurring in an indirect manner *via* modulation of APCs.

In the following section, we will elaborate in detail the different mechanisms of how DC–NK crosstalk influences the emerging T cell response.

NK Cells Promoting T Cell Immunity

Natural killer cells can be beneficial for mounting a T cell response by modulating DC function. NK cells act by different mechanisms depending on the DC subsets and the prevailing cytokine environment. In contrast to immature DCs (iDCs), which are found at steady state in peripheral tissues and in the blood, mature DCs (mDCs) migrate to the LN in a CCR7-dependent manner and have the ability to potently stimulate naive T cells. The conversion from iDCs to mDC occurs *via* sensing PAMPs *via* PRRs (57). mDCs are characterized by high level expression of costimulatory molecules, e.g., CD80, CD83, CD86, and MHC class II/human leukocyte antigen-DR (HLA-DR), as well by the secretion of pro-inflammatory cytokines such as IL-12 (58). To reveal interactions between DCs and NK cells at the early T cell priming phase, most studies with human cells used monocyte-derived DCs. iDCs are obtained by culturing monocytes in the presence of granulocyte macrophage colony-stimulating factor (GM-CSF) and IL-4. DC maturation is induced by the addition of LPS, IL-1, type I IFN, or TNF (59).

Human *in vitro* experiments revealed that DCs promote NK cell activity *via* the secretion of cytokines such as type I IFN, IL-12, or TNF. In return, NK/DC interaction leads to NKp30 engagement, resulting in TNF and IFNγ production by the NK cells, which further promotes DC maturation (58, 60, 61). Besides cytokines (IL-12, TNF, and IFNγ), mDCs can promote IFNγ production, CD69 expression, and cytotoxic functions of NK cells in a cell-to-cell contact-dependent manner. Ligands for the activating receptor NKG2D on NK cells, MICA and B, are expressed on human monocyte-derived DCs and contribute to CD69 expression on NK cells (58, 62, 63).

This bidirectional crosstalk can secondarily influence T cell differentiation. Specifically, the differentiation of naive CD4 T cells into IFNγ secreting T_H1 T cells is promoted by NK cells, shown by transwell experiments *in vitro* using human myeloid blood DCs (64). Similarly, NK cells can also directly shape, possibly without direct DC crosstalk, the quality of T cell responses *via* their cytokine secretion. A murine *in vivo* study demonstrated that after activated NK cells migrated to the LN in a CXCR3-dependent manner where they served as an early source of IFNγ that was essential for Th1 polarization of naive CD4 T cells (56). In a mouse model of *Leishmania major* infection, the blockade of TGFβ signaling in NK cells led to increased IFNγ secretion by NK cells, promoting the differentiation of naive CD4 T cells into Th1 cells, leading to improved pathogen control (65).

Another mechanism how NK cells indirectly influence the emerging T cell response during the priming phase of T cells *via* enhancing cross-presentation. NK cells promote DC cross-presentation by killing target cells, leading to the release of antigens that can be taken up by DCs and presented *via* MHC-I to T cells, resulting in an increased CD8 T cell response. Specifically, transfer of allogenic B cells in mice resulted in NK-mediated killing that promoted endocytosis of apoptotic bodies by $CD8^+$ DCs and MHC I presentation of respective antigens (66). Similarly, *in vivo* killing of OVA-expressing splenocytes by NK cells resulted in better priming of CD8 as well as CD4 T cells due to NK-mediated release of antigen. Here, NK-mediated cytotoxicity promoted CD4 T cell responses, which was crucial to support CD8 T cell responses and strong IgG responses (67). Also in human cells, NK cell-secreted IFNγ and TNF induce cross-presentation of tumor cell-derived antigens by monocyte-derived DCs, promoting the induction of a tumor-specific CD8 T cell response (68).

Several studies have reported that activated NK cells specifically kill iDCs, while sparing mDCs. Albeit the direct killing of iDC might have a negative impact on the emerging T cell response, it is more likely that the killing of iDCs might be beneficial for robust T cell immunity. This DC selection, known as DC editing, might serve as a quality control to guarantee the survival of DCs expressing sufficient costimulation molecules needed for successful T cell priming or to ensure that effective T cell priming is stopped after resolution of infection-induced inflammation (61, 69).

Human NK cells, activated with IL-2 *in vitro*, killed, and produced IFNγ when encountering iDCs. In contrast, lysis of mDCs occurred only when inhibitory ligands (HLA molecules) were blocked, showing that mDCs protect themselves against NK cell attack by KIR engagement. NKp30 engagement was mainly responsible for detection and elimination of iDCs. However, NKp30 was also shown to induce maturation and favor DC activation, and it is currently unclear how this differential role of NKp30 in NK cells is regulated (28, 69). The *in vitro* finding that human iDCs are preferably lysed by NK cells was confirmed in a tumor mouse model *in vivo*. Impaired tumor-specific T cell immunity was related to the missing deletion of iDCs in the absence of NK cells (70). NK cells can kill iDCs due to lacking expression of the inhibitory ligand Qa-1 in contrast to mDCs that upregulate its expression. Interestingly, iDCs were also killed under inflammatory conditions (71).

By contrary, it was also reported that iDCs can promote T cell immunity: in a DC-based vaccine model in mice prevention of TRAIL-mediated iDC killing by NK cells enhanced antigen-specific T cell responses (72). However, these studies relied on adoptive transfer of *in vitro* cultivated DCs (LPS treated, mature, monocyte-derived DCs and untreated, immature, DCs). To what extent these *in vitro* cultivated iDCs resemble *in vivo* iDCs remains to be demonstrated (69, 73).

In summary, NK cells improve maturation, effector functions, and cross-presentation of DCs and are thereby able to promote T cell responses. Moreover, NK cells can secrete cytokines during T cell priming, which promote the differentiation of naive CD4 T cells to Th1 cells. The elimination of iDCs might be a possible control mechanism to select immunogenic DC, which provide sufficient costimulation needed for the effective stimulation of naive T cells (**Figure 1**).

NK Cells Impairing T Cell Immunity

Negative impact of NK cell on T cells during priming can be either mediated by directly affecting T cells or indirectly *via* DC modulation.

Even though killing of iDCs might have a positive effect on T cell priming, direct killing of mature APCs, especially DCs, results in diminished antigen availability and thereby represents a mechanism how NK cells can negatively regulate T cell responses.

An *in vivo* mouse study revealed that in the absence of NK cells APCs show improved capacity to stimulate CD8, but not CD4 T cells early during LCMV infection and this enhanced virus-specific CD8 T cell response resulted in viral clearance of a usually persistent infection. The beneficial effects could only be observed when NK cells were depleted within first 2 days of infection and were not due to enhanced costimulation provided by APCs, but most likely due to increased numbers of APCs in NK cell-depleted mice (74). In the context of MCMV infection in C57BL/6 mice, the MCMV protein m157, that is directly recognized by L49H on NK cells, triggered NK-mediated killing of MCMV-infected DCs. The resulting lower numbers of antigen-presenting DCs in this setting translated into impaired T cell immunity and hence protracted lytic replication (75). Furthermore, in MCMV infection, proinflammatory cytokines, such as IL-12, IFNγ, and TNF, and NK cell-activating receptors, such as NKG2D and NCR1, were shown to promote a bidirectional NK/DC crosstalk that supported CD4 T cell priming. Interestingly, IL-10 negatively regulated this crosstalk, such that only in absence of IL-10, NK cells supported DCs to more effectively prime MCMV-specific CD4 T cells, resulting in enhanced virus control (76).

In vitro- and *in vivo*-activated murine NK cells can acquire MHC-II proteins from interacting DCs *via* membrane transfer in a process known as trogocytosis. However, MHC-II- and costimulatory molecule-expressing NK cells did not promote but rather inhibited proliferation of CD4 T cells in presence of DCs, suggesting that they competed with antigen-presenting DCs for engagement of CD4 T cells (77).

In contrast, human MHC-II-expressing NK cells that also express costimulatory molecules typical for DCs, such as CD80/CD86, are capable to induce CD4 T cell proliferation (78–81). Thus, the crosstalk between DCs and NK cells can either support or impair initial activation of T cells, raising the question which factors decide about the outcome of the interaction. First of all, depending on the nature of the cytokines NK cells are exposed

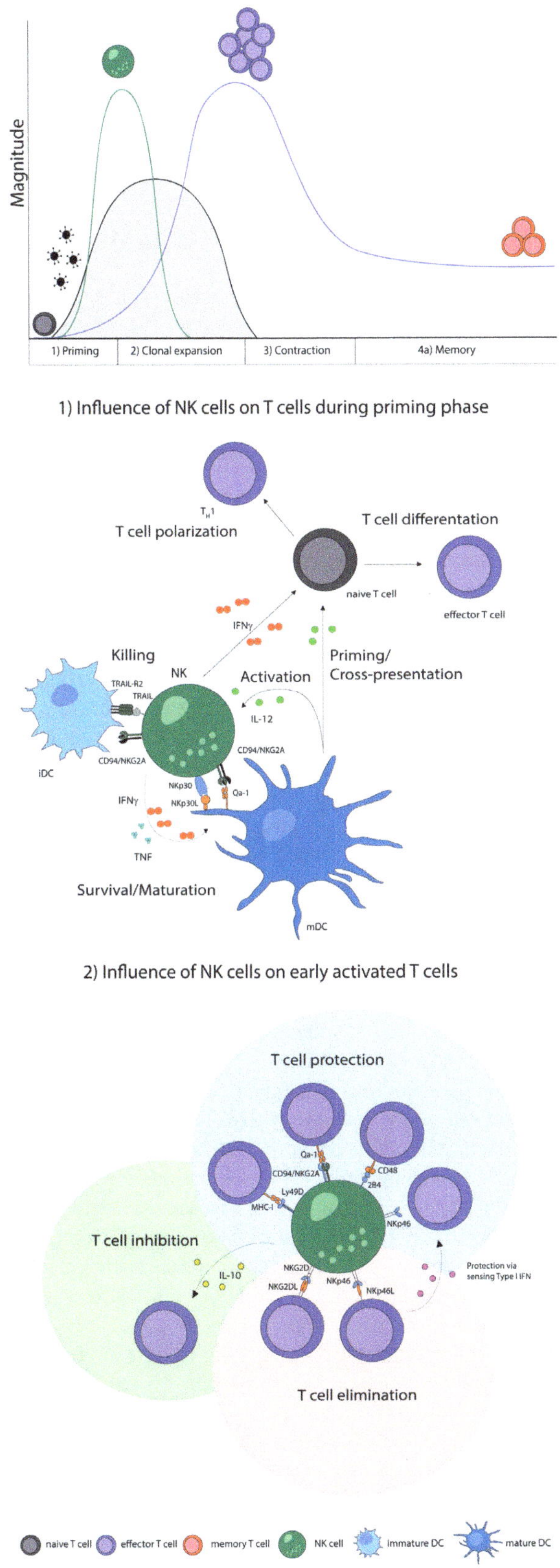

FIGURE 1 | NK–T cell interaction during early phases of viral infections.

(Continued)

FIGURE 1 | Continued

Naive T cells (gray) are primed by DCs, expand, and differentiate into effector T cells (purple), contract and form memory T cells (red). Activation of NK cells and T cells show a temporal and microanatomical overlap early during infection allowing T cell regulation by NK cells. (1) Early during infection NK cells can directly regulate T cells *via* the secretion of cytokines. Indirectly, NK cells modulate DC numbers and function, which affect T cell responses. The bidirectional NK–DC crosstalk is achieved by cytokine secretion or *via* direct cell-to-cell contact, resulting in either positive or negative consequences for T cells. (2) Early activated T cells can be recognized and inhibited (green shading) or eliminated (red shading) by NK cells *via* cytokines or directly in a cell-to-cell-dependent manner. T cells shield themselves by upregulating ligands for inhibitory NK cell receptors (blue shading). Abbreviations: IFNγ, interferon γ; TNF, tumor necrosis factor; IL-12, interleukin 12; IL-10, interleukin 10; type I IFN, type I interferon; NKp30L and NKp46L, ligands for the NK cell-activating natural cytotoxicity receptors; NKp46 and NKp30, NK cell-activating natural cytotoxicity receptors; NKG2DL, ligands for activating NK cell receptor; NKG2D, activating receptor of NK cells; TRAIL-R2, tumor necrosis factor receptor; TRAIL, ligand for tumor necrosis factor receptor; CD94/NKG2A, Ly49D, and 2B4, inhibitory NK cell receptors; Qa-1, MHC-I, and CD48, inhibitory receptor ligands; T_H1, T helper cell type 1; iDC, immature dendritic cell; mDC, mature dendritic cell.

to, they can either promote or inhibit the polarization of T cells. While IL-12 and IL-2 can promote the maturation of DCs that are capable of priming IFNγ secreting T_H1 T cells, regulatory cytokines, such as IL-10, lead to a reduction of the NK–DC cross-talk and to a decreased priming of virus-specific T cells (76, 82).

Moreover, another decisive parameter is the NK:DC ratio. While in a situation of a low NK:DC ratio, NK cells preferentially promote DC activation and enhance secretion of cytokines by the DCs, a high NK:DC ratio favors NK-mediated killing of autologous DCs (61). Besides this ratio also the density of receptors on the NK cell surface may decide about their impact in shaping T cell responses. For example, NK cells that express a high density of NCR1 showed an increased cytotoxicity against NK sensitive tumor cell lines compared with NK cells expressing lower levels of NCR1. However, if the density of activating receptors on NK cells is also decisive for their ability to directly kill T cells or APCs remains to be investigated (83).

Taken together, NK cells have the potential to regulate T cell responses during their priming phase directly *via* proinflammatory cytokines, such as IFNγ, as well as indirectly by eliminating or promoting APCs numbers and function by various mechanisms (**Figure 1**). Future studies are essential to elucidate the factors which decide whether NK cells promote DC maturation, leading to an enhanced T cell response due to efficient priming of naive T cells or whether NK cells eliminate DCs which leads to reduced T cell immunity.

Early Activation and Clonal Expansion

The following paragraph focuses on effector T cells; how NK cells influence Tregs and *vice versa* is summarized in **Box 1**. After encountering its cognate peptide presented by an APC in presence of costimulation and cytokines, a T cell rapidly undergoes clonal expansion. NK cells have the ability to recognize and directly kill early activated T cells and can thereby determine the quality and magnitude of T cell responses, which can influence the course of infection (84–90).

BOX 1 | Treg–NK cell interaction.

Regulatory T cells (Tregs) are characterized by the expression of the master transcription factor Foxp3 and are able to control effector T cells responses by inhibiting innate and adaptive immunity (91). Lately, several studies reported that Tregs influence NK cells. In *Scurfy* mice, that fail to develop Tregs, NK cells are hyperproliferative and show enhanced effector functions, hinting toward an interaction between Tregs and NK cells – or alternatively, that NK cells are activated as a consequence of the increased proinflammatory milieu presenting in absence of Tregs (92). Also, under homeostatic conditions, Treg depletion results in hyperreactivity of NK cells in terms of CD69 expression and cytolytic functions and IFNγ secretion (93). In line with increased activation in the absence of Tregs, NK cells show an increased missing self-responsiveness toward target cells and rapidly respond to weak stimulation. This hyperactivation depends on high levels of available IL-2 produced by CD4 T cells after Treg depletion. Thereby, low levels of IL-2, as in the presence of Tregs, can serve as a limiting factor restricting NK cell activity (94). This was confirmed in a genetic model of type I diabetes in mice in which Treg depletion led to infiltration and proliferation of NK cells in the pancreas, accompanied by the secretion of IFNγ that promoted disease progression. In this setting, the lack of IL-2 consumption by Tregs led to the elevated levels of IL-2, promoting NK cell activity, and was not due to higher numbers of CD4 T cells secreting IL-2 (95). Interestingly, upon MCMV infection, the restricting effect of Tregs on NK cells was abolished and virus titers between Treg-sufficient and Treg-depleted mice were comparable. Thus, the Treg-mediated suppression of NK cells might be overridden by availability of high IL-2 levels during viral infections (93).

However, in hepatitis C virus (HCV) infection, NK cells were suppressed by Tregs, which was shown in a human *in vitro* study. In HCV, NK cells are known to kill activated hepatic stellate cells (HSCs), which promote fibrosis. NK cell-mediated killing could be inhibited by Tregs in a cell–cell contact-dependent manner involving the cytotoxic T lymphocyte antigen 4 (CTLA-4). Interestingly, the preincubation of HSCs with Tregs resulted in reduced expression of activating ligands, such as MICA/B on HSCs, leading to suppression of NK cell effector functions (96). Not only CD4 Tregs interact with NK cells but also regulatory DCs (regDCs), which is a special DC subset inducing tolerance under certain physiological settings. The interaction between regDCs and NK cells induced an alternative activation state in NK cells, characterized by low levels of IFNγ secretion. This suppression was attributed to the secretion of IL-10 by DCs and, surprisingly, by the engagement of the activating receptor NKp46 on NK cells (97). This opposing effect of NKp46 might be due to the fact that naive NK cells were used in this DC coculture, while other studies used activated NK cells or cocultured NK cells in presence of virus-infected DCs (98, 99). Collectively, these studies show that NK cells can be controlled reciprocally by regulatory immune cells, in particular by Tregs. This notion might be relevant in the setting of diseases in which Tregs are increased, such as in chronic viral infections or cancer (100–103).

In mice, IL-2-activated NK cells specifically detect and kill activated CD4 and CD8 T cells in a perforin-dependent manner. Activated NK cells discriminate between activated and naive T cells *via* the activating receptor NKG2D (84). Similarly, human-activated T cells are susceptible to NK-mediated killing by upregulating ligands for NKG2D. Furthermore, human T cells also upregulate ligands for the activating receptor DNAM-1, such as PVR, upon stimulation with superantigen. Only T cells in the S and G2/M phases expressed PVR which is in line with the fact that NK cells preferentially kill proliferating T cells (104). These human *in vitro* studies showed that autologous T cells can be killed by NK cells *via* DNAM-1 and NKG2D in a granule exocytosis-mediated manner, which might serve as a mechanism to control T cell responses (85, 104).

In vivo, the absence of NK cells prevented establishment of a chronic LCMV infection in mice (86, 87). Mechanistically, some studies suggested that CD8 T cells can be directly killed by NK cells in a NKG2D- and perforin-dependent manner, while others reported that NK cells killed quite selectively activated CD4 T cells, which subsequently led to reduced CD8 T cell numbers and function. The NK cell-mediated killing of CD4 T cells required perforin; however, the exact mechanism how NK cells recognize activated CD4 T cells was not revealed (86, 87, 105). The differences in mode of action of NK cells during LCMV infection reported in these studies might be based on the use of different viral strains and infection doses.

Natural killer cells can regulate T cells also in a contact-independent manner *via* the secretion of IL-10, leading to the suppression of allergen and Ag-induced T cell proliferation (106, 107). In LCMV infection, IL-10 secreted by NK cells can restrict the magnitude of CD8 T cell responses during persistent viral infection and can thereby limit immunopathology (108).

As activated T cells are prone to be recognized and eliminated by NK cells, T cells have evolved mechanisms to protect themselves against NK-mediated cytolytic attack. One mechanism how T cells can protect themselves is *via* sensing type I IFN because virus-specific CD4 and CD8 T cells deficient for type-I interferon receptor *(Ifnar)* were effectively eliminated by NK cells in a perforin-dependent manner upon acute LCMV infection, demonstrating that sensing of type I IFN-induced specific downstream pathways in virus-specific T cells leading to protection from NK cell recognition and hence promoted survival and expansion (89, 90). Indeed, *Ifnar*-deficient activated T cells express ligands for the activating receptor NCR1 on their surface and are recognized *via* NCR1 on NK cells and are killed by the release of perforin. Virus-specific wild-type T cells, which were able to sense type I IFN, safeguarded themselves and resisted the NK cell attack due to absent expression of NCR1 ligands (90).

Interestingly, not only activated T cells need to evade NK cell-mediated killing but also NK cells themselves. Recently, it was reported that *Ifnar*-deficient NK cells were also killed by activated NK cells in a perforin-dependent manner. NK cell killing of *Ifnar*-deficient NK cells was not NCR1, but NKG2D dependent (109). The independent findings that type I IFN sensing is associated with absent expression of NCR1 or NKG2D ligands on the surface of different immune cells suggests that type I IFN downstream signaling pathways, including STAT 1 signaling, seem to provide conserved protective mechanisms of how activated immune cells avoid NK-mediated killing. In future studies, it will be important to gain a more detailed knowledge about the regulation of expression of ligands for activating or inhibitory NK cell receptors on activated immune cells.

Besides the upregulation of activating ligands on T cells, the absence of inhibitory ligands is also a mechanism how T cells can be rendered targets for NK cell-mediated lysis early during activation. On this line, *Ifnar*-sufficient virus-specific T cells showed higher expression levels of classical and non-classical MHC-I molecules, ligands for inhibitory NK cell receptors, compared with *Ifnar*-deficient cells (89).

In addition, NLCR5, a transcription modulator promoting MHC-I expression, was shown to protect T cells against NK cell attack, as *NLRC5*-deficient T cells became targets for NK cells under inflammatory conditions due to strongly reduced MHC-I expression (110). Furthermore, the absence of the inhibitory receptor 2B4 on NK cells was associated with a prolonged persistence of LCMV because virus-specific CD8 T cells were killed by NK cells. Even though CD8 T cells expressed high levels of MHC-I, they could not shield themselves from NK cell-mediated attack, showing that the protective features of MHC-I and CD48 are non-redundant (111, 112). Of note, 2B4 can act as a coactivating receptor on human NK cells, in particular, in combination with CD16 (113) or in synergy with NKG2D, NKp46, and DNAM-1, resulting in enhanced cytotoxicity of naive human NK cells (114, 115). However, if 2B4 as a coactivating receptor on NK cells impacts on T cell regulation remains to be demonstrated.

Natural killer cells were also shown to diminish the number of CD4 T_{FH} cells and thereby to affect B cell responses. During an acute LCMV infection, NK cells killed CD4 T_{FH} cells in a perforin-dependent manner, resulting in a weak germinal center (GC) response and reduced titers of virus-specific antibodies (116). NK depletion early during a chronic LCMV infection enhanced the levels of virus-specific antibodies, thereby contributing to control of the infection. Even though the underlying mechanisms were not revealed so far, elevated numbers of CD4 T_{FH} cells suggested that NK cells promote viral persistence by suppressing not only CD8 T immunity but also CD4 T_{FH} cells and consequently humoral immunity (117).

Taken together, NK cells have the potential to shape T cell responses early during activation, mostly by curtailing T cell responses and humoral immunity (**Figure 1**). Why would such regulation of T and B cell responses by NK cells exist? One situation in which this regulation has physiological relevance is chronic viral infections when NK cell-dependent killing of T cells and suppression of T cell functions may prevent T cell driven immunopathology. Furthermore, infection with intermediate LCMV doses normally leads to detrimental immunopathology due to a strong T cell response, which is not sufficiently inhibited by NK cells. Depletion of NK cells, in this model, leads to an early enhanced T cell response that mediates control of the infection (86).

Finally, NK cells may ensure that only "correctly" activated CD8 T cells, which have received all three activation signals, survive. "Incorrectly" activated T cells (i.e., those that might have only perceived one or two signals) might become targets for NK cells. This hypothesis originates from the observation that LCMV-specific T cells that have not sensed signal 3 (i.e., type I IFN) are exquisite targets for NK cells. Such incorrectly activated T cell might potentially be harmful for the host, for instance, in case of autoreactive T cells. This is, however, at the moment, pure speculation that needs to be addressed experimentally.

Contraction and Memory Formation

The peak in the life of a T cell is around day 8 after antigen activation before the population of expanded T cells contracts massively due to apoptosis of antigen-specific effector T cells. The decline of T cell numbers is necessary to allow the organism to respond to new pathogens and to avoid immunopathology (118). NK cells do not seem to be involved in the contraction phase because virus-specific T cells decline in NK-depleted mice to the same extent as in NK-sufficient mice after acute LCMV infection (90, 105). After the contraction phase, few antigen-specific T cells differentiate into long-lived memory cells, which are dependent on IL-7 and IL-15 for homeostasis (119, 120).

Even though the decision about T cell differentiation occurs early during T cell activation and not during the contraction phase, the effects of NK cells on long-term memory formation will be discussed briefly. At the peak of T cell expansion, the population of antigen-specific T cell can be divided into SLECs and MPECs, with the latter fueling the pool of long-lived memory cells. In the absence or presence of NK cells, virus-specific T cells differentiated into MPECs and SLECs, indicating that NK cells do not affect the establishment of both effector subpopulations. Hence, establishment of T cell memory occurs both in presence and absence of NK cells during the priming phase in the context of an acute LCMV infection - albeit total numbers of memory cells might be slightly reduced in presence of NK cells due to slightly reduced peak expansion (88, 90, 116). In mice, the lack of Qa-1 on T cells, which binds to the inhibitory receptor CD94/NKG2A on NK cells, led to reduced survival of CD4 T cells. Furthermore, memory formation was drastically reduced, suggesting that Qa-1 plays a role in CD4 T cell survival and memory development (121). In an OVA vaccination model in mice, memory CD8 T cell responses were increased in the absence of NK cells, which was associated with a better control of tumor growth (105).

Natural killer–DC interactions not only occur during primary immune responses but also during recall responses, which was shown in a secondary infection model using different OVA-expressing pathogens. Upon secondary infection with *Listeria monocytogenes*, memory CD8 T cells were activated by IL-12 and CXCL9 secreting $XCR1^+$ DCs, leading to improved reactivation of memory CD8 T cells. Upon recall, NK cells were rapidly activated and provided an early source of IFNγ, boosting $XCR1^+$ DC functions, which resulted in a stronger recall CD8 T cell response (122).

Collectively, there are some indications that NK cells have the potential to shape the differentiation of memory T cells - although such impact may actually take place during the priming period - and decide thereby about the quantity of memory T cells.

Disease

In the following part, we provide an overview about NK–T cell interactions in specific disease settings. In particular, we emphasize on how NK cells affect T cells during chronic viral infections and how chronic viral infections can modulate NK–T cell interaction. We also extend the discussion on the role of NK–T cell interaction in autoimmune disorders.

Role of NK Cells during Chronic Virus Infections

During chronic active viral infections, "classical" memory T cell formation is impaired. Instead, virus-specific CD8 T cells are maintained as a population of actively cycling cells that depends on recurrent activation by cognate antigen (123). Functionally, the virus-specific CD8 T cells are markedly impaired in their effector functions, predominantly in their ability to produce inflammatory cytokines such as IFNγ and TNF. This hypofunctional state is termed "exhaustion," and its genesis is related to recurrent antigen exposure and the expression of a number of coinhibitory receptors that downregulate signal transduction from the TCR (123). Even though the depletion of NK cells early during a chronic LCMV infection leads to morbidity and mortality, due to enhanced antiviral T cell immunity, NK depletion 2 or 3 weeks after the peak of cytotoxic CD8 T cells is associated with improved antigen-specific CD8 T cell response and better virus control and led to only weak immunopathological symptoms (86, 124).

In human immunodeficiency virus-1 (HIV-1) infection, NK cells are activated and upregulate the activating receptor NKp44, while CD4 T cells express a cellular ligand for NKp44 (NKp44L). The expression of NKp44L is induced by the HIV-1 envelope gp41 protein and renders CD4 T cells highly susceptible to NK-mediated lysis. Interestingly, NKp44L expression was only observed in uninfected CD4 T cells. In infected CD4 T cells, the HIV-derived protein Nef inhibited the upregulation of NKp44L on the surface. Thus, the downregulation of NKp44-activating ligands on infected CD4 T cells might serve as an immune escape mechanism, while NKp44 ligand expression on non-infected CD4 T cells may contribute to the overall depletion of CD4 T cells over the course of infection (125). In addition, Nef and the late viral factor Vpu induced the downregulation of PVR, which is a ligand for the activating receptor DNAM-1 on NK cells. The blockade of DNAM-1 and NKG2D together reduced killing of HIV-1-infected cells by NK cells, indicating that both DNAM-1 and NKG2D are involved in the regulation of NK cell recognition of HIV-1-infected cells (126). Another study in HIV-1-infected patients revealed that plasmacytoid DCs secreted type I IFN which induced NK cells to kill HIV-infected CD4 T cells with the engagement of NKp46 and NKG2D. In this setting, NK-mediated lysis represents an antiviral immune response contributing to control the infection (127). Chronic hepatitis B virus (CHB) infection is also characterized by activated NK cells and a decrease of T cell functions or clonal deletion of virus-specific T cells. Studies using human samples derived from CHB patients revealed that virus-specific CD8 T cells in the liver are NK cell targets as they expressed higher levels of TRAIL-R2 which engages TRAIL on NK cells, leading to NK-mediated killing. Surprisingly, only HBV-specific T cells were affected by NK regulation *via* the TRAIL-R2/TRAIL pathway, but not other virus-specific T cells such as Epstein–Barr virus (EBV) or *Cytomegalovirus* (CMV)-specific T cells (128). NK cells and T cells derived from CHB patients are both functionally impaired. Nucleos(t)ide analog (NUC) therapy can improve T- and NK-cell responses and resulted in a reduction of an inflammatory phenotype of NK cells derived from CHB patients, characterized by reduced expression of TRAIL, CD38, and Ki-67. The change to a quiescent phenotype in NK cells was accompanied by the restoration of effector functions of HBV-specific T cells under NUC therapy. NK depletion and blockade of TRAIL and NKG2D further ameliorated the HBV-specific T cell functions (129). These studies clearly demonstrate that NK cells play a role in regulating HBV-specific T cells during chronic HBV infection. The regulation of HBV-specific T cells by NK cells might be under certain conditions beneficial for the host due to diminished immunopathological consequences. The complex interplay of NK cells and T cells in CHB was reviewed in Ref. (130).

EBV, which belongs to the *herpesviridae* family, causes a persistent latent infection and has a prevalence of over 90% in the population. While, in most individuals, the infection occurs without symptoms, in around 10% the infection results in infectious mononucleosis (IM), which is related to massive expansion of EBV-specific CD8 T cells (131). A study using humanized mice for EBV infection presented the hypothesis that NK cells could limit T cell expansion by directly killing T cells and thereby preventing the development of IM. It is therefore speculated that EBV seronegative individuals that have low numbers of NK cells are more prone to develop IM (132).

Natural killer cells are likewise involved in regulating T cell responses specific for MCMV, also a member of the *herpesviridae* family. During MCMV infection, NK cells kill activated CD4 T cells in the salivary gland. The elimination of activated CD4 T cells was dependent on TRAIL and led to a prolongation of the lytic virus replication; however, it prevented virus-induced autoimmunity at the same time (133). This shows that NK cells have, indeed, a regulatory role which has the ability to restrict strong adaptive immune responses in the setting of persistent viral infections (**Figure 2**).

Role of NK Cells in Autoimmune Disease

Natural killer cells are also involved in regulating autoreactive T cells. This was demonstrated in the mouse model for multiple sclerosis, experimental autoimmune encephalomyelitis (EAE). Depletion of NK cells was associated with a more severe form of EAE, characterized by the occurrence of relapses. The exacerbation of the disease was due to increased T cell proliferation and cytokine production (134). Besides the TRAIL–TRAILR pathway inducing apoptosis in autoreactive T cells, blockade of the Qa-1/CD94-NKG2A axis led to an amelioration of EAE due to the elimination of autoreactive CD4 T cells by NK cells in a perforin-dependent manner (121, 133).

The human CD56^{bright} NK cell subset is most prominent in peripheral lymphoid organs compared with the CD56^{dim} subset which mostly circulates. CD56^{bright} cells derived from healthy individuals could suppress autologous CD4 T cell proliferation under inflammatory conditions *via* the activation of NCRs and the secretion of granzyme. In comparison, CD56^{bright} derived from MS patients exhibited reduced granzyme secretion due to higher expression levels of the inhibitory ligand HLA-E on CD4 T cells and not due to altered NCR expression on NK cells (135). In addition, enrichment of CNS-resident NK cells in EAE had beneficial effects on the disease progression by suppressing T_H17 cells (136). Paradoxically, the absence of IFNγ in a transgenic mouse model led to more severe symptoms in EAE, uveitis, and arthritis. In the presence of IFNγ, NK cells migrated to the LN where they interacted with DCs promoting IL-27 production, leading to the suppression of disease-inducing T_H17 T cells and therefore to protection from autoimmunity (137). Studies in MS patients revealed that the reduced cytolytic functions of NK cells are due to an impaired interaction between NK cells and CD4 T cells *via* DNAM-1-PVR and 2B4-CD48.

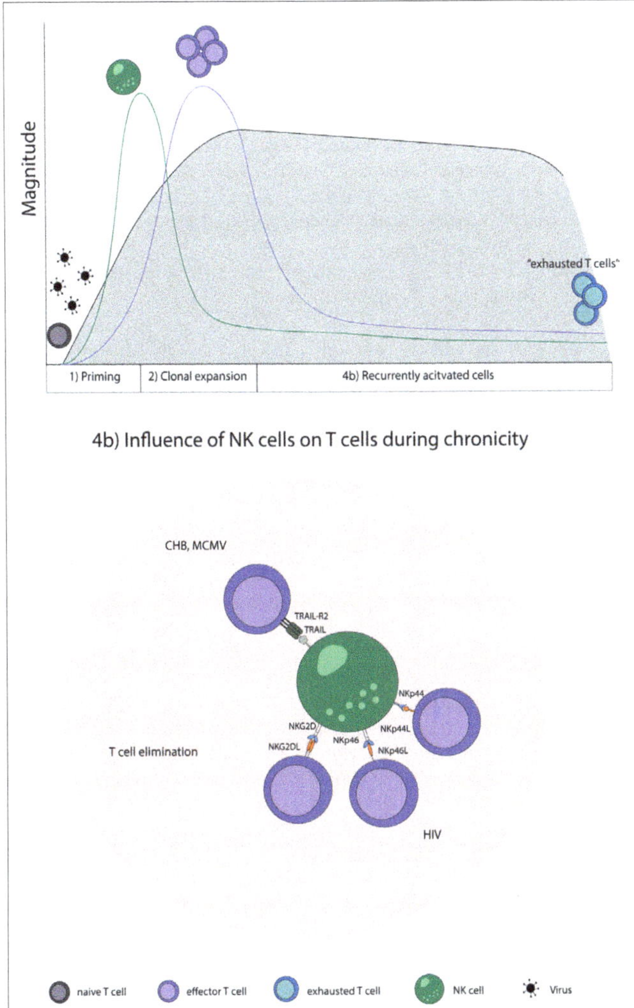

FIGURE 2 | NK–T cell interaction in chronic virus infections. Naive T cells (gray) are primed by DCs, expand, and initially form effector T cells (purple) that transition during a chronic infection T cells to "exhausted T cells" (blue) due to constant restimulation by cognate antigen. (4b) NK cells recognize and eliminate T cells during a chronic state of viral infections by a direct cell-to-cell contact. Abbreviations: NKp44L and NKp46L, ligands for the NK cell-activating natural cytotoxicity receptors; NKp44 and NKp46, NK cell-activating natural cytotoxicity receptors; NKG2DL, ligands for activating NK cell receptor; NKG2D, activating receptor of NK cells; TRAIL-R2, tumor necrosis factor receptor; TRAIL, ligand for tumor necrosis factor receptor superfamily; CHB, chronic hepatitis B virus; MCMV, murine *Cytomegalovirus*; HIV, human immunodeficiency virus.

Thus, the dysregulation of NK–T cell interaction in MS may present an attractive therapeutic target. Indeed, therapeutic immune modulation targeting IL-2Rα (CD25) by daclizumab was shown to selectively expand CD56^{bright} NK cells, activate NK cell functions, and induced upregulation of PVR on the surface of CD4 T cells, rendering these susceptible for NK cell-mediated killing (138–140).

The activating receptor NKp46 is known to be associated with the induction of type I diabetes. Using knockout mice, in which NKp46 is ablated, revealed that these mice are less prone to develop diabetes. Mechanistically, NK cells degranulated after binding *via* NKp46 to beta islet cells of the pancreas. The ligand for NKp46 expressed on beta islet cells is still unknown (141, 142). Whether the lack of NKp46 has also influence on autoreactive T cells in diabetes type I remains to be shown.

Moreover, NKp46 is also involved in graft-versus-host disease (GVHD) because the absence of NKp46 in GVHD led to rapid mortality. However, it has remained elusive if NKp46 is needed for direct killing of host-reactive T cells, or whether regulation occurs indirectly *via* APCs in the context of GVHD (143). Furthermore, NK cells expressing the activating receptor Ly49D rapidly killed allogenic DCs in a murine skin transplantation model through the release of perforin. The absence of allogenic DCs inhibited alloreactive CD8 T cell responses emerging in the draining LN. In addition, NK cells could limit alloreactive CD4 T cells *via* the activating receptor Ly49D (144). These studies demonstrate that NK cells play a role in GVHD *via* the activating receptors Ly49D and NKp46.

CONCLUDING REMARKS

Even though NK cells are mainly known as killer cells of the innate immune system, there is more and more evidence that NK cells can shape the adaptive immune system by influencing T cells in different stages of their lifespan. During T cell priming, NK cells indirectly alter T cell responses by affecting DCs. Reciprocally, DCs are also able to modulate NK cells and this bidirectional interaction affects the emerging T cell response. More detailed insight into the detailed mechanisms of DC/NK interaction will be important to tailor T cell immunity in the context of vaccination or toleration.

The absence of NK cells induces alterations in the early phase of T cell responses, including direct attack of T cells. The detailed mechanisms of this direct regulation are, however, still being defined. Since NK cells are regulated *via* the net balance of signals perceived by their activating and inhibiting receptors, more insights into the regulation of NK receptor ligand expression on activated T cells is required.

Since T cells have evolved several mechanisms to shield themselves against NK-mediated killing during early activation, a more detailed molecular understanding about this shielding process is important. Such knowledge might also be useful to understand whether and how autoreactive T cells can be rendered targets for NK cell attack.

AUTHOR CONTRIBUTIONS

All authors listed have made substantial, direct, and intellectual contribution to the work and approved it for publication.

ACKNOWLEDGMENTS

The authors are grateful to the members of the Oxenius group for helpful discussions and critical reading of the manuscript. This work was supported by the ETH and the Swiss National Science Foundation (Grant 310030_146140 to AO).

REFERENCES

1. Lucas M, Schachterle W, Oberle K, Aichele P, Diefenbach A. Dendritic cells prime natural killer cells by trans-presenting interleukin 15. *Immunity* (2007) 26(4):503–17. doi:10.1016/j.immuni.2007.03.006
2. Fehniger TA, Cai SF, Cao X, Bredemeyer AJ, Presti RM, French AR, et al. Acquisition of murine NK cell cytotoxicity requires the translation of a pre-existing pool of granzyme B and perforin mRNAs. *Immunity* (2007) 26(6):798–811. doi:10.1016/j.immuni.2007.04.010
3. Kang SJ, Liang HE, Reizis B, Locksley RM. Regulation of hierarchical clustering and activation of innate immune cells by dendritic cells. *Immunity* (2008) 29(5):819–33. doi:10.1016/j.immuni.2008.09.017
4. Gidlund M, Orn A, Wigzell H, Senik A, Gresser I. Enhanced NK cell activity in mice injected with interferon and interferon inducers. *Nature* (1978) 273(5665):759–61. doi:10.1038/273759a0
5. DeBlaker-Hohe DF, Yamauchi A, Yu CR, Horvath-Arcidiacono JA, Bloom ET. IL-12 synergizes with IL-2 to induce lymphokine-activated cytotoxicity and perforin and granzyme gene expression in fresh human NK cells. *Cell Immunol* (1995) 165(1):33–43. doi:10.1006/cimm.1995.1184
6. Takeda K, Tsutsui H, Yoshimoto T, Adachi O, Yoshida N, Kishimoto T, et al. Defective NK cell activity and Th1 response in IL-18-deficient mice. *Immunity* (1998) 8(3):383–90. doi:10.1016/S1074-7613(00)80543-9
7. Biron CA, Nguyen KB, Pien GC, Cousens LP, Salazar-Mather TP. Natural killer cells in antiviral defense: function and regulation by innate cytokines. *Annu Rev Immunol* (1999) 17:189–220. doi:10.1146/annurev.immunol.17.1.189
8. Hart OM, Athie-Morales V, O'Connor GM, Gardiner CM. TLR7/8-mediated activation of human NK cells results in accessory cell-dependent IFN-gamma production. *J Immunol* (2005) 175(3):1636–42. doi:10.4049/jimmunol.175.3.1636
9. Schmidt KN, Leung B, Kwong M, Zarember KA, Satyal S, Navas TA, et al. APC-independent activation of NK cells by the toll-like receptor 3 agonist double-stranded RNA. *J Immunol* (2004) 172(1):138–43. doi:10.4049/jimmunol.172.1.138
10. Sivori S, Falco M, Della Chiesa M, Carlomagno S, Vitale M, Moretta L, et al. CpG and double-stranded RNA trigger human NK cells by toll-like receptors: induction of cytokine release and cytotoxicity against tumors and dendritic cells. *Proc Natl Acad Sci U S A* (2004) 101(27):10116–21. doi:10.1073/pnas.0403744101
11. Lanier LL. NK cell receptors. *Annu Rev Immunol* (1998) 16:359–93. doi:10.1146/annurev.immunol.16.1.359
12. Moretta A, Bottino C, Vitale M, Pende D, Cantoni C, Mingari MC, et al. Activating receptors and coreceptors involved in human natural killer cell-mediated cytolysis. *Annu Rev Immunol* (2001) 19:197–223. doi:10.1146/annurev.immunol.19.1.197
13. Pegram HJ, Andrews DM, Smyth MJ, Darcy PK, Kershaw MH. Activating and inhibitory receptors of natural killer cells. *Immunol Cell Biol* (2011) 89(2):216–24. doi:10.1038/icb.2010.78
14. Vivier E, Ugolini S, Blaise D, Chabannon C, Brossay L. Targeting natural killer cells and natural killer T cells in cancer. *Nat Rev Immunol* (2012) 12(4):239–52. doi:10.1038/nri3174
15. Diefenbach A, Raulet DH. Strategies for target cell recognition by natural killer cells. *Immunol Rev* (2001) 181:170–84. doi:10.1034/j.1600-065X.2001.1810114.x
16. Vivier E, Tomasello E, Baratin M, Walzer T, Ugolini S. Functions of natural killer cells. *Nat Immunol* (2008) 9(5):503–10. doi:10.1038/ni1582
17. Raulet DH. Roles of the NKG2D immunoreceptor and its ligands. *Nat Rev Immunol* (2003) 3(10):781–90. doi:10.1038/nri1199
18. Groh V, Bahram S, Bauer S, Herman A, Beauchamp M, Spies T. Cell stress-regulated human major histocompatibility complex class I gene expressed in gastrointestinal epithelium. *Proc Natl Acad Sci U S A* (1996) 93(22):12445–50. doi:10.1073/pnas.93.22.12445
19. Jinushi M, Takehara T, Kanto T, Tatsumi T, Groh V, Spies T, et al. Critical role of MHC class I-related chain A and B expression on IFN-alpha-stimulated dendritic cells in NK cell activation: impairment in chronic hepatitis C virus infection. *J Immunol* (2003) 170(3):1249–56. doi:10.4049/jimmunol.170.3.1249
20. Groh V, Rhinehart R, Randolph-Habecker J, Topp MS, Riddell SR, Spies T. Costimulation of CD8alphabeta T cells by NKG2D via engagement by MIC induced on virus-infected cells. *Nat Immunol* (2001) 2(3):255–60. doi:10.1038/85321
21. Welte SA, Sinzger C, Lutz SZ, Singh-Jasuja H, Sampaio KL, Eknigk U, et al. Selective intracellular retention of virally induced NKG2D ligands by the human *Cytomegalovirus* UL16 glycoprotein. *Eur J Immunol* (2003) 33(1):194–203. doi:10.1002/immu.200390022
22. Shibuya A, Campbell D, Hannum C, Yssel H, Franz-Bacon K, McClanahan T, et al. DNAM-1, a novel adhesion molecule involved in the cytolytic function of T lymphocytes. *Immunity* (1996) 4(6):573–81. doi:10.1016/S1074-7613(00)70060-4
23. Takai Y, Miyoshi J, Ikeda W, Ogita H. Nectins and nectin-like molecules: roles in contact inhibition of cell movement and proliferation. *Nat Rev Mol Cell Biol* (2008) 9(8):603–15. doi:10.1038/nrm2457
24. Sakisaka T, Takai Y. Biology and pathology of nectins and nectin-like molecules. *Curr Opin Cell Biol* (2004) 16(5):513–21. doi:10.1016/j.ceb.2004.07.007
25. Chan CJ, Smyth MJ, Martinet L. Molecular mechanisms of natural killer cell activation in response to cellular stress. *Cell Death Differ* (2014) 21(1):5–14. doi:10.1038/cdd.2013.26
26. Piédavent-Salomon M, Willing A, Engler JB, Steinbach K, Bauer S, Eggert B, et al. Multiple sclerosis associated genetic variants of CD226 impair regulatory T cell function. *Brain* (2015) 138(Pt 11):3263–74. doi:10.1093/brain/awv256
27. Biassoni R, Pessino A, Bottino C, Pende D, Moretta L, Moretta A. The murine homologue of the human NKp46, a triggering receptor involved in the induction of natural cytotoxicity. *Eur J Immunol* (1999) 29(3):1014–20. doi:10.1002/(SICI)1521-4141(199903)29:03<1014::AID-IMMU1014>3.0.CO;2-O
28. Kruse PH, Matta J, Ugolini S, Vivier E. Natural cytotoxicity receptors and their ligands. *Immunol Cell Biol* (2014) 92(3):221–9. doi:10.1038/icb.2013.98
29. Rosental B, Brusilovsky M, Hadad U, Oz D, Appel MY, Afergan F, et al. Proliferating cell nuclear antigen is a novel inhibitory ligand for the natural cytotoxicity receptor NKp44. *J Immunol* (2011) 187(11):5693–702. doi:10.4049/jimmunol.1102267
30. Baychelier F, Sennepin A, Ermonval M, Dorgham K, Debré P, Vieillard V. Identification of a cellular ligand for the natural cytotoxicity receptor NKp44. *Blood* (2013) 122(17):2935–42. doi:10.1182/blood-2013-03-489054
31. Mandelboim O, Lieberman N, Lev M, Paul L, Arnon TI, Bushkin Y, et al. Recognition of haemagglutinins on virus-infected cells by NKp46 activates lysis by human NK cells. *Nature* (2001) 409(6823):1055–60. doi:10.1038/35059110
32. Arase H, Mocarski ES, Campbell AE, Hill AB, Lanier LL. Direct recognition of *Cytomegalovirus* by activating and inhibitory NK cell receptors. *Science* (2002) 296(5571):1323–6. doi:10.1126/science.1070884
33. Raulet DH, Gasser S, Gowen BG, Deng W, Jung H. Regulation of ligands for the NKG2D activating receptor. *Annu Rev Immunol* (2013) 31:413–41. doi:10.1146/annurev-immunol-032712-095951
34. Warren HS, Smyth MJ. NK cells and apoptosis. *Immunol Cell Biol* (1999) 77(1):64–75. doi:10.1046/j.1440-1711.1999.00790.x
35. Clynes RA, Towers TL, Presta LG, Ravetch JV. Inhibitory Fc receptors modulate in vivo cytotoxicity against tumor targets. *Nat Med* (2000) 6(4):443–6. doi:10.1038/74704
36. Long EO, Kim HS, Liu D, Peterson ME, Rajagopalan S. Controlling natural killer cell responses: integration of signals for activation and inhibition. *Annu Rev Immunol* (2013) 31:227–58. doi:10.1146/annurev-immunol-020711-075005
37. Höglund P, Ohlén C, Carbone E, Franksson L, Ljunggren HG, Latour A, et al. Recognition of beta 2-microglobulin-negative (beta 2m-) T-cell blasts by natural killer cells from normal but not from beta 2m- mice: nonresponsiveness controlled by beta 2m- bone marrow in chimeric mice. *Proc Natl Acad Sci U S A* (1991) 88(22):10332–6. doi:10.1073/pnas.88.22.10332
38. Liao NS, Bix M, Zijlstra M, Jaenisch R, Raulet D. MHC class I deficiency: susceptibility to natural killer (NK) cells and impaired NK activity. *Science* (1991) 253(5016):199–202. doi:10.1126/science.1853205
39. Held W, Kijima M, Angelov G, Bessoles S. The function of natural killer cells: education, reminders and some good memories. *Curr Opin Immunol* (2011) 23(2):228–33. doi:10.1016/j.coi.2010.11.008
40. Crouse J, Xu HC, Lang PA, Oxenius A. NK cells regulating T cell responses: mechanisms and outcome. *Trends Immunol* (2015) 36(1):49–58. doi:10.1016/j.it.2014.11.001

41. Crome SQ, Lang PA, Lang KS, Ohashi PS. Natural killer cells regulate diverse T cell responses. *Trends Immunol* (2013) 34(7):342–9. doi:10.1016/j.it.2013.03.002
42. von Burg N, Turchinovich G, Finke D. Maintenance of immune homeostasis through ILC/T cell interactions. *Front Immunol* (2015) 6:416. doi:10.3389/fimmu.2015.00416
43. Diefenbach A. Innate lymphoid cells in the defense against infections. *Eur J Microbiol Immunol (Bp)* (2013) 3(3):143–51. doi:10.1556/EuJMI.3.2013.3.1
44. Banchereau J, Briere F, Caux C, Davoust J, Lebecque S, Liu YJ, et al. Immunobiology of dendritic cells. *Annu Rev Immunol* (2000) 18:767–811. doi:10.1146/annurev.immunol.18.1.767
45. Lenschow DJ, Walunas TL, Bluestone JA. CD28/B7 system of T cell costimulation. *Annu Rev Immunol* (1996) 14:233–58. doi:10.1146/annurev.immunol.14.1.233
46. Borst J, Hendriks J, Xiao Y. CD27 and CD70 in T cell and B cell activation. *Curr Opin Immunol* (2005) 17(3):275–81. doi:10.1016/j.coi.2005.04.004
47. Croft M, So T, Duan W, Soroosh P. The significance of OX40 and OX40L to T-cell biology and immune disease. *Immunol Rev* (2009) 229(1):173–91. doi:10.1111/j.1600-065X.2009.00766.x
48. Alderson MR, Smith CA, Tough TW, Davis-Smith T, Armitage RJ, Falk B, et al. Molecular and biological characterization of human 4-1BB and its ligand. *Eur J Immunol* (1994) 24(9):2219–27. doi:10.1002/eji.1830240943
49. Anderson DM, Maraskovsky E, Billingsley WL, Dougall WC, Tometsko ME, Roux ER, et al. A homologue of the TNF receptor and its ligand enhance T-cell growth and dendritic-cell function. *Nature* (1997) 390(6656):175–9. doi:10.1038/36593
50. Mescher MF, Curtsinger JM, Agarwal P, Casey KA, Gerner M, Hammerbeck CD, et al. Signals required for programming effector and memory development by CD8+ T cells. *Immunol Rev* (2006) 211:81–92. doi:10.1111/j.0105-2896.2006.00382.x
51. Blattman JN, Antia R, Sourdive DJ, Wang X, Kaech SM, Murali-Krishna K, et al. Estimating the precursor frequency of naive antigen-specific CD8 T cells. *J Exp Med* (2002) 195(5):657–64. doi:10.1084/jem.20001021
52. Wiesel M, Walton S, Richter K, Oxenius A. Virus-specific CD8 T cells: activation, differentiation and memory formation. *APMIS* (2009) 117(5–6):356–81. doi:10.1111/j.1600-0463.2009.02459.x
53. Kallies A. Distinct regulation of effector and memory T-cell differentiation. *Immunol Cell Biol* (2008) 86(4):325–32. doi:10.1038/icb.2008.16
54. Ferlazzo G, Thomas D, Lin SL, Goodman K, Morandi B, Muller WA, et al. The abundant NK cells in human secondary lymphoid tissues require activation to express killer cell Ig-like receptors and become cytolytic. *J Immunol* (2004) 172(3):1455–62. doi:10.4049/jimmunol.172.3.1455
55. Fehniger TA, Cooper MA, Nuovo GJ, Cella M, Facchetti F, Colonna M, et al. CD56bright natural killer cells are present in human lymph nodes and are activated by T cell-derived IL-2: a potential new link between adaptive and innate immunity. *Blood* (2003) 101(8):3052–7. doi:10.1182/blood-2002-09-2876
56. Martín-Fontecha A, Thomsen LL, Brett S, Gerard C, Lipp M, Lanzavecchia A, et al. Induced recruitment of NK cells to lymph nodes provides IFN-gamma for T(H)1 priming. *Nat Immunol* (2004) 5(12):1260–5. doi:10.1038/ni1138
57. Joffre O, Nolte MA, Spörri R, Reis e Sousa C. Inflammatory signals in dendritic cell activation and the induction of adaptive immunity. *Immunol Rev* (2009) 227:234–47. doi:10.1111/j.1600-065X.2008.00718.x
58. Gerosa F, Baldani-Guerra B, Nisii C, Marchesini V, Carra G, Trinchieri G. Reciprocal activating interaction between natural killer cells and dendritic cells. *J Exp Med* (2002) 195(3):327–33. doi:10.1084/jem.20010938
59. Wehner R, Dietze K, Bachmann M, Schmitz M. The bidirectional crosstalk between human dendritic cells and natural killer cells. *J Innate Immun* (2011) 3(3):258–63. doi:10.1159/000323923
60. Vitale M, Della Chiesa M, Carlomagno S, Pende D, Aricò M, Moretta L, et al. NK-dependent DC maturation is mediated by TNFalpha and IFNgamma released upon engagement of the NKp30 triggering receptor. *Blood* (2005) 106(2):566–71. doi:10.1182/blood-2004-10-4035
61. Piccioli D, Sbrana S, Melandri E, Valiante NM. Contact-dependent stimulation and inhibition of dendritic cells by natural killer cells. *J Exp Med* (2002) 195(3):335–41. doi:10.1084/jem.20010934
62. Nishioka Y, Nishimura N, Suzuki Y, Sone S. Human monocyte-derived and CD83(+) blood dendritic cells enhance NK cell-mediated cytotoxicity. *Eur J Immunol* (2001) 31(9):2633–41. doi:10.1002/1521-4141(200109)31:9<2633::AID-IMMU2633>3.0.CO;2-2
63. Amakata Y, Fujiyama Y, Andoh A, Hodohara K, Bamba T. Mechanism of NK cell activation induced by coculture with dendritic cells derived from peripheral blood monocytes. *Clin Exp Immunol* (2001) 124(2):214–22. doi:10.1046/j.1365-2249.2001.01550.x
64. Wehner R, Löbel B, Bornhäuser M, Schäkel K, Cartellieri M, Bachmann M, et al. Reciprocal activating interaction between 6-sulfo LacNAc+ dendritic cells and NK cells. *Int J Cancer* (2009) 124(2):358–66. doi:10.1002/ijc.23962
65. Laouar Y, Sutterwala FS, Gorelik L, Flavell RA. Transforming growth factor-beta controls T helper type 1 cell development through regulation of natural killer cell interferon-gamma. *Nat Immunol* (2005) 6(6):600–7. doi:10.1038/ni1197
66. Iyoda T, Shimoyama S, Liu K, Omatsu Y, Akiyama Y, Maeda Y, et al. The CD8+ dendritic cell subset selectively endocytoses dying cells in culture and in vivo. *J Exp Med* (2002) 195(10):1289–302. doi:10.1084/jem.20020161
67. Krebs P, Barnes MJ, Lampe K, Whitley K, Bahjat KS, Beutler B, et al. NK-cell-mediated killing of target cells triggers robust antigen-specific T-cell-mediated and humoral responses. *Blood* (2009) 113(26):6593–602. doi:10.1182/blood-2009-01-201467
68. Deauvieau F, Ollion V, Doffin AC, Achard C, Fonteneau JF, Verronese E, et al. Human natural killer cells promote cross-presentation of tumor cell-derived antigens by dendritic cells. *Int J Cancer* (2015) 136(5):1085–94. doi:10.1002/ijc.29087
69. Ferlazzo G, Tsang ML, Moretta L, Melioli G, Steinman RM, Münz C. Human dendritic cells activate resting natural killer (NK) cells and are recognized via the NKp30 receptor by activated NK cells. *J Exp Med* (2002) 195(3):343–51. doi:10.1084/jem.20011149
70. Morandi B, Mortara L, Chiossone L, Accolla RS, Mingari MC, Moretta L, et al. Dendritic cell editing by activated natural killer cells results in a more protective cancer-specific immune response. *PLoS One* (2012) 7(6):e39170. doi:10.1371/journal.pone.0039170
71. Persson CM, Assarsson E, Vahlne G, Brodin P, Chambers BJ. Critical role of Qa1b in the protection of mature dendritic cells from NK cell-mediated killing. *Scand J Immunol* (2008) 67(1):30–6. doi:10.1111/j.1365-3083.2007.02034.x
72. Hayakawa Y, Screpanti V, Yagita H, Grandien A, Ljunggren HG, Smyth MJ, et al. NK cell TRAIL eliminates immature dendritic cells in vivo and limits dendritic cell vaccination efficacy. *J Immunol* (2004) 172(1):123–9. doi:10.4049/jimmunol.172.1.123
73. Della Chiesa M, Vitale M, Carlomagno S, Ferlazzo G, Moretta L, Moretta A. The natural killer cell-mediated killing of autologous dendritic cells is confined to a cell subset expressing CD94/NKG2A, but lacking inhibitory killer Ig-like receptors. *Eur J Immunol* (2003) 33(6):1657–66. doi:10.1002/eji.200390042
74. Cook KD, Whitmire JK. The depletion of NK cells prevents T cell exhaustion to efficiently control disseminating virus infection. *J Immunol* (2013) 190(2):641–9. doi:10.4049/jimmunol.1202448
75. Andrews DM, Estcourt MJ, Andoniou CE, Wikstrom ME, Khong A, Voigt V, et al. Innate immunity defines the capacity of antiviral T cells to limit persistent infection. *J Exp Med* (2010) 207(6):1333–43. doi:10.1084/jem.20091193
76. Mandaric S, Walton SM, Rülicke T, Richter K, Girard-Madoux MJ, Clausen BE, et al. IL-10 suppression of NK/DC crosstalk leads to poor priming of MCMV-specific CD4 T cells and prolonged MCMV persistence. *PLoS Pathog* (2012) 8(8):e1002846. doi:10.1371/journal.ppat.1002846
77. Nakayama M, Takeda K, Kawano M, Takai T, Ishii N, Ogasawara K. Natural killer (NK)-dendritic cell interactions generate MHC class II-dressed NK cells that regulate CD4+ T cells. *Proc Natl Acad Sci U S A* (2011) 108(45):18360–5. doi:10.1073/pnas.1110584108
78. Phillips JH, Le AM, Lanier LL. Natural killer cells activated in a human mixed lymphocyte response culture identified by expression of Leu-11 and class II histocompatibility antigens. *J Exp Med* (1984) 159(4):993–1008. doi:10.1084/jem.159.4.993
79. Hanna J, Gonen-Gross T, Fitchett J, Rowe T, Daniels M, Arnon TI, et al. Novel APC-like properties of human NK cells directly regulate T cell activation. *J Clin Invest* (2004) 114(11):1612–23. doi:10.1172/JCI22787
80. Zingoni A, Sornasse T, Cocks BG, Tanaka Y, Santoni A, Lanier LL. Cross-talk between activated human NK cells and CD4+ T cells via OX40-OX40

ligand interactions. *J Immunol* (2004) 173(6):3716–24. doi:10.4049/jimmunol.173.6.3716

81. Spits H, Lanier LL. Natural killer or dendritic: what's in a name? *Immunity* (2007) 26(1):11–6. doi:10.1016/j.immuni.2007.01.004
82. Agaugué S, Marcenaro E, Ferranti B, Moretta L, Moretta A. Human natural killer cells exposed to IL-2, IL-12, IL-18, or IL-4 differently modulate priming of naive T cells by monocyte-derived dendritic cells. *Blood* (2008) 112(5):1776–83. doi:10.1182/blood-2008-02-135871
83. Sivori S, Pende D, Bottino C, Marcenaro E, Pessino A, Biassoni R, et al. NKp46 is the major triggering receptor involved in the natural cytotoxicity of fresh or cultured human NK cells. Correlation between surface density of NKp46 and natural cytotoxicity against autologous, allogeneic or xenogeneic target cells. *Eur J Immunol* (1999) 29(5):1656–66. doi:10.1002/(SICI)1521-4141(199905)29:05<1656::AID-IMMU1656>3.0.CO;2-1
84. Rabinovich BA, Li J, Shannon J, Hurren R, Chalupny J, Cosman D, et al. Activated, but not resting, T cells can be recognized and killed by syngeneic NK cells. *J Immunol* (2003) 170(7):3572–6. doi:10.4049/jimmunol.170.7.3572
85. Cerboni C, Zingoni A, Cippitelli M, Piccoli M, Frati L, Santoni A. Antigen-activated human T lymphocytes express cell-surface NKG2D ligands via an ATM/ATR-dependent mechanism and become susceptible to autologous NK-cell lysis. *Blood* (2007) 110(2):606–15. doi:10.1182/blood-2006-10-052720
86. Waggoner SN, Cornberg M, Selin LK, Welsh RM. Natural killer cells act as rheostats modulating antiviral T cells. *Nature* (2012) 481(7381):394–8. doi:10.1038/nature10624
87. Lang PA, Lang KS, Xu HC, Grusdat M, Parish IA, Recher M, et al. Natural killer cell activation enhances immune pathology and promotes chronic infection by limiting CD8+ T-cell immunity. *Proc Natl Acad Sci U S A* (2012) 109(4):1210–5. doi:10.1073/pnas.1118834109
88. Soderquest K, Walzer T, Zafirova B, Klavinskis LS, Polić B, Vivier E, et al. Cutting edge: CD8+ T cell priming in the absence of NK cells leads to enhanced memory responses. *J Immunol* (2011) 186(6):3304–8. doi:10.4049/jimmunol.1004122
89. Xu HC, Grusdat M, Pandyra AA, Polz R, Huang J, Sharma P, et al. Type I interferon protects antiviral CD8+ T cells from NK cell cytotoxicity. *Immunity* (2014) 40(6):949–60. doi:10.1016/j.immuni.2014.05.004
90. Crouse J, Bedenikovic G, Wiesel M, Ibberson M, Xenarios I, Von Laer D, et al. Type I interferons protect T cells against NK cell attack mediated by the activating receptor NCR1. *Immunity* (2014) 40(6):961–73. doi:10.1016/j.immuni.2014.05.003
91. Rudensky AY. Regulatory T cells and Foxp3. *Immunol Rev* (2011) 241(1):260–8. doi:10.1111/j.1600-065X.2011.01018.x
92. Ghiringhelli F, Ménard C, Terme M, Flament C, Taieb J, Chaput N, et al. CD4(+) CD25(+) regulatory T cells inhibit natural killer cell functions in a transforming growth factor-beta-dependent manner. *J Exp Med* (2005) 202(8):1075–85. doi:10.1084/jem.20051511
93. Lindenberg M, Solmaz G, Puttur F, Sparwasser T. Mouse *Cytomegalovirus* infection overrules T regulatory cell suppression on natural killer cells. *Virol J* (2014) 11:145. doi:10.1186/1743-422X-11-145
94. Gasteiger G, Hemmers S, Firth MA, Le Floc'h A, Huse M, Sun JC, et al. IL-2-dependent tuning of NK cell sensitivity for target cells is controlled by regulatory T cells. *J Exp Med* (2013) 210(6):1167–78. doi:10.1084/jem.20122462
95. Sitrin J, Ring A, Garcia KC, Benoist C, Mathis D. Regulatory T cells control NK cells in an insulitic lesion by depriving them of IL-2. *J Exp Med* (2013) 210(6):1153–65. doi:10.1084/jem.20122248
96. Langhans B, Alwan AW, Krämer B, Glässner A, Lutz P, Strassburg CP, et al. Regulatory CD4(+) T cells modulate the interaction between NK cells and hepatic stellate cells by acting on either cell type. *J Hepatol* (2015) 62(2):398–404. doi:10.1016/j.jhep.2014.08.038
97. Spallanzani RG, Torres NI, Avila DE, Ziblat A, Iraolagoitia XL, Rossi LE, et al. Regulatory dendritic cells restrain NK Cell IFN-gamma production through mechanisms involving NKp46, IL-10, and MHC class I-specific inhibitory receptors. *J Immunol* (2015) 195(5):2141–8. doi:10.4049/jimmunol.1403161
98. Magri G, Muntasell A, Romo N, Sáez-Borderías A, Pende D, Geraghty DE, et al. NKp46 and DNAM-1 NK-cell receptors drive the response to human *Cytomegalovirus*-infected myeloid dendritic cells overcoming viral immune evasion strategies. *Blood* (2011) 117(3):848–56. doi:10.1182/blood-2010-08-301374
99. Draghi M, Pashine A, Sanjanwala B, Gendzekhadze K, Cantoni C, Cosman D, et al. NKp46 and NKG2D recognition of infected dendritic cells is necessary for NK cell activation in the human response to influenza infection. *J Immunol* (2007) 178(5):2688–98. doi:10.4049/jimmunol.178.5.2688
100. Claassen MA, de Knegt RJ, Tilanus HW, Janssen HL, Boonstra A. Abundant numbers of regulatory T cells localize to the liver of chronic hepatitis C infected patients and limit the extent of fibrosis. *J Hepatol* (2010) 52(3):315–21. doi:10.1016/j.jhep.2009.12.013
101. Cabrera R, Tu Z, Xu Y, Firpi RJ, Rosen HR, Liu C, et al. An immunomodulatory role for CD4(+)CD25(+) regulatory T lymphocytes in hepatitis C virus infection. *Hepatology* (2004) 40(5):1062–71. doi:10.1002/hep.20454
102. Curiel TJ, Coukos G, Zou L, Alvarez X, Cheng P, Mottram P, et al. Specific recruitment of regulatory T cells in ovarian carcinoma fosters immune privilege and predicts reduced survival. *Nat Med* (2004) 10(9):942–9. doi:10.1038/nm1093
103. Marshall NA, Christie LE, Munro LR, Culligan DJ, Johnston PW, Barker RN, et al. Immunosuppressive regulatory T cells are abundant in the reactive lymphocytes of Hodgkin lymphoma. *Blood* (2004) 103(5):1755–62. doi:10.1182/blood-2003-07-2594
104. Ardolino M, Zingoni A, Cerboni C, Cecere F, Soriani A, Iannitto ML, et al. DNAM-1 ligand expression on Ag-stimulated T lymphocytes is mediated by ROS-dependent activation of DNA-damage response: relevance for NK-T cell interaction. *Blood* (2011) 117(18):4778–86. doi:10.1182/blood-2010-08-300954
105. Soderquest K, Powell N, Luci C, van Rooijen N, Hidalgo A, Geissmann F, et al. Monocytes control natural killer cell differentiation to effector phenotypes. *Blood* (2011) 117(17):4511–8. doi:10.1182/blood-2010-10-312264
106. Deniz G, Erten G, Kücüksezer UC, Kocacik D, Karagiannidis C, Aktas E, et al. Regulatory NK cells suppress antigen-specific T cell responses. *J Immunol* (2008) 180(2):850–7. doi:10.4049/jimmunol.180.2.850
107. Perona-Wright G, Mohrs K, Szaba FM, Kummer LW, Madan R, Karp CL, et al. Systemic but not local infections elicit immunosuppressive IL-10 production by natural killer cells. *Cell Host Microbe* (2009) 6(6):503–12. doi:10.1016/j.chom.2009.11.003
108. Lee SH, Kim KS, Fodil-Cornu N, Vidal SM, Biron CA. Activating receptors promote NK cell expansion for maintenance, IL-10 production, and CD8 T cell regulation during viral infection. *J Exp Med* (2009) 206(10):2235–51. doi:10.1084/jem.20082387
109. Madera S, Rapp M, Firth MA, Beilke JN, Lanier LL, Sun JC. Type I IFN promotes NK cell expansion during viral infection by protecting NK cells against fratricide. *J Exp Med* (2016) 213(2):225–33. doi:10.1084/jem.20150712
110. Ludigs K, Jandus C, Utzschneider DT, Staehli F, Bessoles S, Dang AT, et al. NLRC5 shields T lymphocytes from NK-cell-mediated elimination under inflammatory conditions. *Nat Commun* (2016) 7:10554. doi:10.1038/ncomms10554
111. Waggoner SN, Taniguchi RT, Mathew PA, Kumar V, Welsh RM. Absence of mouse 2B4 promotes NK cell-mediated killing of activated CD8(+) T cells, leading to prolonged viral persistence and altered pathogenesis. *J Clin Invest* (2010) 120(6):1925–38. doi:10.1172/JCI41264
112. McNerney ME, Lee KM, Kumar V. 2B4 (CD244) is a non-MHC binding receptor with multiple functions on natural killer cells and CD8+ T cells. *Mol Immunol* (2005) 42(4):489–94. doi:10.1016/j.molimm.2004.07.032
113. Kim HS, Das A, Gross CC, Bryceson YT, Long EO. Synergistic signals for natural cytotoxicity are required to overcome inhibition by c-Cbl ubiquitin ligase. *Immunity* (2010) 32(2):175–86. doi:10.1016/j.immuni.2010.02.004
114. Bryceson YT, March ME, Ljunggren HG, Long EO. Synergy among receptors on resting NK cells for the activation of natural cytotoxicity and cytokine secretion. *Blood* (2006) 107(1):159–66. doi:10.1182/blood-2005-04-1351
115. Bryceson YT, March ME, Barber DF, Ljunggren HG, Long EO. Cytolytic granule polarization and degranulation controlled by different receptors in resting NK cells. *J Exp Med* (2005) 202(7):1001–12. doi:10.1084/jem.20051143
116. Rydyznski C, Daniels KA, Karmele EP, Brooks TR, Mahl SE, Moran MT, et al. Generation of cellular immune memory and B-cell immunity is impaired by natural killer cells. *Nat Commun* (2015) 6:6375. doi:10.1038/ncomms7375

117. Cook KD, Kline HC, Whitmire JK. NK cells inhibit humoral immunity by reducing the abundance of CD4+ T follicular helper cells during a chronic virus infection. *J Leukoc Biol* (2015) 98(2):153–62. doi:10.1189/jlb.4HI1214-594R
118. Harty JT, Badovinac VP. Shaping and reshaping CD8+ T-cell memory. *Nat Rev Immunol* (2008) 8(2):107–19. doi:10.1038/nri2251
119. Schluns KS, Kieper WC, Jameson SC, Lefrançois L. Interleukin-7 mediates the homeostasis of naive and memory CD8 T cells in vivo. *Nat Immunol* (2000) 1(5):426–32. doi:10.1038/80868
120. Tan JT, Ernst B, Kieper WC, LeRoy E, Sprent J, Surh CD. Interleukin (IL)-15 and IL-7 jointly regulate homeostatic proliferation of memory phenotype CD8+ cells but are not required for memory phenotype CD4+ cells. *J Exp Med* (2002) 195(12):1523–32. doi:10.1084/jem.20020066
121. Lu L, Ikizawa K, Hu D, Werneck MB, Wucherpfennig KW, Cantor H. Regulation of activated CD4+ T cells by NK cells via the Qa-1-NKG2A inhibitory pathway. *Immunity* (2007) 26(5):593–604. doi:10.1016/j.immuni.2007.03.017
122. Alexandre YO, Ghilas S, Sanchez C, Le Bon A, Crozat K, Dalod M. XCR1+ dendritic cells promote memory CD8+ T cell recall upon secondary infections with *Listeria monocytogenes* or certain viruses. *J Exp Med* (2016) 213(1):75–92. doi:10.1084/jem.20142350
123. Wherry EJ. T cell exhaustion. *Nat Immunol* (2011) 12(6):492–9. doi:10.1038/ni.2035
124. Waggoner SN, Daniels KA, Welsh RM. Therapeutic depletion of natural killer cells controls persistent infection. *J Virol* (2014) 88(4):1953–60. doi:10.1128/JVI.03002-13
125. Vieillard V, Strominger JL, Debre P. NK cytotoxicity against CD4+ T cells during HIV-1 infection: a gp41 peptide induces the expression of an NKp44 ligand. *Proc Natl Acad Sci U S A* (2005) 102(31):10981–6. doi:10.1073/pnas.0504315102
126. Matusali G, Potestà M, Santoni A, Cerboni C, Doria M. The human immunodeficiency virus type 1 Nef and Vpu proteins downregulate the natural killer cell-activating ligand PVR. *J Virol* (2012) 86(8):4496–504. doi:10.1128/JVI.05788-11
127. Tomescu C, Mavilio D, Montaner LJ. Lysis of HIV-1-infected autologous CD4+ primary T cells by interferon-alpha-activated NK cells requires NKp46 and NKG2D. *AIDS* (2015) 29(14):1767–73. doi:10.1097/QAD.0000000000000777
128. Peppa D, Gill US, Reynolds G, Easom NJ, Pallett LJ, Schurich A, et al. Up-regulation of a death receptor renders antiviral T cells susceptible to NK cell-mediated deletion. *J Exp Med* (2013) 210(1):99–114. doi:10.1084/jem.20121172
129. Boni C, Lampertico P, Talamona L, Giuberti T, Invernizzi F, Barili V, et al. Natural killer cell phenotype modulation and natural killer/T-cell interplay in nucleos(t)ide analogue-treated hepatitis e antigen-negative patients with chronic hepatitis B. *Hepatology* (2015) 62(6):1697–709. doi:10.1002/hep.28155
130. Maini MK, Peppa D. NK cells: a double-edged sword in chronic hepatitis B virus infection. *Front Immunol* (2013) 4:57. doi:10.3389/fimmu.2013.00057
131. Callan MF, Annels N, Steven N, Tan L, Wilson J, McMichael AJ, et al. T cell selection during the evolution of CD8(+) T cell memory in vivo. *Eur J Immunol* (1998) 28(12):4382–90. doi:10.1002/(SICI)1521-4141(199812)28:12<4382::AID-IMMU4382>3.3.CO;2-Q
132. Chijioke O, Müller A, Feederle R, Barros MH, Krieg C, Emmel V, et al. Human natural killer cells prevent infectious mononucleosis features by targeting lytic Epstein-Barr virus infection. *Cell Rep* (2013) 5(6):1489–98. doi:10.1016/j.celrep.2013.11.041
133. Schuster IS, Wikstrom ME, Brizard G, Coudert JD, Estcourt MJ, Manzur M, et al. TRAIL+ NK cells control CD4+ T cell responses during chronic viral infection to limit autoimmunity. *Immunity* (2014) 41(4):646–56. doi:10.1016/j.immuni.2014.09.013
134. Zhang B, Yamamura T, Kondo T, Fujiwara M, Tabira T. Regulation of experimental autoimmune encephalomyelitis by natural killer (NK) cells. *J Exp Med* (1997) 186(10):1677–87. doi:10.1084/jem.186.10.1677
135. Laroni A, Armentani E, Kerlero de Rosbo N, Ivaldi F, Marcenaro E, Sivori S, et al. Dysregulation of regulatory CD56bright NK cells/T cells interactions in multiple sclerosis. *J Autoimmun* (2016). doi:10.1016/j.jaut.2016.04.003
136. Hao J, Liu R, Piao W, Zhou Q, Vollmer TL, Campagnolo DI, et al. Central nervous system (CNS)-resident natural killer cells suppress Th17 responses and CNS autoimmune pathology. *J Exp Med* (2010) 207(9):1907–21. doi:10.1084/jem.20092749
137. Chong WP, van Panhuys N, Chen J, Silver PB, Jittayasothorn Y, Mattapallil MJ, et al. NK-DC crosstalk controls the autopathogenic Th17 response through an innate IFN-gamma-IL-27 axis. *J Exp Med* (2015) 212(10):1739–52. doi:10.1084/jem.20141678
138. Gross CC, Schulte-Mecklenbeck A, Rünzi A, Kuhlmann T, Posevitz-Fejfár A, Schwab N, et al. Impaired NK-mediated regulation of T-cell activity in multiple sclerosis is reconstituted by IL-2 receptor modulation. *Proc Natl Acad Sci U S A* (2016) 113(21):E2973–82. doi:10.1073/pnas.1524924113
139. Bielekova B, Catalfamo M, Reichert-Scrivner S, Packer A, Cerna M, Waldmann TA, et al. Regulatory CD56(bright) natural killer cells mediate immunomodulatory effects of IL-2Ralpha-targeted therapy (daclizumab) in multiple sclerosis. *Proc Natl Acad Sci U S A* (2006) 103(15):5941–6. doi:10.1073/pnas.0601335103
140. Jiang W, Chai NR, Maric D, Bielekova B. Unexpected role for granzyme K in CD56bright NK cell-mediated immunoregulation of multiple sclerosis. *J Immunol* (2011) 187(2):781–90. doi:10.4049/jimmunol.1100789
141. Gur C, Porgador A, Elboim M, Gazit R, Mizrahi S, Stern-Ginossar N, et al. The activating receptor NKp46 is essential for the development of type 1 diabetes. *Nat Immunol* (2010) 11(2):121–8. doi:10.1038/ni.1834
142. Gur C, Enk J, Kassem SA, Suissa Y, Magenheim J, Stolovich-Rain M, et al. Recognition and killing of human and murine pancreatic beta cells by the NK receptor NKp46. *J Immunol* (2011) 187(6):3096–103. doi:10.4049/jimmunol.1101269
143. Ghadially H, Ohana M, Elboim M, Gazit R, Gur C, Nagler A, et al. NK cell receptor NKp46 regulates graft-versus-host disease. *Cell Rep* (2014) 7(6):1809–14. doi:10.1016/j.celrep.2014.05.011
144. Laffont S, Seillet C, Ortaldo J, Coudert JD, Guéry JC. Natural killer cells recruited into lymph nodes inhibit alloreactive T-cell activation through perforin-mediated killing of donor allogeneic dendritic cells. *Blood* (2008) 112(3):661–71. doi:10.1182/blood-2007-10-120089
145. Gazit R, Gruda R, Elboim M, Arnon TI, Katz G, Achdout H, et al. Lethal influenza infection in the absence of the natural killer cell receptor gene Ncr1. *Nat Immunol* (2006) 7(5):517–23. doi:10.1038/ni1322
146. Jarahian M, Watzl C, Fournier P, Arnold A, Djandji D, Zahedi S, et al. Activation of natural killer cells by Newcastle disease virus hemagglutinin-neuraminidase. *J Virol* (2009) 83(16):8108–21. doi:10.1128/JVI.00211-09
147. Jarahian M, Fiedler M, Cohnen A, Djandji D, Hämmerling GJ, Gati C, et al. Modulation of NKp30- and NKp46-mediated natural killer cell responses by poxviral hemagglutinin. *PLoS Pathog* (2011) 7(8):e1002195. doi:10.1371/journal.ppat.1002195
148. Arnon TI, Lev M, Katz G, Chernobrov Y, Porgador A, Mandelboim O. Recognition of viral hemagglutinins by NKp44 but not by NKp30. *Eur J Immunol* (2001) 31(9):2680–9. doi:10.1002/1521-4141(200109)31:9<2680::AID-IMMU2680>3.0.CO;2-A
149. Arnon TI, Achdout H, Levi O, Markel G, Saleh N, Katz G, et al. Inhibition of the NKp30 activating receptor by pp65 of human *Cytomegalovirus*. *Nat Immunol* (2005) 6(5):515–23. doi:10.1038/ni1190
150. Alter G, Suscovich TJ, Teigen N, Meier A, Streeck H, Brander C, et al. Single-stranded RNA derived from HIV-1 serves as a potent activator of NK cells. *J Immunol* (2007) 178(12):7658–66. doi:10.4049/jimmunol.178.12.7658
151. Townsley E, O'Connor G, Cosgrove C, Woda M, Co M, Thomas SJ, et al. Interaction of a dengue virus NS1-derived peptide with the inhibitory receptor KIR3DL1 on natural killer cells. *Clin Exp Immunol* (2016) 183(3):419–30. doi:10.1111/cei.12722

10

Dysregulation of Chemokine/Chemokine Receptor Axes and NK Cell Tissue Localization during Diseases

Giovanni Bernardini[1,2], Fabrizio Antonangeli[1], Valentina Bonanni[1] and Angela Santoni[2,3]*

[1] Department of Molecular Medicine, Sapienza University, Rome, Italy, [2] IRCCS NEUROMED – Mediterranean Neurological Institute, Isernia, Italy, [3] Department of Molecular Medicine, Istituto Pasteur-Fondazione Cenci Bolognetti, Rome, Italy

Edited by:
Emanuela Marcenaro,
Università di Genova, Italy

Reviewed by:
Christine Susanne Falk,
Hannover Medical School, Germany
Nicholas D. Huntington,
Walter and Eliza Hall Institute of Medical Research, Australia

***Correspondence:**
Giovanni Bernardini
giovanni.bernardini@uniroma1.it

Chemokines are small chemotactic molecules that play key roles in physiological and pathological conditions. Upon signaling *via* their specific receptors, chemokines regulate tissue mobilization and trafficking of a wide array of immune cells, including natural killer (NK) cells. Current research is focused on analyzing changes in chemokine/chemokine receptor expression during various diseases to interfere with pathological trafficking of cells or to recruit selected cell types to specific tissues. NK cells are a heterogeneous lymphocyte population comprising several subsets endowed with distinct functional properties and mainly representing distinct stages of a linear development process. Because of their different functional potential, the type of subset that accumulates in a tissue drives the final outcome of NK cell-regulated immune response, leading to either protection or pathology. Correspondingly, chemokine receptors, including CXCR4, CXCR3, and CX_3CR1, are differentially expressed by NK cell subsets, and their expression levels can be modulated during NK cell activation. At first, this review will summarize the current knowledge on the contribution of chemokines to the localization and generation of NK cell subsets in homeostasis. How an inappropriate chemotactic response can lead to pathology and how chemokine targeting can therapeutically affect tissue recruitment/localization of distinct NK cell subsets will also be discussed.

Keywords: chemokine receptors, CXCR4, CXCR3, NK cell subsets, multiple sclerosis, multiple myeloma, cross-inhibition, migration

INTRODUCTION

Natural killer (NK) cells are innate lymphocytes that play a key role in the immune response to tumors and infections through their ability to kill transformed or infected cells and to produce immunoregulatory cytokines and chemokines. Activation of NK cell effector functions can be achieved through a complex integration of inhibitory and activation signals provided by membrane expressed receptors and/or through cytokine stimulation (1).

Natural killer cells are widely distributed into different tissues such as the bone marrow (BM), liver, thymus, lymph node, and uterus, thus contributing to immune surveillance in homeostasis, and can be further recruited into tissues in pathological conditions (2, 3). While tissue-resident NK cells have been identified in uterus, liver, and skin, conventional NK cells continuously traffic and localize into tissue through a combination of stimuli able to promote their mobilization from storage compartments to blood circulation and their entry and retention into tissue (4–6).

Chemokines are a family of more than 50 small proteins, mostly secreted, that accomplish their function by interacting with heterotrimeric G protein-coupled receptors (GPCR). Chemokine binding promotes a conformational change in the receptor, triggering intracellular signals that drive cell polarization, migration, and adhesion, thus resulting in the induction of leukocyte trafficking and homing (3). Besides leukocyte chemotaxis, chemokines can affect a number of other leukocyte functions and are recognized as important regulators of the immune response (3). Chemokines may be grouped according to their modality of expression and function, as inflammatory or homeostatic (7).

CHEMOKINE RECEPTOR EXPRESSION ON NK CELL SUBSETS

A number of evidence indicates that the differential functional properties underlying NK cell-mediated protective effect in pathological conditions can be attributed to distinct NK cell populations endowed with distinct expression patterns of activating and inhibitory as well as homing receptors. Two major subsets of mature NK cells were identified in human peripheral blood with respect to the neural cell adhesion molecule CD56 and the low affinity receptor for IgG, FcγRIII CD16. The $CD56^{high}CD16^{low}$ subset accounts for around 10% of circulating $CD56^{+}$ NK cells and exerts immunomodulatory effects producing large amount of cytokines such as IFNγ in response to activation, whereas $CD56^{low}CD16^{high}$ cells are the major cytotoxic population representing the majority of circulating $CD56^{+}$ NK cells [for a review on human NK subsets, see Ref. (8)]. Although the exact relationship between these NK cell subsets still remains unclear, evidence suggest that $CD56^{low}$ NK cells originate from $CD56^{high}$ NK cells (9–12).

Natural killer cell subsets display a differential pattern of chemokine receptor expression (**Figure 1**). $CD56^{high}$ NK cells are targeted to lymph nodes *via* CCR7, preferentially express CXCR3 and have higher CXCR4 expression levels as compared with $CD56^{low}$ cells. $CD56^{low}$ NK cells uniquely express CXCR1, ChemR23, and CX_3CR1 (**Figure 1**) (see **Table 1** for a list of ligands of chemoattractant receptors expressed by human and mouse NK cell subsets) (13–18). More recently, a new $CD56^{low}CD16^{low}$ subset has been identified and found to be prominent in the BM of healthy pediatric donors, to display potent killing and IFNγ producing capacity, and to expresses higher levels of CXCR4 and

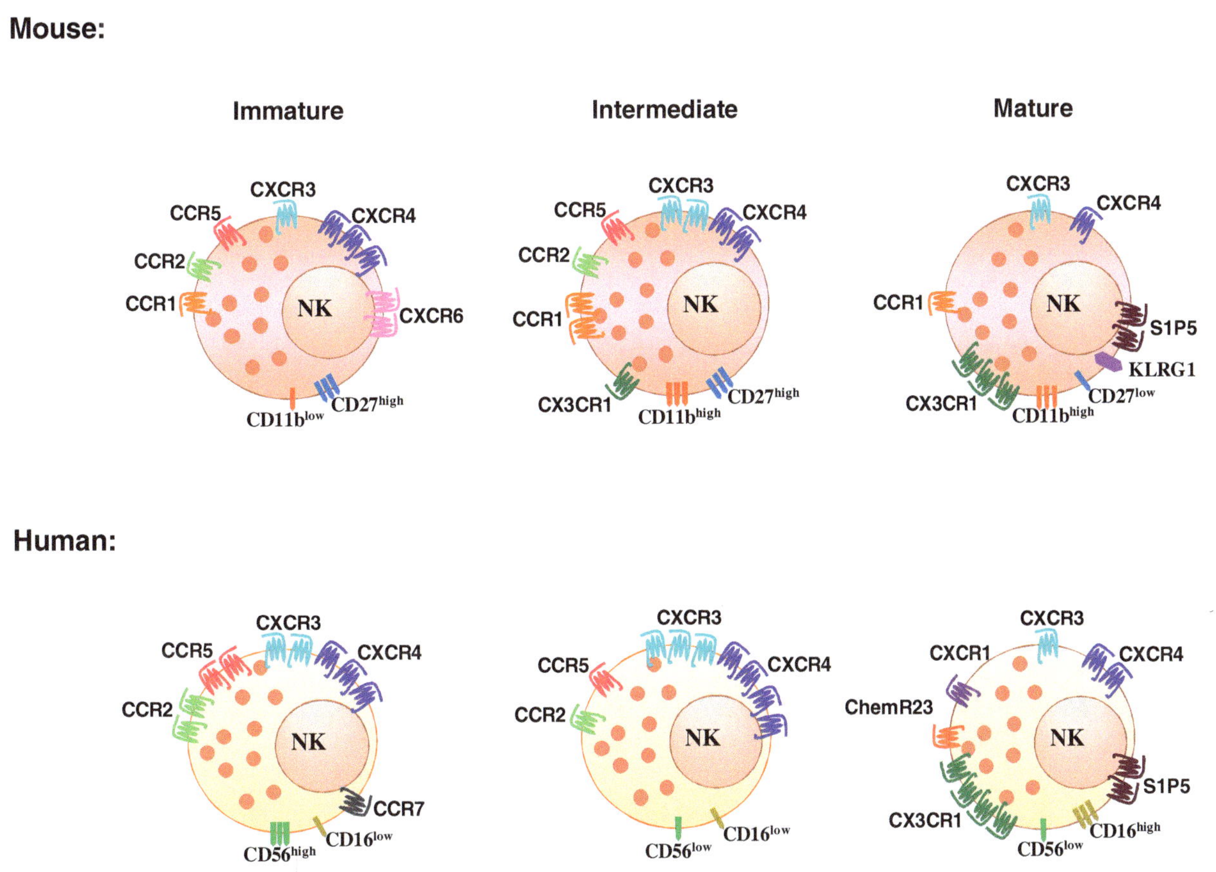

FIGURE 1 | Differential chemoattractant receptor expression by NK cell subsets in human and mouse.

TABLE 1 | Chemokine receptor expression by NK cells and their respective ligands.

Chemokine receptor	Chemokine ligand
CCR1	CCL3/MIP-1α, CCL5/RANTES, CCL7/MCP-3, CCL9/CCL10/MIP-1γ, CCL14/HCC-1, CCL15/HCC-2, CCL16/HCC-4, CCL23/MPIF-1
CCR2	CCL2/MCP-1, CCL7/MCP-3, CCL12, CCL13/MCP-4, CCL16/HCC-4
CCR5	CCL3/MIP-1α, CCL4/MIP-1β, CCL5/RANTES, CCL8/MCP-2, CCL14/HCC-1
CCR7	CCL19/MIP-3β/ELC, CCL21/SLC
CXCR1	CXCL8/IL-8
CXCR3	CXCL9/Mig, CXCL10/IP-10, CXCL11/I-TAC
CXCR4	CXCL12/SDF-1
CXCR6	CXCL16/SR-PSOX
CX_3CR1	CX3CL1/fractalkine
Chemoattractant receptors	**Ligand**
S1P5	S1P
ChemR23, CCRL2	Chemerin

The previous chemokine name follows the one currently used.

CXCR3 compared with the other subsets (17). In addition, other subsets related to NK cell maturation, including cells coexpressing CD57, a member of the glucuronyl-transferase gene family or the L-selectin within the $CD56^{low}$ NK cells are currently active field of investigation (19–21).

Mouse NK cells do not express the murine ortholog of CD56, but four different developmentally related subsets have been identified on the basis of the expression levels of the integrin chain CD11b and of a member of the TNF receptor superfamily, CD27: $CD11b^{low}CD27^{low}$, $CD11b^{low}CD27^{high}$, $CD11b^{high}CD27^{high}$, and $CD11b^{high}CD27^{low}$ (22, 23). The inhibitory receptor KLRG1 is acquired by the most mature subset and identifies NK cells with reduced effector functions (24). KLRG1 coexpression with the chemokine receptor CX_3CR1 identifies an even later maturation stage with unique functional properties (25). Likewise, other NK cell subsets can be defined according to the expression of selected chemokine receptors: the prevalent expression of CXCR3 on $CD56^{high}$ was related to expression of the receptor on mouse $CD27^{high}$ NK cells and more in general to stronger proliferative and cytokine production capacity (26, 27). In addition, CXCR6 was shown to identify an NK cell population resident in liver displaying in humans a $CD56^{high}CD16^{-}CD57^{-}$ phenotype, and expressing TRAIL in diseased liver, and in the mouse a $DX5^{-}TRAIL^{+}$ phenotype (28, 29).

CHEMOKINE RECEPTOR INTERPLAY FOR EFFICIENT MIGRATION IN COMPLEX CHEMOTACTIC ENVIRONMENT

Leukocytes express multiple chemokine receptors not only to robustly and promptly infiltrate tissues upon activation but also to navigate through the network provided by multiple competing chemotactic gradients perceived into tissue. Indeed, discrimination of a dominant chemoattractant is required to localize in the correct microenvironment and/or to make a decision on whether to stay or to leave a tissue (30, 31). The ability to preferentially respond to selected chemoattractants was at first demonstrated in neutrophils and was postulated to require heterologous receptor desensitization that is the transinhibition of a chemokine receptor resulting from the activation of second messenger-dependent kinases by a dominant chemoattractant receptor (32–34). In this regard, triggering CXCR2 by KC was shown to be required for neutrophil egress from the BM not only by promoting neutrophil migration but also by inhibiting CXCR4-mediated BM retention (35). Afterward, our group showed that CXCR4 heterologous desensitization is also associated with NK cell egress from BM into circulation (36, 37). On the other hand, the sphingophospholipid chemoattractant sphingosine-1-phosphate (S1P) promotes NK cell egress from BM under steady state without inducing CXCR4 heterologous desensitization (38). CX3CL1, a ligand for CX_3CR1 constitutively expressed in BM, acts similarly on a small highly differentiated subset which poorly expresses CXCR4 (25, 39). While homeostatic chemokines may be sufficient for NK cell egress under steady state, heterologous receptor desensitization could be a mechanism to rapidly switch NK cell responsiveness promoted by inflammatory chemokines, to promptly facilitate BM NK cell availability in circulation.

Besides desensitization, CXCR4-mediated transinhibition can occur through other mechanisms, including receptor heterodimerization and G protein scavenging (40). In regard to heterodimerization, it is becoming increasingly clear that receptor dimers are constitutively formed, and that ligand binding to one receptor can reorganize receptor complexes thus affecting different aspects of the associated chemokine receptor activity, including ligand affinity, the activated signaling cascades, and the receptor internalization (41–43). For example, CXCR3/CXCR4-heterodimerization was shown to reduce the binding affinity of CXCR4 for its ligand (44).

Apart from the mechanism involved, cross-regulation promoted by CXCR3 was shown to be relevant in several pathological conditions in mouse disease models. For example, O'Boyle and coworkers demonstrated CXCR4 and CCR5 inhibition on T cells by using a mimetic of CXCL10, a CXCR3 ligand. The use of this mimetic in a humanized mouse air-pouch model demonstrated reduced trafficking of T cells toward synovial fluids from patients with active rheumatoid arthritis, indicating that the triggering of a single chemokine receptor can control the immune response in chronic inflammatory conditions where CXCL12 is produced at high levels (45, 46). More recently, our observations in a mouse model of multiple myeloma led us to hypothesize that increased expression of CXCR3 ligands in the tumor BM microenvironment constitutes a new mechanism to avoid tumor infiltration by NK cells: ligand-induced CXCR3 activation on $KLRG1^{-}$ NK cells resulted in cross-desensitization of CXCR4 that together with the coincident down-modulation of CXCL12 protein levels promotes NK cell egress from BM into blood circulation. The final outcome of this process is the reduction of the localization of this subset at the tumor site (39). On the other hand, when studying human plasmacytoid dendritic cells (pDC), a positive cooperative interaction was observed between the two receptors (47). This observation, together with a consistent adjacent expression of

CXCL12 with CXCR3 ligands in human tissues, led the authors to hypothesize that the cooperation between CXCR3 ligands and CXCL12 controls the tissue recruitment of pDCs.

Considering the highlighted importance of CXCR3/CXCR4 interplay, hereafter we will document the critical role of CXCR3 and of CXCR4 receptor/ligand axes in the regulation of NK cell-mediated function in pathologies (summarized in **Table 2**).

CXCR4 RECEPTOR/LIGAND AXIS

CXCL12 displays a constitutive but restricted expression pattern *in situ*, with selective expression by CXCL12 abundant reticular (CAR) cells and osteoblasts in BM, by subpopulations of neuronal and endothelial cells in the brain, by dermal endothelial cells, and by invading trophoblast cells and lymph node high endothelial venules (48–54). Correspondingly, CXCR4/CXCL12 axis was reported to regulate NK cell functions in several physiological processes. It was shown that CXCL12 regulates the positioning in the BM of selected NK cell subsets at various stages of maturation; in addition, during pregnancy, human peripheral blood CD56^{high}CD16^{-} NK cells can be recruited by CXCL12 and migrate to the uterus (36, 48).

Considering the key role of CXCL12 in the localization of NK cells in BM, subversion of the CXCR4/CXCL12 axis has been hypothesized to represent a mechanism of immune evasion from NK-mediated immune surveillance in neuroblastoma and multiple myeloma (39, 55). TGF-β1 produced by neuroblastoma cell lines was shown to upregulate the surface expression of CXCR4 and CXCR3 on both CD56^{high} and CD56^{low} NK cells, while it downregulated CX_3CR1 in the CD56^{low} subset. Increased CXCR3 and reduced CX_3CR1 expression was observed also in peripheral blood NK cells of stage 4 neuroblastoma patients, and it may represent an attempt to avoid NK cell cytolytic subset recruitment to the tumor site, while promoting the enrichment of immature and poorly cytotoxic CD56^{high} subset in tumor leukocyte infiltrates (55, 56). Similarly, accumulation of CD56^{high} NK cells was shown to occur in several tumors and may be related to the responsiveness of this subset to a combination of tissue-expressed chemokines: by analyzing the chemokine expression pattern of various normal solid tissues, Carrega and coworkers documented that some tissues are clearly oriented to recruit CD56^{high} cells when CXCL12 is coexpressed with other chemokines (LN, colorectal, stomach, and liver tissues), while anatomic compartments with the lowest proportion of CD56^{high} within NK cells (lung and breast tissues) display a chemokine expression profile favoring CD56^{low} NK cell recruitment (57). Similarly, decreased levels of CXCL12 in BM plasma samples from a cohort of patients with active multiple myeloma as compared with a premalignant stage support the hypothesis of a reduced NK cell surveillance based on reduced BM recruitment (39).

The key role of CXCR4 expression for NK cell homeostasis is also highlighted by genetic defects responsible for altered or lost CXCR4 function. For example, a more abundant proportion of circulating GPI^{-}CD56^{high} NK cells in paroxysmal nocturnal hemoglobinuria patients, a disease caused by dampened biosynthesis of glycosylphosphatidylinositol (GPI)-linked protein, was associated with reduced responsiveness of this population to the CXCR4 ligand CXCL12 (58, 59). In addition, deficiency of the transcription factor GATA2 is characterized by several hematological and non-hematological abnormalities among which NK cell cytopenia, with almost complete absence of circulating CD56^{high} subset (60). Interestingly, CD56^{low} NK cells display reduced CXCR4 surface expression levels and reduced chemotaxis to CXCL12 that was attributed decreased expression of filamin A and β-arrestin-1, two proteins regulating CXCR4 cell surface expression and endocytosis (61–63). On the other hand,

TABLE 2 | Influence of dysregulation of chemokine receptor/ligand axes in pathologies on NK cell subsets.

NK cell subset	Chemokine-related alterations	Disease	Effects	Reference
CD56^{low}	↑CXCR4 ↓CX_3CR1	Neuroblastoma	Less recruitment to tumor site – tumor immune evasion	(55)
KLRG1^{-}	↓CXCL12 ↓CXCR3 ↑CXCL9/10	Multiple myeloma	Less recruitment to tumor site – tumor immune evasion	(39)
NK cells	↑CXCR4	Multiple sclerosis	Enhanced chemotaxis toward CXCL12 – neuroprotective role	(69)
CD56^{high}	↓CXCR4	Paroxysmal nocturnal hemoglobinuria	Less secondary lymphoid tissue relocation	(58)
CD56^{low}	↓CXCR4	GATA2 deficiency	Reduced chemotaxis toward CXCL12	(63)
NK cells	↑CXCR4	WHIM	Enhanced chemotaxis toward CXCL12	(38, 64, 65)
CD27^{high}	↑CXCL9/10	Cowpox virus infection	Increased recruitment to lymph nodes Protection from infection	(77)
NK cells	↑CXCL10 Truncated form	HCV infection	Less recruitment to liver – immune response dampening	(81)
CD56^{low}	↑CXCL9/10	Primary biliary cirrhosis	Autologous cytotoxicity – tissue damage	(82–85)
CD56^{high}	↑CXCR3 ↑CXCL10	Psoriatic skin	Local inflammation – disease progression	(86)
CD56^{high}	↑CXCR3 ligands	Periprosthetic osteolysis	Local inflammation – immunoregulatory role	(87)

↑ indicates up-modulation/increase frequency and ↓ indicates down-modulation/reduced frequency.

NK cells from patients affected by warts, hypogammaglobulinemia, infections, and myelokathexis (WHIM) syndrome display enhanced responsiveness to CXCL12 (38, 64, 65). Similarly, NK cells from mice displaying the most common mutation of the *CXCR4* gene associated with WHIM syndrome show enhanced migration to CXCL12 that is linked to impaired CXCR4 desensitization and internalization after CXCL12 stimulation. Possibly for this reason, NK cell distribution is altered, with CD11b^{low} and CD11b^{high}CD27^{high} NK cells accumulating in the BM.

Although *CXCR4* mutation in WHIM syndrome was not associated with any NK cell-related disease, a selective defect of CXCR4 internalization after CXCL12 binding underlies a new rare immune deficiency documented in two cases of disseminated *Mycobacterium avium* infection, where a marked reduction in the number of circulating NK cells as well as neutrophils and B cells was observed (66).

An altered pattern of CXCL12 expression in brain has been reported in multiple sclerosis (67, 68), suggesting the involvement of the CXCR4/CXCL12 axis in the leukocyte infiltration that characterizes this pathology. In this regard, while the interference with CXCR4/CXCL12 axis often leads to reduced NK cell protection in pathological conditions, Serrano-Pertierra and colleagues have found increased NK cell chemotaxis in response to CXCL12 in multiple sclerosis patients in the remitting phase and in clinically isolated syndrome patients with respect to relapsing multiple sclerosis patients and healthy controls. This finding has been associated with higher frequencies of NK cells expressing CXCR4 in the blood of the former patients' cohorts (69). The enhanced NK cell migration in patients with a less active disease course supports the idea of a neuroprotective role for NK cells in multiple sclerosis (70). Unfortunately, NK cells have been studied as a whole, and the authors agree it would be of interest to analyze NK cell subsets, also considering that the size of the circulating CD56^{high} NK cell pool is significantly associated with clinical remissions and that expansion of this population is associated with amelioration of diseases in response to therapy (71–75).

CXCR3 RECEPTOR/LIGAND AXES

CXCR3 ligands are expressed at low levels in homeostatic conditions, but their expression can be upregulated in both the hematopoietic and non-hematopoietic compartment by IFN-γ and some related cytokines. Several studies in humans and mice reveal that NK cells can promote adaptive immune response by modulating dendritic cell (DC) function and T helper cell polarization (76). This important function is linked to CXCR3-mediated NK cell recruitment into draining lymph node in several conditions. In mouse, in accordance with higher and preferential expression levels of CXCR3, the NK cell population mostly affected by CXCR3 function is the CD27^{high} subset that colonizes draining LN following DC vaccination, cowpox virus infection, and during tumor growth (77, 78). Several studies have correlated high numbers of tumor-infiltrating NK cells with a good prognosis for cancer patients and with tumor cell clearance in mouse tumor models. This has been related to the IFN-γ promotion of CXCL9 and CXCL10 production by tumor-infiltrating leukocytes, leading to the CXCR3-mediated recruitment of mouse CD27^{high} NK cells, the population of NK cells with the higher functional potential (79).

The influence of CXCR3–CXCL10 axis on NK cell function was documented also in human pathologies. Upregulation of CXCR3 ligands in multiple myeloma patients with active disease corresponded to marked down-modulation of CXCR3 expression levels by BM NK cells, an event that was linked to reduction of NK cell localization in the BM in multiple myeloma-bearing mice. In addition, high CXCL9 and CXCL10 serum levels were associated with several established prognostic parameters and predicted poor overall survival (39, 80).

CXCL10/CXCR3 axis is involved in hepatic trafficking of NK cells, which also represent an important component of the intrahepatic lymphocyte pool and has been implicated in the pathogenesis of chronic hepatitis C virus (HCV) infection. The CD56^{high}CXCR3^{+} NK cells display the strongest activity against hepatic stellate cells, thus regulating liver fibrosis. Although expanded in HCV-infected patients, the CD56^{high}CXCR3^{+} NK cell subset display impaired functions that may be linked to HCV-associated liver fibrosis (27). Elevated levels of CXCL10 were found in serum of patients and were predictive of the failure to respond to HCV therapy. Nevertheless, in a recent study, it has been reported that CXCL10 in the serum of HCV patients may not be biologically active, representing a truncated form that can bind to CXCR3 without signaling. In the presence of higher levels of this CXCL10 antagonist, NK cells might fail to migrate to the infected liver and accumulate instead in the peripheral circulation (81).

Chemokines play an important role in destruction of the biliary tract (82) by recruiting cells of the immune system, including NK cells. As such, liver NK cells have been reported to express the chemokine receptors CX_3CR1 and CXCR3 (83). Among the principal chemokines involved in hepatic immune cell migration, CXCL9 and CXCL10 are both increased in serum of patients with primary biliary cirrhosis (PBC) compared with normal individuals and are preferentially expressed in the portal areas, corresponding to CD56^{low}CD16^{+} NK cell liver infiltration increased numbers of CD56^{+} cells located around the destroyed small bile ducts (84, 85).

Several reports documented that CD56^{high} NK cells also infiltrate inflamed skin in a CXCL10-dependent fashion. Ottaviani et al. have shown that psoriatic keratinocytes display an enhanced capacity to produce CXCL10, and CD56^{high}CD16^{-} NK cells showed an upregulation of CXCR3, in comparison to CD56^{low}CD16^{+} NK cells (86). CD56^{high}CXCR3^{+}CCR5^{+} cells produced IFNγ after IL2 stimulation that in turn potentiates activation of keratinocytes and upregulates HLA class-I. These findings would suggest that CD56^{+} NK cells are recruited in psoriatic skin through a mechanism involving the CXCL10/CXCR3 axes and that, once in the skin, they may contribute to the disease progression by inducing local inflammation and amplifying T cell autoimmune reactivity.

Similar to psoriatic skin-infiltrating NK cells, NK cells in the synovial tissue of osteoarthritis patients are CCR5^{+}CXCR3^{+}. High levels of CXCR3 and CCR5 ligands present in synovial fluids after revision surgery, as well as evidence of particle-induced chemokine production by macrophages (87), suggest a mechanism for recruitment of a subset functionally corresponding to $CD56^{high}CD16^{-}$ NK cells during periprosthetic osteolysis. The majority of synovial tissue-infiltrating NK cells express a combination of surface receptors consistent with a non-cytotoxic phenotype similar to blood.

CONCLUSION

The correct localization of NK cells into tissues has a fundamental role in several aspects of NK cell-mediated immune responses *in vivo*. Thus, identification of the key mediators regulating NK cell tissue recruitment is a critical step in the optimization of current cancer immunotherapy protocols or in the treatment of inflammatory diseases.

When NK cell tissue accumulation is important, the therapeutic enhancement of expression of selected chemokines that attract NK cells specifically is a valuable approach to increase the penetration and/or local activation and differentiation of NK cells at the tumor site. Ectopic expression of chemokines/chemoattractants known to preferentially attract effector lymphocytes, including CXCR3 ligands as well as CX_3CL1 and chemerin, was shown to positively affect the antitumor nature of tumor-infiltrating lymphocytes with a large proportion of NK cells (88–91). Nevertheless, high concentrations of attracting chemokines do not always imply increased NK cell migration, as shown by Halama and coworkers in colorectal cancer tissue where $NKp46^{+}$ NK cells are poorly infiltrated, despite high local chemokine levels (92).

An alternative and highly novel strategy to improve NK cell migration to target tissues is to promote or optimize the expression of chemoattractant receptors on NK cells to be used for adoptive immunotherapy. The expression of chemokine receptors and the corresponding NK cell chemotactic response can be modulated upon cytokine-mediated activation thus suggesting that they may better home to tumor sites where their corresponding ligands are expressed (93–96). In addition, NK cells *ex vivo* engineered to express chemokine receptors by gene transfer or by trogocytosis are under investigation for their better tissue homing and function (97–101). Conversely, a number of new clinical trials for immune-mediated diseases based on the use of chemokine receptor antagonists are ongoing and will help to understand the therapeutic potential of these important targets for NK cell-promoted pathologies (102). Finally, the emerging role of chemoattractant receptor interplay in the regulation of immune cell response may also lead to the discovery of molecules able to block chemokine receptor cross-inhibition thus allowing to unleash the full chemotactic potential of important NK cell receptors, such as CXCR4.

AUTHOR CONTRIBUTIONS

GB, FA, VB, and AS contributed equally to writing and critically revised the paper.

REFERENCES

1. Yokoyama WM, Kim S, French AR. The dynamic life of natural killer cells. *Annu Rev Immunol* (2004) 22:405–29. doi:10.1146/annurev.immunol.22.012703.104711
2. Gajewski TF, Fuertes M, Spaapen R, Zheng Y, Kline J. Molecular profiling to identify relevant immune resistance mechanisms in the tumor microenvironment. *Curr Opin Immunol* (2011) 23(2):286–92. doi:10.1016/j.coi.2010.11.013
3. Zlotnik A, Yoshie O. The chemokine superfamily revisited. *Immunity* (2012) 36(5):705–16. doi:10.1016/j.immuni.2012.05.008
4. Sojka DK, Tian Z, Yokoyama WM. Tissue-resident natural killer cells and their potential diversity. *Semin Immunol* (2014) 26(2):127–31. doi:10.1016/j.smim.2014.01.010
5. Bernardini G, Sciume G, Santoni A. Differential chemotactic receptor requirements for NK cell subset trafficking into bone marrow. *Front Immunol* (2013) 4:12. doi:10.3389/fimmu.2013.00012
6. Gregoire C, Chasson L, Luci C, Tomasello E, Geissmann F, Vivier E, et al. The trafficking of natural killer cells. *Immunol Rev* (2007) 220:169–82. doi:10.1111/j.1600-065X.2007.00563.x
7. Mantovani A. The chemokine system: redundancy for robust outputs. *Immunol Today* (1999) 20(6):254–7. doi:10.1016/S0167-5699(99)01469-3
8. Cooper MA, Fehniger TA, Caligiuri MA. The biology of human natural killer-cell subsets. *Trends Immunol* (2001) 22(11):633–40. doi:10.1016/S1471-4906(01)02060-9
9. Chan A, Hong DL, Atzberger A, Kollnberger S, Filer AD, Buckley CD, et al. CD56bright human NK cells differentiate into CD56dim cells: role of contact with peripheral fibroblasts. *J Immunol* (2007) 179(1):89–94. doi:10.4049/jimmunol.179.1.89
10. Ouyang Q, Baerlocher G, Vulto I, Lansdorp PM. Telomere length in human natural killer cell subsets. *Ann N Y Acad Sci* (2007) 1106:240–52. doi:10.1196/annals.1392.001
11. Romagnani C, Juelke K, Falco M, Morandi B, D'Agostino A, Costa R, et al. CD56brightCD16- killer Ig-like receptor- NK cells display longer telomeres and acquire features of CD56dim NK cells upon activation. *J Immunol* (2007) 178(8):4947–55. doi:10.4049/jimmunol.178.8.4947
12. Huntington ND, Legrand N, Alves NL, Jaron B, Weijer K, Plet A, et al. IL-15 trans-presentation promotes human NK cell development and differentiation in vivo. *J Exp Med* (2009) 206(1):25–34. doi:10.1084/jem.20082013
13. Berahovich RD, Lai NL, Wei Z, Lanier LL, Schall TJ. Evidence for NK cell subsets based on chemokine receptor expression. *J Immunol* (2006) 177(11):7833–40. doi:10.4049/jimmunol.177.11.7833
14. Campbell JJ, Qin S, Unutmaz D, Soler D, Murphy KE, Hodge MR, et al. Unique subpopulations of CD56+ NK and NK-T peripheral blood lymphocytes identified by chemokine receptor expression repertoire. *J Immunol* (2001) 166(11):6477–82. doi:10.4049/jimmunol.166.11.6477
15. Fehniger TA, Cooper MA, Nuovo GJ, Cella M, Facchetti F, Colonna M, et al. CD56bright natural killer cells are present in human lymph nodes and are activated by T cell-derived IL-2: a potential new link between adaptive and innate immunity. *Blood* (2003) 101(8):3052–7. doi:10.1182/blood-2002-09-2876
16. Parolini S, Santoro A, Marcenaro E, Luini W, Massardi L, Facchetti F, et al. The role of chemerin in the colocalization of NK and dendritic cell subsets into inflamed tissues. *Blood* (2007) 109(9):3625–32. doi:10.1182/blood-2006-08-038844
17. Stabile H, Nisti P, Morrone S, Pagliara D, Bertaina A, Locatelli F, et al. Multifunctional human CD56 low CD16 low natural killer cells are the prominent subset in bone marrow of both healthy pediatric donors and

leukemic patients. *Haematologica* (2015) 100(4):489–98. doi:10.3324/haematol.2014.116053
18. Vitale M, Della Chiesa M, Carlomagno S, Romagnani C, Thiel A, Moretta L, et al. The small subset of CD56brightCD16- natural killer cells is selectively responsible for both cell proliferation and interferon-gamma production upon interaction with dendritic cells. *Eur J Immunol* (2004) 34(6):1715–22. doi:10.1002/eji.200425100
19. Lanier LL, Le AM, Phillips JH, Warner NL, Babcock GF. Subpopulations of human natural killer cells defined by expression of the Leu-7 (HNK-1) and Leu-11 (NK-15) antigens. *J Immunol* (1983) 131(4):1789–96.
20. Juelke K, Killig M, Luetke-Eversloh M, Parente E, Gruen J, Morandi B, et al. CD62L expression identifies a unique subset of polyfunctional CD56dim NK cells. *Blood* (2010) 116(8):1299–307. doi:10.1182/blood-2009-11-253286
21. Lopez-Verges S, Milush JM, Pandey S, York VA, Arakawa-Hoyt J, Pircher H, et al. CD57 defines a functionally distinct population of mature NK cells in the human CD56dimCD16+ NK-cell subset. *Blood* (2010) 116(19):3865–74. doi:10.1182/blood-2010-04-282301
22. Chiossone L, Chaix J, Fuseri N, Roth C, Vivier E, Walzer T. Maturation of mouse NK cells is a 4-stage developmental program. *Blood* (2009) 113(22):5488–96. doi:10.1182/blood-2008-10-187179
23. Hayakawa Y, Huntington ND, Nutt SL, Smyth MJ. Functional subsets of mouse natural killer cells. *Immunol Rev* (2006) 214:47–55. doi:10.1111/j.1600-065X.2006.00454.x
24. Huntington ND, Tabarias H, Fairfax K, Brady J, Hayakawa Y, Degli-Esposti MA, et al. NK cell maturation and peripheral homeostasis is associated with KLRG1 up-regulation. *J Immunol* (2007) 178(8):4764–70. doi:10.4049/jimmunol.178.8.4764
25. Sciume G, De Angelis G, Benigni G, Ponzetta A, Morrone S, Santoni A, et al. CX3CR1 expression defines 2 KLRG1+ mouse NK-cell subsets with distinct functional properties and positioning in the bone marrow. *Blood* (2011) 117(17):4467–75. doi:10.1182/blood-2010-07-297101
26. Marquardt N, Wilk E, Pokoyski C, Schmidt RE, Jacobs R. Murine CXCR3+CD27bright NK cells resemble the human CD56bright NK-cell population. *Eur J Immunol* (2010) 40(5):1428–39. doi:10.1002/eji.200940056
27. Eisenhardt M, Glassner A, Kramer B, Korner C, Sibbing B, Kokordelis P, et al. The CXCR3(+)CD56Bright phenotype characterizes a distinct NK cell subset with anti-fibrotic potential that shows dys-regulated activity in hepatitis C. *PLoS One* (2012) 7(7):e38846. doi:10.1371/journal.pone.0038846
28. Paust S, Gill HS, Wang BZ, Flynn MP, Moseman EA, Senman B, et al. Critical role for the chemokine receptor CXCR6 in NK cell-mediated antigen-specific memory of haptens and viruses. *Nat Immunol* (2010) 11(12):1127–35. doi:10.1038/ni.1953
29. Stegmann KA, Robertson F, Hansi N, Gill U, Pallant C, Christophides T, et al. CXCR6 marks a novel subset of T-bet(lo)Eomes(hi) natural killer cells residing in human liver. *Sci Rep* (2016) 6:26157. doi:10.1038/srep26157
30. Broxmeyer HE, Kim CH. Regulation of hematopoiesis in a sea of chemokine family members with a plethora of redundant activities. *Exp Hematol* (1999) 27(7):1113–23. doi:10.1016/S0301-472X(99)00045-4
31. Zabel BA, Rott A, Butcher EC. Leukocyte chemoattractant receptors in human disease pathogenesis. *Annu Rev Pathol* (2015) 10:51–81. doi:10.1146/annurev-pathol-012513-104640
32. Campbell JJ, Foxman EF, Butcher EC. Chemoattractant receptor cross talk as a regulatory mechanism in leukocyte adhesion and migration. *Eur J Immunol* (1997) 27(10):2571–8. doi:10.1002/eji.1830271016
33. Foxman EF, Kunkel EJ, Butcher EC. Integrating conflicting chemotactic signals. The role of memory in leukocyte navigation. *J Cell Biol* (1999) 147(3):577–88. doi:10.1083/jcb.147.3.577
34. Ali H, Richardson RM, Haribabu B, Snyderman R. Chemoattractant receptor cross-desensitization. *J Biol Chem* (1999) 274(10):6027–30. doi:10.1074/jbc.274.10.6027
35. Martin C, Burdon PC, Bridger G, Gutierrez-Ramos JC, Williams TJ, Rankin SM. Chemokines acting via CXCR2 and CXCR4 control the release of neutrophils from the bone marrow and their return following senescence. *Immunity* (2003) 19(4):583–93. doi:10.1016/S1074-7613(03)00263-2
36. Bernardini G, Sciume G, Bosisio D, Morrone S, Sozzani S, Santoni A. CCL3 and CXCL12 regulate trafficking of mouse bone marrow NK cell subsets. *Blood* (2008) 111(7):3626–34. doi:10.1182/blood-2007-08-106203
37. Ponzetta A, Sciume G, Benigni G, Antonangeli F, Morrone S, Santoni A, et al. CX3CR1 regulates the maintenance of KLRG1+ NK cells into the bone marrow by promoting their entry into circulation. *J Immunol* (2013) 191(11):5684–94. doi:10.4049/jimmunol.1300090
38. Mayol K, Biajoux V, Marvel J, Balabanian K, Walzer T. Sequential desensitization of CXCR4 and S1P5 controls natural killer cell trafficking. *Blood* (2011) 118(18):4863–71. doi:10.1182/blood-2011-06-362574
39. Ponzetta A, Benigni G, Antonangeli F, Sciume G, Sanseviero E, Zingoni A, et al. Multiple myeloma impairs bone marrow localization of effector natural killer cells by altering the chemokine microenvironment. *Cancer Res* (2015) 75(22):4766–77. doi:10.1158/0008-5472.CAN-15-1320
40. Vischer HF, Watts AO, Nijmeijer S, Leurs R. G protein-coupled receptors: walking hand-in-hand, talking hand-in-hand? *Br J Pharmacol* (2011) 163(2):246–60. doi:10.1111/j.1476-5381.2011.01229.x
41. Kramp BK, Sarabi A, Koenen RR, Weber C. Heterophilic chemokine receptor interactions in chemokine signaling and biology. *Exp Cell Res* (2011) 317(5):655–63. doi:10.1016/j.yexcr.2010.11.014
42. Contento RL, Molon B, Boularan C, Pozzan T, Manes S, Marullo S, et al. CXCR4-CCR5: a couple modulating T cell functions. *Proc Natl Acad Sci U S A* (2008) 105(29):10101–6. doi:10.1073/pnas.0804286105
43. Sohy D, Yano H, de Nadai P, Urizar E, Guillabert A, Javitch JA, et al. Hetero-oligomerization of CCR2, CCR5, and CXCR4 and the protean effects of "selective" antagonists. *J Biol Chem* (2009) 284(45):31270–9. doi:10.1074/jbc.M109.054809
44. Watts AO, van Lipzig MM, Jaeger WC, Seeber RM, van Zwam M, Vinet J, et al. Identification and profiling of CXCR3-CXCR4 chemokine receptor heteromer complexes. *Br J Pharmacol* (2013) 168(7):1662–74. doi:10.1111/bph.12064
45. Bradfield PF, Amft N, Vernon-Wilson E, Exley AE, Parsonage G, Rainger GE, et al. Rheumatoid fibroblast-like synoviocytes overexpress the chemokine stromal cell-derived factor 1 (CXCL12), which supports distinct patterns and rates of CD4+ and CD8+ T cell migration within synovial tissue. *Arthritis Rheum* (2003) 48(9):2472–82. doi:10.1002/art.11219
46. O'Boyle G, Fox CR, Walden HR, Willet JD, Mavin ER, Hine DW, et al. Chemokine receptor CXCR3 agonist prevents human T-cell migration in a humanized model of arthritic inflammation. *Proc Natl Acad Sci U S A* (2012) 109(12):4598–603. doi:10.1073/pnas.1118104109
47. Vanbervliet B, Bendriss-Vermare N, Massacrier C, Homey B, de Bouteiller O, Briere F, et al. The inducible CXCR3 ligands control plasmacytoid dendritic cell responsiveness to the constitutive chemokine stromal cell-derived factor 1 (SDF-1)/CXCL12. *J Exp Med* (2003) 198(5):823–30. doi:10.1084/jem.20020437
48. Hanna J, Wald O, Goldman-Wohl D, Prus D, Markel G, Gazit R, et al. CXCL12 expression by invasive trophoblasts induces the specific migration of CD16- human natural killer cells. *Blood* (2003) 102(5):1569–77. doi:10.1182/blood-2003-02-0517
49. Krug A, Uppaluri R, Facchetti F, Dorner BG, Sheehan KC, Schreiber RD, et al. IFN-producing cells respond to CXCR3 ligands in the presence of CXCL12 and secrete inflammatory chemokines upon activation. *J Immunol* (2002) 169(11):6079–83. doi:10.4049/jimmunol.169.11.6079
50. Nagasawa T. CXCL12/SDF-1 and CXCR4. *Front Immunol* (2015) 6:301. doi:10.3389/fimmu.2015.00301
51. Pablos JL, Amara A, Bouloc A, Santiago B, Caruz A, Galindo M, et al. Stromal-cell derived factor is expressed by dendritic cells and endothelium in human skin. *Am J Pathol* (1999) 155(5):1577–86. doi:10.1016/S0002-9440(10)65474-0
52. Petit I, Szyper-Kravitz M, Nagler A, Lahav M, Peled A, Habler L, et al. G-CSF induces stem cell mobilization by decreasing bone marrow SDF-1 and up-regulating CXCR4. *Nat Immunol* (2002) 3(7):687–94. doi:10.1038/ni813
53. Stumm R, Culmsee C, Schafer MK, Krieglstein J, Weihe E. Adaptive plasticity in tachykinin and tachykinin receptor expression after focal cerebral ischemia is differentially linked to gabaergic and glutamatergic cerebrocortical circuits and cerebrovenular endothelium. *J Neurosci* (2001) 21(3):798–811.
54. Tham TN, Lazarini F, Franceschini IA, Lachapelle F, Amara A, Dubois-Dalcq M. Developmental pattern of expression of the alpha chemokine stromal cell-derived factor 1 in the rat central nervous system. *Eur J Neurosci* (2001) 13(5):845–56. doi:10.1046/j.0953-816x.2000.01451.x

55. Castriconi R, Dondero A, Bellora F, Moretta L, Castellano A, Locatelli F, et al. Neuroblastoma-derived TGF-beta1 modulates the chemokine receptor repertoire of human resting NK cells. *J Immunol* (2013) 190(10):5321–8. doi:10.4049/jimmunol.1202693
56. Carrega P, Morandi B, Costa R, Frumento G, Forte G, Altavilla G, et al. Natural killer cells infiltrating human nonsmall-cell lung cancer are enriched in CD56 bright CD16(-) cells and display an impaired capability to kill tumor cells. *Cancer* (2008) 112(4):863–75. doi:10.1002/cncr.23239
57. Carrega P, Bonaccorsi I, Di Carlo E, Morandi B, Paul P, Rizzello V, et al. CD56(bright)perforin(low) noncytotoxic human NK cells are abundant in both healthy and neoplastic solid tissues and recirculate to secondary lymphoid organs via afferent lymph. *J Immunol* (2014) 192(8):3805–15. doi:10.4049/jimmunol.1301889
58. El-Sherbiny YM, Kelly RJ, Hill A, Doody GM, Hillmen P, Cook GP. Altered natural killer cell subset homeostasis and defective chemotactic responses in paroxysmal nocturnal hemoglobinuria. *Blood* (2013) 122(11):1887–90. doi:10.1182/blood-2013-06-507574
59. Szpurka H, Schade AE, Jankowska AM, Maciejewski JP. Altered lipid raft composition and defective cell death signal transduction in glycosylphosphatidylinositol anchor-deficient PIG-A mutant cells. *Br J Haematol* (2008) 142(3):413–22. doi:10.1111/j.1365-2141.2008.07203.x
60. Mace EM, Hsu AP, Monaco-Shawver L, Makedonas G, Rosen JB, Dropulic L, et al. Mutations in GATA2 cause human NK cell deficiency with specific loss of the CD56(bright) subset. *Blood* (2013) 121(14):2669–77. doi:10.1182/blood-2012-09-453969
61. Clift IC, Bamidele AO, Rodriguez-Ramirez C, Kremer KN, Hedin KE. beta-Arrestin1 and distinct CXCR4 structures are required for stromal derived factor-1 to downregulate CXCR4 cell-surface levels in neuroblastoma. *Mol Pharmacol* (2014) 85(4):542–52. doi:10.1124/mol.113.089714
62. Gomez-Mouton C, Fischer T, Peregil RM, Jimenez-Baranda S, Stossel TP, Nakamura F, et al. Filamin A interaction with the CXCR4 third intracellular loop regulates endocytosis and signaling of WT and WHIM-like receptors. *Blood* (2015) 125(7):1116–25. doi:10.1182/blood-2014-09-601807
63. Maciejewski-Duval A, Meuris F, Bignon A, Aknin ML, Balabanian K, Faivre L, et al. Altered chemotactic response to CXCL12 in patients carrying GATA2 mutations. *J Leukoc Biol* (2016) 99(6):1065–76. doi:10.1189/jlb.5MA0815-388R
64. Gulino AV, Moratto D, Sozzani S, Cavadini P, Otero K, Tassone L, et al. Altered leukocyte response to CXCL12 in patients with warts hypogammaglobulinemia, infections, myelokathexis (WHIM) syndrome. *Blood* (2004) 104(2):444–52. doi:10.1182/blood-2003-10-3532
65. Balabanian K, Lagane B, Pablos JL, Laurent L, Planchenault T, Verola O, et al. WHIM syndromes with different genetic anomalies are accounted for by impaired CXCR4 desensitization to CXCL12. *Blood* (2005) 105(6):2449–57. doi:10.1182/blood-2004-06-2289
66. Doncker AV, Balabanian K, Bellanne-Chantelot C, de Guibert S, Revest M, Bachelerie F, et al. Two cases of disseminated *Mycobacterium avium* infection associated with a new immunodeficiency syndrome related to CXCR4 dysfunctions. *Clin Microbiol Infect* (2011) 17(2):135–9. doi:10.1111/j.1469-0691.2010.03187.x
67. Calderon TM, Eugenin EA, Lopez L, Kumar SS, Hesselgesser J, Raine CS, et al. A role for CXCL12 (SDF-1alpha) in the pathogenesis of multiple sclerosis: regulation of CXCL12 expression in astrocytes by soluble myelin basic protein. *J Neuroimmunol* (2006) 177(1–2):27–39. doi:10.1016/j.jneuroim.2006.05.003
68. Krumbholz M, Theil D, Cepok S, Hemmer B, Kivisakk P, Ransohoff RM, et al. Chemokines in multiple sclerosis: CXCL12 and CXCL13 up-regulation is differentially linked to CNS immune cell recruitment. *Brain* (2006) 129(Pt 1):200–11. doi:10.1093/brain/awh680
69. Serrano-Pertierra E, Blanco-Gelaz MA, Oliva-Nacarino P, Martinez-Camblor P, Villafani J, Lopez-Larrea C, et al. Increased natural killer cell chemotaxis to CXCL12 in patients with multiple sclerosis. *J Neuroimmunol* (2015) 282:39–44. doi:10.1016/j.jneuroim.2015.03.007
70. Lunemann JD, Munz C. Do natural killer cells accelerate or prevent autoimmunity in multiple sclerosis? *Brain* (2008) 131(Pt 7):1681–3. doi:10.1093/brain/awn132
71. Kastrukoff LF, Lau A, Wee R, Zecchini D, White R, Paty DW. Clinical relapses of multiple sclerosis are associated with 'novel' valleys in natural killer cell functional activity. *J Neuroimmunol* (2003) 145(1–2):103–14. doi:10.1016/j.jneuroim.2003.10.001
72. Bielekova B, Catalfamo M, Reichert-Scrivner S, Packer A, Cerna M, Waldmann TA, et al. Regulatory CD56(bright) natural killer cells mediate immunomodulatory effects of IL-2Ralpha-targeted therapy (daclizumab) in multiple sclerosis. *Proc Natl Acad Sci U S A* (2006) 103(15):5941–6. doi:10.1073/pnas.0601335103
73. Rodriguez-Martin E, Picon C, Costa-Frossard L, Alenda R, Sainz de la Maza S, Roldan E, et al. Natural killer cell subsets in cerebrospinal fluid of patients with multiple sclerosis. *Clin Exp Immunol* (2015) 180(2):243–9. doi:10.1111/cei.12580
74. Vandenbark AA, Huan J, Agotsch M, La Tocha D, Goelz S, Offner H, et al. Interferon-beta-1a treatment increases CD56bright natural killer cells and CD4+CD25+ Foxp3 expression in subjects with multiple sclerosis. *J Neuroimmunol* (2009) 215(1–2):125–8. doi:10.1016/j.jneuroim.2009.08.007
75. Martinez-Rodriguez JE, Lopez-Botet M, Munteis E, Rio J, Roquer J, Montalban X, et al. Natural killer cell phenotype and clinical response to interferon-beta therapy in multiple sclerosis. *Clin Immunol* (2011) 141(3):348–56. doi:10.1016/j.clim.2011.09.006
76. Martin-Fontecha A, Thomsen LL, Brett S, Gerard C, Lipp M, Lanzavecchia A, et al. Induced recruitment of NK cells to lymph nodes provides IFN-gamma for T(H)1 priming. *Nat Immunol* (2004) 5(12):1260–5. doi:10.1038/ni1138
77. Pak-Wittel MA, Yang L, Sojka DK, Rivenbark JG, Yokoyama WM. Interferon-gamma mediates chemokine-dependent recruitment of natural killer cells during viral infection. *Proc Natl Acad Sci U S A* (2013) 110(1):E50–9. doi:10.1073/pnas.1220456110
78. Watt SV, Andrews DM, Takeda K, Smyth MJ, Hayakawa Y. IFN-gamma-dependent recruitment of mature CD27(high) NK cells to lymph nodes primed by dendritic cells. *J Immunol* (2008) 181(8):5323–30. doi:10.4049/jimmunol.181.8.5323
79. Wendel M, Galani IE, Suri-Payer E, Cerwenka A. Natural killer cell accumulation in tumors is dependent on IFN-gamma and CXCR3 ligands. *Cancer Res* (2008) 68(20):8437–45. doi:10.1158/0008-5472.CAN-08-1440
80. Bolomsky A, Schreder M, Hubl W, Zojer N, Hilbe W, Ludwig H. Monokine induced by interferon gamma (MIG/CXCL9) is an independent prognostic factor in newly diagnosed myeloma. *Leuk Lymphoma* (2016) 21:1–10. doi:10.3109/10428194.2016.1151511
81. Riva A, Laird M, Casrouge A, Ambrozaitis A, Williams R, Naoumov NV, et al. Truncated CXCL10 is associated with failure to achieve spontaneous clearance of acute hepatitis C infection. *Hepatology* (2014) 60(2):487–96. doi:10.1002/hep.27139
82. Ishibashi H, Shimoda S. [Pathogenesis of biliary tract injury in primary biliary cirrhosis]. *Nihon Rinsho Meneki Gakkai Kaishi* (2012) 35(6):455–62. doi:10.2177/jsci.35.455
83. Chuang YH, Lian ZX, Cheng CM, Lan RY, Yang GX, Moritoki Y, et al. Increased levels of chemokine receptor CXCR3 and chemokines IP-10 and MIG in patients with primary biliary cirrhosis and their first degree relatives. *J Autoimmun* (2005) 25(2):126–32. doi:10.1016/j.jaut.2005.08.009
84. Gao B, Bertola A. Natural killer cells take two tolls to destruct bile ducts. *Hepatology* (2011) 53(4):1076–9. doi:10.1002/hep.24275
85. Shimoda S, Selmi C, Gershwin ME. Fractalkine and other chemokines in primary biliary cirrhosis. *Int J Hepatol* (2012) 2012:102839. doi:10.1155/2012/102839
86. Ottaviani C, Nasorri F, Bedini C, de Pita O, Girolomoni G, Cavani A. CD56brightCD16(-) NK cells accumulate in psoriatic skin in response to CXCL10 and CCL5 and exacerbate skin inflammation. *Eur J Immunol* (2006) 36(1):118–28. doi:10.1002/eji.200535243
87. Nakashima Y, Sun DH, Trindade MC, Chun LE, Song Y, Goodman SB, et al. Induction of macrophage C-C chemokine expression by titanium alloy and bone cement particles. *J Bone Joint Surg Br* (1999) 81(1):155–62. doi:10.1302/0301-620X.81B1.8884
88. Ju DW, Tao Q, Cheng DS, Zhang W, Zhang M, Hamada H, et al. Adenovirus-mediated lymphotactin gene transfer improves therapeutic efficacy of cytosine deaminase suicide gene therapy in established murine colon carcinoma. *Gene Ther* (2000) 7(4):329–38. doi:10.1038/sj.gt.3301082
89. Lode HN, Xiang R, Dreier T, Varki NM, Gillies SD, Reisfeld RA. Natural killer cell-mediated eradication of neuroblastoma metastases to bone marrow by targeted interleukin-2 therapy. *Blood* (1998) 91(5):1706–15.

90. Nokihara H, Yanagawa H, Nishioka Y, Yano S, Mukaida N, Matsushima K, et al. Natural killer cell-dependent suppression of systemic spread of human lung adenocarcinoma cells by monocyte chemoattractant protein-1 gene transfection in severe combined immunodeficient mice. *Cancer Res* (2000) 60(24):7002–7.
91. Pachynski RK, Zabel BA, Kohrt HE, Tejeda NM, Monnier J, Swanson CD, et al. The chemoattractant chemerin suppresses melanoma by recruiting natural killer cell antitumor defenses. *J Exp Med* (2012) 209(8):1427–35. doi:10.1084/jem.20112124
92. Halama N, Braun M, Kahlert C, Spille A, Quack C, Rahbari N, et al. Natural killer cells are scarce in colorectal carcinoma tissue despite high levels of chemokines and cytokines. *Clin Cancer Res* (2011) 17(4):678–89. doi:10.1158/1078-0432.CCR-10-2173
93. Inngjerdingen M, Damaj B, Maghazachi AA. Expression and regulation of chemokine receptors in human natural killer cells. *Blood* (2001) 97(2):367–75. doi:10.1182/blood.V97.2.367
94. Mailliard RB, Alber SM, Shen H, Watkins SC, Kirkwood JM, Herberman RB, et al. IL-18-induced CD83+CCR7+ NK helper cells. *J Exp Med* (2005) 202(7):941–53. doi:10.1084/jem.20050128
95. Moustaki A, Argyropoulos KV, Baxevanis CN, Papamichail M, Perez SA. Effect of the simultaneous administration of glucocorticoids and IL-15 on human NK cell phenotype, proliferation and function. *Cancer Immunol Immunother* (2011) 60(12):1683–95. doi:10.1007/s00262-011-1067-6
96. Sechler JM, Barlic J, Grivel JC, Murphy PM. IL-15 alters expression and function of the chemokine receptor CX3CR1 in human NK cells. *Cell Immunol* (2004) 230(2):99–108. doi:10.1016/j.cellimm.2004.10.001
97. Burga RA, Nguyen T, Zulovich J, Madonna S, Ylisastigui L, Fernandes R, et al. Improving efficacy of cancer immunotherapy by genetic modification of natural killer cells. *Cytotherapy* (2016). doi:10.1016/j.jcyt.2016.05.018
98. Carlsten M, Levy E, Karambelkar A, Li L, Reger R, Berg M, et al. Efficient mRNA-based genetic engineering of human NK cells with high-affinity CD16 and CCR7 augments rituximab-induced ADCC against lymphoma and targets NK cell migration toward the lymph node-associated chemokine CCL19. *Front Immunol* (2016) 7:105. doi:10.3389/fimmu.2016.00105
99. Marcenaro E, Cantoni C, Pesce S, Prato C, Pende D, Agaugue S, et al. Uptake of CCR7 and acquisition of migratory properties by human KIR+ NK cells interacting with monocyte-derived DC or EBV cell lines: regulation by KIR/HLA-class I interaction. *Blood* (2009) 114(19):4108–16. doi:10.1182/blood-2009-05-222265
100. Muller N, Michen S, Tietze S, Topfer K, Schulte A, Lamszus K, et al. Engineering NK cells modified with an EGFRvIII-specific chimeric antigen receptor to overexpress CXCR4 improves immunotherapy of CXCL12/SDF-1alpha-secreting glioblastoma. *J Immunother* (2015) 38(5):197–210. doi:10.1097/CJI.0000000000000082
101. Somanchi SS, Somanchi A, Cooper LJ, Lee DA. Engineering lymph node homing of ex vivo-expanded human natural killer cells via trogocytosis of the chemokine receptor CCR7. *Blood* (2012) 119(22):5164–72. doi:10.1182/blood-2011-11-389924
102. Allegretti M, Cesta MC, Locati M. Allosteric modulation of chemoattractant receptors. *Front Immunol* (2016) 7:170. doi:10.3389/fimmu.2016.00170

11

Decreased Human Leukocyte Antigen-G Expression by miR-133a Contributes to Impairment of Proinvasion and Proangiogenesis Functions of Decidual NK Cells

Wenwei Guo[1,2†], Liang Fang[2†], Bo Li[1], Xifeng Xiao[1], Shuqiang Chen[1], Jun Wang[1], Fang Yang[1], Lihua Chen[2] and Xiaohong Wang[1]**

[1] Department of Gynecology and Obstetrics, Tangdu Hospital, The Fourth Military Medical University, Xi'an, China,
[2] Department of Immunology, The Fourth Military Medical University, Xi'an, China

Edited by:
Emanuela Marcenaro,
Università di Genova, Italy

Reviewed by:
Cai Zhang,
Shandong University, China
Gunnur Deniz,
Istanbul University, Turkey

***Correspondence:**
Lihua Chen
chenlh@fmmu.edu.en;
Xiaohong Wang
wangxh-99919@163.com

†These authors have contributed equally to this work.

Human leukocyte antigen (HLA)-G plays a crucial role in conferring fetal–maternal tolerance and ensuring a successful pregnancy. CD56^{bright} natural killer (NK) cells accumulate at the maternal decidua in large numbers during pregnancy and are found in direct contact with fetal trophoblasts. There are increasing evidences that decidual NK (dNK) cells are crucial for pregnancy. However, the regulation of dNK cells is mostly unknown. Here, we provide evidences that the secretion function of dNK cells in recurrent spontaneous abortion was impaired, which led to the impairment of the proinvasion and proangiogenesis functions of dNK cells. Decreased HLA-G expression induced by the transfection of miR-133a mimics in HTR-8/SVneo affected the secretory functions of dNK cells. Thus, our data revealed that the functions of dNK cells could be suppressed by the decreased expression of HLA-G and suggest a possible mechanism of recurrent miscarriage.

Keywords: miR-133a, human leukocyte antigen-G, decidua, nature killer cell, killer immunoglobulin-like receptor 2DL4, recurrent spontaneous abortion

INTRODUCTION

Recurrent spontaneous abortion (RSA) is defined as two or more consecutive spontaneous abortions with the same partner (1). However, the causes of RSA are complicated and mostly unknown. In addition to chromosomal abnormalities, anatomical anomalies, and endocrine disorders, immunologic dysfunction is generally considered the most important cause leading to RSA (2, 3). Whether gravid or inchoate, there is a unique combination of human leukocyte antigen (HLA) expression on extravillous trophoblasts (EVTs) and maternal leukocytes in decidual tissues at the maternal–fetal interface. EVT expresses the polymorphic non-classical HLA class I antigens HLA-E, HLA-F, HLA-G, and HLA-C (4, 5). Of these, HLA-G is the most frequently studied and plays important roles in inducing immune tolerance to maintain pregnancy (6–8).

In humans, approximately 40% of the decidual cells are leukocytes, and more than 70% are CD56^{bright}CD16^{-} natural killer (NK) cells (9), which are a minority of the blood NK population. Only about 10% of lymphocytes are CD56^{dim}CD16^{+} NK cells in decidual cells which are the main population in peripheral blood and display cytotoxic activity toward the EVTs (10). Compared to peripheral blood $CD56^{dim}CD16^{+}$ NK cells (pNK), $CD56^{bright}CD16^{-}$ NK cells have a higher secretory ability and lower cytotoxicity (11). Therefore, these $CD56^{bright}CD16^{-}$ NK cells are defined as decidual NK (dNK) cells (12). Compared to pNK cells, dNK cells express higher levels of certain inhibitory receptors that can recognize non-classical HLA class I antigens on EVTs (13). Then, these receptors transfer inhibitory signals to reduce the cytotoxicity of dNK cells (12). dNK cells also express CD94/NKG2A and killer immunoglobulin-like receptor (KIR) that could recognize HLA-C and HLA-G/E, respectively. The interaction between dNK cells and such HLA molecules would contribute to maintain pregnancy successfully (14). Recent studies have shown that dNK cells play an important role in early pregnancy (15–19). A report by Hanna et al. indicated that dNK cells possess the unique ability to regulate crucial placental developmental processes, including trophoblast invasion and vascular growth, *via* the production of interleukin (IL)-8, interferon induced protein (IP)-10, and vascular endothelial growth factor (VEGF) (15). Moreover, NK cells belong to the group 1 innate lymphoid cells (ILC1). The human decidua tissues also contain the group 3 ILC which express CD56 and NCR but lack CD94/NKG2A and KIR. These NCR^{+} ILC3 cells could also produce IL-8, which may be involved in the trophoblast invasion and neo-angiogenesis process (20, 21).

Killer immunoglobulin-like receptor 2DL4 (KIR2DL4), a specific receptor for HLA-G, is expressed on human NK cells (22). KIR2DL4 has an immunoreceptor tyrosine-based inhibitory motif (ITIM) and shows weak inhibition of dNK cell when bound to HLA-G. Therefore, it has been suggested that NK cells expressing KIR2DL4 might be involved in the maintenance of pregnancy by recognizing HLA-G (23, 24). The receptors of HLA-G also include immunoglobulin-like transcript 2 (ILT2) and ILT4. Both ILT2 and ILT4 contain three ITIMs and are inhibitory receptors expressed on leukocytes including NK cells (25).

MicroRNAs (miRNAs) are a class of non-protein-coding RNAs that are estimated to regulate 30% of all genes in animals by binding to specific sites in the 3′ untranslated region (UTR) (26, 27). In studying the mechanisms underlying RSA, we found that miR-133a was greatly overexpressed in RSA villi compared to villi from induced abortion (IA) patients. Multi-software prediction and real-time PCR confirmed that miR-133a was most likely to bind to the HLA-G 3′UTR, as established in our previous study (28). Therefore, this study was designed to confirm that miR-133a negatively regulates HLA-G expression to influence dNK function *via* KIR2DL4 in RSA patients.

MATERIALS AND METHODS

Human Samples

Maternal decidual tissues were obtained from 12 patients with RSAs and from 11 patients with IA, excluding chromosomal and anatomic abnormalities as causes for abortion. Maternal peripheral blood mononuclear cells were prepared from six patients with IA and RSA. The normal and RSA samples were aged between 7 and 12 weeks of gestation, and all samples were obtained from the Tangdu Hospital of the Fourth Military Medical University. Written informed consent was obtained from each subject, and this study was approved by the Institutional Review Board, Tang Du Hospital, Fourth Military Medical University. The decidual tissues were placed into cold PBS immediately and quickly transported to the laboratory.

Human Decidual Cell Isolation, dNK Purification, and Culture

For the isolation of decidual cells, an enzymatic dispersion method was used, as described previously (17, 29). Briefly, the decidual tissues were washed in PBS twice and then cut into small pieces. The decidual tissues were digested with 1 mg/ml collagenase type IV (MP Biomedical, Santa Ana, CA, USA) and 0.01 mg/ml DNase I (MP Biomedical, Santa Ana, CA, USA) in RPMI 1640 medium (HyClone Cell Culture, Carlsbad, CA, USA) in 37°C for 40 min. The suspensions were filtered through a nylon mesh, and the supernatants were discarded after centrifugation. The decidual mononuclear cells were isolated by density gradient centrifugation with Ficoll and were then immediately used for flow cytometry analysis and further study. CD56 Microbeads (Miltenyi Biotec, Bergisch Gladbach, Germany) were used for the positive selection of dNK cells from the decidual mononuclear cells. For excluding ILC3 cells, dNK cells were stained with PE-conjugated anti-CD94 antibodies at 4°C for 30 min for fluorescence-activated cell sorting (FACS), and >97% purity was obtained for $CD56^{+}CD16^{-}CD94^{+}$ cells. These cells were used for the subsequent experiments. The $CD56^{+}CD94^{+}$ dNK cells were cultured in RPMI 1640 medium containing 10% fetal bovine serum (FBS; Gibco, Grand Island, NY, USA) and 100 IU/ml penicillin/streptomycin plus 20 ng/ml IL-15 (PeproTech, Rocky Hill, NJ, USA) for 24 h, and culture supernatants were harvested for the subsequent experiments. When $CD56^{+}CD94^{+}$ dNK cells were cocultured with HTR-8/SVneo cells, there were no IL-15 stimulation. Peripheral blood samples from IA and RSA were centrifuged (800 × *g*, room temperature) for 30 min on Ficoll gradients (MP Biomedical, Santa Ana, CA, USA) to collect mononuclear cells.

Flow Cytometry

The following conjugated mouse anti-human antibodies were used for FACS analysis: CD45-APC, CD3-Percep/Cy5.5, CD94-Percp/Cy5.5, CD56-FITC, CD16-PE, NKp44-APC, NKp46-APC, NKG2A-APC, NKG2D-APC, ILT-2-APC, ILT-4-APC, and KIR2DL4-APC. These antibodies were purchased from Biolegend (San Diego, CA, USA). IL-8-APC, IP-10-APC, VEGF-APC, placental growth factor (PLGF)-APC, and interferon (IFN)-γ-APC were purchased from BD Biosciences (San Jose, CA, USA). Homologous IgG antibodies conjugated with the same fluorescent dye were used as negative controls. Approximately $1–3 \times 10^{6}$ decidua mononuclear cells were resuspended in 100 μl PBS containing 5% NaN_3 and 5% FBS. The antibodies targeting specific surface markers for NK cell staining was performed according to the manufacturer's protocol and incubated in the dark for 30 min at 4°C. After fixation and permeabilization with Fixation and Permeabilization solution (BD Biosciences, San Jose, CA, USA) for 30 min, cells were stained with antibodies targeting specific intracellular cytokines at 4°C for 30 min. The cells were resuspended in 300 μl PBS containing 5% FBS, and then, 100,000 cells per sample were collected for analysis using a standard FACS Calibur flow cytometer (BD Biosciences, San Jose,

CA, USA) and analyzed using the FlowJo software (TreeStar, San Carlos, CA, USA).

Real-time PCR

Real-time PCR was used to measure the mRNA levels of interleukin (IL)-8, IFN-inducible protein (IP)-10, VEGF, PLGF, and IFN-γ in decidual tissues. Total RNA was extracted from 1 cm^3 decidual tissue or 1 × 10^6 dNK cells using the Trizol reagent (Invitrogen, Carlsbad, CA, USA). The cDNA was synthesized from 500 ng of total RNA using PrimeScript RT Master Mix in a total volume of 10 μl at 37°C for 15 min, 85°C for 5 s (TaKaRa, Japan) and 4°C for storage. Next, real-time PCR was run based on the detected fluorescence with SYBR Premix Ex Taq (TaKaRa). The GAPDH mRNA levels were used in the mRNA levels of the standard. PCR was performed using specific primers (**Table 1**). The PCR reaction involved preliminary denaturation at 95°C for 3 min, followed by 40 cycles of denaturation at 95°C for 5 s, annealing at 60°C for 30 s, and elongation at 72°C for 10 s. The RNA samples were assessed for purity and concentration using spectrophotometry (Nanodrop-2000, Thermo Scientific).

Multiplex Cytokine Assays

The secretory cytokines of the decidual tissues and dNK cells were detected using a multiplex cytokine assay (Magpix, Bio-Rad Laboratories, Veenendaal, The Netherlands) according to the manufacturer's instructions. The Human Magnetic Luminex Screening 5 Plex Assay kit (R&D systems, Minneapolis, MN, USA) contained beads conjugated with monoclonal antibodies for IL-8, IP-10, VEGF, PLGF, and IFN-γ. Cytokine assays sensitivity and measurement ranges have shown in Table S1 in Supplementary Material.

Matrigel Invasion Assay

HTR-8/SVneo is the human first-trimester trophoblast cell line (from University of Toronto, ON, Canada) and has the similar function to EVTs. We determined the ability of dNK cells to induce the invasion of HTR-8/SVneo cells *in vitro* by Transwell (Millipore, Billerica, MA, USA) assays. First, 200 μl of HTR-8/SVneo cells at a concentration of 5 × 10^5 cells/ml were plated in invasion chambers, which were immersed in 24-well cell culture plates containing 500 μl RPMI 1640 with 10% FBS (control) or the supernatant of dNK cells from the IA or RSA samples.

TABLE 1 | Details of primer sequences for real-time PCR.

Primer set	Sequence (5′–3′)
GAPDH	F:GCACCGTCAAGGCTGAGAAC R:TGGTGAAGACGCCAGTGGA
Interleukin-8	F:TTTCAGAGACAGCAGAGCACACAA R:CACACAGAGCTGCAGAAATCAGG
IP-10	F:GGCCATCAAGAATTTACTGAAAGCA R:TCTGTGTGGTCCATCCTTGGAA
Vascular endothelial growth factor	F:GAGCCTTGCCTTGCTGCTCTA R:CACCAGGGTCTCGATTGGATG
Placental growth factor	F:GAGACGGCCAATGTCACCA R:GCTGAGAGAACGTCAGCTCCA
Interferon-γ	F:CTTTAAAGATGACCAGAGCATCCAA R:GGCGACAGTTCAGCCATCAC

The plates were cultured at 37°C and 5% CO_2. After 24 h, using cotton swabs, non-invading cells were removed from the top of the Matrigel (BD Biosciences, San Jose, CA, USA). The cells that had invaded the Matrigel were fixed using 4% paraformaldehyde and dyed using 0.1% crystal violet. Non-overlapping fields at 100× magnification were analyzed using a light microscope.

Tube Formation Assay

The human umbilical vein endothelial cells (HUVECs) are isolated from pooled donors. Experiments on HUVECs were carried out at the 4–6 passages. The ability of HUVECs to form network-like structures *in vitro* was examined. 25 μl of HUVECs at a concentration of 1.2 × 10^6 cells/ml was placed in 96-well plate pre-coated with 50 μl/well Matrigel (BD Biosciences, Bedford, MA, USA), and 25 μl of supernatants of dNK cells from the two groups were added to 96-well plate and incubated at 37°C, 5% CO_2. The HUVECs were photographed microscopically to evaluate the extent of tube formation after 6 h. Non-overlapping fields at 40× magnification were analyzed using a light microscope.

HTR-8/SVneo Cell Line Culture and Transfection

HTR-8/SVneo cells were cultured in RPMI 1640 medium containing 10% FBS. The transfection methods were used as described previously (30). Briefly, 1 × 10^6 HTR-8/SVneo cells were plated in 6-well plates and transfected with miR-133a mimics, inhibitor, or negative control using Lipofectamine 2000 (Invitrogen, Carlsbad, CA, USA) according to the manufacturer's instructions, when the cells reached 50–60% confluence.

Western Blot Analysis

The methods of western blot analysis were described previously (28). Briefly, HTR-8/SVneo cells transfected with miRNAs were washed by PBS and lysed by RIPA lysis buffer. Samples consisting of 50 μg total protein were loaded onto an SDS-PAGE gel (P0012AC, Beyotime Biotechnology, China) and transferred electrophoretically to nitrocellulose membranes (LC2000, Invitrogen). After blocking with Tris-buffered saline with Tween-20 (TBST) containing 5% milk powder, the membranes were incubated with the appropriate primary antibody against HLA-G (MEM-G/1, ab7759; Abcam, Cambridge, UK, 1:500) or tubulin (ab6161, Abcam, Cambridge, UK, 1:2,000), anti-mouse IgG horseradish peroxidase (HRP)-conjugated secondary antibody (GE Healthcare, Piscataway, NJ, USA) was added for 1 h.

The blots were developed using Immobilon Western HRP Substrate Luminol Reagent (Millipore, Billerica, MA, USA). The quantification of HLA-G relative to tubulin expression within each sample was determined using the QuantiOne imaging software (Bio-Rad, USA).

Measurement of Soluble HLA-G Production in Culture Supernatants

The methodology to measure HLA-G using ELISA was described previously. Culture supernatants of HTR-8/SVneo cells transfected with miR-133a were harvested. Soluble HLA-G was quantified using MEM-G/9 (ab7758, Abcam, Cambridge, UK), which recognizes the most abundant soluble isoforms and anti-human β2-microglobulin as capture and detection antibodies, respectively (31). The plates were incubated for 30 min with substrate and absorbance was measured at 450 nm using a Benchmark microplate reader (Bio-Rad, USA).

Cocultures of dNK Cells and HTR-8/SVneo

The coculture system of fresh isolated $CD56^{+}CD94^{+}$ dNK cells (5×10^5) of IA patients and HTR-8/SVneo cells (5×10^5) transfected with miR-133a relative sequence (1:1 ratio) was established and then seeded onto 24-well culture plates in 1 ml RPMI 1640 medium containing 10% FBS without IL-15. In addition, the fresh isolated dNK cells of IA patients with HTR-8/SVneo cells in RPMI 1640 medium containing 10% FBS plus blocking antibodies for KIR2DL4 (clone mab33) or isotype controls at a concentration of 10 μg/ml were cocultured.

Statistical Analysis

The data from independent experiments were presented as the mean ± SD. Differences between two groups were analyzed by Student's *t*-test, and multiple groups were analyzed by the one-way analysis of variance. A *P*-value < 0.05 was considered to be statistically significant. All analyses were performed using SPSS 19.0 software (SPSS, Chicago, IL, USA).

RESULTS

Identification of Human Normal and Abnormal First-Trimester dNK Cells

To characterize the phenotypes of dNK cells in patients with IA and patients with RSAs, flow cytometry analysis was performed with CD45, CD3, CD56, and CD16 mAbs. To exclude the confounding fluorescent signals from other cells, only $CD45^{+}$ leukocytes were examined, and $CD3^{-}CD56^{bright}CD16^{-}$ NK cells were regarded as dNK cells (**Figure 1A**). In both IA and RSA samples, the $CD56^{bright}CD16^{-}$ NK cells were the dominant lympholeukocyte cell type in the decidua (~70%) (**Figure 1B**). The percentage of dNK cells exhibited no significant difference between the groups ($P > 0.05$, **Figure 1C**).

Lower KIR2DL4 Expression in dNK Cells in the RSA Group than in the IA Group

Cell surface receptors, including NKp44, NKp46, NKG2D, NKG2A, ILT2, ILT4, and KIR2DL4 on dNK cells in both groups, were subsequently investigated by flow cytometry. Among these receptors, ILT2, ILT4, and KIR2DL4 were the ligands of HLA-G. Six samples of the RSA group and six samples of the IA group were measured. We found that the expression of NKp44, NKp46, NKG2D, ILT2, and ILT4 on dNK cells showed no significant differences between the groups ($P > 0.05$, **Figures 2A–C**). However, the RSA group expressed lower KIR2DL4 and NKG2A than did the IA group in terms of both the percentage and mean fluorescence intensity ($P < 0.05$, **Figures 2A–C**). Considering KIR2DL4 resides predominantly in endosomes, we examined the KIR2DL4 expression with intracellular staining and found that the RSA group also expressed lower KIR2DL4 in endosomes ($P < 0.05$, **Figures 2D,E**). Additionally, KIR2DL4 expression on $CD56^{dim}$ and $CD56^{bright}$ peripheral blood NK (pNK) cells were detected both in the IA and the RSA group. The results demonstrated that $CD56^{dim}$ pNK cells scarcely expressed KIR2DL4, while $CD56^{bright}$ pNK cells have low expression of KIR2DL4 in both two groups ($P > 0.05$, **Figures 2F–H**).

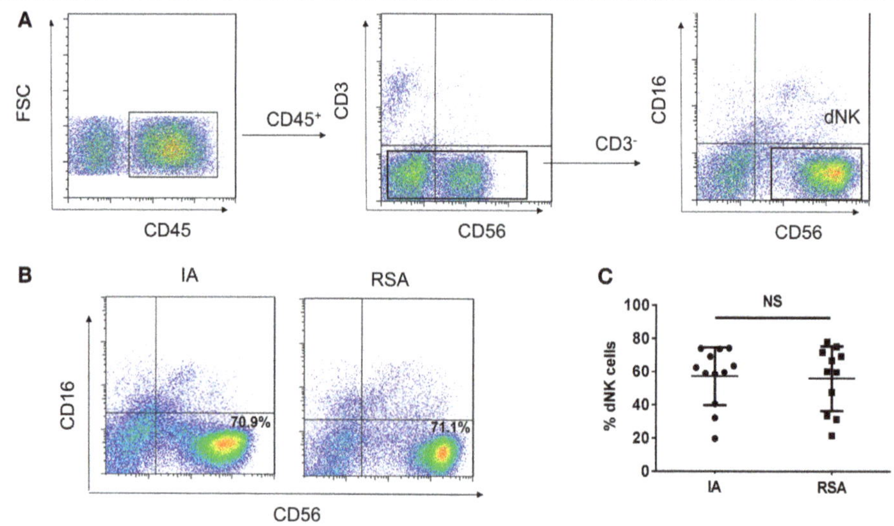

FIGURE 1 | $CD56^{bright}CD16^{-}$ natural killer (NK) cells present in large numbers in human deciduas of induced abortion (IA) group and recurrent spontaneous abortion (RSA) group. **(A)** Representative density plots showing the $CD56^{bright}$ NK (dNK) cells in gated $CD45^{+}CD3^{-}CD56^{+}CD16^{-}$ NK cells isolated from decidua in the first trimester of IA group and RSA group. **(B)** Representative density plots showing the decidual NK (dNK) cell phenotypes of the two groups. **(C)** Percentages of dNK cells in gated decidual lymphocytes. $n = 12$ and 11 for IA and RSA, respectively. The data in B are presented as the means ± SD. Student's *t*-test.

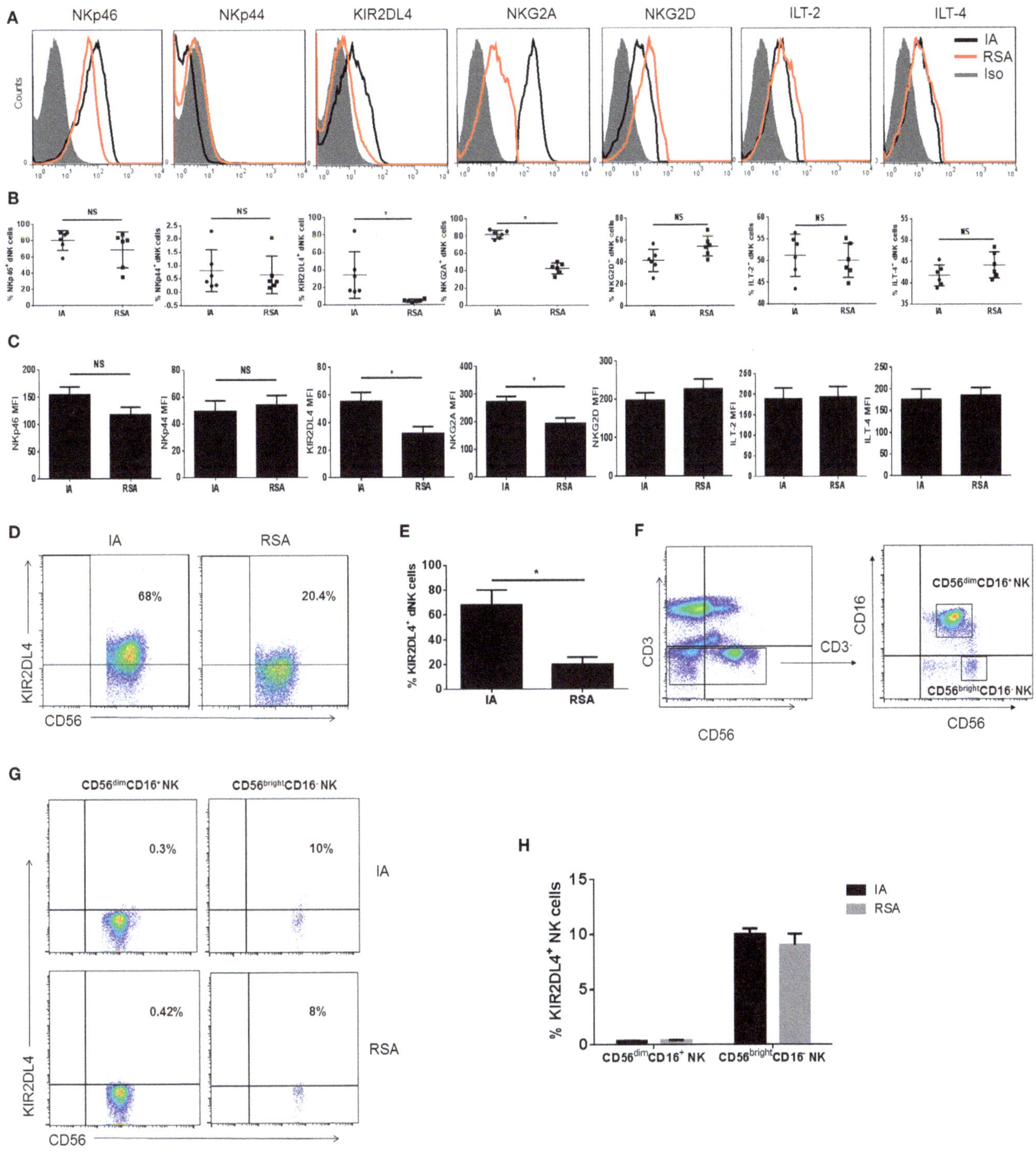

FIGURE 2 | Expression of natural killer (NK) cell receptors in decidual NK (dNK) cells of induced abortion (IA) and recurrent spontaneous abortion (RSA). **(A)** Expressions of NKp46, NKp44, killer immunoglobulin-like receptor 2DL4 (KIR2DL4), NKG2A, NKG2D, immunoglobulin-like transcript 2 (ILT2), and ILT4 on the $CD3^-CD56^+CD16^-$ dNK cells of the two groups. The gray-filled histograms represent background staining of corresponding isotype-matched control. **(B)** Percentage of NKp46, NKp44, KIR2DL4, NKG2A, NKG2D, ILT2, and ILT4 on $CD3^-CD56^+CD16^-$ dNK cells of IA and RSA. **(C)** Mean fluorescence intensity (MFI) of these receptors on $CD3^-CD56^{bright}CD16^-$ dNK cells of IA and RSA. **(D)** Representative density plots showing the KIR2DL4 expression on gated $CD3^-CD56^{bright}CD16^-$ dNK cells of IA and RSA with intracellular staining. **(E)** Percentage of KIR2DL4 expression on gated $CD3^-CD56^{bright}CD16^-$ dNK cells of two groups with intracellular staining. **(F)** Representative density plots showing $CD56^{bright}$ and $CD56^{dim}$ NK cells in gated $CD56^+CD3^-$ NK cells isolated from peripheral blood (pNK) in the first trimester of IA group. **(G)** Representative density plots showing the KIR2DL4 expression on gated $CD3^-CD56^{bright}$ and $CD3^-CD56^{dim}$ pNK cells of IA and RSA. **(H)** Percentage of KIR2DL4 expression on gated $CD3^-CD56^{bright}$ and $CD3^-CD56^{dim}$ pNK cells of IA and RSA. $n = 6$ for each group. The data are presented as the means ± SD. $^*P < 0.01$ Student's t-test.

Differences in Secretion Functions of dNK Cells in the RSA and IA Groups

Owing to the ability of dNK cells to secrete cytokines that may affect pregnancy, the profile of the cytokine secretion of dNK cells in the two groups was examined using a multiplex cytokine assay. The $CD56^{bright}$ dNK cells were purified with immunomagnetic beads, and $CD56^{+}CD94^{+}$ dNK cells were further purified by FCM and more than 97% of dNK were $CD94^{+}CD56^{+}CD16^{-}$ NK cells in two groups (**Figure 3C**). We first examined the mRNA levels of certain cytokines, including IL-8, IP-10, VEGF, PLGF, and IFN-γ, in the decidual tissues and dNK cells in each group. We found that the IL-8 and IP-10 levels were higher in the decidual tissues of the RSA group than in the IA group but that the VEGF level in RSA was lower than that in IA ($P < 0.05$, **Figure 3A**). However, the mRNA levels of IP-10 and VEGF in $CD56^{+}CD94^{+}$ dNK cells were much lower in the RSA group ($P < 0.05$, **Figure 3D**). Then, the supernatants of decidual tissue homogenates and the supernatants of $CD56^{+}CD94^{+}$ dNK cells, which were cultured for 24 h, were harvested and used for the analysis of these cytokines. All $CD56^{+}CD94^{+}$ dNK cell cultures were supplemented with 20 ng/ml IL-15. We found significantly higher levels of IP-10 and VEGF in both tissue homogenate supernatants and dNK cell supernatants in the IA group ($P < 0.05$, **Figures 3B,E**). Meanwhile, all these cytokines in $CD94^{+}CD56^{+}CD16^{-}$ NK cells were detected by intracellular staining and analyzed by FCM. The results also demonstrated that $IP\text{-}10^{+}$ $CD56^{+}$ dNK cells and $VEGF^{+}$ $CD56^{+}$ dNK cells were significantly lower in the RSA group than the IA group ($P < 0.05$, **Figures 3F,G**).

Difference in the Proinvasive and Proangiogenesis Ability of dNK Cells in the RSA and IA Groups

To examine whether the variation of cytokines changed the proinvasive ability of dNK cells, we conducted the Transwell assay. The HUVEC tube formation assay was done to test the

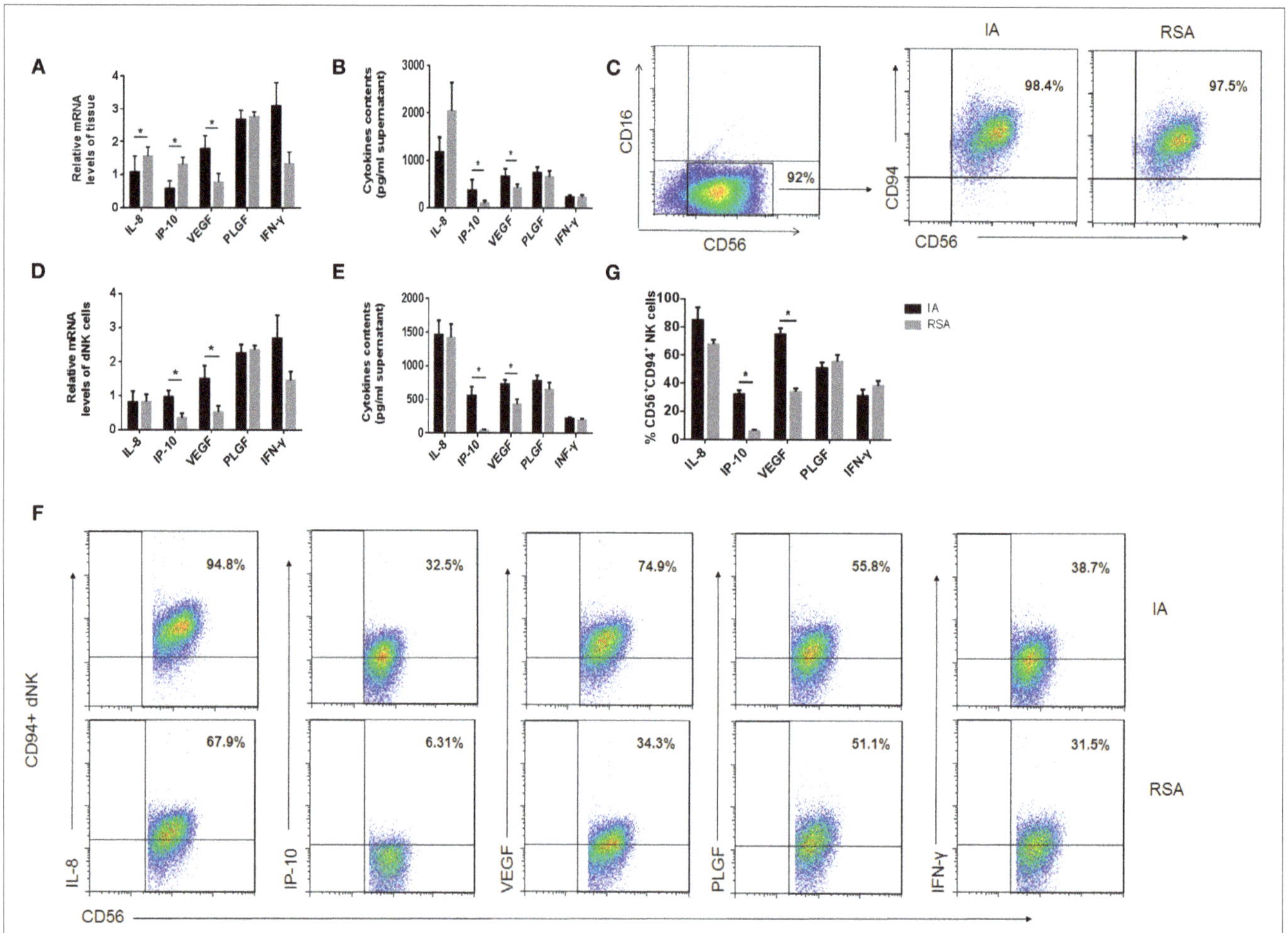

FIGURE 3 | Cytokine production by human decidual NK (dNK) cells of the two groups. **(A)** mRNA levels and **(B)** protein levels of cytokines in decidual tissues of the two groups. **(C)** Representative purity analysis of sorted $CD56^{+}CD16^{-}$ natural killer (NK) cells and $CD56^{+}CD94^{+}$ NK cells in induced abortion (IA) and recurrent spontaneous abortion (RSA). The purified $CD56^{+}CD94^{+}$ NK cells were subsequently used for real-time PCR and coculture. **(D)** mRNA levels of cytokines in fresh purified $CD56^{+}CD94^{+}$ dNK cells and **(E)** protein levels of cytokines in $CD56^{+}CD94^{+}$ dNK cultured 24 h after purification. **(F)** Representative density plots showing the cytokines in $CD56^{+}CD94^{+}$ dNK cells of IA and RSA with intracellular staining. **(G)** Percentage of cytokine expression in $CD56^{+}CD94^{+}$ dNK cells of IA and RSA, $n = 6$ for each group. The data are presented as the means ± SD. $^{*}P < 0.05$ Student's *t*-test.

proangiogenesis property of dNK cells. The CD56^{+}CD94^{+} dNK culture supernatants of the two groups were harvested and cocultured with HTR-8/SVneo cells to perform the Matrigel invasion assay. At the same time, the supernatants were cocultured with HUVECs to perform the tube formation assay. We confirmed that the CD56^{+}CD94^{+} dNK supernatants of the RSA group significantly reduced the invasive activity of HTR-8/SVneo cells compared with the IA group ($P < 0.05$, **Figures 4A,B**). The results of the tube formation assay also revealed that CD56^{+}CD94^{+} dNK supernatants of the RSA group significantly reduced the tube formation ability of HUVECs ($P < 0.05$, **Figures 4C,D**). All dNK cells were cultured with 20 ng/ml IL-15 for 24 h before harvesting the supernatants.

Decreased HLA-G Expression by miR-133a Affected dNK Function

In our previous study, we found that miR-133a was greatly overexpressed in RSA villi compared to those from IA patients, and we confirmed that miR-133a was most likely to bind to the HLA-G 3′UTR (28). To examine whether the downregulation of HLA-G by miR-133a affected the function of dNK cells, we prepared a coculture system involving HTR-8/SVneo cells transfected with miR-133a and dNK cells isolated from IA patients. We first identified the efficiency of the miR-133a transfection with real-time PCR (**Figure 5A**). Next, we confirmed the downregulation of HLA-G by miR-133a using western blot (**Figures 5B,C**). The culture supernatants were harvested 24 h later after transfection and sHLA-G concentrations were detected by ELISA. We confirmed that sHLA-G could also be downregulated by miR-133a (**Figure 5D**). To investigate the role of HLA-G in controlling dNK secretion functions, fresh CD56^{+}CD94^{+} dNK cells were cocultured with HTR-8/SVneo cells transfected with miR-133a. The cell culture medium was harvested, and the cytokines were analyzed (**Figures 5E–I**). The results of the assay showed that CD56^{+}CD94^{+} dNK cells cocultured with HTR-8/SVneo cells transfected with miR-133a mimics significantly decreased the levels of IL-8, IP-10, and VEGF ($P < 0.05$, **Figures 5E–G**). The tube formation assay revealed that the supernatants of the CD56^{+}CD94^{+} dNK cells cocultured with HTR-8/SVneo cells transfected with miR-133a mimics significantly reduced the tube formation ability of HUVECs ($P < 0.05$, **Figures 5J,K**). The Transwell assay also showed that the CD56^{+}CD94^{+} dNK cells cocultured with HTR-8/SVneo cells transfected with miR-133a mimics significantly reduced the migration ability of the HTR-8/SVneo cells ($P < 0.05$, **Figures 5L,M**).

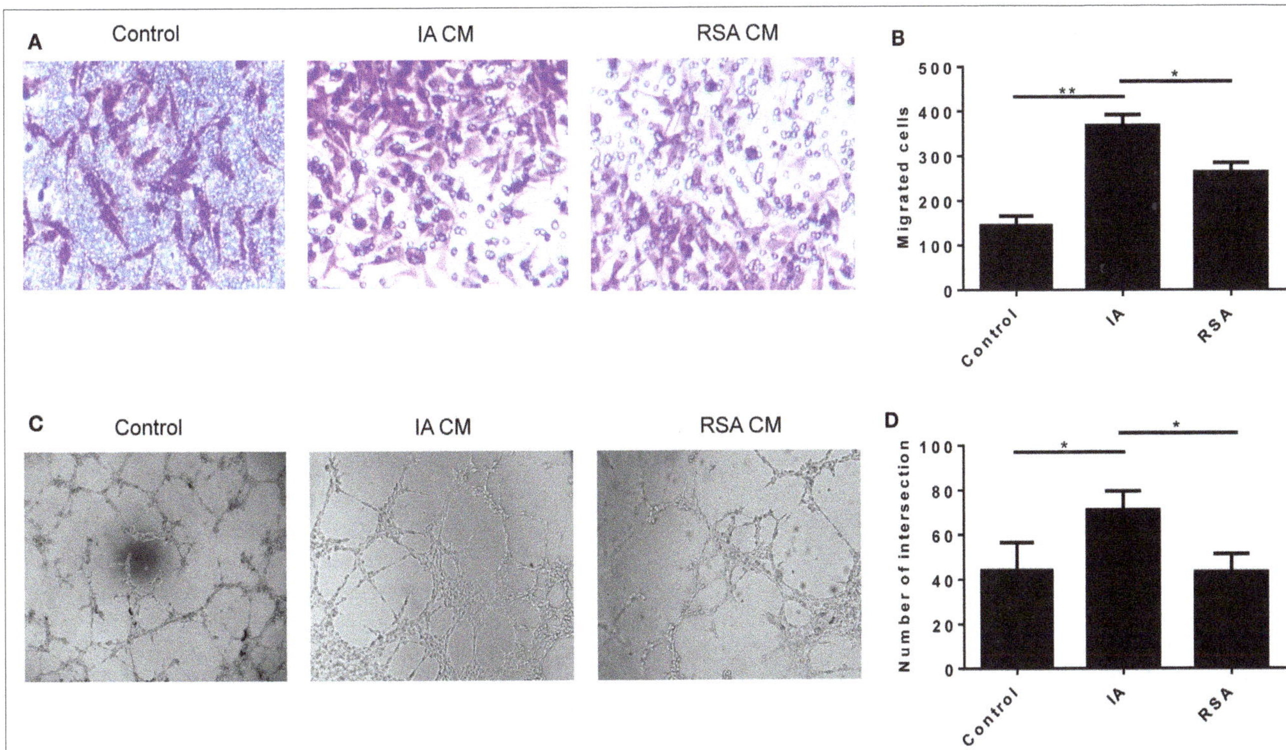

FIGURE 4 | Decidual NK (dNK) cells of recurrent spontaneous abortion (RSA) reduce HTR-8/SVneo invasion and human umbilical vein endothelial cell (HUVEC) tube formation. **(A,B)** Transwell assay showed CD56^{+}CD94^{+}dNK cells culture supernatants of the two groups affects the number of invading HTR-8/SVneo cells. **(A)** Representative images of invading HTR-8/SVneo cells which were dyed with crystal violet. Magnification, 100×. **(B)** Average number of HTR-8/SVneo cells invading across the Matrigel toward dNK cells of different groups. **(C,D)** Formation by HUVECs of tube structures *in vitro* in conditioned supernatants obtained from CD56^{+}CD94^{+} dNK cells of different groups. Values shown on each frame indicate average number of intersection. Magnification, 40×. The data are representative of three experiments. The data in panels **(B,D)** are presented as the means ± SD. $^{**}P < 0.01$ versus control and induced abortion (IA) group. $^{*}P < 0.05$ versus IA and RSA group. Student's *t*-test.

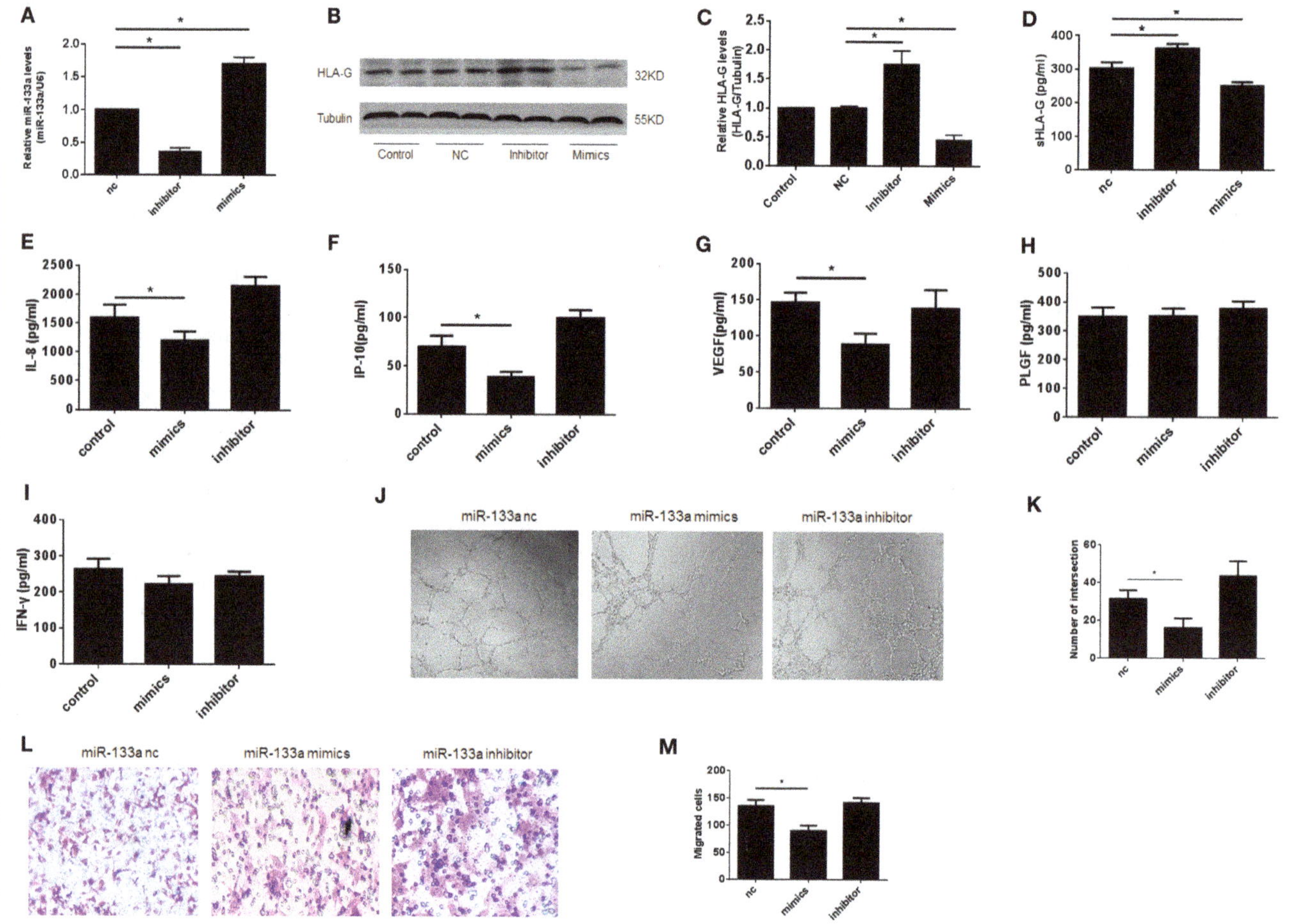

FIGURE 5 | Decreased human leukocyte antigen (HLA)-G expression by miR-133a affected the function of decidual NK (dNK), including proinvasion and proangiogenesis. Overexpression of miR-133a reduced HLA-G expression in HTR-8/SVneo cells. **(A)** Relative miR-133a expression after transfection with mimics, inhibitor, and negative control (nc) in HTR-8/SVneo cells by qRT-PCR. **(B,C)** Relative HLA-G protein levels after transfection with miR-133a by western blot. The data are representative of three experiments. **(D)** Soluble HLA-G concentration in supernatant of HTR-8/SVneo cells transfected with miR-133a for 24 h. Cell culture supernatants were analyzed for **(E)** interleukin (IL)-8, **(F)** IP-10, **(G)** vascular endothelial growth factor (VEGF), **(H)** placental growth factor (PLGF), and **(I)** INF-γ. **(J,K)** Tube formation by human umbilical vein endothelial cells *in vitro* in conditioned supernatants obtained from CD56⁺CD94⁺ dNK cells cocultured with HTR-8/SVneo transfected with miR-133a negative control (nc), inhibitor, or mimics. Magnification, 40×. **(L,M)** Number of invading HTR-8/SVneo cells in conditioned supernatants obtained from CD56⁺CD94⁺ dNK cocultured with HTR-8/SVneo transfected with miR-133a nc, inhibitor, or mimics. Magnification, 100×. The data are representative of three experiments. The data are presented as the means ± SD. $^{*}P < 0.05$. Student's *t*-test.

KIR2DL4 Was Involved in the Impairment of dNK Cell Function

To examine whether KIR2DL4 was involved in the impairment dNK cell function, we cocultured $CD56^+CD94^+$ dNK cells with HTR-8/SVneo cells plus anti-KIR2DL4 antibodies or isotypes controls at a concentration of 10 μg/ml. The results showed reduced IL-8, IP-10, and VEGF in the presence of blocking antibodies for KIR2DL4 ($P < 0.01$, **Figure 6A**). We found that the anti-KIR2DL4 antibodies reduced the invasive activity of HTR-8/SVneo cells compared with the control group ($P < 0.05$, **Figures 6B,C**). In addition, tube formation by HUVECs was observed after 6 h with supernatant stimulation. The results showed the significantly reduced tube formation capacity following the blockade of KIR2DL4 by Ab ($P < 0.05$, **Figures 6D,E**).

DISCUSSION

During pregnancy, there is complicated immunological regulation at the maternal–fetal interface that is necessary for a successful pregnancy. The dNK cells are the major component of the immune cells present in the decidua between $HLA\text{-}G^+$ extravillous cytotrophoblasts that are essential for early pregnancy. Increasing research has suggested that HLA-G and dNK cells have unique features and play important roles at the maternal–fetal interface. However, the specific mechanisms are still unknown.

Our present study demonstrated that decreased HLA-G expression by miR-133a transfection played an important role in the regulation of the secretory functions of dNK cells. We found that there were no significant differences between RSA and IA patients regarding the percentage of dNK cells in decidual lymphocytes. However, regarding to the receptors of dNK cells,

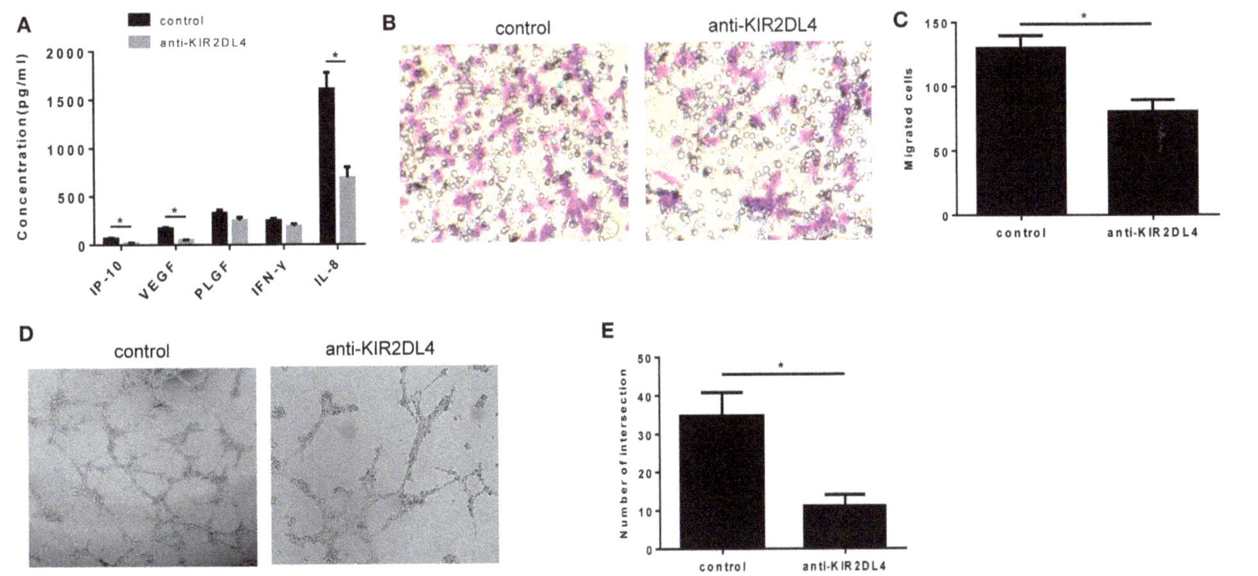

FIGURE 6 | The impairment of proinvasion and proangiogenesis function of decidual NK (dNK) after blocking killer immunoglobulin-like receptor 2DL4 (KIR2DL4). **(A)** Cell culture supernatants were analyzed for IP-10, vascular endothelial growth factor (VEGF), placental growth factor (PLGF), INF-γ, and interleukin (IL)-8. The data are representative of three experiments. **(B,C)** Number of invading HTR-8/SVneo cells in conditioned supernatants obtained from CD56$^+$CD94$^+$ dNK cells cocultured with HTR-8/SVneo cells in the presence of anti-KIR2DL4 Ab or isotype. **(D,E)** Formation by human umbilical vein endothelial cells of tube structures *in vitro* in conditioned supernatants obtained from CD56$^+$CD94$^+$ dNK cells cocultured with HTR-8/SVneo in the presence of anti-KIR2DL4 antibody or isotype. The data in panels **(C,E)** are presented as the means ± SD. $^*P < 0.05$. Student's *t*-test.

especially KIR2DL4, the expression level was lower in RSA than in IA. This finding was consistent with a previous report (32). Although KIR2DL4 has a long cytoplasmic domain and contains ITIMs, which can transduce inhibitory signals to NK cells, it could enhance dNK activity of cytokine and chemokine secretion when bound to HLA-G. KIR2DL4 could also activate endosomal signaling for a proinflammatory or proangiogenic response (22). To date, why the dNK cells of RSA patients have lower KIR2DL4 is still unknown. We speculated that hematological and immunological homeostasis changed the uterine microenvironment so as to disturb the differentiation of dNK cells in RSA patients.

As we know, dNK cells are CD56^{bright}CD16$^-$ NK cells which overexpress CD94 have strong secretory functions. Considering the human decidua contain group 3 ILC which express CD56 and NCR but lack CD94/NKG2A and KIR, and these NCR$^+$ ILC3 cells could also produce IL-8. We separated CD56$^+$CD94$^+$ dNK cells by flow cytometry for excluding ILC3 cells. We confirmed that CD94$^+$CD56$^+$ NK cells could secret certain cytokines, such as IL-8, IP-10, and VEGF, etc. Thus, we examined whether the decreased expression of KIR2DL4 could influence dNK cells' secretion capability. We found that the mRNA levels of IL-8 and IP-10 were increased, whereas the VEGF levels were decreased in the decidual tissues of the RSA group. However, the protein levels of IL-8 and VEGF were in accordance with the mRNA levels in the decidual tissues from the RSA group, whereas the protein levels of IP-10 were not the same as the mRNA levels. In addition, IP-10 and VEGF were decreased in dNK cells of RSA at both the mRNA and protein levels. Although IL-8 seemed to be increased in dNK cells of RSA, there was no significant difference. IL-8 is a chemokine that is produced by decidual cells and dNK cells (33) and can bind to two receptors, IL-8RA (CXCR1) and IL-8RB (CXCR2) (34). Both IL-8RA and IL-8RB were detected on the EVTs (35). IL-8 may lead to the release of matrix metalloproteinase-2 and MMP-9, thereby increasing the EVT and HTR-8/SVneo cell invasion (35, 36). Decreased expression of IL-8 is associated with RSA (37). Our study demonstrated that IL-8 was increased in decidual tissues but was reduced in dNK cells, so we thought that IL-8 was produced not only by dNK cells but by other decidual cells. The higher expression of IL-8 in decidual tissues might have been due to other decidual cells' compensatory mechanisms. However, our conclusion still needs to be further studied due to the limited sample size of our study. IP-10 is also one of the chemokines secreted by endometrial stromal and glandular cells (38). IP-10 can promote trophoblast invasion, but its mechanism is still unknown. We considered the supernatants of cultured dNK cells from the RSA group had weakened the invasion ability of HTR-8/SVneo cells compared with the IA group due to the decreased levels of IL-8 and IP-10, which were both proinvasion cytokines. VEGF is a strong angiogenic factor that is secreted by the decidua and was proven to be secreted by dNK cells (15, 39). VEGF is crucial for vascular growth, which is one of the processes for placental development (40). Lower VEGF expression at the maternal–fetal interface was reported to be related to RSA (41). VEGF and VEGF soluble receptor-1 (sFlt-1) in serum were both increased in RSA because of placental ischemia/hypoxia and endothelial dysfunction (42). However, another study showed that reduced serum VEGF may contribute to RSA (43). Our results demonstrated that VEGF was decreased in both the decidual tissues and dNK cells of the RSA group, which may affect the tube formation of HUVECs. In our

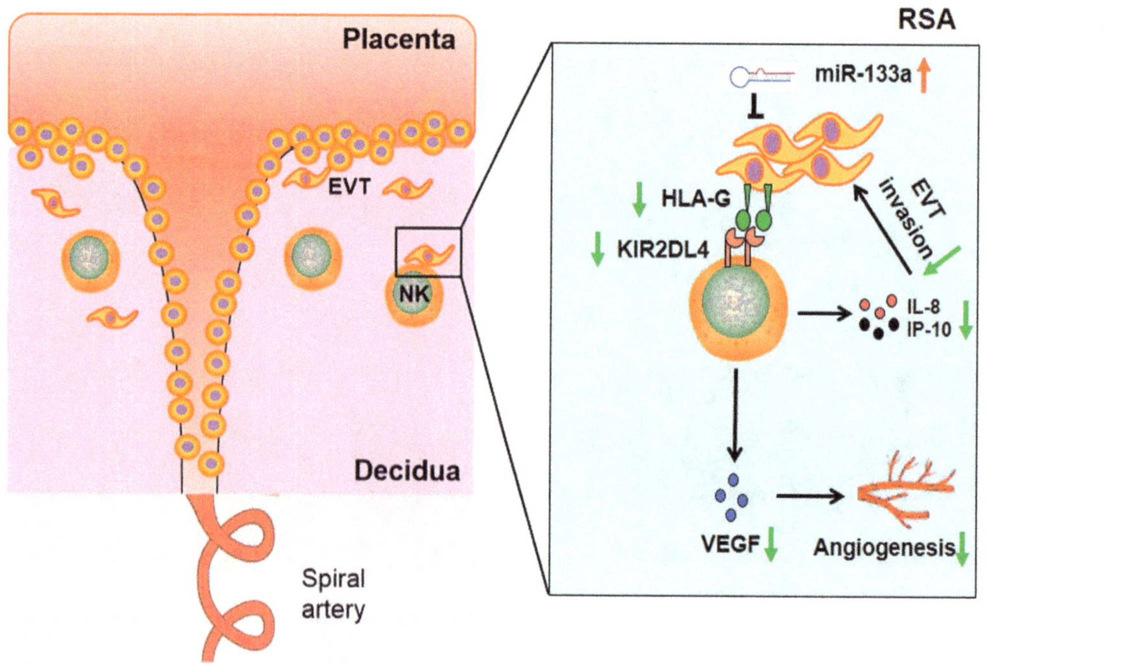

FIGURE 7 | Proposed scheme depicting the hypothesis that decreased human leukocyte antigen (HLA)-G expression impaired functions of decidual NK cells in recurrent spontaneous abortion (RSA).

studies, the decreased expression of VEGF was related to a lower tube formation ability of HUVECs.

Numerous studies have indicated that HLA-G+ extravillous cytotrophoblasts contacted by dNK cells play a crucial role in the maintenance of pregnancy (6, 44, 45). KIR2DL4, as a particular receptor for HLA-G on dNK, may play a mediation role in modulating the secretion ability of dNK cells. Our previous study found that miR-133a was highly expressed in the villi of the RSA group (28). Multi-software prediction and real-time PCR confirmed that miR-133a could bind to the 3′UTR of HLA-G (28). It was also confirmed that HLA-G (all HLA-G isoform and soluble HLA-G) was downregulated by miR-133a in HTR-8/SVneo cells. The dNK cells of IA patients, when cocultured with HTR-8/SVneo cells transfected with miR-133a mimics, decreased the expression of IL-8, IP-10, and VEGF. These cytokines in the coculture supernatants plus the anti-KIR2DL4 antibody were also significantly decreased. The results suggested that low levels of HLA-G or KIR2DL4 may influence the secretion of dNK cells in a way that affected the invasion capability of HTR-8/SVneo cells and tube formation of HUVECs. The mechanism of HLA-G influencing the secretion of dNK cells was reported that soluble HLA-G stimulate resting NK cells on endocytosis into endosomes and combined with KIR2DL4 which stimulated senescent NK cells to promote vascular remodeling and angiogenesis probably by secreting certain cytokines, such as IL-8 (16). KIR2DL4 is reported to be related to vascular remodeling and breast cancer invasion (16, 46). We know that vascular remodeling and trophoblast cell invasion are two key processes of early pregnancy. Trophoblast cells have invasion abilities just like tumor cells. Our study found that KIR2DL4 could also influence trophoblast cell lines' invasion, demonstrating that KIR2DL4 on dNK is crucial for HLA-G interfaced dNK cells and the maintenance of a successful pregnancy.

In conclusion, our current study showed that lower KIR2DL4 expression on dNK cells of RSA group could downregulate the proinvasion and proangiogenic cytokine secretion of dNK cells. Furthermore, it was indicated that decreased HLA-G expression by miR-133a in the trophoblast cell line, HTR-8/SVneo could influence the secretion ability of dNK cells when bound to KIR2DL4. The decreased cytokines could affect trophoblast invasion and angiogenesis (**Figure 7**). Our findings provide a possible mechanism of RSA and a basis for further study; in addition, this study may provide a potential drug target for therapy of RSA.

ETHICS STATEMENT

This study was approved by the Institutional Review Board, Tang Du Hospital, Fourth Military Medical University. All subjects gave written informed consent in accordance with the Declaration of Helsinki.

AUTHOR CONTRIBUTIONS

The mainly task of the research, including experiment and sorting results, was completed by WG. LC and XW were guiding the research and offering the funds supporting collectively. LF contributed to the quality control of the experiment and the article writing. BL, XX, SC, and JW contributed to offer the experiment site and technical supporting selflessly. Especially, gratitude is due to FY for her assistance in specimen collection.

ACKNOWLEDGMENTS

XW and LC are supported by grant No. 81370710 and 81571531, respectively, financed by the National Natural Science Foundation of China.

REFERENCES

1. Practice Committee of the American Society for Reproductive Medicine. Evaluation and treatment of recurrent pregnancy loss: a committee opinion. *Fertil Steril* (2012) 98(5):1103–11. doi:10.1016/j.fertnstert.2012.06.048
2. Ford HB, Schust DJ. Recurrent pregnancy loss: etiology, diagnosis, and therapy. *Rev Obstet Gynecol* (2009) 2(2):76–83.
3. Pandey MK, Rani R, Agrawal S. An update in recurrent spontaneous abortion. *Arch Gynecol Obstet* (2005) 272(2):95–108. doi:10.1007/s00404-004-0706-y
4. Dahl M, Hviid TV. Human leucocyte antigen class Ib molecules in pregnancy success and early pregnancy loss. *Hum Reprod Update* (2012) 18(1):92–109. doi:10.1093/humupd/dmr043
5. Djurisic S, Hviid TV. HLA class Ib molecules and immune cells in pregnancy and preeclampsia. *Front Immunol* (2014) 5:652. doi:10.3389/fimmu.2014.00652
6. Tilburgs T, Evans JH, Crespo AC, Strominger JL. The HLA-G cycle provides for both NK tolerance and immunity at the maternal-fetal interface. *Proc Natl Acad Sci U S A* (2015) 112(43):13312–7. doi:10.1073/pnas.1517724112
7. Zidi I, Rizzo R, Bouaziz A, Laaribi AB, Zidi N, Di Luca D, et al. sHLA-G1 and HLA-G5 levels are decreased in Tunisian women with multiple abortion. *Hum Immunol* (2016) 77(4):342–5. doi:10.1016/j.humimm.2016.01.019
8. Rebmann V, Konig L, Nardi Fda S, Wagner B, Manvailer LF, Horn PA. The potential of HLA-G-bearing extracellular vesicles as a future element in HLA-G immune biology. *Front Immunol* (2016) 7:173. doi:10.3389/fimmu.2016.00173
9. Vivier E, Raulet DH, Moretta A, Caligiuri MA, Zitvogel L, Lanier LL, et al. Innate or adaptive immunity? The example of natural killer cells. *Science* (2011) 331(6013):44–9. doi:10.1126/science.1198687
10. Giuliani E, Parkin KL, Lessey BA, Young SL, Fazleabas AT. Characterization of uterine NK cells in women with infertility or recurrent pregnancy loss and associated endometriosis. *Am J Reprod Immunol* (2014) 72(3):262–9. doi:10.1111/aji.12259
11. Moretta L, Montaldo E, Vacca P, Del Zotto G, Moretta F, Merli P, et al. Human natural killer cells: origin, receptors, function, and clinical applications. *Int Arch Allergy Immunol* (2014) 164(4):253–64. doi:10.1159/000365632
12. Mori M, Bogdan A, Balassa T, Csabai T, Szekeres-Bartho J. The decidua-the maternal bed embracing the embryo-maintains the pregnancy. *Semin Immunopathol* (2016) 38(6):635–49. doi:10.1007/s00281-016-0574-0
13. Lima PD, Zhang J, Dunk C, Lye SJ, Croy BA. Leukocyte driven-decidual angiogenesis in early pregnancy. *Cell Mol Immunol* (2014) 11(6):522–37. doi:10.1038/cmi.2014.63
14. Parham P, Moffett A. Variable NK cell receptors and their MHC class I ligands in immunity, reproduction and human evolution. *Nat Rev Immunol* (2013) 13(2):133–44. doi:10.1038/nri3370
15. Hanna J, Goldman-Wohl D, Hamani Y, Avraham I, Greenfield C, Natanson-Yaron S, et al. Decidual NK cells regulate key developmental processes at the human fetal-maternal interface. *Nat Med* (2006) 12(9):1065–74. doi:10.1038/nm1452
16. Rajagopalan S, Long EO. Cellular senescence induced by CD158d reprograms natural killer cells to promote vascular remodeling. *Proc Natl Acad Sci U S A* (2012) 109(50):20596–601. doi:10.1073/pnas.1208248109
17. Fu B, Li X, Sun R, Tong X, Ling B, Tian Z, et al. Natural killer cells promote immune tolerance by regulating inflammatory TH17 cells at the human maternal-fetal interface. *Proc Natl Acad Sci U S A* (2013) 110(3):E231–40. doi:10.1073/pnas.1206322110
18. Zhang J, Dunk CE, Kwan M, Jones RL, Harris LK, Keating S, et al. Human dNK cell function is differentially regulated by extrinsic cellular engagement and intrinsic activating receptors in first and second trimester pregnancy. *Cell Mol Immunol* (2015) 14(2):203–13. doi:10.1038/cmi.2015.66
19. Li YH, Zhou WH, Tao Y, Wang SC, Jiang YL, Zhang D, et al. The galectin-9/Tim-3 pathway is involved in the regulation of NK cell function at the maternal-fetal interface in early pregnancy. *Cell Mol Immunol* (2016) 13(1):73–81. doi:10.1038/cmi.2014.126
20. Vacca P, Montaldo E, Vitale C, Croxatto D, Moretta L, Mingari MC. MSC and innate immune cell interactions: a lesson from human decidua. *Immunol Lett* (2015) 168(2):170–4. doi:10.1016/j.imlet.2015.05.006
21. Doisne JM, Balmas E, Boulenouar S, Gaynor LM, Kieckbusch J, Gardner L, et al. Composition, development, and function of uterine innate lymphoid cells. *J Immunol* (2015) 195(8):3937–45. doi:10.4049/jimmunol.1500689
22. Rajagopalan S, Long EO. KIR2DL4 (CD158d): an activation receptor for HLA-G. *Front Immunol* (2012) 3:258. doi:10.3389/fimmu.2012.00258
23. Rajalingam R. Overview of the killer cell immunoglobulin-like receptor system. *Methods Mol Biol* (2012) 882:391–414. doi:10.1007/978-1-61779-842-9_23
24. Brusilovsky M, Radinsky O, Yossef R, Campbell KS, Porgador A. Carbohydrate-mediated modulation of NK cell receptor function: structural and functional influences of heparan sulfate moieties expressed on NK cell surface. *Front Oncol* (2014) 4:185. doi:10.3389/fonc.2014.00185
25. Shiroishi M, Tsumoto K, Amano K, Shirakihara Y, Colonna M, Braud VM, et al. Human inhibitory receptors Ig-like transcript 2 (ILT2) and ILT4 compete with CD8 for MHC class I binding and bind preferentially to HLA-G. *Proc Natl Acad Sci U S A* (2003) 100(15):8856–61. doi:10.1073/pnas.1431057100
26. Lewis BP, Burge CB, Bartel DP. Conserved seed pairing, often flanked by adenosines, indicates that thousands of human genes are microRNA targets. *Cell* (2005) 120(1):15–20. doi:10.1016/j.cell.2004.12.035
27. Lim LP, Lau NC, Garrett-Engele P, Grimson A, Schelter JM, Castle J, et al. Microarray analysis shows that some microRNAs downregulate large numbers of target mRNAs. *Nature* (2005) 433(7027):769–73. doi:10.1038/nature03315
28. Wang X, Li B, Wang J, Lei J, Liu C, Ma Y, et al. Evidence that miR-133a causes recurrent spontaneous abortion by reducing HLA-G expression. *Reprod Biomed Online* (2012) 25(4):415–24. doi:10.1016/j.rbmo.2012.06.022
29. Fu Q, Tao Y, Piao H, Du MR, Li DJ. Trophoblasts and decidual stromal cells regulate decidual NK cell functions via interaction between collagen and LAIR-1. *Am J Reprod Immunol* (2014) 71(4):368–78. doi:10.1111/aji.12211
30. Ma Y, Gong J, Liu Y, Guo W, Jin B, Wang X, et al. MicroRNA-30c promotes natural killer cell cytotoxicity via up-regulating the expression level of NKG2D. *Life Sci* (2016) 151:174–81. doi:10.1016/j.lfs.2016.03.012
31. Rebmann V, Lemaoult J, Rouas-Freiss N, Carosella ED, Grosse-Wilde H. Report of the wet workshop for quantification of soluble HLA-G in essen, 2004. *Hum Immunol* (2005) 66(8):853–63. doi:10.1016/j.humimm.2005.05.003
32. Yan WH, Lin A, Chen BG, Zhou MY, Dai MZ, Chen XJ, et al. Possible roles of KIR2DL4 expression on uNK cells in human pregnancy. *Am J Reprod Immunol* (2007) 57(4):233–42. doi:10.1111/j.1600-0897.2007.00469.x
33. Sharma S, Godbole G, Modi D. Decidual control of trophoblast invasion. *Am J Reprod Immunol* (2016) 75(3):341–50. doi:10.1111/aji.12466
34. Holmes WE, Lee J, Kuang WJ, Rice GC, Wood WI. Structure and functional expression of a human interleukin-8 receptor. *Science* (1991) 253(5025):1278–80. doi:10.1126/science.1840701
35. De Oliveira LG, Lash GE, Murray-Dunning C, Bulmer JN, Innes BA, Searle RF, et al. Role of interleukin 8 in uterine natural killer cell regulation of extravillous trophoblast cell invasion. *Placenta* (2010) 31(7):595–601. doi:10.1016/j.placenta.2010.04.012
36. Jovanovic M, Stefanoska I, Radojcic L, Vicovac L. Interleukin-8 (CXCL8) stimulates trophoblast cell migration and invasion by increasing levels of matrix metalloproteinase (MMP)2 and MMP9 and integrins alpha5 and beta1. *Reproduction* (2010) 139(4):789–98. doi:10.1530/REP-09-0341
37. Pitman H, Innes BA, Robson SC, Bulmer JN, Lash GE. Altered expression of interleukin-6, interleukin-8 and their receptors in decidua of women with sporadic miscarriage. *Hum Reprod* (2013) 28(8):2075–86. doi:10.1093/humrep/det233
38. Dominguez F, Martinez S, Quinonero A, Loro F, Horcajadas JA, Pellicer A, et al. CXCL10 and IL-6 induce chemotaxis in human trophoblast cell lines. *Mol Hum Reprod* (2008) 14(7):423–30. doi:10.1093/molehr/gan032

39. Andraweera PH, Dekker GA, Roberts CT. The vascular endothelial growth factor family in adverse pregnancy outcomes. *Hum Reprod Update* (2012) 18(4):436–57. doi:10.1093/humupd/dms011
40. Red-Horse K, Zhou Y, Genbacev O, Prakobphol A, Foulk R, McMaster M, et al. Trophoblast differentiation during embryo implantation and formation of the maternal-fetal interface. *J Clin Invest* (2004) 114(6):744–54. doi:10.1172/JCI22991
41. Banerjee P, Ghosh S, Dutta M, Subramani E, Khalpada J, Roychoudhury S, et al. Identification of key contributory factors responsible for vascular dysfunction in idiopathic recurrent spontaneous miscarriage. *PLoS One* (2013) 8(11):e80940. doi:10.1371/journal.pone.0080940
42. Pang L, Wei Z, Li O, Huang R, Qin J, Chen H, et al. An increase in vascular endothelial growth factor (VEGF) and VEGF soluble receptor-1 (sFlt-1) are associated with early recurrent spontaneous abortion. *PLoS One* (2013) 8(9):e75759. doi:10.1371/journal.pone.0075759
43. Almawi WY, Saldanha FL, Mahmood NA, Al-Zaman I, Sater MS, Mustafa FE. Relationship between VEGFA polymorphisms and serum VEGF protein levels and recurrent spontaneous miscarriage. *Hum Reprod* (2013) 28(10):2628–35. doi:10.1093/humrep/det308
44. Tilburgs T, Crespo AC, van der Zwan A, Rybalov B, Raj T, Stranger B, et al. Human HLA-G+ extravillous trophoblasts: immune-activating cells that interact with decidual leukocytes. *Proc Natl Acad Sci U S A* (2015) 112(23):7219–24. doi:10.1073/pnas.1507977112
45. Rajagopalan S. HLA-G-mediated NK cell senescence promotes vascular remodeling: implications for reproduction. *Cell Mol Immunol* (2014) 11(5):460–6. doi:10.1038/cmi.2014.53
46. Ueshima C, Kataoka TR, Hirata M, Furuhata A, Suzuki E, Toi M, et al. The killer cell Ig-like receptor 2DL4 expression in human mast cells and its potential role in breast cancer invasion. *Cancer Immunol Res* (2015) 3(8):871–80. doi:10.1158/2326-6066.CIR-14-0199

12

Host Immune Responses in HIV-1 Infection: The Emerging Pathogenic Role of Siglecs and their Clinical Correlates

*Joanna Mikulak[1,2], Clara Di Vito[1], Elisa Zaghi[1] and Domenico Mavilio[1,3]**

[1] *Unit of Clinical and Experimental Immunology, Humanitas Clinical and Research Center, Rozzano, Italy,* [2] *Istituto di Ricerca Genetica e Biomedica, UOS di Milano, Consiglio Nazionale delle Ricerche (UOS/IRGB/CNR), Rozzano, Italy,* [3] *Department of Medical Biotechnologies and Translational Medicine (BioMeTra), University of Milan, Milan, Italy*

Edited by:
Eric Vivier, Centre d'immunologie de Marseille-Luminy, France

Reviewed by:
Vincent Vieillard, Centre National de la Recherche Scientifique (CNRS), France
Catharina C. Gross, University of Münster, Germany
Jacques Zimmer, Centre de Recherche Public de la Santé, Luxembourg

***Correspondence:**
Domenico Mavilio
domenico.mavilio@unimi.it

A better understanding of the mechanisms employed by HIV-1 to escape immune responses still represents one of the major tasks required for the development of novel therapeutic approaches targeting a disease still lacking a definitive cure. Host innate immune responses against HIV-1 are key in the early phases of the infection as they could prevent the development and the establishment of two hallmarks of the infection: chronic inflammation and viral reservoirs. Sialic acid-binding immunoglobulin-like lectins (Siglecs) belong to a family of transmembrane proteins able to dampen host immune responses and set appropriate immune activation thresholds upon ligation with their natural ligands, the sialylated carbohydrates. This immune-modulatory function is also targeted by many pathogens that have evolved to express sialic acids on their surface in order to escape host immune responses. HIV-1 envelope glycoprotein 120 (gp120) is extensively covered by carbohydrates playing active roles in life cycle of the virus. Indeed, besides forming a protecting shield from antibody recognition, this coat of N-linked glycans interferes with the folding of viral glycoproteins and enhances virus infectivity. In particular, the sialic acid residues present on gp120 can bind Siglec-7 on natural killer and monocytes/macrophages and Siglec-1 on monocytes/macrophages and dendritic cells. The interactions between these two members of the Siglec family and the sialylated glycans present on HIV-1 envelope either induce or increase HIV-1 entry in conventional and unconventional target cells, thus contributing to viral dissemination and disease progression. In this review, we address the impact of Siglecs in the pathogenesis of HIV-1 infection and discuss how they could be employed as clinic and therapeutic targets.

Keywords: host-pathogen interactions, NK cells, monocytes/macrophages, dendritic cells

INTRODUCTION

Immunoglobulin-like receptors constitutively present on cellular surface play a key role in the immune recognition of both cell targets and pathogens due to their high plasticity that makes them easily adaptable to adopt an infinite range of molecular structures. In this regard, glycans expressed by several microbes are recognized by several immunoglobulin type-I lectins, including sialic acid-binding immunoglobulin-like lectins (Siglecs). Siglecs belong to a family of transmembrane proteins that display a variable numbers of C2-set immunoglobulin domains (from 1 in CD33 to 16 in Siglec-1), an amino-terminal V-set immunoglobulin domain that recognizes sialic acid residues (1–3) and a conserved arginine residue that binds carbohydrates. Fifteen different functional Siglecs have been identified in humans, including Siglec-12 that has lost the ability to bind sialic acids. They can be classified in two main subgroups: the first one comprises the evolutionary conserved Siglecs-1, -2, -4, and -15 that differ from the second cluster of rapidly evolving Siglecs. This latter type-I lectin subgroup includes Siglec-3, -5 to -11, -14, and -16 sharing structural similarities with the first identified CD33/Siglec-3 (4–6). Siglecs are constitutively expressed mostly on cells derived from the hematopoietic lineage and belonging to innate immunity, such as monocytes/macrophages, neutrophils, basophils, eosinophils, dendritic cells (DCs), and natural killer (NK) cells (5). In contrast, adaptive immune cells such as T lymphocytes have very low or undetectable surface levels of these type-I lectins (7). Of note, the expression of distinct member of Siglec family can be either restricted to specific immune cells (i.e., Siglec-1 and -11 constitutively present only on macrophages) (5, 8, 9) or shared by several myeloid and lymphoid populations (i.e., Siglec-7 and -9 expressed on NK cells, monocytes and macrophages or Siglec-5 on monocytes, B cells, and neutrophils) (5, 10–12).

In regard to Siglec binding to their putative ligands, each one of these lectin-type molecules has a preferential specificity for sialic acid residues, and the recognition depends on both glycan structures and glycan modifications associated with sulfation, acetylation, and glycosylation reactions (13, 14). Sialic acids comprise a family of more than 50 carbohydrates that share a nine-carbon backbone (C1-9) representing the terminal unit of glycoproteins and glycolipids (gangliosides) phylogenetic conserved among different species including humans (15). Siglec-binding sites are usually masked at the cellular surface due to *cis* interactions with highly expressed low affinity sialic acids. Their unmasking in response to cellular activation or sialidase treatment cleaves the *cis* low affinity ligands and allows free interactions in *trans* with highly glycosylated ligands. Nonetheless, *trans* interactions can also occur when Siglecs encounter other cells or pathogens expressing higher affinity ligands competing with *cis* (16, 17). Following the binding with these sugars, Siglecs regulate and control inflammatory responses by setting appropriate thresholds of immune activation (5, 18–20). Indeed, Siglec cytosolic tails mostly include immune-receptor tyrosine-based inhibitory motifs (ITIMs) or ITIM-like domains, whose engagement inhibits leukocyte proliferation, induces apoptosis, modulates endocytosis and the production of inflammatory cytokines.

In the context of host–pathogen interactions, sialic acids and their glycan residues expressed on the envelopes of several pathogens serve as recognition molecules interacting with lectin-type molecules such as Siglecs. This is one of the established mechanisms of viral entry in immune cells, as demonstrate with highly pathogenic A, B, and C strains of influenza that encode hemagglutinin enriched with sialic acid residues (21, 22). The engagement of Siglecs in response to their recognition of microbial glycan residues has been first described as a mechanism to dampen host immune responses following the encounter with inflammatory stimuli (23–26). Later on, it became evident that pathogens took advantage of the Siglec recognition by evolving and expressing sialic acids on their surface in order to escape host inflammatory responses (20, 27). More recently, a novel and emerging working hypothesis postulates that CD33-related Siglecs also evolved in response to pathogen manipulation by associating their intracellular domains with either immune-receptor tyrosine-based activation motifs (ITAM) or unconventional activating from of ITIMs in Siglec-7 (17, 20, 28–30).

Several lines of evidence indicate that also HIV-1 employs sialic acid associated-mechanisms to facilitate its viral entry in target cells. Indeed, HIV-1 envelope (env) is equipped with an N-terminal glycoprotein 120 (gp120) heavily conjugated with sialic acids residues that, besides forming a protecting shield from antibody (Ab) recognition, also enhance virus infectivity and dissemination by binding sugar-binding receptors expressed on host immune cells (31–39). Indeed, several human lectin-type receptors including galectins (40, 41), defensins (42–44), and others (45–51) have been shown to bind HIV-1 env by recognizing glycans expressed on gp120. The identification and characterization of these sialic acid-binding receptors also explain the different susceptibility to HIV-1 to infect cells expressing different repertoires of Siglecs (52). Although these lectin-type receptors were identified in the early 1980s, only recently their interactions with HIV-1 have been disclosed. In particular, Siglec-1 (53–55) and Siglec-7 (52, 56) have been reported to bind HIV-1 envelope, thus inducing and/or facilitating HIV-1 infection and disease progression. This review provides an updated summary in regard the pathogenic role of Siglecs and also discusses how they could be employed as clinic and therapeutic targets.

KINETICS OF SIGLECS DURING THE COURSE OF HIV-1 INFECTION

Human NK cells are known as effector-cells capable of killing potentially dangerous non-self targets (i.e., viral-infected, tumor-transformed, or allogeneic cells) and produce pro-inflammatory cytokines such as interferon-γ (IFN-γ) upon activation and in the absence of a prior sensitization to specific antigens. NK cell recognition of autologous cell targets is mediated by a large family of inhibitory NK cell receptors (iNKRs) including killer cell immunoglobulin-like receptors and C-type lectins that recognize different alleles of major histocompatibility complex of class I (MHC-I) (i.e., missing-self hypothesis). Decreased expression or absence of self-MHC-I on viral-infected or tumor-transformed or allogeneic cells trigger NK cell effector functions *via* the engagement of large family of activating NKRs (aNKRs) that bind their putative ligands expressed on cellular targets (57–59). Even though the inhibitory signals are considered dominant over the activating ones, the dynamic balance between iNKRs and aNKRs is key for the regulation of NK-cell-mediated clearance of viruses and tumors (60–62). NK cells are $CD3^{neg}/CD19^{neg}/CD14^{neg}$ lymphocytes whose phenotype is defined by the expression of CD56 and CD16 (56, 59). In healthy subjects, the largest subset of circulating NK cells (up to 90%) is $CD56^{dim}/CD16^{pos}$, has a high cytolytic potential, and can also produce high levels of pro-inflammatory and antiviral cytokines upon stimulation. The second circulating subset of immune-regulatory CD56^{bright}/CD16$^{dim-neg}$ NK cells (up to 10–15%) is able to secrete high amount of pro-inflammatory cytokines while displaying poor cytotoxicity (59, 63, 64).

The first studies assessing the frequencies and the functions of NK cells in HIV-1 infected patients with high levels of ongoing viral replication reported an overall decrease of both their percentage and absolute number in the peripheral blood. Either apoptosis or cellular migration to peripheral tissues has been proposed as possible mechanisms driving the depletion of circulating NK cells during the course of HIV-1 infection (65, 66). Only later, it became evident that HIV-1 viremia does not reduce the absolute number of circulating NK cells but rather induce a pathologic redistribution of their subsets since from the early stages of the infection. Indeed, others and we identified and characterized an anergic population of CD56^{neg}/CD16^{pos} (CD56^{neg}) NK cells expanded in viremic patients and counterbalancing the reduction of the CD56^{dim}/CD16^{pos} cytolytic NK cell subset. An early depletion of the immune-regulatory CD56^{bright}/CD16^{neg} NK cells has been also described (64, 67–69). This subset of CD56^{neg} NK cells in HIV-1 infection is characterized by an imbalanced repertoire of NKRs associated with a remarkably impaired cytotoxic activity and secretion of antiviral cytokines/chemokines (68, 70). These pathological features explain, at least in part, the impairments of CD56^{neg} NK cells in the clearance of both tumor-transformed and viral-infected cell targets as well as their defective interactions with autologous DCs (71–73). The mechanisms driving the expansion of this pathologic CD56^{neg} NK cells during the course of HIV-1 infection remain unknown, although several lines of evidence indicate that this phenomenon is not HIV-1 specific. Indeed, high frequencies of CD56^{neg} NK cells have been reported in other viral infections [such as human cytomegalovirus or hepatitis C virus (HCV)] and autoimmune diseases (such as myasthenia gravis or dermatomyositis) (64). Hence, it is conceivable to hypothesize that persistent inflammation characterizing several human chronic immunological diseases might be one of the factors inducing the expansion of CD56^{neg} NK cells.

The down-modulation of CD56 is not the only NK surface marker whose surface level is altered by high levels of ongoing HIV-1 replication. In fact, also the expressions as well as the functional relevance of several aNKRS and iNKRs are affected by HIV-1 viremia (62, 74). Indeed, these markedly dysfunctional CD56^{neg} NK cell subsets express high levels of iNKRS and low levels of aNKRs, a phenomenon that contributes, at least in part, to a defective control of HIV-1 replication and to disease progression (64, 72). The HIV-1-induced pathologic modulation of CD56 and NKR surface expression does not occur during the acute phase of the infection but require a chronic exposure of NK cells to chronic viral replication (75). In contrast, the down-modulation of Siglec-7, a iNKRs constitutively expressed on the majority of NK cells, is detectable since from the earliest phases of HIV-1 infection and precedes the phenotypic changes of CD56 and NKRs on NK cells (56). This phenomenon makes it possible to identify two distinct subsets present in different stages of HIV-1 infection: the Siglec-7^{neg}/CD56^{dim} NK population only present in the early phases of HIV-1 infection and the Siglec-7^{neg}/CD56^{neg} NK cell subset that become detectable several months later when the disease enters in its chronic phase. All these HIV-1-induced phenotypic changes are reversible following the administration of a successful antiretroviral therapy (ART) and the restoration of Siglec-7 expression on NK cells is much faster compared to that of CD56, aNKRs, and iNKRs. The impact of viral replication in inducing these phenotypic abnormalities is also confirmed by the experimental evidence showing that NK cells from those HIV-1-infected patients with a low/undetectable levels of viral replication that do not progress to AIDS (i.e., long-term non-progressors or LTNP) are similar and undistinguishable from the ones of uninfected healthy individuals (56). Although Siglec-7 is an iNKRs containing ITIM-like domains in its cytosolic tail, it has been also reported that Siglec-7^{pos} NK cells are more functionally relevant compared to their Siglec-7^{neg} counterparts. Indeed, the expression of Siglec-7 in NK cells is associated with higher levels of aNKRs (i.e., CD16, NKp30, NKp46) and lower levels of iNKRs (i.e., CD94/NKG2A, CD158b) (76). This likely explains the fact that NK cells still retain a certain degree of antiviral and antitumor functions in the acute phases of HIV-1 infection, while become anergic and highly dysfunctional only in the chronic stages of the disease featuring the expansion of the Siglec-7^{neg}/CD56^{neg} NK cells (56).

A soluble form of Siglec-7 (sSiglec-7) resulted to be increased in the sera of HIV-1-infected patients and could serve as an additional clinical biomarker to monitor disease progression. In fact, higher serum levels of sSiglec-7 directly correlate with HIV-1 viremia and parallel the expansion of Siglec-7^{neg} NK cells in HIV-1-infected patients, thus suggesting that the continuous exposure to ongoing viral replication induce the shedding of Siglec-7 receptor from NK cell surface (52, 56). Among circulating immune cells constitutively expressing this C-lectin-type molecule (i.e., NK cells and monocytes) (5), the main source of sSiglec-7 seems to be NK cells because a subset of Siglec-7^{neg} monocytes has never been reported during the course of HIV-1 infection (52). Similarly, it has been recently shown that also the stimulation *in vitro* of NK cells with HCV induces both a significant release of sSiglec-7 in culture supernatant and a decrease of Siglec-7 expression from NK cell surface (77). Further investigations are required to identify the cellular and molecular mechanisms that certain viruses employ to induce a shedding of this C lectin-type receptor selectively on NK cells during the acute phases of the infection.

Siglec-1 is a sialoadhesin (i.e., an adhesion molecule present on cell surface and binding sialic acid residues) constitutively expressed on monocytes and macrophages whose expression is remarkably increased on HIV-1 infected monocytes (54, 78, 79). These higher amount of Siglec-1 on CD14^{pos} cells are induced by IFN-α and IFN-γ, occur rapidly after infection, enhance their susceptibility to infection, and are maintained during the following chronic phases of HIV-1 disease. In particular, the expression of Siglec-1 on monocytes appears to be higher in AIDS patients compared to that of asymptomatic HIV-1-infected individuals and correlates with plasma viral load. These findings suggest that high frequencies of Siglec-1^{pos}/CD14^{pos} monocytes are associated with disease progression (80). In addition, higher surface levels of Siglec-1 on monocytes correlate with immune activation as the treatment of these cells with IFNs or bacterial compounds such as lipopolysaccharide increases the surface amount of this type-I lectin. This is consistent with the condition of chronic inflammation and immune activation induced by the presence of HIV-1 viremia that also triggers the expansion of a subset of CD14^{pos}/CD16^{pos}-activated monocytes (54, 78, 81–84). Hence, although having different kinetics during the course of HIV-1 infection, both the expressions of Siglec-1 and Siglec-7 are greatly influenced by viral replication and can be used as clinical tools to identify and monitor acute and chronic clinical stages of the infection.

SIGLECS IN THE PATHOGENESIS OF HIV-1 INFECTION

In addition to CD4^{pos} T lymphocytes that represent the main cell target of HIV-1 infection, the virus is able to also infect immune cells of myeloid origin such as DCs and tissue-resident macrophages (71, 85). These two latter antigen-presenting cells express both CD4 and the chemokine co-receptors CCR5 and CXCR4, which are required for a productive HIV-1 infection (86). In particular, different Siglecs have been reported to play key roles in both binding the virus and facilitating its viral entry in monocytes and macrophages due to the fact that HIV-1 env-gp120 is extensively covered by carbohydrates first processed in the endoplasmic reticulum and then further modified in the *trans*-Golgi apparatus (31–35). The high degree of sialylation on HIV-1 envelope, other than representing a protecting shield that hamper Ab recognition of viral antigens, is also highly permissive and facilitates HIV-1 infection of target cells (33, 87–89). Indeed, given the high avidity and affinity of Siglecs for sialic acid residues, the viral env-gp120 (in particular within the V1/V2 domain) represents an important Siglec-binding site (90). Different from CD4^{pos} T cells though, monocytes and macrophage are resistant to HIV-1-mediated cytopathic effects as there is no depletion of these cells either in peripheral blood or tissues of HIV-1-infected patients. Moreover, macrophage can serve as cellular storage of accumulating virions in their sub-cellular regions that are termed "virion containing compartment" and whose origin is still unclear (91, 92). The identification of VVCs gave rise among the scientific community to the "Trojan horse" metaphor meaning that the virus can be hidden in infected macrophages that, in turn, can disseminate infectious particles when interplaying with CD4^{pos} T cells and/or in response to environmental signals (i.e., cytokines or damage-associated molecular patterns). Several experimental evidence confirmed the existence of this "Trojan horse" in *in vitro* infected monocytes/macrophages, in the brain of AIDS patients with HIV encephalitis (93, 94) as well as in macaques infected with Simian immunodeficiency virus (SIV) (95, 96). It is still being debated whether HIV-1 infection of macrophages leads to the establishment of an additional reservoirs of latently infected cells contributing to boost viral replication and disease progression in patients undergoing ART interruption (97). In supports to this hypothesis, it has been reported that the functional polarization of monocyte-derived macrophages (MDM) into M1 cells upon exposure to pro-inflammatory signals (i.e., IFN-γ and TNF-α) leads to a state of quasi-latency upon HIV-1 infection (98). Moreover, recent evidence demonstrates that M2 macrophages are more permissive to HIV-infection/replication compared to their M1 counterparts. This is associated with the fact that the stimulation with macrophage colony-stimulating factor induces both *in vitro* and *in vivo* the expression on macrophage of several cellular receptors required for viral entry, including CD4, CCR5, and Siglec-1 (90, 99).

The first evidence of a direct contribution of Siglecs to HIV-1 pathogenesis has been reported in monocytes with the Siglec-1-mediated uptake of HIV-1 in a sialic acid-dependent mechanism (54). Indeed, Siglec-1 can bind HIV-1 and promote a *trans*-infection of monocytes. Unlike the other Siglecs in which the sialic acid binding sites could be masked by ligands expressed in *cis*, Siglec-1 is equipped with a larger extracellular domain that allow both cell-to-cell and cell-to-virus interactions. The discovery of this mechanism made it possible to understand that Siglec-1 is able to directly interact with HIV-1 and contributes to the spreading of HIV-1 infection (5). Moreover, Siglec-1 on macrophages can also function as an adhesion molecule carrying sialic acid conjugates to T cells (100, 101). Hence, the main current working hypothesis postulates that monocytes and macrophages interplay at the same time with viral envelope and target cells *via* Siglec-1 that bridges HIV-1 in close proximity to CD4 and chemokine co-receptors (**Figure 1**). This interaction would overcome the electrostatic repulsive forces created by the negative charges present on both virus envelope and cell surface, thereby unmasking the receptors involved in the internalization of HIV-1 and promoting viral entry. In support of this hypothesis, several studies showed that removal of sialic acids from cell surface or viral envelope with neuraminidase treatment increases the susceptibility of target cells to be infected and form cellular syncytia (33, 39, 102, 103). Moreover, a recent report demonstrated in an *in vivo* rodent model of HIV infection that the virus can disseminate over long distance in a cell-free manner in the blood and lymph and can be captured by Siglec-1^{pos} macrophages mainly located at the interface between lymphoid tissues and fluids, thus allowing and facilitating the trans-infection of lymphocytes (104).

Siglec-1 also mediates HIV-1 *trans*-infection of mature DCs (mDCs) and contributes to viral dissemination *via* the establishment of immunological synapses between mDCs and T cells in the early steps of the infection at mucosal sites (53). In this regard, intra-vaginal inoculation of SIV recruit plasmacytoid DCs to lymphoid tissue and increases tissue levels of IFN-α. In turn, this pro-inflammatory cytokine induces the maturation of DCs and their ability to mediate HIV-1 *trans*-infection of CD4^{pos} T cells and of themselves *via* Siglec-1 (105, 106). The interactions between Siglec-1 expressed on mDCs and sialic acids residues present on viral membrane or exosomes also represents a host immune mechanism employed by mDCs to process and present antigens to CD4^{pos} T cells. HIV-1 evolved to take advantage of this physiologic pathway and used mDCs as a carrier of infection (53). Recently, it has been proposed that HIV-1 particles do not only parasite mDCs in their cellular compartments, but virions are also retained on surface-connected plasma membrane invaginations (107) (**Figure 1**).

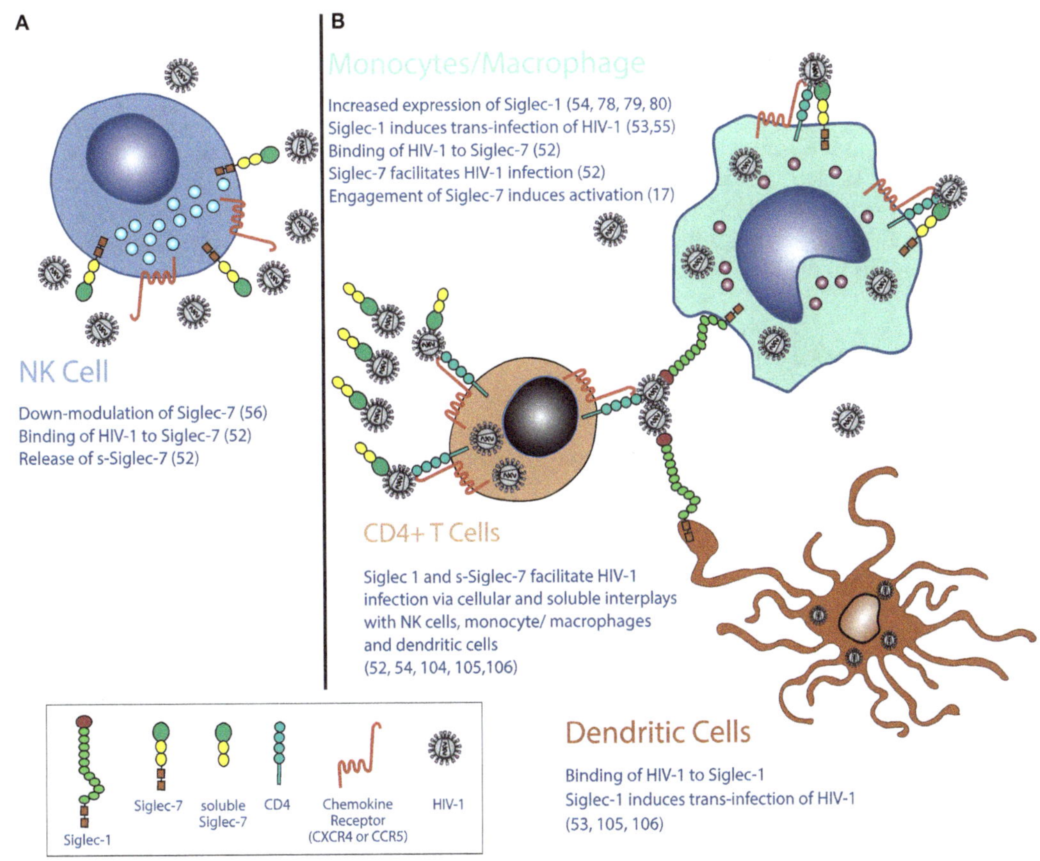

FIGURE 1 | Siglecs in the pathogenesis and clinical outcomes of HIV-1 infection. (A) Chronic HIV-1 infection leads to the downregulation of Siglec-7 expression on natural killer (NK) cells and increases plasma levels of its soluble form of Siglec-7 (sSiglec-7). **(B)** Moreover, HIV-1 monocytes/macrophages induces higher expressions of Siglec-1, CD4, and chemokine co-receptors (CCR5 and CXCR4) on monocytes and macrophages, while do not affects the constitutive high surface levels of Siglec-7 on the same cells. Siglec-1 is also expressed on mature dendritic cells (mDCs), monocytes, and macrophages, and high levels of HIV-1 viremia increase even more its surface expression on the latter two cell types. Due to their distribution on immune cells and to their ability to bind the viral envelope glycoprotein 120 heavily conjugated with sialic acids residues, Siglec-7 and Siglec-1 are highly relevant in the pathogenesis of HIV-1 infection. Indeed, after the binding of HIV-1 to Siglec-1, both mDCs and monocytes/macrophages retain virions with different mechanisms, a phenomenon enhancing the HIV-1 infection of CD4^{pos} lymphocytes. Moreover, the HIV-1 uptake by Siglec-1 and Siglec-7 further contribute to the spreading of infection by using both DCs and monocyte/macrophages as cellular-infected carriers of the virus. The shedding of Siglec-7 from NK cells in the presence of high levels of HIV-1 viremia also significantly increases HIV-1 viral entry in Siglec-7^{neg}/CD4^{pos} T cells. Finally, the engagement of Siglec-7 on monocytes and macrophages promotes cellular activation in monocytes and macrophages and contributes to the establishment of persistent inflammation in HIV-1-infected patients. These abnormalities in Siglec phenotype and functions in HIV-1 infection can be used as clinical tools to monitor disease progression or can be targeted to develop innovative therapeutic approaches to limit viral spreading and disease progression.

Monocyte-derived macrophages express different Siglecs that, other than binding HIV-1, have been also reported to enhance the susceptibility of MDMs to be infected by the same virus (52, 55). Siglec-1 leads this process by serving as the most effective MDM surface receptor up-taking HIV-1 envelope, even though its degree of expression on these cells is significantly lower compared to that of Siglec-3 and -9. As previously mentioned, this is likely due to the fact that Siglec-1 has a larger size with 17 immunoglobulin domains size and is present in an un-masked state, while Siglec-3 and -9 are masked on MDM cell surface and display a great affinity toward gp120 only after neuraminidase treatment (55). It has been recently reported that also Siglec-7 acts as a membrane-bound molecule that contributes to enhance HIV-1 infection of MDMs (52). Several studies showed that both monocytes and macrophages constitutively express high levels of Siglec-7 (5, 10, 20, 24, 25, 52), while only one report claiming a very low expression of this type-I molecule on MDMs (53). Similar to Siglec-1, Siglec-7 binds gp120 present on viral envelope and increases the viral entry of HIV-1 in MDMs. Indeed, the masking of Siglec-7 greatly reduce the amount of HIV-RNA in MDMs, thus decreasing their susceptibility to infection (52) (**Figure 1**).

Siglec-7 has been recently reported of being able to increase the susceptibility to HIV-1 infection even its soluble form (52). Indeed, Siglec-7-Fc fusion protein significantly increases HIV-1 viral entry in Siglec-7^{neg}/CD4^{pos} T cells. This phenomenon is highly relevant in the course of active and chronic HIV-1 infection since AIDS patients show increased serum levels of sSiglec-7. Again, although a direct experimental evidence is still lacking, it has been hypothesized that sSiglec-7 is shed from the cellular surface on NK cells since high levels of HIV-1 replication induce an early decrement of Siglec-7 specifically on this population and not on in monocytes/macrophages and CD8^{pos} T cells, which are the other two immune cells constitutively expressing this type-I lectin (10, 52, 56). Even though the mechanisms, the kinetic, and the pathogenic relevance of the sSiglec-7-mediated infection of CD4^{pos} T cells are still unknown, it is conceivable to hypothesize that this c-lectin-type molecule could serve as a highly efficient and T-trophic carrier of the virus in the bloodstream and tissue sites (**Figure 1**).

SIGLEC-7 AND INFLAMMATION IN HIV-1 INFECTION

Since from the very beginning of HIV-1 pandemic, it became clear that the occurrence of a persistent inflammation is a hallmark of the disease together with the establishment of viral reservoirs (108, 109). Chronic inflammation also leads to a long-term activation of immune system even in the presence of ART and low levels of viremia (110). Several HIV-1-mediated mechanisms have been reported to induce an aberrant state of long-lasting inflammation, including the pathologic translocation of commensal micro-flora from the mucosal interfaces to both lamina propria and mesenteric lymph nodes in response to the massive depletion of mucosal CD4^{pos} T occurring during the acute phases of the infection (111). Macrophages, other then constitutively expressing several members of the Siglec family, also represent the sentinels of the host innate immune system against pathogens at mucosal interface (5, 112). In the case of HIV-1 infection, the persistent engagement of inflammatory pathway in macrophages chronically challenged by the virus itself or by several opportunistic pathogens likely contributes to induce a chronic state of immune activation. Indeed, it has been reported that Siglec-7 on MDMs is able to bind different pathogens either expressing sialic acid residues (i.e., HIV-1, K1 strain of *Escherichia coli* and *Candida albicans)* or not expressing syalilated glycans (i.e., K12 strain of *E. coli* and *Zymosan* yeast particles) (17, 52). Interestingly and in contrast with the reported inhibitory function of Siglec-7 in regulating the homeostasis and/or the effector functions of T and NK cells (113–115), the engagement of this type-I lectin on monocytes induces the production of several pro-inflammatory cytokines such as IL-6, IL-1α, CCL4, IL-8, and TNF-α (17). As a matter of fact, although coupled with ITIMs in its cytoplasmc domain, Siglec-7 promotes activation of monocytes *via* the engagement of an unconventional activating signal transduction pathway of these inhibitory motifs. It is well-known that ITIMs usually induce the phosphorylation of Src family kinases that, in turn, recruit phosphotyrosine phosphatase (PTP) such as Src homology region 2 domain-containing phosphatase-1 (SHP-1) and SHP-2. These PTPs switches off cellular pathways by dephosphorylating tyrosine residues and also prevents the activation signals that originate from receptors coupled with ITAMs (116). Recently, the interpretation of this dichotomy in signal transduction has been revised due to the identification of new transmembrane receptors containing ITIMs. Indeed, several studies demonstrated that, in some circumstances and depending on the cellular context, ITIMs can propagate activation signals and ITAMs can mediate inhibition (117–124). One example is given by the dendritic-cell-associated C-type lectin 2, an ITIM bearing receptor, whose cross-linking with an anti-DCAL-2 monoclonal Ab (mAb) recruits a SHP-2 domain that, instead of negatively regulating cell signaling, is involved in activating the ERK pathway and in inducing the production of IL-10, TNF-α, IL-6, and CCL5 (117). This phenomenon is also occurring with Siglec-7 isoform selectively expressed on monocyte and bearing "activating" ITIMs inducing the phosphorylation of ERK. This latter well-known signaling pathway triggers various transcription factors modulating many genes involved in the immune activation (125). Therefore, the encounter of different pathogens by Siglec-7 expressed on monocytes and macrophages likely provide an additional mechanism contributing to the appearance of persistent inflammation during the course of HIV infection.

Besides inducing the production of pro-inflammatory cytokines and chemokines in monocytes and macrophages, the cross-linking of Siglec-7 also induces an up-modulation of ICAM-1 and CD49e. These two adhesion molecules are known to play a relevant role in the trans-endothelial migration of leukocytes during inflammatory responses following their binding with either the β2 integrin lymphocyte function-associated antigen (LFA-1) or macrophage 1 antigen expressed on endothelial cells. Moreover, the interaction between ICAM-1 and LFA-1 is able to deliver a co-stimulatory signal to T cells (126). Therefore, the engagement of Siglec-7 has also the potential to regulate the migration of other inflammatory cells toward tissue sites, where they deliver co-stimulatory signals. This is highly relevant at gut and cervix mucosal interfaces, as these two tissue sites highly enriched of macrophages represent the main gates of HIV-1 entrance (127, 128). The activation and the migration of monocytes to tissues can lead to cellular activation and plastic polarization toward pro-inflammatory M1 or anti-inflammatory M2 macrophages on the basis of different signals and stimuli delivered within the local microenvironment (112). Further investigations are required to better understand the impact of HIV-1 binding to Siglecs constitutively expressed on macrophages in the modulation of their plasticity toward a classic or alternative polarization, a process that is known to be highly affected and dysfunctional during the course of HIV-1 infection (85). This is highly relevant in the context of the synergic interplays by NK cells and macrophages aimed to induce and optimize the links between innate and adaptive immune responses. In particular, the NK cell-macrophage cross talk occurring through mechanisms requiring both cell-to-cell contacts and soluble mediators is key in the context of host–pathogen interactions (129–131) and during the course of viral infections (132). Indeed, it has been recently reported that M1 macrophages can prime the activation and the effector functions of NK cells that, in turn, are able to rescue M2 toward pro-inflammatory M1 polarization (133). This virtuous loop appears to be completed disrupted during the course of chronic and active HIV-1 infections that is characterized by a macrophage deactivation or M2 polarization and by the expansion of anergic CD56^{neg} NK cells also impaired in the priming of autologous DCs (73, 75, 85). The cellular and molecular mechanisms behind the appearance of markedly dysfunctional NK cells and macrophages are still unknown and not even the selective down-modulation of Siglec-7 on NK cells has been clarified. Another important pathogenic element to elucidate is why the chronic HIV-1 replication induces the polarization of M2 macrophages that are also more susceptible to the infection through the engagement of several surface receptors including Siglec-7 (52, 90, 99). Finally, we do no know why HIV-1 binding to Siglec-7 does not induce a the polarization of macrophage toward M1 with a consequent priming of autologous NK cell able to better function as antiviral effectors. The elucidation of all these points will help us to better understand the basis of innate immune dysfunctions during the course of HIV-1 infection.

SIGLECS AS POSSIBLE THERAPEUTIC TARGETS IN HIV-1 INFECTION

Targeting Siglecs as novel immunotherapy approaches has been proposed soon after the identification of Siglec-3 (CD33) and Siglec-2 (CD22) as biological markers of myeloid leukemias and B cell lymphomas, respectively (4). Indeed, the anti-Siglec-3 mAb Gemtuzumab ozogamicin is approved for treatment of acute myeloid leukemia, while Epratuzumab, the mAb targeting Siglec-2, is currently in clinical trials for treatment of acute lymphoblastic leukemia, follicular non-Hodgkins lymphoma, and autoimmune diseases (134–136). The identification of novel members within the Siglec family and the recent new insights disclosing the pathogenic relevance of this type-I lectins in several model of human diseases shed new light for developing innovative therapeutic approaches targeting the immune-modulatory functions of these transmembrane receptors. In particular, the existence of peculiar functions exerted by several Siglecs in distinct immune cell populations make these lectin-type molecules suitable candidates for therapies modulating host–pathogen interactions and cellular signal transduction.

In the context of HIV-1 infection, other then using Siglec-1 and Siglec-7 as possible clinical biomarkers, we can take advantage of their ability of binding HIV-1 envelope and enhance/induce the susceptibility to infections of both conventional and unconventional cell targets (52–55). Indeed, the use of pharmaceutical compounds or mAbs blocking the interactions of HIV-1 with either Siglec-1 or Siglec-7 can potentially limit HIV-1 dissemination in several cellular compartments (i.e., monocytes/macrophages, DCs, and CD4^{pos} T cells). This approach is particularly relevant during acute infection in order to damp the ability of HIV-1 to infect immune cells, to slow down the CD4^{pos} T cell depletion especially at mucosal interfaces and to put a break to the early establishment of immune activation and viral reservoirs. Furthermore, inhibiting HIV-1 uptake by Siglec-1 would also limit the infection of circulating monocytes that upregulate the surface expression of this type-I lectin in the acute phases of the infection (78, 79). Another possible therapeutic target is sSiglec-7 that is increased in the sera of AIDS patients and enhances the degree of infection of CD4^{pos} T cells (52). The elucidation of the mechanism(s) inducing the down-modulation of Siglec-7 on NK cell surface and inducing the shedding of its soluble form in AIDS patients' sera will certainly make it possible to develop strategies that will limit the ability of sSiglec-7 to spread the infection due to its capacity to carrier HIV-1 in periphery and enhance the susceptibility to infection of Siglec-7^{neg}/CD4^{pos} T cell targets.

In the context of developing a vaccine strategy, it has been also reported that Siglec-1 can be targeted to improve antigen presentation to T cells on the basis of its effective and rapid endocytic activity following interaction with pathogens (137, 138). Further investigations are needed to assess whether such approach can be used for HIV-1. Finally, it is also possible to take advantage of the immune-modulatory potential of Siglecs by employing their natural ligands. In this regard, it has been recently reported that high-affinity synthetized glycans targeting Siglec-2 expressed on B cells compete in *cis* with the natural ligands and induce endocytosis and killing of B cells (139). A similar approach could be used to inhibit the interactions between Siglecs and HIV-1. Particularly, attractive from a pharmaceutical perspective would be the use of lyposomal nanoparticles bearing Siglec glycan ligands (140).

CONCLUDING REMARKS

Among the several mechanisms that HIV-1 has developed to evade host immune defenses there is the heavily conjugation on Gp120 of viral envelope with sialic acids residues that bind Siglec-1 and -7 mostly expressed on NK cells, macrophages, and DCs. These interactions are relevant for HIV-1 survival, serve as molecular mimicry for host immune responses, avoid immune attack, and facilitate viral entry. Indeed, the binding of HIV-1 to these type-I lectins induce and/or enhance the infections of target cells is associated with inflammatory pathways and correlated with disease progression (**Figure 1**). These novel insights disclosing the key roles played by Siglecs in the pathogenesis of HIV-1 infection open new avenues also for clinical perspectives. Indeed, this transmembrane type-I lectin could be either used as biological marker of disease or therapeutically targeted to limit viral dissemination and the establishment of viral reservoirs. Siglec-7 and Siglec-1 dysfunctions in HIV-1 infection are also highly relevant to advance our knowledge in regard to the physiology and physiopathology of NK cells either alone or in the context of their interplays with both monocytes/macrophages and DCs.

AUTHOR CONTRIBUTIONS

JM and DM wrote the manuscript and compiled the citations. CV and EZ contributed to write the manuscript and drew the figures.

REFERENCES

1. Powell LD, Varki A. I-type lectins. *J Biol Chem* (1995) 270(24):14243–6. doi:10.1074/jbc.270.24.14243
2. Angata T, Brinkman-Van der Linden E. I-type lectins. *Biochim Biophys Acta* (2002) 1572(2–3):294–316. doi:10.1016/S0304-4165(02)00316-1
3. Varki A, Angata T. Siglecs – the major subfamily of I-type lectins. *Glycobiology* (2006) 16(1):1R–27R. doi:10.1093/glycob/cwj008
4. O'Reilly MK, Paulson JC. Siglecs as targets for therapy in immune-cell-mediated disease. *Trends Pharmacol Sci* (2009) 30(5):240–8. doi:10.1016/j.tips.2009.02.005
5. Crocker PR, Paulson JC, Varki A. Siglecs and their roles in the immune system. *Nat Rev Immunol* (2007) 7(4):255–66. doi:10.1038/nri2056
6. Cao H, Lakner U, de Bono B, Traherne JA, Trowsdale J, Barrow AD. SIGLEC16 encodes a DAP12-associated receptor expressed in macrophages that evolved from its inhibitory counterpart SIGLEC11 and has functional and non-functional alleles in humans. *Eur J Immunol* (2008) 38(8):2303–15. doi:10.1002/eji.200738078
7. Razi N, Varki A. Masking and unmasking of the sialic acid-binding lectin activity of CD22 (Siglec-2) on B lymphocytes. *Proc Natl Acad Sci U S A* (1998) 95(13):7469–74. doi:10.1073/pnas.95.13.7469
8. O'Neill AS, van den Berg TK, Mullen GE. Sialoadhesin – a macrophage-restricted marker of immunoregulation and inflammation. *Immunology* (2013) 138(3):198–207. doi:10.1111/imm.12042
9. Angata T, Kerr SC, Greaves DR, Varki NM, Crocker PR, Varki A. Cloning and characterization of human Siglec-11. A recently evolved signaling molecule that can interact with SHP-1 and SHP-2 and is expressed by tissue macrophages, including brain microglia. *J Biol Chem* (2002) 277(27):24466–74. doi:10.1074/jbc.M202833200
10. Nicoll G, Ni J, Liu D, Klenerman P, Munday J, Dubock S, et al. Identification and characterization of a novel siglec, siglec-7, expressed by human natural killer cells and monocytes. *J Biol Chem* (1999) 274(48):34089–95. doi:10.1074/jbc.274.48.34089
11. Zhang JQ, Nicoll G, Jones C, Crocker PR. Siglec-9, a novel sialic acid binding member of the immunoglobulin superfamily expressed broadly on human blood leukocytes. *J Biol Chem* (2000) 275(29):22121–6. doi:10.1074/jbc.M002788200
12. Cornish AL, Freeman S, Forbes G, Ni J, Zhang M, Cepeda M, et al. Characterization of siglec-5, a novel glycoprotein expressed on myeloid cells related to CD33. *Blood* (1998) 92(6):2123–32.
13. Yamaji T, Teranishi T, Alphey MS, Crocker PR, Hashimoto Y. A small region of the natural killer cell receptor, Siglec-7, is responsible for its preferred binding to alpha 2,8-disialyl and branched alpha 2,6-sialyl residues. A comparison with Siglec-9. *J Biol Chem* (2002) 277(8):6324–32. doi:10.1074/jbc.M110146200
14. Hartnell A, Steel J, Turley H, Jones M, Jackson DG, Crocker PR. Characterization of human sialoadhesin, a sialic acid binding receptor expressed by resident and inflammatory macrophage populations. *Blood* (2001) 97(1):288–96. doi:10.1182/blood.V97.1.288
15. Schauer R. Sialic acids as regulators of molecular and cellular interactions. *Curr Opin Struct Biol* (2009) 19(5):507–14. doi:10.1016/j.sbi.2009.06.003
16. Crocker PR, Varki A. Siglecs in the immune system. *Immunology* (2001) 103(2):137–45. doi:10.1046/j.0019-2805.2001.01241.x
17. Varchetta S, Brunetta E, Roberto A, Mikulak J, Hudspeth KL, Mondelli MU, et al. Engagement of Siglec-7 receptor induces a pro-inflammatory response selectively in monocytes. *PLoS One* (2012) 7(9):e45821. doi:10.1371/journal.pone.0045821
18. McMillan SJ, Crocker PR. CD33-related sialic-acid-binding immunoglobulin-like lectins in health and disease. *Carbohydr Res* (2008) 343(12):2050–6. doi:10.1016/j.carres.2008.01.009
19. Crocker PR, Redelinghuys P. Siglecs as positive and negative regulators of the immune system. *Biochem Soc Trans* (2008) 36(Pt 6):1467–71. doi:10.1042/BST0361467
20. Cao H, Crocker PR. Evolution of CD33-related siglecs: regulating host immune functions and escaping pathogen exploitation? *Immunology* (2011) 132(1):18–26. doi:10.1111/j.1365-2567.2010.03368.x
21. Neu U, Bauer J, Stehle T. Viruses and sialic acids: rules of engagement. *Curr Opin Struct Biol* (2011) 21(5):610–8. doi:10.1016/j.sbi.2011.08.009
22. Skehel JJ, Wiley DC. Receptor binding and membrane fusion in virus entry: the influenza hemagglutinin. *Annu Rev Biochem* (2000) 69:531–69. doi:10.1146/annurev.biochem.69.1.531
23. Jacobs T, Erdmann H, Fleischer B. Molecular interaction of Siglecs (sialic acid-binding Ig-like lectins) with sialylated ligands on *Trypanosoma cruzi*. *Eur J Cell Biol* (2010) 89(1):113–6. doi:10.1016/j.ejcb.2009.10.006
24. von Gunten S, Bochner BS. Basic and clinical immunology of Siglecs. *Ann N Y Acad Sci* (2008) 1143:61–82. doi:10.1196/annals.1443.011
25. Avril T, Floyd H, Lopez F, Vivier E, Crocker PR. The membrane-proximal immunoreceptor tyrosine-based inhibitory motif is critical for the inhibitory signaling mediated by Siglecs-7 and -9, CD33-related Siglecs expressed on human monocytes and NK cells. *J Immunol* (2004) 173(11):6841–9. doi:10.4049/jimmunol.173.11.6841
26. Ravetch JV, Lanier LL. Immune inhibitory receptors. *Science* (2000) 290(5489):84–9. doi:10.1126/science.290.5489.84
27. Pillai S, Netravali IA, Cariappa A, Mattoo H. Siglecs and immune regulation. *Annu Rev Immunol* (2012) 30:357–92. doi:10.1146/annurev-immunol-020711-075018
28. Turnbull IR, Colonna M. Activating and inhibitory functions of DAP12. *Nat Rev Immunol* (2007) 7(2):155–61. doi:10.1038/nri2014
29. Takaki R, Watson SR, Lanier LL. DAP12: an adapter protein with dual functionality. *Immunol Rev* (2006) 214:118–29. doi:10.1111/j.1600-065X.2006.00466.x
30. Hamerman JA, Lanier LL. Inhibition of immune responses by ITAM-bearing receptors. *Sci STKE* (2006) 2006(320):re1. doi:10.1126/stke.3202006re1
31. Geyer H, Holschbach C, Hunsmann G, Schneider J. Carbohydrates of human immunodeficiency virus. Structures of oligosaccharides linked to the envelope glycoprotein 120. *J Biol Chem* (1988) 263(24):11760–7.
32. Kozarsky K, Penman M, Basiripour L, Haseltine W, Sodroski J, Krieger M. Glycosylation and processing of the human immunodeficiency virus type 1 envelope protein. *J Acquir Immune Defic Syndr* (1989) 2(2):163–9.
33. Hu H, Shioda T, Moriya C, Xin X, Hasan MK, Miyake K, et al. Infectivities of human and other primate lentiviruses are activated by desialylation of the virion surface. *J Virol* (1996) 70(11):7462–70.
34. Kornfeld R, Kornfeld S. Assembly of asparagine-linked oligosaccharides. *Annu Rev Biochem* (1985) 54:631–64. doi:10.1146/annurev.bi.54.070185.003215
35. Olofsson S, Eriksson S, Karlsson A, Oberg B. The HIV replication inhibitor 3'-fluoro-3'-deoxythymidine blocks sialylation of N-linked oligosaccharides. *Antiviral Res* (1992) 19(1):71–80. doi:10.1016/0166-3542(92)90057-C
36. Quinones-Kochs MI, Buonocore L, Rose JK. Role of N-linked glycans in a human immunodeficiency virus envelope glycoprotein: effects on protein function and the neutralizing antibody response. *J Virol* (2002) 76(9):4199–211. doi:10.1128/JVI.76.9.4199-4211.2002
37. Wolk T, Schreiber M. N-Glycans in the gp120 V1/V2 domain of the HIV-1 strain NL4-3 are indispensable for viral infectivity and resistance against antibody neutralization. *Med Microbiol Immunol* (2006) 195(3):165–72. doi:10.1007/s00430-006-0016-z
38. Pantophlet R, Burton DR. GP120: target for neutralizing HIV-1 antibodies. *Annu Rev Immunol* (2006) 24:739–69. doi:10.1146/annurev.immunol.24.021605.090557
39. Sun J, Barbeau B, Sato S, Tremblay MJ. Neuraminidase from a bacterial source enhances both HIV-1-mediated syncytium formation and the virus binding/entry process. *Virology* (2001) 284(1):26–36. doi:10.1006/viro.2001.0889
40. Mercier S, St-Pierre C, Pelletier I, Ouellet M, Tremblay MJ, Sato S. Galectin-1 promotes HIV-1 infectivity in macrophages through stabilization of viral adsorption. *Virology* (2008) 371(1):121–9. doi:10.1016/j.virol.2007.09.034
41. Ouellet M, Mercier S, Pelletier I, Bounou S, Roy J, Hirabayashi J, et al. Galectin-1 acts as a soluble host factor that promotes HIV-1 infectivity through stabilization of virus attachment to host cells. *J Immunol* (2005) 174(7):4120–6. doi:10.4049/jimmunol.174.7.4120
42. Furci L, Sironi F, Tolazzi M, Vassena L, Lusso P. Alpha-defensins block the early steps of HIV-1 infection: interference with the binding of gp120 to CD4. *Blood* (2007) 109(7):2928–35. doi:10.1182/blood-2006-05-024489

43. Feng Z, Dubyak GR, Lederman MM, Weinberg A. Cutting edge: human beta defensin 3 - a novel antagonist of the HIV-1 coreceptor CXCR4. *J Immunol* (2006) 177(2):782–6. doi:10.4049/jimmunol.177.2.782
44. Munk C, Wei G, Yang OO, Waring AJ, Wang W, Hong T, et al. The theta-defensin, retrocyclin, inhibits HIV-1 entry. *AIDS Res Hum Retroviruses* (2003) 19(10):875–81. doi:10.1089/088922203322493049
45. Ezekowitz RA, Kuhlman M, Groopman JE, Byrn RA. A human serum mannose-binding protein inhibits in vitro infection by the human immunodeficiency virus. *J Exp Med* (1989) 169(1):185–96. doi:10.1084/jem.169.1.185
46. de Witte L, Nabatov A, Pion M, Fluitsma D, de Jong MA, de Gruijl T, et al. Langerin is a natural barrier to HIV-1 transmission by Langerhans cells. *Nat Med* (2007) 13(3):367–71. doi:10.1038/nm1541
47. Geijtenbeek TB, Kwon DS, Torensma R, van Vliet SJ, van Duijnhoven GC, Middel J, et al. DC-SIGN, a dendritic cell-specific HIV-1-binding protein that enhances trans-infection of T cells. *Cell* (2000) 100(5):587–97. doi:10.1016/S0092-8674(00)80694-7
48. Nguyen DG, Hildreth JE. Involvement of macrophage mannose receptor in the binding and transmission of HIV by macrophages. *Eur J Immunol* (2003) 33(2):483–93. doi:10.1002/immu.200310024
49. de Witte L, Bobardt M, Chatterji U, Degeest G, David G, Geijtenbeek TB, et al. Syndecan-3 is a dendritic cell-specific attachment receptor for HIV-1. *Proc Natl Acad Sci U S A* (2007) 104(49):19464–9. doi:10.1073/pnas.0703747104
50. Lambert AA, Gilbert C, Richard M, Beaulieu AD, Tremblay MJ. The C-type lectin surface receptor DCIR acts as a new attachment factor for HIV-1 in dendritic cells and contributes to trans- and cis-infection pathways. *Blood* (2008) 112(4):1299–307. doi:10.1182/blood-2008-01-136473
51. Mikulak J, Teichberg S, Arora S, Kumar D, Yadav A, Salhan D, et al. DC-specific ICAM-3-grabbing nonintegrin mediates internalization of HIV-1 into human podocytes. *Am J Physiol Renal Physiol* (2010) 299(3):F664–73. doi:10.1152/ajprenal.00629.2009
52. Varchetta S, Lusso P, Hudspeth K, Mikulak J, Mele D, Paolucci S, et al. Sialic acid-binding Ig-like lectin-7 interacts with HIV-1 gp120 and facilitates infection of CD4pos T cells and macrophages. *Retrovirology* (2013) 10:154. doi:10.1186/1742-4690-10-154
53. Izquierdo-Useros N, Lorizate M, Puertas MC, Rodriguez-Plata MT, Zangger N, Erikson E, et al. Siglec-1 is a novel dendritic cell receptor that mediates HIV-1 trans-infection through recognition of viral membrane gangliosides. *PLoS Biol* (2012) 10(12):e1001448. doi:10.1371/journal.pbio.1001315
54. Rempel H, Calosing C, Sun B, Pulliam L. Sialoadhesin expressed on IFN-induced monocytes binds HIV-1 and enhances infectivity. *PLoS One* (2008) 3(4):e1967. doi:10.1371/journal.pone.0001967
55. Zou Z, Chastain A, Moir S, Ford J, Trandem K, Martinelli E, et al. Siglecs facilitate HIV-1 infection of macrophages through adhesion with viral sialic acids. *PLoS One* (2011) 6(9):e24559. doi:10.1371/journal.pone.0024559
56. Brunetta E, Fogli M, Varchetta S, Bozzo L, Hudspeth KL, Marcenaro E, et al. The decreased expression of Siglec-7 represents an early marker of dysfunctional natural killer-cell subsets associated with high levels of HIV-1 viremia. *Blood* (2009) 114(18):3822–30. doi:10.1182/blood-2009-06-226332
57. Ljunggren HG, Karre K. In search of the 'missing self': MHC molecules and NK cell recognition. *Immunol Today* (1990) 11(7):237–44. doi:10.1016/0167-5699(90)90097-S
58. Vivier E, Raulet DH, Moretta A, Caligiuri MA, Zitvogel L, Lanier LL, et al. Innate or adaptive immunity? The example of natural killer cells. *Science* (2011) 331(6013):44–9. doi:10.1126/science.1198687
59. Cooper MA, Fehniger TA, Caligiuri MA. The biology of human natural killer-cell subsets. *Trends Immunol* (2001) 22(11):633–40. doi:10.1016/S1471-4906(01)02060-9
60. Moretta A, Bottino C, Vitale M, Pende D, Cantoni C, Mingari MC, et al. Activating receptors and coreceptors involved in human natural killer cell-mediated cytolysis. *Annu Rev Immunol* (2001) 19:197–223. doi:10.1146/annurev.immunol.19.1.197
61. Lanier LL. NK cell recognition. *Annu Rev Immunol* (2005) 23:225–74. doi:10.1146/annurev.immunol.23.021704.115526
62. Fauci AS, Mavilio D, Kottilil S. NK cells in HIV infection: paradigm for protection or targets for ambush. *Nat Rev Immunol* (2005) 5(11):835–43. doi:10.1038/nri1711
63. De Maria A, Bozzano F, Cantoni C, Moretta L. Revisiting human natural killer cell subset function revealed cytolytic CD56(dim)CD16+ NK cells as rapid producers of abundant IFN-gamma on activation. *Proc Natl Acad Sci U S A* (2011) 108(2):728–32. doi:10.1073/pnas.1012356108
64. Lugli E, Marcenaro E, Mavilio D. NK cell subset redistribution during the course of viral infections. *Front Immunol* (2014) 5:390. doi:10.3389/fimmu.2014.00390
65. Iannello A, Debbeche O, Samarani S, Ahmad A. Antiviral NK cell responses in HIV infection: II. Viral strategies for evasion and lessons for immunotherapy and vaccination. *J Leukoc Biol* (2008) 84(1):27–49. doi:10.1189/jlb.0907649
66. Scott-Algara D, Paul P. NK cells and HIV infection: lessons from other viruses. *Curr Mol Med* (2002) 2(8):757–68. doi:10.2174/1566524023361781
67. Hu PF, Hultin LE, Hultin P, Hausner MA, Hirji K, Jewett A, et al. Natural killer cell immunodeficiency in HIV disease is manifest by profoundly decreased numbers of CD16+CD56+ cells and expansion of a population of CD16dimCD56- cells with low lytic activity. *J Acquir Immune Defic Syndr Hum Retrovirol* (1995) 10(3):331–40. doi:10.1097/00042560-199511000-00005
68. Mavilio D, Benjamin J, Daucher M, Lombardo G, Kottilil S, Planta MA, et al. Natural killer cells in HIV-1 infection: dichotomous effects of viremia on inhibitory and activating receptors and their functional correlates. *Proc Natl Acad Sci U S A* (2003) 100(25):15011–6. doi:10.1073/pnas.2336091100
69. Alter G, Teigen N, Davis BT, Addo MM, Suscovich TJ, Waring MT, et al. Sequential deregulation of NK cell subset distribution and function starting in acute HIV-1 infection. *Blood* (2005) 106(10):3366–9. doi:10.1182/blood-2005-03-1100
70. Mavilio D, Benjamin J, Kim D, Lombardo G, Daucher M, Kinter A, et al. Identification of NKG2A and NKp80 as specific natural killer cell markers in rhesus and pigtailed monkeys. *Blood* (2005) 106(5):1718–25. doi:10.1182/blood-2004-12-4762
71. Brunetta E, Hudspeth KL, Mavilio D. Pathologic natural killer cell subset redistribution in HIV-1 infection: new insights in pathophysiology and clinical outcomes. *J Leukoc Biol* (2010) 88(6):1119–30. doi:10.1189/jlb.0410225
72. Fogli M, Mavilio D, Brunetta E, Varchetta S, Ata K, Roby G, et al. Lysis of endogenously infected CD4+ T cell blasts by rIL-2 activated autologous natural killer cells from HIV-infected viremic individuals. *PLoS Pathog* (2008) 4(7):e1000101. doi:10.1371/journal.ppat.1000101
73. Mavilio D, Lombardo G, Kinter A, Fogli M, La Sala A, Ortolano S, et al. Characterization of the defective interaction between a subset of natural killer cells and dendritic cells in HIV-1 infection. *J Exp Med* (2006) 203(10):2339–50. doi:10.1084/jem.20060894
74. De Maria A, Fogli M, Costa P, Murdaca G, Puppo F, Mavilio D, et al. The impaired NK cell cytolytic function in viremic HIV-1 infection is associated with a reduced surface expression of natural cytotoxicity receptors (NKp46, NKp30 and NKp44). *Eur J Immunol* (2003) 33(9):2410–8. doi:10.1002/eji.200324141
75. Mavilio D, Lombardo G, Benjamin J, Kim D, Follman D, Marcenaro E, et al. Characterization of CD56-/CD16+ natural killer (NK) cells: a highly dysfunctional NK subset expanded in HIV-infected viremic individuals. *Proc Natl Acad Sci U S A* (2005) 102(8):2886–91. doi:10.1073/pnas.0409872102
76. Shao JY, Yin WW, Zhang QF, Liu Q, Peng ML, Hu HD, et al. Siglec-7 defines a highly functional natural killer cell subset and inhibits cell-mediated activities. *Scand J Immunol* (2016) 84(3):182–90. doi:10.1111/sji.12455
77. Varchetta S, Mele D, Lombardi A, Oliviero B, Mantovani S, Tinelli C, et al. Lack of Siglec-7 expression identifies a dysfunctional natural killer cell subset associated with liver inflammation and fibrosis in chronic HCV infection. *Gut* (2016) 65(12):1998–2006. doi:10.1136/gutjnl-2015-310327
78. van der Kuyl AC, van den Burg R, Zorgdrager F, Groot F, Berkhout B, Cornelissen M. Sialoadhesin (CD169) expression in CD14+ cells is upregulated early after HIV-1 infection and increases during disease progression. *PLoS One* (2007) 2(2):e257. doi:10.1371/journal.pone.0000257

79. Pulliam L, Sun B, Rempel H. Invasive chronic inflammatory monocyte phenotype in subjects with high HIV-1 viral load. *J Neuroimmunol* (2004) 157(1–2):93–8. doi:10.1016/j.jneuroim.2004.08.039
80. Pino M, Erkizia I, Benet S, Erikson E, Fernandez-Figueras MT, Guerrero D, et al. HIV-1 immune activation induces Siglec-1 expression and enhances viral trans-infection in blood and tissue myeloid cells. *Retrovirology* (2015) 12:37. doi:10.1186/s12977-015-0160-x
81. Abel PM, McSharry C, Galloway E, Ross C, Severn A, Toner G, et al. Heterogeneity of peripheral blood monocyte populations in human immunodeficiency virus-1 seropositive patients. *FEMS Microbiol Immunol* (1992) 5(5–6):317–23. doi:10.1111/j.1574-6968.1992.tb05916.x
82. Locher C, Vanham G, Kestens L, Kruger M, Ceuppens JL, Vingerhoets J, et al. Expression patterns of Fc gamma receptors, HLA-DR and selected adhesion molecules on monocytes from normal and HIV-infected individuals. *Clin Exp Immunol* (1994) 98(1):115–22. doi:10.1111/j.1365-2249.1994.tb06616.x
83. Nockher WA, Bergmann L, Scherberich JE. Increased soluble CD14 serum levels and altered CD14 expression of peripheral blood monocytes in HIV-infected patients. *Clin Exp Immunol* (1994) 98(3):369–74. doi:10.1111/j.1365-2249.1994.tb05499.x
84. Thieblemont N, Weiss L, Sadeghi HM, Estcourt C, Haeffner-Cavaillon N. CD14lowCD16high: a cytokine-producing monocyte subset which expands during human immunodeficiency virus infection. *Eur J Immunol* (1995) 25(12):3418–24. doi:10.1002/eji.1830251232
85. Herbein G, Varin A. The macrophage in HIV-1 infection: from activation to deactivation? *Retrovirology* (2010) 7:33. doi:10.1186/1742-4690-7-33
86. Persidsky Y, Gendelman HE. Mononuclear phagocyte immunity and the neuropathogenesis of HIV-1 infection. *J Leukoc Biol* (2003) 74(5):691–701. doi:10.1189/jlb.0503205
87. Scanlan CN, Offer J, Zitzmann N, Dwek RA. Exploiting the defensive sugars of HIV-1 for drug and vaccine design. *Nature* (2007) 446(7139):1038–45. doi:10.1038/nature05818
88. Wei X, Decker JM, Wang S, Hui H, Kappes JC, Wu X, et al. Antibody neutralization and escape by HIV-1. *Nature* (2003) 422(6929):307–12. doi:10.1038/nature01470
89. Go EP, Chang Q, Liao HX, Sutherland LL, Alam SM, Haynes BF, et al. Glycosylation site-specific analysis of clade C HIV-1 envelope proteins. *J Proteome Res* (2009) 8(9):4231–42. doi:10.1021/pr9002728
90. Jobe O, Trinh HV, Kim J, Alsalmi W, Tovanabutra S, Ehrenberg PK, et al. Effect of cytokines on Siglec-1 and HIV-1 entry in monocyte-derived macrophages: the importance of HIV-1 envelope V1V2 region. *J Leukoc Biol* (2016) 99(6):1089–106. doi:10.1189/jlb.2A0815-361R
91. Berre S, Gaudin R, Cunha de Alencar B, Desdouits M, Chabaud M, Naffakh N, et al. CD36-specific antibodies block release of HIV-1 from infected primary macrophages and its transmission to T cells. *J Exp Med* (2013) 210(12):2523–38. doi:10.1084/jem.20130566
92. Gaudin R, Berre S, Cunha de Alencar B, Decalf J, Schindler M, Gobert FX, et al. Dynamics of HIV-containing compartments in macrophages reveal sequestration of virions and transient surface connections. *PLoS One* (2013) 8(7):e69450. doi:10.1371/journal.pone.0069450
93. Gartner S, Markovits P, Markovitz DM, Betts RF, Popovic M. Virus isolation from and identification of HTLV-III/LAV-producing cells in brain tissue from a patient with AIDS. *JAMA* (1986) 256(17):2365–71. doi:10.1001/jama.256.17.2365
94. Koenig S, Gendelman HE, Orenstein JM, Dal Canto MC, Pezeshkpour GH, Yungbluth M, et al. Detection of AIDS virus in macrophages in brain tissue from AIDS patients with encephalopathy. *Science* (1986) 233(4768):1089–93. doi:10.1126/science.3016903
95. Chakrabarti L, Hurtrel M, Maire MA, Vazeux R, Dormont D, Montagnier L, et al. Early viral replication in the brain of SIV-infected rhesus monkeys. *Am J Pathol* (1991) 139(6):1273–80.
96. Clements JE, Babas T, Mankowski JL, Suryanarayana K, Piatak M Jr, Tarwater PM, et al. The central nervous system as a reservoir for simian immunodeficiency virus (SIV): steady-state levels of SIV DNA in brain from acute through asymptomatic infection. *J Infect Dis* (2002) 186(7):905–13. doi:10.1086/343768
97. Deeks SG, Autran B, Berkhout B, Benkirane M, Cairns S, Chomont N, et al. Towards an HIV cure: a global scientific strategy. *Nat Rev Immunol* (2012) 12(8):607–14. doi:10.1038/nri3262
98. Cassetta L, Kajaste-Rudnitski A, Coradin T, Saba E, Della Chiara G, Barbagallo M, et al. M1 polarization of human monocyte-derived macrophages restricts pre and postintegration steps of HIV-1 replication. *AIDS* (2013) 27(12):1847–56. doi:10.1097/QAD.0b013e328361d059
99. Kutza J, Fields K, Grimm TA, Clouse KA. Inhibition of HIV replication and macrophage colony-stimulating factor production in human macrophages by antiretroviral agents. *AIDS Res Hum Retroviruses* (2002) 18(9):619–25. doi:10.1089/088922202760019310
100. van den Berg TK, Breve JJ, Damoiseaux JG, Dopp EA, Kelm S, Crocker PR, et al. Sialoadhesin on macrophages: its identification as a lymphocyte adhesion molecule. *J Exp Med* (1992) 176(3):647–55. doi:10.1084/jem.176.3.647
101. van den Berg TK, Nath D, Ziltener HJ, Vestweber D, Fukuda M, van Die I, et al. Cutting edge: CD43 functions as a T cell counterreceptor for the macrophage adhesion receptor sialoadhesin (Siglec-1). *J Immunol* (2001) 166(6):3637–40. doi:10.4049/jimmunol.166.6.3637
102. Stamatos NM, Gomatos PJ, Cox J, Fowler A, Dow N, Wohlhieter JA, et al. Desialylation of peripheral blood mononuclear cells promotes growth of HIV-1. *Virology* (1997) 228(2):123–31. doi:10.1006/viro.1996.8373
103. Means RE, Desrosiers RC. Resistance of native, oligomeric envelope on simian immunodeficiency virus to digestion by glycosidases. *J Virol* (2000) 74(23):11181–90. doi:10.1128/JVI.74.23.11181-11190.2000
104. Sewald X, Ladinsky MS, Uchil PD, Beloor J, Pi R, Herrmann C, et al. Retroviruses use CD169-mediated trans-infection of permissive lymphocytes to establish infection. *Science* (2015) 350(6260):563–7. doi:10.1126/science.aab2749
105. Sabado RL, O'Brien M, Subedi A, Qin L, Hu N, Taylor E, et al. Evidence of dysregulation of dendritic cells in primary HIV infection. *Blood* (2010) 116(19):3839–52. doi:10.1182/blood-2010-03-273763
106. Puryear WB, Akiyama H, Geer SD, Ramirez NP, Yu X, Reinhard BM, et al. Interferon-inducible mechanism of dendritic cell-mediated HIV-1 dissemination is dependent on Siglec-1/CD169. *PLoS Pathog* (2013) 9(4):e1003291. doi:10.1371/journal.ppat.1003291
107. Akiyama H, Ramirez NG, Gudheti MV, Gummuluru S. CD169-mediated trafficking of HIV to plasma membrane invaginations in dendritic cells attenuates efficacy of anti-gp120 broadly neutralizing antibodies. *PLoS Pathog* (2015) 11(3):e1004751. doi:10.1371/journal.ppat.1004751
108. Pantaleo G, Fauci AS. New concepts in the immunopathogenesis of HIV infection. *Annu Rev Immunol* (1995) 13:487–512. doi:10.1146/annurev.iy.13.040195.002415
109. Finzi D, Blankson J, Siliciano JD, Margolick JB, Chadwick K, Pierson T, et al. Latent infection of CD4+ T cells provides a mechanism for lifelong persistence of HIV-1, even in patients on effective combination therapy. *Nat Med* (1999) 5(5):512–7. doi:10.1038/8394
110. Chun TW, Fauci AS. HIV reservoirs: pathogenesis and obstacles to viral eradication and cure. *AIDS* (2012) 26(10):1261–8. doi:10.1097/QAD.0b013e328353f3f1
111. Brenchley JM, Price DA, Schacker TW, Asher TE, Silvestri G, Rao S, et al. Microbial translocation is a cause of systemic immune activation in chronic HIV infection. *Nat Med* (2006) 12(12):1365–71. doi:10.1038/nm1511
112. Sica A, Mantovani A. Macrophage plasticity and polarization: in vivo veritas. *J Clin Invest* (2012) 122(3):787–95. doi:10.1172/JCI59643
113. Falco M, Biassoni R, Bottino C, Vitale M, Sivori S, Augugliaro R, et al. Identification and molecular cloning of p75/AIRM1, a novel member of the sialoadhesin family that functions as an inhibitory receptor in human natural killer cells. *J Exp Med* (1999) 190(6):793–802. doi:10.1084/jem.190.6.793
114. Vitale C, Romagnani C, Falco M, Ponte M, Vitale M, Moretta A, et al. Engagement of p75/AIRM1 or CD33 inhibits the proliferation of normal or leukemic myeloid cells. *Proc Natl Acad Sci U S A* (1999) 96(26):15091–6. doi:10.1073/pnas.96.26.15091
115. Ikehara Y, Ikehara SK, Paulson JC. Negative regulation of T cell receptor signaling by Siglec-7 (p70/AIRM) and Siglec-9. *J Biol Chem* (2004) 279(41):43117–25. doi:10.1074/jbc.M403538200
116. Daeron M, Jaeger S, Du Pasquier L, Vivier E. Immunoreceptor tyrosine-based inhibition motifs: a quest in the past and future. *Immunol Rev* (2008) 224:11–43. doi:10.1111/j.1600-065X.2008.00666.x

117. Chen CH, Floyd H, Olson NE, Magaletti D, Li C, Draves K, et al. Dendritic-cell-associated C-type lectin 2 (DCAL-2) alters dendritic-cell maturation and cytokine production. *Blood* (2006) 107(4):1459–67. doi:10.1182/blood-2005-08-3264
118. Robinson MJ, Sancho D, Slack EC, LeibundGut-Landmann S, Reis e Sousa C. Myeloid C-type lectins in innate immunity. *Nat Immunol* (2006) 7(12):1258–65. doi:10.1038/ni1417
119. Barrow AD, Trowsdale J. You say ITAM and I say ITIM, let's call the whole thing off: the ambiguity of immunoreceptor signalling. *Eur J Immunol* (2006) 36(7):1646–53. doi:10.1002/eji.200636195
120. Pinheiro da Silva F, Aloulou M, Benhamou M, Monteiro RC. Inhibitory ITAMs: a matter of life and death. *Trends Immunol* (2008) 29(8):366–73. doi:10.1016/j.it.2008.05.001
121. Faure M, Long EO. KIR2DL4 (CD158d), an NK cell-activating receptor with inhibitory potential. *J Immunol* (2002) 168(12):6208–14. doi:10.4049/jimmunol.168.12.6208
122. Blank U, Launay P, Benhamou M, Monteiro RC. Inhibitory ITAMs as novel regulators of immunity. *Immunol Rev* (2009) 232(1):59–71. doi:10.1111/j.1600-065X.2009.00832.x
123. Barrow AD, Astoul E, Floto A, Brooke G, Relou IA, Jennings NS, et al. Cutting edge: TREM-like transcript-1, a platelet immunoreceptor tyrosine-based inhibition motif encoding costimulatory immunoreceptor that enhances, rather than inhibits, calcium signaling via SHP-2. *J Immunol* (2004) 172(10):5838–42. doi:10.4049/jimmunol.172.10.5838
124. Peterson ME, Long EO. Inhibitory receptor signaling via tyrosine phosphorylation of the adaptor Crk. *Immunity* (2008) 29(4):578–88. doi:10.1016/j.immuni.2008.07.014
125. May LT, Hill SJ. ERK phosphorylation: spatial and temporal regulation by G protein-coupled receptors. *Int J Biochem Cell Biol* (2008) 40(10):2013–7. doi:10.1016/j.biocel.2008.04.001
126. Chirathaworn C, Kohlmeier JE, Tibbetts SA, Rumsey LM, Chan MA, Benedict SH. Stimulation through intercellular adhesion molecule-1 provides a second signal for T cell activation. *J Immunol* (2002) 168(11):5530–7. doi:10.4049/jimmunol.168.11.5530
127. Lackner AA, Lederman MM, Rodriguez B. HIV pathogenesis: the host. *Cold Spring Harb Perspect Med* (2012) 2(9):a007005. doi:10.1101/cshperspect.a007005
128. Shacklett BL. Immune responses to HIV and SIV in mucosal tissues: 'location, location, location'. *Curr Opin HIV AIDS* (2010) 5(2):128–34. doi:10.1097/COH.0b013e328335c178
129. Nedvetzki S, Sowinski S, Eagle RA, Harris J, Vely F, Pende D, et al. Reciprocal regulation of human natural killer cells and macrophages associated with distinct immune synapses. *Blood* (2007) 109(9):3776–85. doi:10.1182/blood-2006-10-052977
130. Lapaque N, Walzer T, Meresse S, Vivier E, Trowsdale J. Interactions between human NK cells and macrophages in response to *Salmonella* infection. *J Immunol* (2009) 182(7):4339–48. doi:10.4049/jimmunol.0803329
131. Bellora F, Castriconi R, Dondero A, Reggiardo G, Moretta L, Mantovani A, et al. The interaction of human natural killer cells with either unpolarized or polarized macrophages results in different functional outcomes. *Proc Natl Acad Sci U S A* (2010) 107(50):21659–64. doi:10.1073/pnas.1007654108
132. Romo N, Magri G, Muntasell A, Heredia G, Baia D, Angulo A, et al. Natural killer cell-mediated response to human cytomegalovirus-infected macrophages is modulated by their functional polarization. *J Leukoc Biol* (2011) 90(4):717–26. doi:10.1189/jlb.0311171
133. Mattiola I, Pesant M, Tentorio PF, Molgora M, Marcenaro E, Lugli E, et al. Priming of human resting NK cells by autologous M1 macrophages via the engagement of IL-1beta, IFN-beta, and IL-15 pathways. *J Immunol* (2015) 195(6):2818–28. doi:10.4049/jimmunol.1500325
134. Cowan AJ, Laszlo GS, Estey EH, Walter RB. Antibody-based therapy of acute myeloid leukemia with gemtuzumab ozogamicin. *Front Biosci (Landmark Ed)* (2013) 18:1311–34. doi:10.2741/4181
135. Kochuparambil ST, Litzow MR. Novel antibody therapy in acute lymphoblastic leukemia. *Curr Hematol Malig Rep* (2014) 9(2):165–73. doi:10.1007/s11899-014-0202-9
136. Relan M, Vishwanath S, Shen L, Ambrus JL. Update on the use of biologics in lupus. *Curr Pharm Biotechnol* (2014) 15:516–20. doi:10.2174/1389201015666140804162053
137. Revilla C, Poderoso T, Martinez P, Alvarez B, Lopez-Fuertes L, Alonso F, et al. Targeting to porcine sialoadhesin receptor improves antigen presentation to T cells. *Vet Res* (2009) 40(3):14. doi:10.1051/vetres:2008052
138. Poderoso T, Martinez P, Alvarez B, Handler A, Moreno S, Alonso F, et al. Delivery of antigen to sialoadhesin or CD163 improves the specific immune response in pigs. *Vaccine* (2011) 29(29–30):4813–20. doi:10.1016/j.vaccine.2011.04.076
139. Collins BE, Blixt O, Han S, Duong B, Li H, Nathan JK, et al. High-affinity ligand probes of CD22 overcome the threshold set by cis ligands to allow for binding, endocytosis, and killing of B cells. *J Immunol* (2006) 177(5):2994–3003. doi:10.4049/jimmunol.177.5.2994
140. Chen WC, Kawasaki N, Nycholat CM, Han S, Pilotte J, Crocker PR, et al. Antigen delivery to macrophages using liposomal nanoparticles targeting sialoadhesin/CD169. *PLoS One* (2012) 7(6):e39039. doi:10.1371/journal.pone.0039039

13

Human NK Cell Subsets Redistribution in Pathological Conditions: A Role for CCR7 Receptor

Silvia Pesce[1], *Lorenzo Moretta*[2], *Alessandro Moretta*[1,3]* *and Emanuela Marcenaro*[1,3]

[1] *Dipartimento di Medicina Sperimentale, Università degli Studi di Genova, Genova, Italy,* [2] *Dipartimento di Immunologia, IRCCS Bambino Gesù Ospedale Pediatrico, Rome, Italy,* [3] *CEBR, Università degli Studi di Genova, Genova, Italy*

Edited by:
Laurent Brossay, Brown University, USA

Reviewed by:
Clair Gardiner, Trinity College Dublin, Ireland
Kerry S. Campbell, Fox Chase Cancer Center, USA

***Correspondence:**
Alessandro Moretta
alemoret@unige.it

Innate and adaptive immunity has evolved complex molecular mechanisms regulating immune cell migration to facilitate the dynamic cellular interactions required for its function involving the chemokines and their receptors. One important chemokine receptor in the immune system is represented by CCR7. Together with its ligands CCL19 and CCL21, this chemokine receptor controls different arrays of migratory events, both in innate and adaptive immunity, including homing of CD56^{bright} NK cells, T cells, and DCs to lymphoid compartments, where T cell priming occurs. Only recently, a key role for CCR7 in promoting CD56^{dim} NK cell migration toward lymphoid tissues has been described. Remarkably, this event can influence the shaping and polarization of adaptive T cell responses. In this review, we describe recent progress in understanding the mechanisms and the site where CD56^{dim} KIR^{+} NK cells can acquire the capability to migrate toward lymph nodes. The emerging significance of this event in clinical transplantation is also discussed.

Keywords: human NK cells, CCR7, transplantation, alloreactivity, NK cell subsets, immune checkpoint

INTRODUCTION

NK cells are effector cells of the innate immune system able to recognize and kill stressed, transformed, or virus-infected cells *via* a delicate balance of signals transmitted by activating and inhibitory receptors, and to secrete various effector molecules (1–3).

Two main subsets of human NK cells have been identified, according to the cell surface density of CD56 and expression of CD16 (FcγRIIIa). The CD56^{dim} CD16^{bright} NK cell subset expresses KIR and/or CD94/NKG2A molecules and predominates in peripheral blood (~90% of circulating NK cells), while the CD56^{bright} CD16$^{neg/dim}$ NK cells express CD94/NKG2A (but are KIR negative) and represent only ~10% of circulating NK cells. CD56^{dim} CD16^{bright} NK cells display potent cytolytic activity and produce cytokines following receptor-mediated stimulation (e.g., engagement of activating surface receptors during target cell recognition) (4–6). On the other hand, CD56^{bright} CD16$^{neg/dim}$ NK cells produce cytokines including interferon-γ (IFNγ), tumor necrosis factor-α (TNFα), and granulocyte–macrophage colony-stimulating factor (GM-CSF) and undergo proliferation following stimulation with pro-inflammatory cytokines. Cytolytic activity is acquired only after prolonged cell stimulation (4–6). Notably the CD56^{bright} CD16$^{neg/dim}$ NK cells can undergo differentiation into CD56^{dim} CD16^{bright} NK cells. Moreover this subset can undergo further phenotypic and functional maturation toward terminally differentiated NK cells (7–10).

WHAT DETERMINES NK CELL SUBSET RECRUITMENT TO DIFFERENT ORGANS DURING PHYSIOLOGICAL AND PATHOLOGICAL CONDITIONS?

In bone marrow, NK cell precursors undergo a maturation process that includes the acquisition of effector functions and the expression of chemotactic receptors that will drive their migration from the bone marrow to different organs through the blood stream (11, 12).

The recirculation and the distribution of cells of the immune system to the various organs depend primarily on the release of particular chemokines by organ-specific cell types (13, 14).

NK cells can respond to a large array of chemokines (13), and can be recruited to different district of the body and to sites of inflammation (15, 16). The distribution of NK cells is subset specific. Indeed, the two main NK cell subsets display major functional differences not only for their cytolytic activity and modality of cytokine production but also in their homing capabilities, as shown by their organ-specific localization (16). In particular, the cytolytic CD56^{dim} CD16^{bright} NK cell subset expresses CXCR1, CX3CR1, and ChemR23 chemokine receptors; therefore, it is mainly recruited to inflamed peripheral tissues. In contrast, CD56^{bright} CD16$^{neg/dim}$ NK cells preferentially express CCR7 and are primarily attracted by secondary lymphoid organs (lymph nodes, tonsils, and spleen) (17–20). These cells also express CD62L (L-selectin), which provides important adhesion to endothelial surfaces, required for extravasation of CD56^{bright} NK cells (21). Accordingly, CD56^{bright} NK cells are 10 times more frequent than CD56^{dim} in parafollicular (T-cell) regions of healthy (non-inflamed) lymph nodes, where they can be activated by T-cell-derived IL-2 (19, 22). Therefore, it is likely that the expression of the high-affinity IL-2 receptors on CD56^{bright} NK cells may promote a cross talk between NK and T cells in these lymphoid compartments (19).

It has recently been shown that, in addition to secondary lymphoid compartments (SLCs), CD56^{bright} CD16$^{neg/dim}$ NK cells populate other normal human tissues. These include uterine mucosa, liver, skin, adrenal gland, colorectal, liver, and visceral adipose tissues. On the other hand, tissues such as lung, breast, and sottocutaneous adipose tissue contain preferentially CD56^{dim} CD16^{bright} cells (14, 16, 23).

The specific distribution of the two subsets is mainly reflecting differences in their chemokine receptor repertoires and, as a consequence, in their ability to respond to different chemotactic factors (14, 16, 23).

Remarkably the localization of the two NK cell subsets can change in pathological conditions, e.g., in the presence of tumors (16). Thus, in different tumor types, both migration and homing of NK cells may be altered and even reversed. For example, NK cells present in tumor-infiltrated peripheral tissues are often enriched in CD56^{bright} CD16$^{neg/dim}$ NK cells (24–26); in contrast, an expansion of an unusual subset characterized by a CD56^{dim} CD69^{+} CCR7^{+} KIR^{+} phenotype has been detected in tumor-infiltrated lymph nodes (27). A possible explanation of these findings is that, a different pattern of chemokines, released by cells of the tumor microenvironment, or the acquisition of different/new chemokine receptors by NK cells, may operate an altered recruitment of the two NK cell subsets.

Thus, chemokines/chemokine receptors play a critical role in the regulation of the distribution of NK cell subpopulations in the various tissues, both in normal and the pathological conditions, such as inflammatory processes or tumors.

Notably, also organ-specific features, such as the anatomical structure, the type of cells present in the tissue, the soluble factors released, and the different interactions that NK cells can establish with different cell types, may considerably influence their homing capability.

WHAT CONDITIONS DETERMINE NK CELL ACTIVATION IN INFLAMED TISSUES?

As illustrated above, in the course of pathological conditions, such as inflammation, viral infection, and tumor growth, NK cells are rapidly recruited from peripheral blood into injured tissues, thanks to the interplay between chemokines and their receptors (20, 28–31).

NK cells, once they reach inflammatory sites, need to be activated in order to perform their functional activities. NK cell activation can occur through different types of signals. These include signals delivered by several activating NK receptors (32), upon recognition of specific ligands expressed on tumor cells, or signals generated in response to stimulation *via* toll-like receptors (TLR) that are constitutively expressed by NK cells and allow their responses to products of viral or bacterial origin (3, 33, 34). In addition, the activation of NK cells can occur through various soluble signals, including different cytokines that are provided by other cell types. Indeed, although NK cells can directly recognize cells infected by viruses and/or transformed cells, many recent studies revealed that the microenvironment and the interaction with other cells of the immune system, in particular with DCs, may considerably contribute to ensure an optimal priming of NK cells. For example, the production of cytokines (such as IL-12) by activated DCs improves NK cell proliferation, production of IFNγ, and antitumor cytotoxic activity (35).

In addition, a full NK cell activation allows the establishment of important processes, such as the mechanism of NK-mediated "editing" of DCs (28), which would play a crucial role in controlling the induction of appropriate antigen-specific adaptive immune responses (36, 37). In particular, activated NK cells appear to contribute to the quality control of immature DCs (iDCs) undergoing maturation by selecting DCs most fitting to optimal antigen presentation (36, 38). This process is thought to be crucial for the subsequent DC migration to SLCs and priming of naïve T lymphocytes toward Type 1 (Th1) immune responses.

In this context, it has to be considered that the exposure of NK cells to type II cytokines, such as IL-4 (35), released by other cells of the innate immune system in certain inflammatory microenvironments may result in low levels of cytotoxicity against tumor cells. Importantly, NK cells would become unable to kill iDCs and to release IFNγ and TNFα. This results in the failure of an efficient "editing" process and in the generation of Th0 cell leading to non polarized T cell responses in lymph nodes that compromise downstream antigen-specific, Th1-mediated immune responses (35). On the other hand, IL-4 by driving a type II immune response may be important for the eradication of particular pathogens. The absence of NK cells driving a Th1 response, in this circumstance, may be optimal for an efficient immune response to that particular pathogen (39).

Other studies suggest that NK cells may play not only a beneficial regulatory role in shaping adaptive immune responses but also an inhibition of the adaptive immune cells (e.g., T and B cells), by killing both antigen-presenting cells (APCs) and antigen-specific T cells and by producing anti-inflammatory cytokines, such as IL-10. These events can shape the overall immune response against certain pathogens, which can have consequences for disease pathogenesis, and infection outcome. Continued evaluation of the specific context-dependent roles of NK cells in human infection will be necessary to guide attempts to modulate NK cells in therapy or prevention of infection (40).

Thus, depending on the nature of the pathogens responsible for the invasion of a given tissue and, as a consequence, on the cytokine microenvironment created during the early stages of an inflammatory response, NK cells may differentially contribute to the quality and magnitude of both innate and adaptive immune responses (1, 35, 41).

HOW DO THE PHENOTYPIC PROPERTIES (AND THE EFFECTOR FUNCTIONS) OF NK CELLS CHANGE FOLLOWING RECRUITMENT TO INFLAMED PERIPHERAL TISSUES?

According to the data reported in the previous paragraphs, there is clear evidence that $CD56^{dim}$ $CD16^{+}$ NK cells, which have been recruited into inflamed tissues, in the presence of pro-inflammatory stimuli favoring their interaction with DCs (28, 31, 42–44) may usually influence naïve T cell priming, for instance by the "DC editing program." This mechanism may take place in the periphery and, at least in some instances, does not require the recruitment of tissue-activated NK cells into lymph nodes (35, 41, 45).

On the other hand, the inflammatory microenvironment and the interactions established with other cells at the inflammatory sites may also affect both phenotypic and homing properties of NK cells by generating altered NK cell subsets. For example, there is evidence that, during an immune response, the highly cytotoxic $CD56^{dim}$ KIR^{+} (as well as $CD56^{dim}$ KIR^{-} $NKG2A^{+}$) NK cells, that, different from $CD56^{bright}$ NK cells are CCR7 negative, could be directed from the peripheral (non-lymphoid tissues) toward inflamed lymph nodes. These represent a crucial site for NK cell-mediated immunosurveillance against tumor metastases and the control of viral infections and also contribute to priming of adaptive immune responses (27, 41, 46).

In this context, it has been demonstrated that NK cells exposed to a microenvironment rich of IL-18 may *de novo* express CCR7 and, thus, acquire the ability to respond to the lymph node chemokines CCL19 and CCL21 (47).

IL-18 may be released by APCs in an inflammatory microenvironment triggered by pathogens. In this regard, it has recently been demonstrated that M0 and M2 macrophages express a membrane form of IL-18 (mIL-18), which is released in the course of polarization to M1, induced by bacterial stimuli (LPS or BCG). IL-18 induces NK cells to release large amounts of IFNγ and plays a pivotal role in the expression on their surface of CCR7 receptor (47, 48) (**Figure 1A**).

The acquisition of CCR7 provides $CD56^{dim}$ KIR^{+} NK cells with the potential to reach lymph nodes where they can interact with DCs and T cells and regulate T cell responses (41, 47, 49). On the other hand, the expression of CCR7 in response to IL-18 can occur only in resting NK cells, because cytokine-mediated priming of NK cells induces their activation and a marked downregulation of IL-18Rα expression (50, 51). In light of these data, it is conceivable that other mechanisms may contribute to the acquisition of CCR7 by "activated" $CD56^{dim}$ NK cells. Indeed, recent data revealed that an unexpected crosstalk between NK cells and other cells present in the inflamed niche environment (tumor/infected cells or other immune cells, including DCs and T cells) leads to the *de novo* surface expression of CCR7 by $CD56^{dim}$ KIR^{+} NK cells in an IL-18-independent manner (51–54) (**Figures 1B,C**).

The acquisition of CCR7 by these NK cells requires direct cell-to-cell contact, is detectable within few minutes, and is due to receptor uptake from $CCR7^{+}$ cells. This mechanism is termed "trogocytosis" and is distinct from other mechanisms of intercellular exchanges, such as nanotubes and exosomes (55–58). For example, interaction between NK cells and mature DCs (mDCs) that express surface CCR7 could induce its expression on NK cells (52).

An important aspect is that the acquisition of CCR7 is promoted by non-HLA-specific activating receptors (including NKp46), but, it is negatively regulated by the interaction between inhibitory KIR receptors expressed by NK cells ("acceptor") and HLA class I molecules expressed by $CCR7^{+}$ cells ("donor"). In particular, in the course of infection in a self-environment, KIR^{+} NK cells can acquire CCR7 only when they interact with HLA class I negative $CCR7^{+}$ cells (e.g., target cells that have undergone tumor transformation or viral infection) (**Figure 1B**). On the contrary, in an allogeneic setting characterized by KIR/HLA class I mismatch, KIR^{+} NK cells can acquire CCR7 following interaction with any $CCR7^{+}$ cell, including those expressing high levels of HLA class I, such as mDCs (**Figure 1C**).

This condition is reproduced during KIR-mismatched haploidentical hematopoietic stem cell transplantation (haplo-HSCT) to cure high-risk leukemias (46, 59). Importantly, these events have been recently reproduced and confirmed *in vivo* in a mouse model (53).

Thus, the *de novo* CCR7 expression renders $CD56^{dim}$ KIR^{+} NK cells able to migrate into the SLCs in response to CCL19 and CCL21 chemokines (47, 52, 54). In this compartment, NK cells release IFNγ (directly promoting the development of Th1 responses) and exert their cytotoxic activity against different

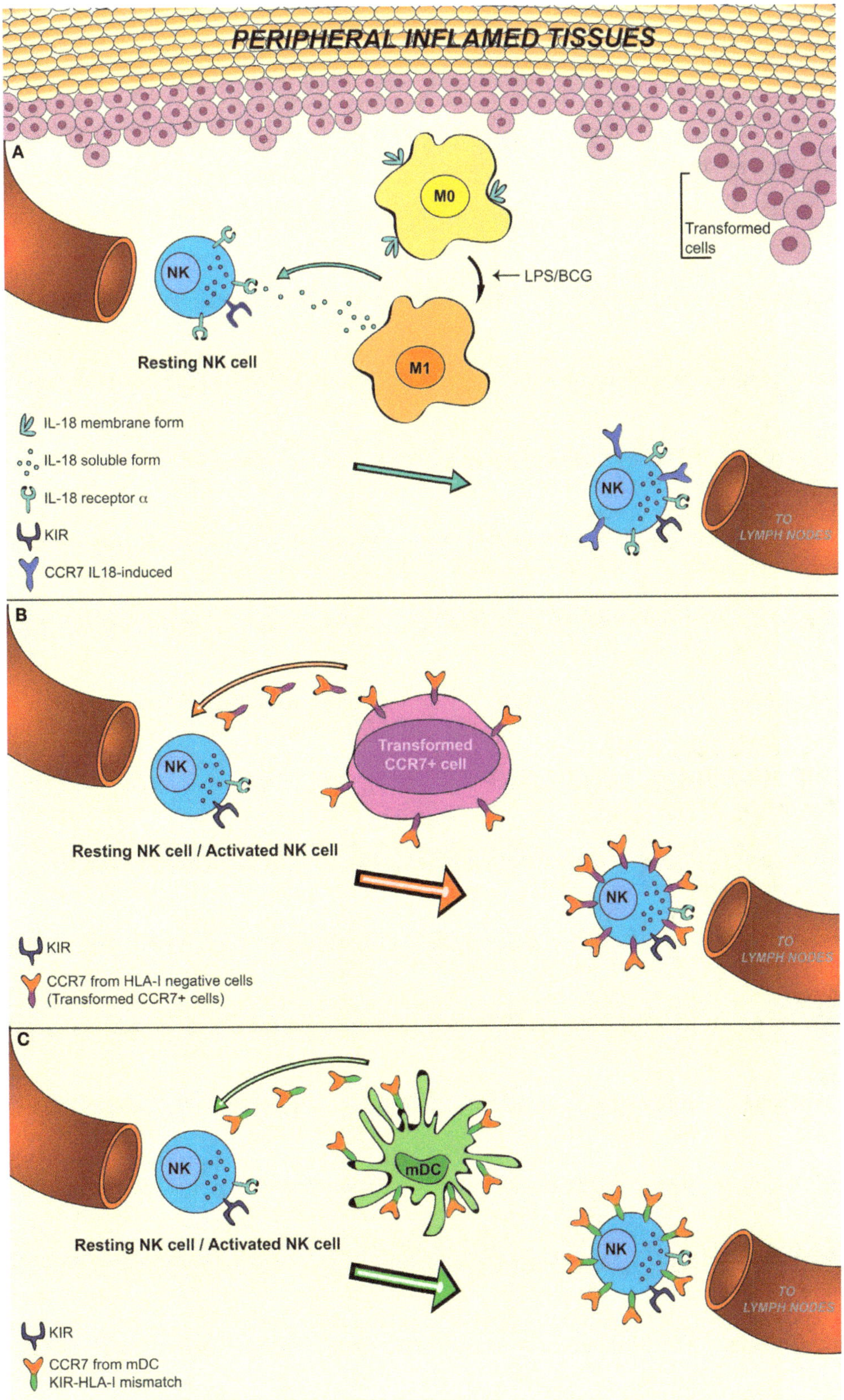

FIGURE 1 | **In this figure, it is shown that CD56^{dim} KIR^{+} NK cells, after recruitment from blood into inflamed peripheral tissues in response to chemokine gradients, may *de novo* express CCR7, a chemokine receptor able to confer immune cells with the ability to migrate to lymph nodes.**

(Continued)

FIGURE 1 | Continued
This event may occur in a microenvironment rich in IL-18, a pro-inflammatory cytokine released by M0 macrophages during differentiation toward M1 following activation. This *de novo* expression can be induced only on a fraction of NK cells characterized by the expression of the IL-18Rα (resting NK cells), whereas it cannot be induced on activated NK cells that express little IL-18Rα **(A)**. Otherwise, the expression of CCR7 on CD56^{dim} NK cells can be induced by a mechanism of trogocytosis on either resting or activated NK cells, regardless of their level of IL-18Rα expression. This mechanism is finely controlled by the specific interaction between KIRs on NK cells and HLA class I molecules on CCR7$^+$ cells. In particular, inhibitory KIRs block this transfer, whereas some activating KIRs are able to strongly promote the CCR7 acquisition by NK cells. This means that in an autologous setting, NK cells may acquire CCR7 only by interacting with HLA-I negative CCR7$^+$ cells, such as transformed cells **(B)**. In contrast, in an allogeneic setting, as in haplo-HSCT, characterized by KIR/HLA class I mismatch, KIR$^+$ alloreactive NK cells can express CCR7 when they interact with allogeneic CCR7$^+$ mDCs **(C)**.

types of targets, including tumor cells or (in an allogeneic system such as in transplantation) other target cells including patient's DCs (46).

In line with these data, Martin-Fontecha et al. showed that murine NK cells can be recruited to lymph nodes in a CXCR3-dependent manner by the injection of mDCs and that these recruitment correlated with the induction of Th1 responses (60). The induction of IFNγ release by NK cells migrated into lymph nodes appears to depend on direct interactions with DCs that release IL-12 and provide IL-15 by trans-presentation (61).

Very recently, it has been shown that NK cells can be genetically reprogramed efficiently using a electroporation method with mRNA coding for the chemokine receptor CCR7, in order to induce the expression of this chemokine receptor on their surface and the consequent NK cell migration toward the lymph node-associated chemokine CCL19 (62).

DOES CCR7$^+$ EXPRESSION ON ALLOREACTIVE KIR$^+$ NK CELLS PLAYS A ROLE IN TRANSPLANTATION?

Alloreactivity in haploidentical haplo-HSCT is operating through the mechanism of "missing self" recognition, provided that the donor expresses a given inhibitory KIR, whose ligand is missing in the recipient's HLA genotype. This combination will lead to a KIR/KIR–ligand mismatch in graft-versus-host (GvH) direction (63–67). In particular, it has been shown that, in T-cell depleted haplo-HSCT, "alloreactive" NK cells kill KIR–ligand mismatched leukemic blasts (graft versus leukemia, GvL), thus contributing to eradication of high-risk acute myeloid or lymphoid leukemias. It is of note that, in this transplantation setting, all cases would be at risk of T cell-mediated alloreactivity both in the HvG and in the GvH direction. However, patients transplanted from an NK "alloreactive donor" benefit from higher rates of engraftment and reduced incidence of graft-versus-host disease (GvHD) (68–71).

Such low rate of GvHD would be consequent to the inefficient priming of alloreactive donor T cells (reconstituted from donor CD34$^+$ HSC precursors) due to the NK cell-mediated killing of recipient's APCs (68). Furthermore, NK cells are able to eliminate residual patient's T cells, thus preventing HvG reactions and thereby promoting engraftment. In agreement with this concept, *in vitro* studies showed that alloreactive NK cell clones are able to kill not only leukemic blasts (71) but also allogeneic mDCs and T cell blasts (59, 72), while sparing other tissues that are common targets for T-cell-mediated GvHD (70). Although some of these target cells might be killed within peripheral tissues, it is conceivable that the elimination of patient's DCs by alloreactive NK cells occurs primarily within lymph nodes, thus preventing GvH reactions. Indeed, it is mainly at these sites that patient's DCs would prime donor's allospecific naïve T cells.

However, since KIR$^+$ NK cells do not express CCR7, it was difficult to explain how donor's KIR-mismatched NK cells could get in close proximity with recipient DCs and kill them within lymph nodes. In this context, our findings on trogocytosis of CCR7 have provided a possible explanation for the mechanism by which KIR-mismatched NK cells can acquire CCR7, migrate to lymph nodes, and kill recipient's DC, thus preventing GvH responses in this compartment (52).

Recent evidence revealed that the expression of activating KIRs, in particular KIR2DS1 (specific for HLA-C2) could improve the clinical outcome in haplo-HSCT. It is interesting that the expression of KIR2DS1 can induce alloreactivity by an otherwise non-alloreactive subset of NK cells expressing NKG2A. Thus, in appropriate donor/recipient pairs, the expression of KIR2DS1 can considerably increase the fraction of alloreactive NK cells (71, 72). In light of these data, we could demonstrate that the activating KIR KIR2DS1 represented an advantage for CCR7$^-$ NK cells favoring the acquisition of this receptor from HLA-C2$^+$ CCR7$^+$ cells. In particular, we showed that NK cell, triggered by this receptor in NK cell clones expressing the KIR2DS1$^+$/NKG2A$^+$ phenotype (isolated from C1$^+$ donors), could by-pass almost completely the inhibition mediated by NKG2A, thus making NKG2A$^+$ cells (which usually are not alloreactive) capable of killing allogeneic target cells expressing HLA-C2$^+$ and HLA-E. Remarkably, this interaction results also in the capture of CCR7 from CCR7$^+$ cells that express HLA-C2, including EBV-transformed B cell lines as well as mDCs or T blasts derived from C2$^+$ donors. We also showed that the ability to migrate in response to CCL19/21 NK cell was directly proportional to the level of expression of CCR7 acquired after coculture with CCR7$^+$ cells (54). Hence, the acquisition of CCR7 on CD56^{dim} KIR$^+$ NK cells is negatively regulated by inhibitory KIRs, but is strongly favored by positive signals delivered by activating KIR (52, 54, 59). In line with these results, it has been shown that also the KIR2DS4-activating receptor (73, 74) may induce the uptake of CCR7 by some KIR2DS4$^+$ NKG2A$^+$ NK cell clones upon interaction with CCR7$^+$ target cells expressing appropriate HLA-C alleles (51).

In conclusion, it is conceivable that the migration of CD56^{dim} KIR$^+$ NK cells to lymph nodes may play an important role not only in the mechanism of Th1 polarization in adaptive immune responses (through the release of IFNγ and the mechanism of

DC "editing") but also in the prevention of GvHD and HvGD in haplo-HSCT.

ARE THERE POTENTIAL CLINICAL APPLICATION SUGGESTED BY THESE FINDINGS?

The data described further support the concept that alloreactive NK cells, appropriately selected according to their phenotypic and functional characteristics, may play a central role in haplo-HSCT, by preventing leukemia relapses and improving engraftment by killing leukemia blasts and host APCs, which are known to initiate T cell-mediated GvHD (**Figure 2**). Notably, in the haplo-HSCT setting, the differentiation of KIR+ alloreactive NK cells from CD34+ HSC precursors may require 6–8 weeks. Therefore, their anti-leukemia effect would occur only after this time period. In case of infections or highly proliferating residual leukemia blasts, this delay may represent a major problem, resulting in leukemia relapses. In addition, in haplo-HSCT patients, due to the extensive T-cell depletion, required to prevent GvHD, there is an increased risk of life-threatening infections. With the aim to improve posttransplantation immune reconstitution, several immunotherapeutic approaches have been applied. For example, mature donor alloreactive NK cells may be infused at transplantation together with HSC and/or at short time intervals after HSCT. These NK cells may rapidly kill leukemic cells, thus anticipating the effect of alloreactive NK cells generated from transplanted HSC (75–77).

Another approach that could improve the efficiency of alloreactive NK cells would be to exploit trogocytosis as a tool to rapidly (and transiently) modify lymphocytes, for adoptive immunotherapy applications (52, 54). For example, pre-incubation of KIR+ NK cells (from an alloreactive donor) with CCR7+ HLA

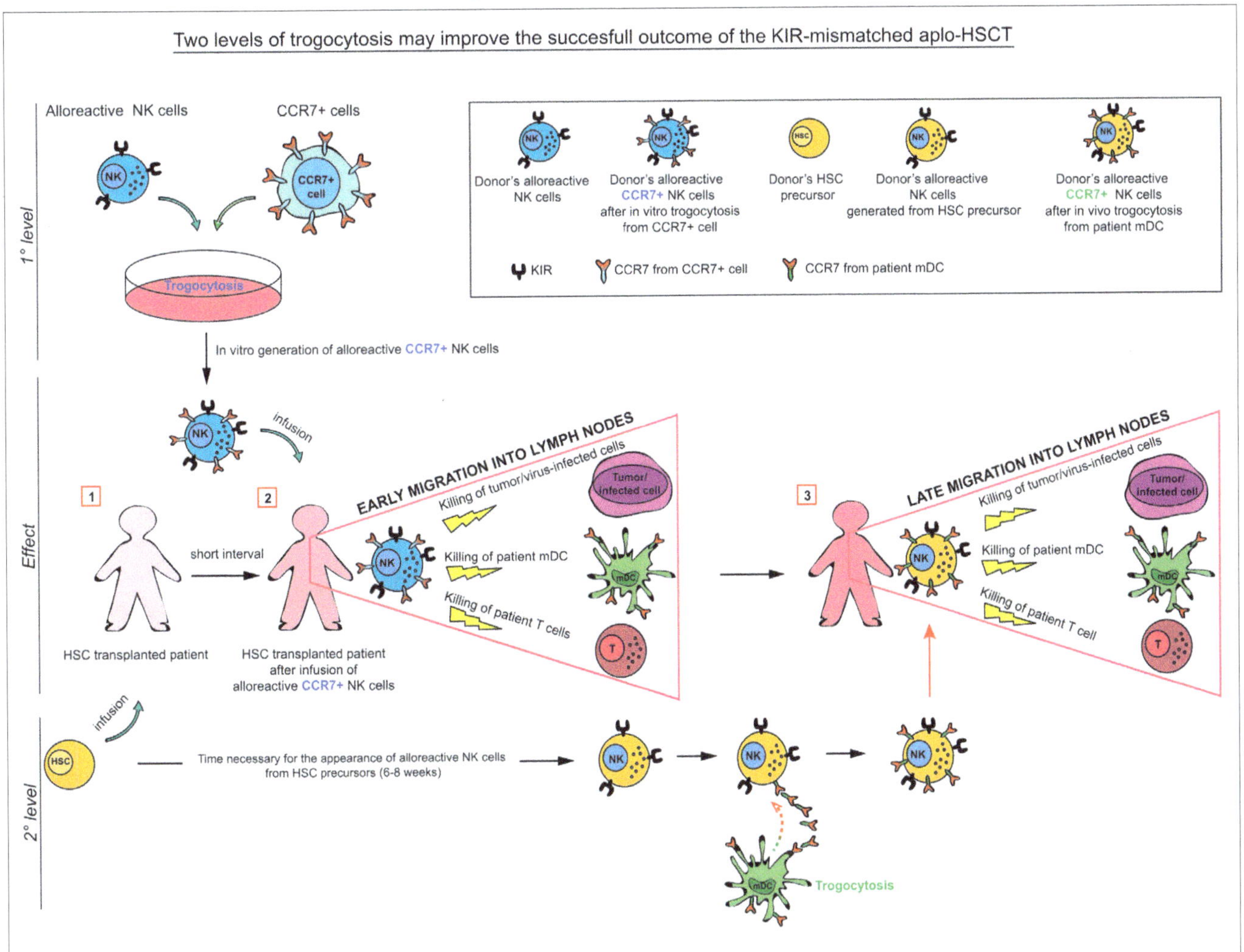

FIGURE 2 | In KIR-mismatched haplo-HSCT, the late appearance of alloreactive (KIR+) NK cells (6–8 weeks after transplant) may result in a delayed anti-leukemia and antiviral infection effect. In order to prevent these risks, a promising approach has been proposed consisting in the infusion of manipulated NK cells at short time intervals after HSCT. In particular, trogocytosis could be used as a tool to rapidly and transiently modify the migratory properties of donor-derived alloreactive NK cells by inducing CCR7 expression on their surface. These alloreactive NK cells, infused during the time required for the maturation of NK cells from HSC, may play a relevant role in preventing leukemic relapse (GvL activity), GvH/HvG reactions (by killing patient's DCs/Tcells), and in controlling infections. After this period, mature alloreactive NK cells generated from HSC acquire CCR7 directly (*in vivo*) from patient's CCR7+ cells, including mDC and T cell blasts.

class I-deficient cells, before their adoptive transfer to the patient, would allow CCR7+ alloreactive NK cells to be rapidly recruited to lymph nodes and to eliminate the leukemia cells residual after conditioning and patient's DC/T cells (78) (**Figure 2**). Importantly, in these immunosuppressed patients, CCR7+ KIR+ NK cells could also provide a rapid and more efficient line of defense against life-threatening infections (**Figure 2**).

In this context, CCR7 acquired by NK cells *via* trogocytosis has been shown to enhance their lymph node homing upon adoptive transfer to athymic nude mice, contributing to eliminate lymph node tumor metastases (53).

Trogocytosis may involve several cell surface molecules and different cell types, including T and B lymphocytes, NK cells, DCs, and tumor cells. It has been proposed that the formation of the immune synapse for capturing target cell membrane fragments by NK cells may be promoted by the interaction of NK receptors with their specific ligands. During this process several molecules (included in the immune synapse?) can be transferred from one cell to another. For example, upon coculture with the CCR7+ 221 cell line, NK cells may acquire, together with CCR7, also CD19 and CD86 (a marker commonly used to assess trogocytosis) (54).

Another interesting possible approach is based on the use of anti-KIR mAbs (their use is now in phase II clinical trials in patients with acute myeloid leukemia or multiple myeloma) (79–83). By blocking inhibitory KIRs, these mAbs, are able to confer alloreactivity to any KIR-2D+ NK cell (in individuals expressing HLA-specific activating KIRs, such as KIR2DS1, it is possible that the use of anti-KIR mAbs may partially modulate the function of an NK cell subset expressing this receptor) (52, 54). Notably, in this setting, all anti-KIR-treated NK cells of a given patient become capable of capturing CCR7 by any autologous CCR7+ cell soon after infusion in the patient. These "pseudo-alloreactive" NK cells could be rapidly routed toward lymph nodes, where they could carry out their functions.

CONCLUSION

The cross talk occurring between NK cells and DCs further supports the concept that NK cells play a critical role in the initiation and regulation of both innate and adaptive immune responses. These cellular interactions allow the establishment of important processes crucial for shaping of adaptive immunity: (a) the "DC editing program" is a process resulting in the selection of DCs with optimal antigen-presenting properties allowing appropriate Th1 responses protective against tumors and infections; (b) the acquisition of CCR7 expression by human KIR+ NK cells allows mature cytolytic NK cells to migrate to lymph nodes and exert antitumor and antiviral activity. Moreover, at these sites, NK cells may modulate Th1 polarization. In addition, in the haplo-HSCT transplantation setting "alloreactive" NK cells, migrated to lymph nodes, can mediate important anti-GvH and anti-HvG responses by killing recipient's allogeneic cells, including lymph node DCs and T cell blasts.

As illustrated in this review, CCR7 expression by KIR+ cytolytic NK cells is based on the capture of this chemokine receptor from allogeneic CCR7+ cells by a mechanism termed trogocytosis. This event is induced by activating KIRs, including KIR2DS1 and KIR2DS4, while it is negatively regulated by inhibitory KIRs and NKG2A. Thus, selection of appropriate donor/recipient pairs may greatly expand the contingent of alloreactive NK cells migrating to lymph nodes in HSCT, by increasing the effectiveness of alloreactive NK cells and the positive outcome of transplantation.

In addition to various protocols of chemotherapy and radiotherapy, fundamental progress in fighting cancer have been recently obtained with immunotherapy, thanks to the use of anti-immune checkpoints monoclonal antibodies, such as anti-CTLA4 and anti-PD-1, that are capable of reversing the function of otherwise exhausted tumor-specific T lymphocytes (84–86). Importantly, similar novel strategies now can be used also to manipulate NK cell function by the use of antibodies targeting their major inhibitory checkpoints, such as anti-KIR (79, 82), anti-NKG2A (87), and anti-PD-1 (86). In addition, novel means to optimize NK cell traffic to crucial sites (e.g., CCR7 acquisition and homing to lymph nodes) could be applied to treatment of different diseases besides tumors. These include viral infections and autoimmune diseases. In conclusion, a promising approach to improve the cure of life-threatening diseases may be based on "reshaping" NK cell phenotypic and functional properties.

AUTHOR CONTRIBUTIONS

All authors listed have made substantial, direct, and intellectual contribution to the work and approved it for publication.

REFERENCES

1. Vivier E, Raulet DH, Moretta A, Caligiuri MA, Zitvogel L, Lanier LL, et al. Innate or adaptive immunity? The example of natural killer cells. *Science* (2011) 331(6013):44–9. doi:10.1126/Science.1198687
2. Lanier LL. NK cell receptors. *Annu Rev Immunol* (1998) 16:359–93. doi:10.1146/annurev.immunol.16.1.359
3. Sivori S, Carlomagno S, Pesce S, Moretta A, Vitale M, Marcenaro E. TLR/NCR/KIR: which one to use and when? *Front Immunol* (2014) 5:105. doi:10.3389/fimmu.2014.00105
4. Cooper MA, Fehniger TA, Caligiuri MA. The biology of human natural killer-cell subsets. *Trends Immunol* (2001) 22(11):633–40. doi:10.1016/S1471-4906(01)02060-9
5. Fauriat C, Ivarsson MA, Ljunggren HG, Malmberg KJ, Michaelsson J. Education of human natural killer cells by activating killer cell immunoglobulin-like receptors. *Blood* (2010) 115(6):1166–74. doi:10.1182/blood-2009-09-245746
6. De Maria A, Bozzano F, Cantoni C, Moretta L. Revisiting human natural killer cell subset function revealed cytolytic CD56(dim)CD16+ NK cells as rapid producers of abundant IFN-gamma on activation. *Proc Natl Acad Sci U S A* (2011) 108(2):728–32. doi:10.1073/pnas.1012356108
7. Chan A, Hong DL, Atzberger A, Kollnberger S, Filer AD, Buckley CD, et al. CD56bright human NK cells differentiate into CD56dim cells: role of contact with peripheral fibroblasts. *J Immunol* (2007) 179(1):89–94. doi:10.4049/jimmunol.179.1.89

8. Bjorkstrom NK, Riese P, Heuts F, Andersson S, Fauriat C, Ivarsson MA, et al. Expression patterns of NKG2A, KIR, and CD57 define a process of CD56dim NK-cell differentiation uncoupled from NK-cell education. *Blood* (2010) 116(19):3853–64. doi:10.1182/blood-2010-04-281675
9. Della Chiesa M, Marcenaro E, Sivori S, Carlomagno S, Pesce S, Moretta A. Human NK cell response to pathogens. *Semin Immunol* (2014) 26(2):152–60. doi:10.1016/j.smim.2014.02.001
10. Moretta L. Dissecting CD56dim human NK cells. *Blood* (2010) 116(19): 3689–91. doi:10.1182/blood-2010-09-303057
11. Lysakova-Devine T, O'Farrelly C. Tissue-specific NK cell populations and their origin. *J Leukoc Biol* (2014) 96(6):981–90. doi:10.1189/jlb.1RU0514-241R
12. Shibuya A. Development and functions of natural killer cells. *Int J Hematol* (2003) 78(1):1–6. doi:10.1007/BF02983233
13. Gregoire C, Chasson L, Luci C, Tomasello E, Geissmann F, Vivier E, et al. The trafficking of natural killer cells. *Immunol Rev* (2007) 220:169–82. doi:10.1111/j.1600-065X.2007.00563.x
14. Carrega P, Bonaccorsi I, Di Carlo E, Morandi B, Paul P, Rizzello V, et al. CD56(bright)perforin(low) noncytotoxic human NK cells are abundant in both healthy and neoplastic solid tissues and recirculate to secondary lymphoid organs via afferent lymph. *J Immunol* (2014) 192(8):3805–15. doi:10.4049/jimmunol.1301889
15. Wald O, Weiss ID, Wald H, Shoham H, Bar-Shavit Y, Beider K, et al. IFN-gamma acts on T cells to induce NK cell mobilization and accumulation in target organs. *J Immunol* (2006) 176(8):4716–29. doi:10.4049/jimmunol.176.8.4716
16. Shi FD, Ljunggren HG, La Cava A, Van Kaer L. Organ-specific features of natural killer cells. *Nat Rev Immunol* (2011) 11(10):658–71. doi:10.1038/nri3065
17. Campbell JJ, Qin S, Unutmaz D, Soler D, Murphy KE, Hodge MR, et al. Unique subpopulations of CD56+ NK and NK-T peripheral blood lymphocytes identified by chemokine receptor expression repertoire. *J Immunol* (2001) 166(11):6477–82. doi:10.4049/jimmunol.166.11.6477
18. Vitale M, Della Chiesa M, Carlomagno S, Romagnani C, Thiel A, Moretta L, et al. The small subset of CD56brightCD16-natural killer cells is selectively responsible for both cell proliferation and interferon-gamma production upon interaction with dendritic cells. *Eur J Immunol* (2004) 34(6):1715–22. doi:10.1002/eji.200425100
19. Ferlazzo G, Thomas D, Lin SL, Goodman K, Morandi B, Muller WA, et al. The abundant NK cells in human secondary lymphoid tissues require activation to express killer cell Ig-like receptors and become cytolytic. *J Immunol* (2004) 172(3):1455–62. doi:10.4049/jimmunol.172.3.1455
20. Parolini S, Santoro A, Marcenaro E, Luini W, Massardi L, Facchetti F, et al. The role of chemerin in the colocalization of NK and dendritic cell subsets into inflamed tissues. *Blood* (2007) 109(9):3625–32. doi:10.1182/blood-2006-08-038844
21. Frey M, Packianathan NB, Fehniger TA, Ross ME, Wang WC, Stewart CC, et al. Differential expression and function of L-selectin on CD56bright and CD56dim natural killer cell subsets. *J Immunol* (1998) 161(1):400–8.
22. Fehniger TA, Cooper MA, Nuovo GJ, Cella M, Facchetti F, Colonna M, et al. CD56bright natural killer cells are present in human lymph nodes and are activated by T cell-derived IL-2: a potential new link between adaptive and innate immunity. *Blood* (2003) 101(8):3052–7. doi:10.1182/blood-2002-09-2876
23. Sharma R, Das A. Organ-specific phenotypic and functional features of NK cells in humans. *Immunol Res* (2014) 58(1):125–31. doi:10.1007/s12026-013-8477-9
24. Belisle JA, Gubbels JA, Raphael CA, Migneault M, Rancourt C, Connor JP, et al. Peritoneal natural killer cells from epithelial ovarian cancer patients show an altered phenotype and bind to the tumour marker MUC16 (CA125). *Immunology* (2007) 122(3):418–29. doi:10.1111/j.1365-2567.2007.02660.x
25. Carlsten M, Norell H, Bryceson YT, Poschke I, Schedvins K, Ljunggren HG, et al. Primary human tumor cells expressing CD155 impair tumor targeting by down-regulating DNAM-1 on NK cells. *J Immunol* (2009) 183(8):4921–30. doi:10.4049/jimmunol.0901226
26. Pesce S, Tabellini G, Cantoni C, Patrizi O, Coltrini D, Rampinelli F, et al. B7-H6-mediated downregulation of NKp30 in NK cells contributes to ovarian carcinoma immune escape. *Oncoimmunology* (2015) 4(4):e1001224. doi:10.1080/2162402X.2014.1001224
27. Ali TH, Pisanti S, Ciaglia E, Mortarini R, Anichini A, Garofalo C, et al. Enrichment of CD56(dim)KIR + CD57 + highly cytotoxic NK cells in tumour-infiltrated lymph nodes of melanoma patients. *Nat Commun* (2014) 5:5639. doi:10.1038/ncomms6639
28. Moretta A. Natural killer cells and dendritic cells: rendezvous in abused tissues. *Nat Rev Immunol* (2002) 2(12):957–64. doi:10.1038/nri956
29. Zitvogel L. Dendritic and natural killer cells cooperate in the control/switch of innate immunity. *J Exp Med* (2002) 195(3):F9–14. doi:10.1084/jem.20012040
30. Cooper MA, Fehniger TA, Fuchs A, Colonna M, Caligiuri MA. NK cell and DC interactions. *Trends Immunol* (2004) 25(1):47–52. doi:10.1016/j.it.2003.10.012
31. Marcenaro E, Ferranti B, Moretta A. NK-DC interaction: on the usefulness of auto-aggression. *Autoimmun Rev* (2005) 4(8):520–5. doi:10.1016/j.autrev.2005.04.015
32. Moretta A, Bottino C, Vitale M, Pende D, Cantoni C, Mingari MC, et al. Activating receptors and coreceptors involved in human natural killer cell-mediated cytolysis. *Annu Rev Immunol* (2001) 19:197–223. doi:10.1146/annurev.immunol.19.1.197
33. Sivori S, Falco M, Della Chiesa M, Carlomagno S, Vitale M, Moretta L, et al. CpG and double-stranded RNA trigger human NK cells by toll-like receptors: induction of cytokine release and cytotoxicity against tumors and dendritic cells. *Proc Natl Acad Sci U S A* (2004) 101(27):10116–21. doi:10.1073/pnas.0403744101
34. Marcenaro E, Ferranti B, Falco M, Moretta L, Moretta A. Human NK cells directly recognize *Mycobacterium bovis* via TLR2 and acquire the ability to kill monocyte-derived DC. *Int Immunol* (2008) 20(9):1155–67. doi:10.1093/intimm/dxn073
35. Marcenaro E, Della Chiesa M, Bellora F, Parolini S, Millo R, Moretta L, et al. IL-12 or IL-4 prime human NK cells to mediate functionally divergent interactions with dendritic cells or tumors. *J Immunol* (2005) 174(7):3992–8. doi:10.4049/jimmunol.174.7.3992
36. Della Chiesa M, Vitale M, Carlomagno S, Ferlazzo G, Moretta L, Moretta A. The natural killer cell-mediated killing of autologous dendritic cells is confined to a cell subset expressing CD94/NKG2A, but lacking inhibitory killer Ig-like receptors. *Eur J Immunol* (2003) 33(6):1657–66. doi:10.1002/eji.200390042
37. Ferlazzo G, Moretta L. Dendritic cell editing by natural killer cells. *Crit Rev Oncog* (2014) 19(1–2):67–75. doi:10.1615/CritRevOncog.2014010827
38. Ferlazzo G, Morandi B, D'Agostino A, Meazza R, Melioli G, Moretta A, et al. The interaction between NK cells and dendritic cells in bacterial infections results in rapid induction of NK cell activation and in the lysis of uninfected dendritic cells. *Eur J Immunol* (2003) 33(2):306–13. doi:10.1002/immu.200310004
39. Sahoo A, Wali S, Nurieva R. T helper 2 and T follicular helper cells: regulation and function of interleukin-4. *Cytokine Growth Factor Rev* (2016) 30:29–37. doi:10.1016/j.cytogfr.2016.03.011
40. Waggoner SN, Reighard SD, Gyurova IE, Cranert SA, Mahl SE, Karmele EP, et al. Roles of natural killer cells in antiviral immunity. *Curr Opin Virol* (2016) 16:15–23. doi:10.1016/j.coviro.2015.10.008
41. Agaugue S, Marcenaro E, Ferranti B, Moretta L, Moretta A. Human natural killer cells exposed to IL-2, IL-12, IL-18, or IL-4 differently modulate priming of naive T cells by monocyte-derived dendritic cells. *Blood* (2008) 112(5):1776–83. doi:10.1182/blood-2008-02-135871
42. Ferlazzo G, Munz C. NK cell compartments and their activation by dendritic cells. *J Immunol* (2004) 172(3):1333–9. doi:10.4049/jimmunol.172.3.1333
43. Gerosa F, Baldani-Guerra B, Nisii C, Marchesini V, Carra G, Trinchieri G. Reciprocal activating interaction between natural killer cells and dendritic cells. *J Exp Med* (2002) 195(3):327–33. doi:10.1084/jem.20010938
44. Piccioli D, Sbrana S, Melandri E, Valiante NM. Contact-dependent stimulation and inhibition of dendritic cells by natural killer cells. *J Exp Med* (2002) 195(3):335–41. doi:10.1084/jem.20010934
45. Moretta A, Marcenaro E, Parolini S, Ferlazzo G, Moretta L. NK cells at the interface between innate and adaptive immunity. *Cell Death Differ* (2008) 15(2):226–33. doi:10.1038/sj.cdd.4402170
46. Moretta L, Locatelli F, Pende D, Marcenaro E, Mingari MC, Moretta A. Killer Ig-like receptor-mediated control of natural killer cell alloreactivity in haploidentical hematopoietic stem cell transplantation. *Blood* (2011) 117(3):764–71. doi:10.1182/blood-2010-08-264085

47. Mailliard RB, Alber SM, Shen H, Watkins SC, Kirkwood JM, Herberman RB, et al. IL-18-induced CD83+CCR7+ NK helper cells. *J Exp Med* (2005) 202(7):941–53. doi:10.1084/jem.20050128
48. Bellora F, Castriconi R, Dondero A, Reggiardo G, Moretta L, Mantovani A, et al. The interaction of human natural killer cells with either unpolarized or polarized macrophages results in different functional outcomes. *Proc Natl Acad Sci U S A* (2010) 107(50):21659–64. doi:10.1073/pnas.1007654108
49. Bajenoff M, Glaichenhaus N, Germain RN. Fibroblastic reticular cells guide T lymphocyte entry into and migration within the splenic T cell zone. *J Immunol* (2008) 181(6):3947–54. doi:10.4049/jimmunol.181.6.3947
50. White MJ, Nielsen CM, McGregor RH, Riley EH, Goodier MR. Differential activation of CD57-defined natural killer cell subsets during recall responses to vaccine antigens. *Immunology* (2014) 142(1):140–50. doi:10.1111/imm.12239
51. Pesce S, Carlomagno S, Moretta A, Sivori S, Marcenaro E. Uptake of CCR7 by KIR2DS4(+) NK cells is induced upon recognition of certain HLA-C alleles. *J Immunol Res* (2015) 2015:754373. doi:10.1155/2015/754373
52. Marcenaro E, Cantoni C, Pesce S, Prato C, Pende D, Agaugue S, et al. Uptake of CCR7 and acquisition of migratory properties by human KIR+ NK cells interacting with monocyte-derived DC or EBV cell lines: regulation by KIR/HLA-class I interaction. *Blood* (2009) 114(19):4108–16. doi:10.1182/blood-2009-05-222265
53. Somanchi SS, Somanchi A, Cooper LJ, Lee DA. Engineering lymph node homing of ex vivo-expanded human natural killer cells via trogocytosis of the chemokine receptor CCR7. *Blood* (2012) 119(22):5164–72. doi:10.1182/blood-2011-11-389924
54. Marcenaro E, Pesce S, Sivori S, Carlomagno S, Moretta L, Moretta A. KIR2DS1-dependent acquisition of CCR7 and migratory properties by human NK cells interacting with allogeneic HLA-C2+ DCs or T-cell blasts. *Blood* (2013) 121(17):3396–401. doi:10.1182/blood-2012-09-458752
55. Joly E, Hudrisier D. What is trogocytosis and what is its purpose? *Nat Immunol* (2003) 4(9):815. doi:10.1038/ni0903-815
56. Davis DM. Intercellular transfer of cell-surface proteins is common and can affect many stages of an immune response. *Nat Rev Immunol* (2007) 7(3):238–43. doi:10.1038/nri2020
57. Mack M, Kleinschmidt A, Bruhl H, Klier C, Nelson PJ, Cihak J, et al. Transfer of the chemokine receptor CCR5 between cells by membrane-derived microparticles: a mechanism for cellular human immunodeficiency virus 1 infection. *Nat Med* (2000) 6(7):769–75. doi:10.1038/77498
58. Hudrisier D, Riond J, Mazarguil H, Gairin JE, Joly E. Cutting edge: CTLs rapidly capture membrane fragments from target cells in a TCR signaling-dependent manner. *J Immunol* (2001) 166(6):3645–9. doi:10.4049/jimmunol.166.6.3645
59. Marcenaro E, Carlomagno S, Pesce S, Della Chiesa M, Moretta A, Sivori S. Role of alloreactive KIR2DS1(+) NK cells in haploidentical hematopoietic stem cell transplantation. *J Leukoc Biol* (2011) 90(4):661–7. doi:10.1189/jlb.0311137
60. Martin-Fontecha A, Thomsen LL, Brett S, Gerard C, Lipp M, Lanzavecchia A, et al. Induced recruitment of NK cells to lymph nodes provides IFN-gamma for T(H)1 priming. *Nat Immunol* (2004) 5(12):1260–5. doi:10.1038/ni1138
61. Lucas M, Schachterle W, Oberle K, Aichele P, Diefenbach A. Dendritic cells prime natural killer cells by trans-presenting interleukin 15. *Immunity* (2007) 26(4):503–17. doi:10.1016/j.immuni.2007.03.006
62. Carlsten M, Levy E, Karambelkar A, Li L, Reger R, Berg M, et al. Efficient mRNA-based genetic engineering of human NK cells with high-affinity CD16 and CCR7 augments rituximab-induced ADCC against lymphoma and targets NK cell migration toward the lymph node-associated chemokine CCL19. *Front Immunol* (2016) 7:105. doi:10.3389/fimmu.2016.00105
63. Moretta A, Bottino C, Vitale M, Pende D, Biassoni R, Mingari MC, et al. Receptors for HLA class-I molecules in human natural killer cells. *Annu Rev Immunol* (1996) 14:619–48. doi:10.1146/annurev.immunol.14.1.619
64. Long EO. Regulation of immune responses through inhibitory receptors. *Annu Rev Immunol* (1999) 17:875–904. doi:10.1146/annurev.immunol.17.1.875
65. Karre K. Natural killer cell recognition of missing self. *Nat Immunol* (2008) 9(5):477–80. doi:10.1038/ni0508-477
66. Moretta A, Bottino C, Pende D, Tripodi G, Tambussi G, Viale O, et al. Identification of four subsets of human CD3-CD16+ natural killer (NK) cells by the expression of clonally distributed functional surface molecules: correlation between subset assignment of NK clones and ability to mediate specific alloantigen recognition. *J Exp Med* (1990) 172(6):1589–98. doi:10.1084/jem.172.6.1589
67. Moretta A, Vitale M, Bottino C, Orengo AM, Morelli L, Augugliaro R, et al. P58 molecules as putative receptors for major histocompatibility complex (MHC) class I molecules in human natural killer (NK) cells. Anti-p58 antibodies reconstitute lysis of MHC class I-protected cells in NK clones displaying different specificities. *J Exp Med* (1993) 178(2):597–604. doi:10.1084/jem.178.2.597
68. Velardi A, Ruggeri L, Moretta A, Moretta L. NK cells: a lesson from mismatched hematopoietic transplantation. *Trends Immunol* (2002) 23(9):438–44. doi:10.1016/S1471-4906(02)02284-6
69. Ruggeri L, Mancusi A, Capanni M, Urbani E, Carotti A, Aloisi T, et al. Donor natural killer cell allorecognition of missing self in haploidentical hematopoietic transplantation for acute myeloid leukemia: challenging its predictive value. *Blood* (2007) 110(1):433–40. doi:10.1182/blood-2006-07-038687
70. Moretta A, Locatelli F, Moretta L. Human NK cells: from HLA class I-specific killer Ig-like receptors to the therapy of acute leukemias. *Immunol Rev* (2008) 224:58–69. doi:10.1111/j.1600-065X.2008.00651.x
71. Pende D, Marcenaro S, Falco M, Martini S, Bernardo ME, Montagna D, et al. Anti-leukemia activity of alloreactive NK cells in KIR ligand-mismatched haploidentical HSCT for pediatric patients: evaluation of the functional role of activating KIR and redefinition of inhibitory KIR specificity. *Blood* (2009) 113(13):3119–29. doi:10.1182/blood-2008-06-164103
72. Sivori S, Carlomagno S, Falco M, Romeo E, Moretta L, Moretta A. Natural killer cells expressing the KIR2DS1-activating receptor efficiently kill T-cell blasts and dendritic cells: implications in haploidentical HSCT. *Blood* (2011) 117(16):4284–92. doi:10.1182/blood-2010-10-316125
73. Bottino C, Sivori S, Vitale M, Cantoni C, Falco M, Pende D, et al. A novel surface molecule homologous to the p58/p50 family of receptors is selectively expressed on a subset of human natural killer cells and induces both triggering of cell functions and proliferation. *Eur J Immunol* (1996) 26(8):1816–24. doi:10.1002/eji.1830260823
74. Katz G, Gazit R, Arnon TI, Gonen-Gross T, Tarcic G, Markel G, et al. MHC class I-independent recognition of NK-activating receptor KIR2DS4. *J Immunol* (2004) 173(3):1819–25. doi:10.4049/jimmunol.173.3.1819
75. Handgretinger R. New approaches to graft engineering for haploidentical bone marrow transplantation. *Semin Oncol* (2012) 39(6):664–73. doi:10.1053/j.seminoncol.2012.09.007
76. Locatelli F, Moretta F, Brescia L, Merli P. Natural killer cells in the treatment of high-risk acute leukaemia. *Semin Immunol* (2014) 26(2):173–9. doi:10.1016/j.smim.2014.02.004
77. Lucarelli B, Merli P, Bertaina V, Locatelli F. Strategies to accelerate immune recovery after allogeneic hematopoietic stem cell transplantation. *Expert Rev Clin Immunol* (2016) 12(3):343–58. doi:10.1586/1744666X.2016.1123091
78. Locatelli F, Pende D, Mingari MC, Bertaina A, Falco M, Moretta A, et al. Cellular and molecular basis of haploidentical hematopoietic stem cell transplantation in the successful treatment of high-risk leukemias: role of alloreactive NK cells. *Front Immunol* (2013) 4:15. doi:10.3389/fimmu.2013.00015
79. Romagne F, Andre P, Spee P, Zahn S, Anfossi N, Gauthier L, et al. Preclinical characterization of 1-7F9, a novel human anti-KIR receptor therapeutic antibody that augments natural killer-mediated killing of tumor cells. *Blood* (2009) 114(13):2667–77. doi:10.1182/blood-2009-02-206532
80. Benson DM Jr, Hofmeister CC, Padmanabhan S, Suvannasankha A, Jagannath S, Abonour R, et al. A phase 1 trial of the anti-KIR antibody IPH2101 in patients with relapsed/refractory multiple myeloma. *Blood* (2012) 120(22):4324–33. doi:10.1182/blood-2012-06-438028
81. Kohrt HE, Thielens A, Marabelle A, Sagiv-Barfi I, Sola C, Chanuc F, et al. Anti-KIR antibody enhancement of anti-lymphoma activity of natural killer cells as monotherapy and in combination with anti-CD20 antibodies. *Blood* (2014) 123(5):678–86. doi:10.1182/blood-2013-08-519199
82. Vey N, Bourhis JH, Boissel N, Bordessoule D, Prebet T, Charbonnier A, et al. A phase 1 trial of the anti-inhibitory KIR mAb IPH2101 for AML in complete remission. *Blood* (2012) 120(22):4317–23. doi:10.1182/blood-2012-06-437558
83. Nijhof IS, Lammerts van Bueren JJ, van Kessel B, Andre P, Morel Y, Lokhorst HM, et al. Daratumumab-mediated lysis of primary multiple myeloma cells is enhanced in combination with the human anti-KIR antibody

IPH2102 and lenalidomide. *Haematologica* (2015) 100(2):263–8. doi:10.3324/haematol.2014.117531

84. Lipson EJ, Sharfman WH, Drake CG, Wollner I, Taube JM, Anders RA, et al. Durable cancer regression off-treatment and effective reinduction therapy with an anti-PD-1 antibody. *Clin Cancer Res* (2013) 19(2):462–8. doi:10.1158/1078-0432.CCR-12-2625
85. Okazaki T, Chikuma S, Iwai Y, Fagarasan S, Honjo T. A rheostat for immune responses: the unique properties of PD-1 and their advantages for clinical application. *Nat Immunol* (2013) 14(12):1212–8. doi:10.1038/ni.2762
86. Pesce S, Greppi M, Tabellini G, Rampinelli F, Parolini S, Olive D, et al. Identification of a subset of human natural killer cells expressing high levels of programmed death 1: a phenotypic and functional characterization. *J Allergy Clin Immunol* (2016). doi:10.1016/j.jaci.2016.04.025
87. Ruggeri L, Urbani E, Andre P, Mancusi A, Tosti A, Topini F, et al. Effects of anti-NKG2A antibody administration on leukemia and normal hematopoietic cells. *Haematologica* (2016) 101(5):626–33. doi:10.3324/haematol.2015.135301

14

Human NK Cell Diversity in Viral Infection: Ramifications of Ramification

*Dara M. Strauss-Albee[1] and Catherine A. Blish[1,2]**

[1] *Stanford Immunology, Stanford University School of Medicine, Stanford, CA, USA,* [2] *Department of Medicine, Stanford University School of Medicine, Stanford, CA, USA*

Edited by:
Emanuela Marcenaro, University of Genova, Italy

Reviewed by:
Todd A. Fehniger, Washington University School of Medicine, USA
Raffaele Badolato, University of Brescia, Italy

***Correspondence:**
Catherine A. Blish
cblish@stanford.edu

Natural killer (NK) cells are a unique lymphocyte lineage with remarkable agility in the rapid destruction of virus-infected cells. They are also the most poorly understood class of lymphocyte. A spectrum of activating and inhibitory receptors at the NK cell surface leads to an unusual and difficult-to-study mechanism of cellular recognition, as well as a very high capacity for diversity at the single-cell level. Here, we review the evidence for the role of NK cells in the earliest stage of human viral infection, and in its prevention. We argue that single-cell diversity is a logical evolutionary adaptation for their position in the immune response and contributes to their ability to kill virus-infected cells. Finally, we look to the future, where emerging single-cell technologies will enable a new generation of rigorous and clinically relevant studies on NK cells accounting for all of their unique and diverse characteristics.

Keywords: natural killer cells, lymphocyte diversity, mass cytometry, viral susceptibility, single-cell technology

Natural killer (NK) cells were discovered in 1975 (1, 2) on the basis of their ability to selectively lyse leukemic cell lines. Originally called "null cells" because they were believed not to express cell surface receptors, they were eventually recognized as a third lineage of lymphocyte distinct from T and B cells (3). Recently, the role of NK cells as the founding member of a growing group known as innate lymphoid cells has also been appreciated (4).

Natural killer cells are a unique lineage with remarkable agility. They rapidly detect and destroy virus-infected, malignant, and stressed cells (5, 6). They are also the most poorly understood class of lymphocyte, due in part to their unusual and difficult-to-study mechanism of cellular recognition. The NK cell surface contains a spectrum of activating and inhibitory receptors (7). It is the integration and balance of signals from these receptors that determine a cell's activation status (8).

This array of receptors also generates the opportunity for vast diversity in the NK repertoire. Examination of this diversity has been limited: studying a large number of parameters on a single cell is technologically challenging. Yet, it may be a critical functional feature of the NK cell repertoire.

In many of the ways in which immune cells are conventionally categorized, NK cells are intermediates. They borrow, share, and combine functional features of other cell types to form their own recognition paradigm. This helps to explain their enigmatic nature and argues that they occupy a unique evolutionary niche. Here, we review this recognition paradigm, with a focus on NK responses to viruses, and argue that single-cell diversity enhances their ability to fulfill this "middleman" role. We also discuss current and future studies, where single-cell technology will allow a much more detailed and nuanced dissection of the roles and promise of NK cells in the antiviral response.

NK CELLS HAVE LIFETIMES OF INTERMEDIATE LENGTH

In comparison to their classic innate and adaptive counterparts, NK cells take an intermediate position in estimates of cellular lifetime (**Figure 1**). In deuterium incorporation studies, human T cells have the longest estimated leukocyte half-lives, at 1–8 years for naive T cells and 1–12 months for memory T cells (9, 10). Estimated half-lives for CD27$^-$ naive B cells, CD27$^+$ memory B cells, and plasma cells are somewhat more limited, at 22, 11, and 40 days, respectively (11, 12). For NK cells, half-lives are shorter, estimated at 7 days (13). Yet, they still outlast their innate counterparts, with monocytes' half-lives estimated at 71 h (14) and neutrophils' reported between 5 and 90 h (15). It is important to note that only actively dividing cells incorporate deuterium, and thus these studies may not account for long-lived, non-dividing cells. They are also based entirely in peripheral blood; cell kinetics in tissues may follow completely different patterns. Yet, with these caveats in mind, these data suggest that NK cells fall midway on the spectrum of cellular lifetime. These estimates may also serve as a reasonable proxy for the amount of time they require to respond to viral infection.

NK CELLS ARE INTERMEDIATE IN THEIR MECHANISM OF RECOGNITION

Cells of the innate system typically recognize non-self entities through specific receptor–ligand interactions. The toll-like receptor (TLR) system and other pattern recognition receptors,

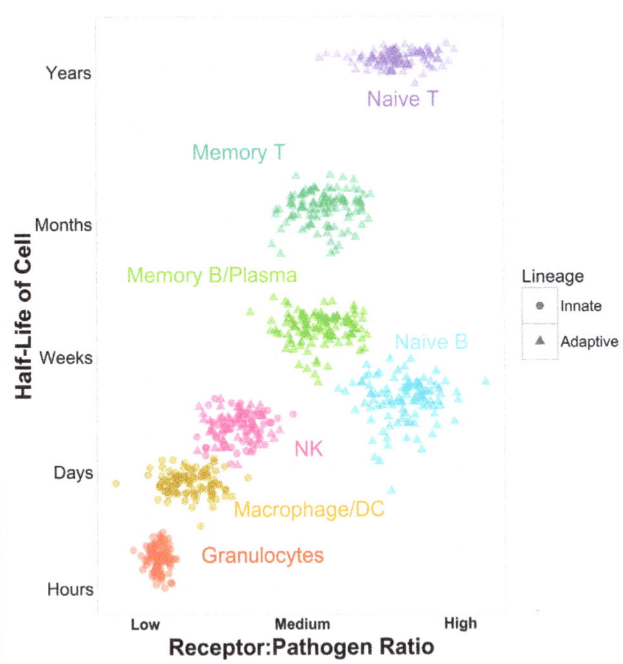

FIGURE 1 | Schematic of NK cells as immune intermediates in cell lineage, ratio of receptors to potential pathogens, and cellular half-life. Half-life values are approximated based on published data (9–15).

including RIG-I-like receptors, NOD-like receptors, and C-type lectin receptors, are designed to recognize specific features of pathogens or other danger signals (16, 17). Recognition is fast and digital, having only on/off states. It triggers phagocytosis, the release of inflammatory cytokines, and regulation of downstream adaptive responses. With 10 TLRs identified in humans to date and other pattern recognition receptor classes in the same order of magnitude or fewer, the innate paradigm is characterized by a low ratio of receptors per potential pathogen (**Figure 1**).

While the innate system has evolved to recognize broad classes of pathogens through a limited set of receptors, the adaptive system takes the converse approach of employing a broad set of receptors that each recognize a very limited set of pathogens. In T and B cells, V(D)J recombination forms specialized, antigen-specific receptors. When triggered in combination with costimulation, these receptors initiate proliferation, cytotoxicity, production of antibodies, and release of cytokines. The total potential diversity of the human T cell receptor (TCR) is estimated at 10^{15}–10^{20} sequences (18, 19). The adaptive system is therefore characterized by a high ratio of receptors to potential pathogens (**Figure 1**).

Natural killer cells take cues from both extremes. They express a spectrum of receptors, which can be either activating or inhibitory, through which they integrate signals to determine their activation status. These include killer immunoglobulin-like receptors (KIR), C-type lectin-like receptors, signaling lymphocyte activation molecule (SLAM) family receptors, and natural cytotoxicity receptors, among others (20). Unlike other cell types, NK activation is analog: increased activating or decreased inhibitory signals can tip the balance toward activation, but typically no specific receptor–ligand interaction is required. In fact, CD16 is the only receptor that has been shown to be capable of activating NK cells independently (21). In addition, NK cells are the only lymphocytes that can be activated purely through soluble signals. Proliferation, cytotoxicity, and a migratory phenotype can be induced by cytokines alone without specific antigen signal or presentation (22, 23).

Thus, the lifespan of NK cells, their mechanism of recognition, and in addition, recent evidence for memory in NK cell responses (24–26) suggest that they play a truly intermediate role between innate and adaptive responses (**Figure 1**).

SINGLE-CELL DIVERSITY: A LOGICAL ADAPTATION FOR MIDDLEMEN

The diversity in a T cell or B cell repertoire is captured in a single receptor. NK cells, by contrast, are diverse at a single-cell level. Each cell expresses a combination of receptors drawn from a vast potential pool. Our early estimates, based on 28 markers, place the total number of combinations in human peripheral blood on the order of 10^4 (27). With deeper sampling, measurement in a greater diversity of tissues, and improvements in technology allowing increased dimensionality, this estimate will certainly increase. A modest increase to measuring 40 NK cell receptors on a single cell would yield 10^{12} theoretical combinations. In comparison, T cells have been estimated to have 2.5×10^7 unique

TCR sequences in humans (28). As mentioned above, their total potential diversity is estimated at 10^{15}–10^{20} sequences (18, 19).

Thus, both theoretically and empirically, NK cells are highly diverse, but several orders of magnitude less so than T cells. Yet, T cell (and B cell) diversity is rigid, modifiable primarily by population-level expansion and contraction. The antigen-specific TCR or BCR, while optimizable via somatic hypermutation or affinity maturation, is pre-formed following exit from the bone marrow or thymus. For adaptive cells tasked with developing a slow but reliable memory response, it is a logical tradeoff to expend energy on selecting a vast but inflexible repertoire.

Natural killer diversity compensates for its lack of breadth with its flexibility, a logical evolutionary adaptation for cells that must act quickly. NK diversity can be modulated with vastly more agility and at many different levels. NK cells can up- or downregulate receptors to adjust repertoire diversity at the timescale of cellular processes rather than cellular division (29), one cell at a time rather than by population. Many NK cell receptors, especially the KIRs, are defined by vast population-level genetic diversity (30). This can be compounded at multiple levels: selective transcription and translation, post-translational modifications, protein trafficking, cytokine signaling, cell–cell interactions, and epigenetic modifications (31–33). NK cell diversity is thus much more responsive to both genetic and environmental determinants. This responsiveness may prove to be a critical feature for maintaining flexibility in the earliest lymphocyte response.

NK CELLS IN VIRAL SUSCEPTIBILITY

Because of their uniquely centered position in the immune system, their associations with improved control of viral infections (34), and recent evidence of their potential for a memory response, much interest has been focused on whether NK cells can be harnessed as part of a strategy for the initial prevention of viral infection (**Table 1**). Prevention is an especially significant goal for chronic incurable diseases, such as HIV. Mechanisms by which NK cells could successfully prevent viral infections include localization of NK cells to barriers of viral entry, efficient cytotoxicity, swift production of IFN-γ for recruitment and activation of downstream adaptive responses, and production of β-chemokines that can block viral entry (35).

Directly studying the immune correlates of human viral susceptibility is challenging. It requires either tracking uninfected people over long periods of time, a difficult logistical hurdle, or

TABLE 1 | Evidence for the role of NK cells in the prevention of viral acquisition.

Virus	Study population	Major finding	Total sample size	Reference
HSV-1	HSV-1-infected adults	The presence of KIR2DL2 or KIR2DS2 by PCR was associated with progression to symptomatic, as compared to asymptomatic, HSV-1 infection	131	(36)
HCV	HCV patients and healthy adult controls	RNA$^+$ HCV patients had increased presence of KIR2DL2 or KIR2DS2 by PCR compared with self-limited RNA$^-$ HCV patients	596	(37)
	Acutely infected HCV patients and healthy adults	NK cells from acutely infected HCV patients produced more IFN-γ and degranulated more than NK cells from healthy controls	39, 44	(38, 39)
	People who inject drugs with or without seroconversion or spontaneous clearance	Relative to chronically infected individuals, homozygosity for KIR2DL3 and its ligand HLA-C1 was more frequent in exposed seronegative individuals or those who spontaneously cleared the virus	1037, 305	(40, 41)
	Acute and chronic HCV patients, patients who naturally resolved infection, and healthy controls	During acute infection, fewer NKp30$^+$, NKp46$^+$, CD161$^+$, and NKG2D$^+$ NK cells were present in individuals who subsequently cleared than those who became chronically infected	57	(42)
	People who inject drugs with or without seroconversion, healthy adults	Higher anti-K562 cytotoxicity and higher NKp30 expression detected in exposed uninfected individuals	33	(43)
	People who inject drugs with or without seroconversion, healthy adults	Presence of KIR2DL3$^+$NKG2A$^-$ NK cells was associated with protection from productive HCV infection	114	(44)
Chikungunya	Chikungunya patients, healthy controls	NK cells from acutely infected chikungunya patients become activated and expand early in infection	55, 143	(45, 46)
EBV	EBV-college students	Increased CD56^{dim}NKG2A$^+$CD57$^+$NK cells detected in peripheral blood during acute infectious mononucleosis	18	(47)
CMV	Solid organ transplant recipients, bone marrow transplant recipients	CD57$^+$NKG2C^{hi} NK cells preferentially respond during CMV reactivation	140, 65	(48, 49)
HIV	South African women: cases acquired HIV, matched controls did not	Decreased pre-infection IFN-γ responses to autologous infected CD4$^+$ T cells were associated with increased acquisition risk	60	(50)
	South African women: cases acquired HIV, matched controls did not	Increased pre-infection NK activation (higher HLA-DR and lower CD38) was associated with increased acquisition risk	81	(51)
	Exposed uninfected intravascular drug users, seroconverters before or after seroconversion, unexposed controls	NK cells from exposed uninfected intravascular drug users showed greater lytic activity and produced more cytokines in response to cell lines than unexposed controls or seroconverters before or after seroconversion	75	(52)
	Kenyan women: cases acquired HIV, matched controls did not	Increased pre-infection NK diversity correlated with increased acquisition risk	36	(53)

infecting people in a controlled environment, an ethical quandary. Several approaches have therefore been used to provide indirect evidence for the importance of NK cells in human viral susceptibility.

First, much has been learned from studies of immunodeficient subjects with deficiencies in NK cell frequency or function. In addition to several mutations known to exert their effects on NK cells in relative isolation, at least 46 primary immunodeficiencies are also associated with NK cell defects. These subjects' overarching and unifying feature is the unusual susceptibility to herpesviruses (54), especially HSV-1, EBV, VSV, and HPV. These findings suggest that NK cells play a non-redundant role in preventing the establishment of these infections.

Second, many studies identify NK-relevant genetic correlates associated with disease-afflicted individuals versus healthy controls. These retrospective studies are not direct measures of acquisition probability, but do provide evidence of possible associations. The presence of KIR2DL2 and KIR2DS2 has been associated with progression to infection in both HSV-1 (36) and HCV (37).

A third approach has been to examine the characteristics and strength of NK responses during acute infection and sometimes draw correlations with disease outcome. In these studies, blood is typically drawn at the time of clinical presentation. This approach provides an imprecise approximation of the earliest events in infection but has the distinct advantage of allowing functional immunological studies. These studies have identified several early-stage markers of NK cell activation in viral infection. In HCV, KIR2DL3 in combination with HLA-C1 at both the genetic and cellular level has emerged as an important correlate of protection from infection (40, 41, 44). In addition, NKG2C$^+$ NK cells have been shown to specifically respond to CMV reactivation in transplant recipients (48, 49).

A fourth type of study measures the immune activity of individuals who remain seronegative despite behavior that grants high probability of viral exposure, providing clues as to what is assumed to be an effective protective immune response. The primary issue with these studies is that this form of resistance is not fully understood, and the highly exposed immune state of these individuals (typically sex workers or people who inject drugs) may not be representative of other routes of acquisition.

Finally, a handful of studies, so far only in the HIV field, have enrolled and tracked large cohorts at risk of infection, banking blood samples before and after detection of infection, and performed functional studies on pre-infection samples. While impressive in scope, their conclusions have unfortunately not been entirely clear. While the ability of NK cells to secrete IFN-γ in response to HIV-infected cells was associated with decreased risk of infection (50), generalized NK cell activation (as measured primarily by expression of CD38 and/or HLA-DR) was associated with increased risk of HIV acquisition (51). Thus, depending on its nature, NK cell activation has been both positively and negatively associated with increased risk of HIV acquisition.

Our recent study took a different approach. In addition to measuring activation status and individual markers, we calculated a diversity score for the NK cell repertoire of each donor before the onset of HIV infection. We showed that no single marker was predictive of HIV acquisition risk, but that higher NK cell diversity was associated with increased risk of HIV acquisition in a cohort of Kenyan women (53). This raises the intriguing possibility that NK repertoire diversification is detrimental to the NK cell response. As it represents an apparent state of ramification and inflexibility, high NK diversity may signal a higher risk of exposure and/or less resistant state of immunity.

All of these studies involve clinical cohorts that are limited in size. The brute-force approach of obtaining additional validation cohorts will be a necessary step in bolstering these results. However, targeted single-cell approaches that incorporate leading edge technologies and account for the increasingly appreciated diversity of the NK repertoire are in rapid development. These approaches hold great promise in bringing a new level of understanding to the role of NK cells in viral susceptibility.

NEW NK CELL-FOCUSED APPROACHES IN SINGLE-CELL TECHNOLOGY

Innovative microchip-based approaches have begun to uncover broad diversity in single-cell function. In a single-cell tracking system, the distribution of average kill times on a per-NK-cell basis was skewed, with a few "serial killers" performing the majority of the killing (55, 56). Furthermore, increased cytotoxicity has been shown to result from simultaneous interaction of NK cells with multiple targets (57) as well as IL-2 activation (23). NKG2A$^+$ NK cells have also been shown to be superior to NKG2A$^-$ NK cells in terms of dynamic migration, conjugation, spreading, and killing. "Serial killers" were also more common in this population (58).

Natural killer cell-focused mass cytometry, which has been pioneered by our group and others (27, 53, 59, 60), also holds great promise in further defining the role of NK diversity in viral acquisition at a single-cell level. This technology will be especially powerful when used in conjunction with humanized mice (61), presenting the next frontier in detailed tracking of all stages of the human NK cell response to viral infection.

Furthermore, single-cell dissection of functional diversity, especially in the context of a viral response, will greatly improve understanding of the processes by which NK cells may contribute to the prevention of infection. The generation of more single-cell data will also help to alleviate the statistical issues associated with small sample sizes in clinical studies.

Many questions remain about the functional consequences of NK cell diversity. Does a differentiated NK cell respond better and faster when re-encountering its initial stimulus? What is the spectrum of antigens that can drive NK cell diversification? Does a diverse NK cell repertoire correspond to a less diverse T and B cell repertoire? In addition, most studies have focused only on peripheral blood, but tissue-specific functions of NK cells are increasingly being appreciated (62).

Together, single-cell studies offer great promise in defining the scope of the human NK repertoire and its significance in the context of viral infection. This understanding will be essential in order to optimally harness the NK cells in the next generation of NK cell-based vaccines and therapeutics.

AUTHOR CONTRIBUTIONS

CB and DSA wrote the manuscript, and approved it for publication.

ACKNOWLEDGMENTS

We thank the members of the Blish laboratory for critical feedback.

REFERENCES

1. Herberman RB, Nunn ME, Lavrin DH. Natural cytotoxic reactivity of mouse lymphoid cells against syngeneic acid allogeneic tumors. I. Distribution of reactivity and specificity. *Int J Cancer* (1975) **16**:216–29. doi:10.1002/ijc.2910160204
2. Kiessling R, Klein E, Wigzell H. "Natural" killer cells in the mouse. I. Cytotoxic cells with specificity for mouse Moloney leukemia cells. Specificity and distribution according to genotype. *Eur J Immunol* (1975) **5**:112–7. doi:10.1002/eji.1830050208
3. Lanier LL, Le AM, Phillips JH, Warner NL, Babcock GF. Subpopulations of human natural killer cells defined by expression of the Leu-7 (HNK-1) and Leu-11 (NK-15) antigens. *J Immunol* (1983) **131**:1789–96.
4. Artis D, Spits H. The biology of innate lymphoid cells. *Nature* (2015) **517**:293–301. doi:10.1038/nature14189
5. Pahl J, Cerwenka A. Tricking the balance: NK cells in anti-cancer immunity. *Immunobiology* (2015):1–10. doi:10.1016/j.imbio.2015.07.012
6. Waggoner SN, Reighard SD, Gyurova IE, Cranert SA, Mahl SE, Karmele EP, et al. Roles of natural killer cells in antiviral immunity. *Curr Opin Virol* (2015) **16**:15–23. doi:10.1016/j.coviro.2015.10.008
7. Lanier LL. NK cell recognition. *Annu Rev Immunol* (2005) **23**:225–74. doi:10.1146/annurev.immunol.23.021704.115526
8. Lanier LL. Up on the tightrope: natural killer cell activation and inhibition. *Nat Immunol* (2008) **9**:495–502. doi:10.1038/ni1581
9. Farber DL, Yudanin NA, Restifo NP. Human memory T cells: generation, compartmentalization and homeostasis. *Nat Rev Immunol* (2014) **14**:24–35. doi:10.1038/nri3567
10. Vrisekoop N, den Braber I, de Boer AB, Ruiter AFC, Ackermans MT, van der Crabben SN, et al. Sparse production but preferential incorporation of recently produced naive T cells in the human peripheral pool. *Proc Natl Acad Sci U S A* (2008) **105**:6115–20. doi:10.1073/pnas.0709713105
11. Bernasconi NL, Traggiai E, Lanzavecchia A. Maintenance of serological memory by polyclonal activation of human memory B cells. *Science* (2002) **298**:2199–202. doi:10.1126/science.1076071
12. Macallan DC, Wallace DL, Zhang Y, Ghattas H, Asquith B, de Lara C, et al. B-cell kinetics in humans: rapid turnover of peripheral blood memory cells. *Blood* (2005) **105**:3633–40. doi:10.1182/blood-2004-09-3740
13. Zhang Y, Wallace DL, de Lara CM, Ghattas H, Asquith B, Worth A, et al. In vivo kinetics of human natural killer cells: the effects of ageing and acute and chronic viral infection. *Immunology* (2007) **121**:258–65. doi:10.1111/j.1365-2567.2007.02573.x
14. Whitelaw DM. Observations on human monocyte kinetics after pulse labeling. *Cell Prolif* (1972) **5**(4):311–7. doi:10.1111/j.1365-2184.1972.tb00369.x/pdf
15. Tak T, Tesselaar K, Pillay J, Borghans JAM, Koenderman L. What's your age again? Determination of human neutrophil half-lives revisited. *J Leukoc Biol* (2013) **94**:595–601. doi:10.1189/jlb.1112571
16. Goubau D, Deddouche S, Sousa CRE. Cytosolic sensing of viruses. *Immunity* (2013) **38**:855–69. doi:10.1016/j.immuni.2013.05.007
17. Takeuchi O, Akira S. Pattern recognition receptors and inflammation. *Cell* (2010) **140**:805–20. doi:10.1016/j.cell.2010.01.022
18. Davis MM, Bjorkman PJ. T-cell antigen receptor genes and T-cell recognition. *Nature* (1988) **334**:395–402. doi:10.1038/334395a0
19. Zarnitsyna VI, Evavold BD, Schoettle LN, Blattman JN, Antia R. Estimating the diversity, completeness, and cross-reactivity of the T cell repertoire. *Front Immunol* (2013) **4**:485. doi:10.3389/fimmu.2013.00485
20. Watzl C, Long EO. Signal transduction during activation and inhibition of natural killer cells. *Curr Protoc Immunol* (2010) **Chapter 11**:Unit11.9B. doi:10.1002/0471142735.im1109bs90
21. Bryceson YT, March ME, Ljunggren H-G, Long EO. Synergy among receptors on resting NK cells for the activation of natural cytotoxicity and cytokine secretion. *Blood* (2006) **107**:159–66. doi:10.1182/blood-2005-04-1351
22. Long EO, Sik Kim H, Liu D, Peterson ME, Rajagopalan S. Controlling natural killer cell responses: integration of signals for activation and inhibition. *Annu Rev Immunol* (2013) **31**:227–58. doi:10.1146/annurev-immunol-020711-075005
23. Olofsson PE, Forslund E, Vanherberghen B, Chechet K, Mickelin O, Ahlin AR, et al. Distinct migration and contact dynamics of resting and IL-2-activated human natural killer cells. *Front Immunol* (2014) **5**:80. doi:10.3389/fimmu.2014.00080
24. Paust S, Gill HS, Wang B-Z, Flynn MP, Moseman EA, Senman B, et al. Critical role for the chemokine receptor CXCR6 in NK cell-mediated antigen-specific memory of haptens and viruses. *Nat Immunol* (2010) **11**:1127–35. doi:10.1038/ni.1953
25. Reeves RK, Li H, Jost S, Blass E, Li H, Schafer JL, et al. Antigen-specific NK cell memory in rhesus macaques. *Nat Immunol* (2015) **16**:927–32. doi:10.1038/ni.3227
26. Sun JC, Beilke JN, Lanier LL. Adaptive immune features of natural killer cells. *Nature* (2009) **457**:557–61. doi:10.1038/nature07665
27. Horowitz A, Strauss-Albee DM, Leipold M, Kubo J, Nemat-Gorgani N, Dogan OC, et al. Genetic and environmental determinants of human NK cell diversity revealed by mass cytometry. *Sci Transl Med* (2013) **5**:208ra145. doi:10.1126/scitranslmed.3006702
28. Arstila TP, Casrouge A, Baron V, Even J, Kanellopoulos J, Kourilsky P. A direct estimate of the human alphabeta T cell receptor diversity. *Science* (1999) **286**:958–61. doi:10.1126/science.286.5441.958
29. Hromadnikova I, Pirkova P, Sedlackova L. Influence of in vitro IL-2 or IL-15 alone or in combination with Hsp-70-derived 14-mer peptide (TKD) on the expression of NK cell activatory and inhibitory receptors. *Mediators Inflamm* (2013) **2013**:1–12. doi:10.1155/2013/405295
30. Parham P. The genetic and evolutionary balances in human NK cell receptor diversity. *Semin Immunol* (2008) **20**:311–6. doi:10.1016/j.smim.2008.10.002
31. Lee J, Zhang T, Hwang I, Kim A, Nitschke L, Kim M, et al. Epigenetic modification and antibody-dependent expansion of memory-like NK cells in human cytomegalovirus-infected individuals. *Immunity* (2015) **42**:431–42. doi:10.1016/j.immuni.2015.02.013
32. Luetke-Eversloh M, Hammer Q, Durek P, Nordström K, Gasparoni G, Pink M, et al. Human cytomegalovirus drives epigenetic imprinting of the IFNG locus in NKG2Chi natural killer cells. *PLoS Pathog* (2014) **10**:e1004441. doi:10.1371/journal.ppat.1004441
33. Schlums H, Cichocki F, Tesi B, Theorell J, Béziat V, Holmes TD, et al. Cytomegalovirus infection drives adaptive epigenetic diversification of NK cells with altered signaling and effector function. *Immunity* (2015) **42**:443–56. doi:10.1016/j.immuni.2015.02.008
34. Jost S, Altfeld M. Control of human viral infections by natural killer cells. *Annu Rev Immunol* (2013) **31**:163–94. doi:10.1146/annurev-immunol-032712-100001
35. Fehniger TA, Herbein G, Yu H, Para MI, Bernstein ZP, O'Brien WA, et al. Natural killer cells from HIV-1+ patients produce C-C chemokines and inhibit HIV-1 infection. *J Immunol* (1998) **161**:6433–8.
36. Estefanía E, Gómez Lozano N, Portero F, De Pablo R, Solís R, Sepúlveda S, et al. Influence of KIR gene diversity on the course of HSV-1 infection: resistance to the disease is associated with the absence of KIR2DL2 and KIR2DS2. *Tissue Antigens* (2007) **70**:34–41. doi:10.1111/j.1399-0039.2007.00844.x
37. Paladino N, Flores AC, Marcos CY, Fainboim H, Theiler G, Arruvito L, et al. Increased frequencies of activating natural killer receptors are associated with liver injury in individuals who do not eliminate hepatitis C virus. *Tissue Antigens* (2007) **69**:109–11. doi:10.1111/j.1399-0039.2006.762_7.x
38. Amadei B, Urbani S, Cazaly A, Fisicaro P, Zerbini A, Ahmed P, et al. Activation of natural killer cells during acute infection with hepatitis C virus. *Gastroenterology* (2010) **138**:1536–45. doi:10.1053/j.gastro.2010.01.006

39. Pelletier S, Drouin C, Bédard N, Khakoo SI, Bruneau J, Shoukry NH. Increased degranulation of natural killer cells during acute HCV correlates with the magnitude of virus-specific T cell responses. *J Hepatol* (2010) **53**:805–16. doi:10.1016/j.jhep.2010.05.013
40. Khakoo SI, Thio CL, Martin MP, Brooks CR, Gao X, Astemborski J, et al. HLA and NK cell inhibitory receptor genes in resolving hepatitis C virus infection. *Science* (2004) **305**:872–4. doi:10.1126/science.1097670
41. Knapp S, Warshow U, Hegazy D, Brackenbury L, Guha IN, Fowell A, et al. Consistent beneficial effects of killer cell immunoglobulin-like receptor 2DL3 and group 1 human leukocyte antigen-C following exposure to hepatitis C virus. *Hepatology* (2010) **51**:1168–75. doi:10.1002/hep.23477
42. Alter G, Jost S, Rihn S, Reyor LL, Nolan BE, Ghebremichael M, et al. Reduced frequencies of NKp30+NKp46+, CD161+, and NKG2D+ NK cells in acute HCV infection may predict viral clearance. *J Hepatol* (2011) **55**:278–88. doi:10.1016/j.jhep.2010.11.030
43. Golden-Mason L, Cox AL, Randall JA, Cheng L, Rosen HR. Increased natural killer cell cytotoxicity and NKp30 expression protects against hepatitis C virus infection in high-risk individuals and inhibits replication in vitro. *Hepatology* (2010) **52**:1581–9. doi:10.1002/hep.23896
44. Thoens C, Berger C, Trippler M, Siemann H, Lutterbeck M, Broering R, et al. KIR2DL3^{+}NKG2A^{-} natural killer cells are associated with protection from productive hepatitis C virus infection in people who inject drugs. *J Hepatol* (2014) **61**:475–81. doi:10.1016/j.jhep.2014.04.020
45. Petitdemange C, Becquart P, Wauquier N, Béziat V, Debré P, Leroy EM, et al. Unconventional repertoire profile is imprinted during acute chikungunya infection for natural killer cells polarization toward cytotoxicity. *PLoS Pathog* (2011) **7**:e1002268. doi:10.1371/journal.ppat.1002268.t001
46. Thanapati S, Das R, Tripathy AS. Phenotypic and functional analyses of NK and NKT-like populations during the early stages of chikungunya infection. *Front Microbiol* (2015) **6**:895. doi:10.3389/fmicb.2015.00895
47. Hendricks DW, Balfour HH, Dunmire SK, Schmeling DO, Hogquist KA, Lanier LL. Cutting edge: NKG2C(hi)CD57+ NK cells respond specifically to acute infection with cytomegalovirus and not Epstein-Barr virus. *J Immunol* (2014) **192**:4492–6. doi:10.4049/jimmunol.1303211
48. Foley B, Cooley S, Verneris MR, Curtsinger J, Luo X, Waller EK, et al. Human cytomegalovirus (CMV)-induced memory-like NKG2C+ NK cells are transplantable and expand *in vivo* in response to recipient CMV antigen. *J Immunol* (2012) **189**:5082–8. doi:10.4049/jimmunol.1201964
49. Lopez-Vergès S, Milush JM, Schwartz BS, Pando MJ, Jarjoura J, York VA, et al. Expansion of a unique CD57+NKG2Chi natural killer cell subset during acute human cytomegalovirus infection. *Proc Natl Acad Sci U S A* (2011) **108**:14725–32. doi:10.1073/pnas.1110900108
50. Naranbhai V, Altfeld M, Abdool Karim Q, Ndung'u T, Abdool Karim SS, Carr WH, et al. Natural killer cell function in women at high risk for HIV acquisition: insights from a microbicide trial. *AIDS* (2012) **26**:1745–53. doi:10.1097/QAD.0b013e328357724f
51. Naranbhai V, Abdool Karim SS, Altfeld M, Samsunder N, Durgiah R, Sibeko S, et al. Innate immune activation enhances HIV acquisition in women, diminishing the effectiveness of tenofovir microbicide gel. *J Infect Dis* (2012) **206**:993–1001. doi:10.1093/infdis/jis465
52. Scott-Algara D, Truong LX, Versmisse P, David A, Luong TT, Nguyen NV, et al. Cutting edge: increased NK cell activity in HIV-1-exposed but uninfected Vietnamese intravascular drug users. *J Immunol* (2003) **171**:5663–7. doi:10.4049/jimmunol.171.11.5663
53. Strauss-Albee DM, Fukuyama J, Liang EC, Yao Y, Jarrell JA, Drake AL, et al. Human NK cell repertoire diversity reflects immune experience and correlates with viral susceptibility. *Sci Transl Med* (2015) **7**:297ra115. doi:10.1126/scitranslmed.aac5722
54. Orange JS. Natural killer cell deficiency. *J Allergy Clin Immunol* (2013) **132**:515–25. doi:10.1016/j.jaci.2013.07.020
55. Choi PJ, Mitchison TJ. Imaging burst kinetics and spatial coordination during serial killing by single natural killer cells. *Proc Natl Acad Sci U S A* (2013) **110**(16):6488–93. doi:10.1073/pnas.1221312110
56. Vanherberghen B, Olofsson PE, Forslund E, Sternberg-Simon M, Khorshidi MA, Pacouret S, et al. Classification of human natural killer cells based on migration behavior and cytotoxic response. *Blood* (2013) **121**:1326–34. doi:10.1182/blood-2012-06-439851
57. Christakou AE, Ohlin M, Vanherberghen B, Khorshidi MA, Kadri N, Frisk T, et al. Live cell imaging in a micro-array of acoustic traps facilitates quantification of natural killer cell heterogeneity. *Integr Biol (Camb)* (2013) **5**:712–9. doi:10.1039/c3ib20253d
58. Forslund E, Sohlberg E, Enqvist M, Olofsson PE, Malmberg KJ, Onfelt B. Microchip-based single-cell imaging reveals that CD56dimCD57-KIR-NKG2A+ NK cells have more dynamic migration associated with increased target cell conjugation and probability of killing compared to CD56dimCD57-KIR-NKG2A-NK cells. *J Immunol* (2015) **195**:3374–81. doi:10.4049/jimmunol.1500171
59. Horowitz A, Guethlein LA, Nemat-Gorgani N, Norman PJ, Cooley S, Miller JS, et al. Regulation of adaptive NK cells and CD8 T cells by HLA-C correlates with allogeneic hematopoietic cell transplantation and with *Cytomegalovirus* reactivation. *J Immunol* (2015) **195**(9):4524–36. doi:10.4049/jimmunol.1401990
60. Strauss-Albee DM, Horowitz A, Parham P, Blish CA. Coordinated regulation of NK receptor expression in the maturing human immune system. *J Immunol* (2014) **193**:4871–9. doi:10.4049/jimmunol.1401821
61. Paust S, Bettini M. Daring to learn from humanized mice. *Blood* (2015) **125**:3829–31. doi:10.1182/blood-2015-04-639435
62. Marquardt N, Béziat V, Nyström S, Hengst J, Ivarsson MA, Kekäläinen E, et al. Cutting edge: identification and characterization of human intrahepatic CD49a+ NK cells. *J Immunol* (2015) **194**:2467–71. doi:10.4049/jimmunol.1402756

15

Natural Killers are made not Born: How to Exploit NK Cells in Lung Malignancies

Paolo Carrega[1,2]* *and Guido Ferlazzo*[1,2,3,4]

[1] Laboratory of Immunology and Biotherapy, Department of Human Pathology, University of Messina, Messina, Italy, [2] Cell Factory Center, University of Messina, Messina, Italy, [3] Cell Therapy Program, University Hospital Policlinico G.Martino, Messina, Italy, [4] Division of Clinical Pathology, University Hospital Policlinico G.Martino, Messina, Italy

Edited by:
Eric Vivier,
Centre d'immunologie de Marseille-Luminy, France

Reviewed by:
Christine Susanne Falk,
Hannover Medical School, Germany
Daniel Olive,
Institut national de la santé et de la recherche médicale (INSERM), France

***Correspondence:**
Paolo Carrega
carregap80@gmail.com

In recent years, progress has been made in the characterization of natural killer (NK) cells in lung malignancies, and we have now gained a better understanding of the frequency, localization, phenotype, and functional status of NK cells infiltrating these tumors. NK cell subset recruited in lung cancer is mainly capable of producing relevant cytokines rather than exerting direct cancer cell killing. Thus, the relevance of NK cells in tumor microenvironment might also go beyond the killing of tumor cells, being NK cells endowed with regulatory functions toward an ample array of immune effectors. Nevertheless, boosting their cytotoxic functions and redirecting the migration of cytotoxic NK cell subset to the tumor site might open new therapeutic avenues for lung cancer. Also, we believe that a deeper investigation into the impact of both conventional (e.g., chemotherapy) or new therapies (e.g., anti-immune checkpoints mAbs) on NK cell homeostasis in lung cancer patients is now required.

Keywords: natural killer cells, lung cancer, tumor microenvironment, cytotoxicity, cytokines

Several lines of evidence support the notion that natural killer (NK) cells play an important role in the control of tumor growth. Notably, along with the large array of results derived from experimental mouse models, a reduced NK cell lytic activity has been associated in human with an increased risk of cancer. Nevertheless, only a very limited number of NK cells can be detected within the majority of human solid tumors.

The discovery that NK cell can easily recognize and lyse tumor cells, immediately translated into hope that NK cells might find a place in the clinic as therapeutic tools. Extensive efforts have been made to exploit NK cells into the clinic and many of these attempts have mainly relayed on knowledge obtained by studying peripheral blood (PB) NK cells. However, the environmental context and cytokine milieu in which NK cells are present, and perceive potential activating signals, may strongly impact on their ultimate effector activity, thus offering one of the possible explanations of the inconsistent outcomes of the majority of NK cell-based therapies. As a matter of fact, several studies have aimed to investigate the biology of NK cells present within the neoplastic tissue and the clinical relevance of this infiltrate: in this context, lung cancer, the main cause of cancer-related mortality, represents a major topic for studying the anti-tumoral role of NK cells.

NK CELLS IN LUNG CANCER AND THEIR CLINICAL SIGNIFICANCE

The impact of NK cells as a potential prognostic marker in resected NSCLC still remains an open question. Although some early studies have demonstrated that the presence of NK cells in the immune infiltrate was associated with a lower risk of relapse and/or longer survival (1, 2), these analyses may now be considered incomplete or not exhaustive because of the use of markers not exclusively expressed by NK cells. Indeed, a more specific analysis of this aspect recently failed to find an association between the presence of NKp46$^+$ NK cells at early stages of the disease with the clinical outcome (3). However, another recent report showed that the number of NK cell infiltrating lung cancer tissue (detected as CD56$^+$CD16$^+$ cells) positively correlated to the survival of lung cancer patients (4).

Although the clinical significance of NK cells in lung cancers seems to require more work, our knowledge about the features of NK cells infiltrating NSCLC has expanded considerably in recent years. A certain consensus has been reached regarding the evidence that NK cells associated to the immune infiltrate of NSCLC are quite different from those normally found in the non-tumor areas of the lungs (5–7), since normal lungs are mainly populated by CD56^{dim} NK cells, whereas tumors are mainly enriched in CD56^{bright} NK cell subset. Overall, the NK cell population found within NSCLC has been described as displaying profound alterations in the expression of relevant NK cell receptors, more specifically, downregulation of activating (NKp30, NKp80, CD16, DNAM-1) and inhibitory (ILT-2 and KIRs) receptors, as well as upregulation of NKG2A when compared to the normal counterpart (3). Functionally, intratumoral NK cells displayed impairments in the ability to both degranulate (8) and produce IFNγ (3) in response to classical NK cell targets, and they may rather acquire a proangiogenic phenotype due to the tumor microenvironment (TM) (9).

Nevertheless, these conclusions were mainly drawn by considering the NSCLC-infiltrating NK cells as a whole population. Recent reports suggested that expression of markers such as CD69, CD103, and CD49a can distinguish tissue resident from circulating NK cells (10). These advances indicate the necessity to revise our previous data and call for a thorough investigation of the various NK cell subsets infiltrating lung tumors. In this regard, high levels of CD69 have been observed on intratumoral NK cells (8) (interpreted as activation marker, at that time) when compared to matched non-tumoral lung tissues, thus suggesting that a substantial amount of NK cells may be actually retained within the TM upon migration (or alternatively upon *in situ* development). Along with CD69, CD103 has also been found to be expressed on a subset of NKp46$^+$ cells within the TM (although potentially co-expressed by ieILC1) [(11) and personal observations by P. Carrega]. These observations propel new questions about the specific contribution of tissue-resident NK cells in lung tumor immunosurveillance as well as about their activity in already established tumors.

A relative abundance of CD56^{bright} NK cells was also observed in pleural effusions (PEs) from different type of primary and metastatic tumors. In contrast with NK cells isolated from solid lung cancer tissues, PE-NK cells express normal levels of both main activating receptors and MHC Class I-specific inhibitory receptors and they rapidly release cytokines upon exposure to neoplastic cells (12). These data further confirm how the microenvironment and cytokine milieu of lung cancer can exert a strong influence on NK cell effector activities (13, 14).

ACCUMULATION OF CD56^{bright} NK CELLS AT THE TUMOR SITE

Overall, current data show that NK cells are very rare within human NSCLC, and this evidence is in accordance with parallel observations in other solid tumors. Similarly to other tumors, NSCLC-infiltrating NK cells resemble PB-CD56^{bright} in their phenotype. These data raise various questions about the actual function of this "regulatory" NK cell subset at the tumor site. Primarily, whether the enrichment of CD56^{bright} NK cells in lung tumors represents a preferential recruitment of these cells from PB or adjacent tissues or rather a local expansion of immature NK cells within the tumor. Recent findings have revealed that tumor microenvironment may play a role in this specific accumulation. In particular, comparison of gene expression data between neoplastic and healthy lung tissues showed a chemokine switch (occurring upon neoplastic transformation) that is in agreement with the accumulation of non-cytotoxic CD56^{bright} NK cells recruited from PB (6). Specifically, variations in the tumor tissues involved a significant downregulation of CXCL2 that can selectively attract CD56^{dim} NK cells and, vice versa, an upregulation of chemokines specific for CCR7 and CXCR3 receptors (i.e., CCL19, CXCL9, and CXCL10), which are, on the contrary, preferentially expressed by CD56^{bright} NK cells. This might represent a further mechanism of cancer immunoediting with implications for both immunosurveillance and tumor escape from NK cell attack. Remarkably, also breast cancer, another tumor type characterized by enrichment in non-cytotoxic CD56^{bright} NK cells, displayed upregulation of genes coding for chemokines attracting this subset, when compared with gene expression profile of healthy breast tissues. Since NSCLC are often associated with the presence of intratumoral tertiary lymphoid structures (15), it is conceivable that these ectopic lymphoid tissues, as well as the establishment of a lymphoid-like stroma within the tumor, might drive the expression of chemokines normally secreted in secondary lymphoid organs (CCL19, CCL21, etc.) and, therefore, preferentially attract CD56^{bright} non-cytotoxic NK cells at the tumor site (**Figure 1**). An experimental approach employing the use of humanized mice xenograft models in which the xenograft develops in the context of a human immune system might potentially help in answering the question on which NK cell subset preferentially migrate to the neoplastic tissues and to shed further light on the mechanisms lying behind their migratory properties.

On the other hand, however, a local expansion of less mature NK cells in the tumor cannot be excluded. In agreement with this hypothesis, it has been reported that a significant amount

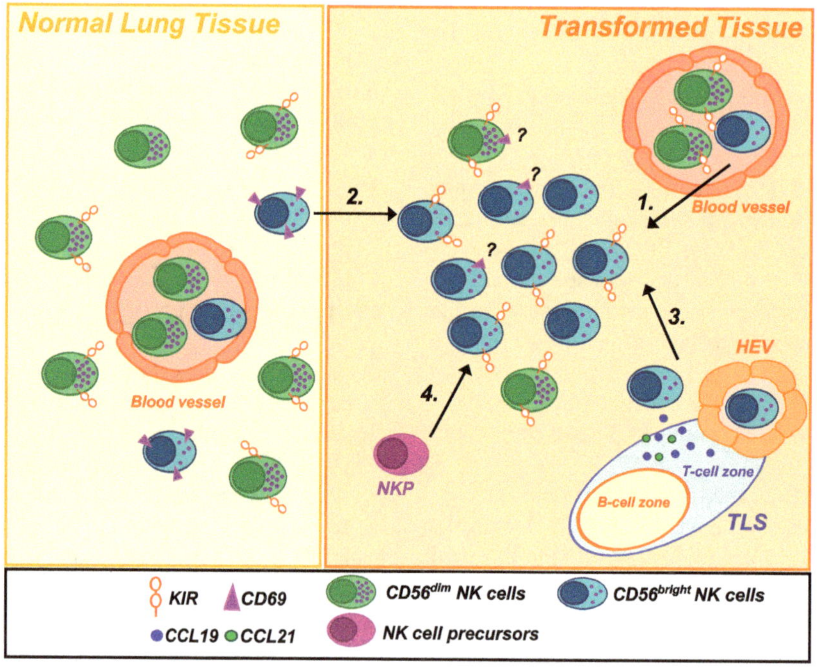

FIGURE 1 | Natural killer (NK) cell subsets in healthy and neoplastic lung tissues. Human healthy lung tissues are mainly populated by CD56^{dim}CD16^{+} NK cells but also present a small subset of CD56^{bright} NK cells expressing CD69, a marker of tissue-residency. Conversely, NSCLC tissues were found to be infiltrated by an NK cell population highly enriched in CD56^{bright}CD16^{neg}Perforin$^{low/neg}$KIR^{+} cells. The reasons for such an accumulation are not clear. They may derive from (1) extravasation of PB NK cells from newly formed blood vessels and/or (2) migration of CD56^{bright} NK cells from the adjacent normal lung tissue in response to upregulated chemokines (CXCL9/CXCL10) in the neoplastic tissues. Moreover, (3) they may be recruited by chemokines (such as CCL19/CCL21) expressed in tertiary lymphoid structures or a lymphoid-like stroma *via* high endothelial venules (HEV), which represent a new gateway for lymphocyte entry into the tumor. Finally, (4) a local expansion from NK cell precursors (NKP) (21)/immature NK cells cannot be excluded. Noteworthy, the ratio of tissue-resident versus circulating/non-residing NK cells in neoplastic tissue has so far not sufficiently been deeply investigated.

of CD56^{bright}perforin^{neg}CD16^{low} NK cells infiltrating lung cancers (and breast cancer) showed upregulation of KIRs on their surface (8, 16). These observations support the hypothesis that TM might, at least in some cases, might favor the expansion of functional intermediate of the normal differentiation between the CD56^{bright} and CD56^{dim} human NK-cell subsets (17, 18). Thus, as previously shown in *in vitro* studies (17), CD56^{bright} NK cells might locally proliferate and eventually differentiate into the cytotoxic counterpart under adequate stimuli obtained within TM.

It is noteworthy that the analysis of perforin expression results critical in this context for identifying the two main distinct subsets of NK cells, since the intratumoral CD56^{bright}CD16^{neg} NK cell phenotype could also represents the result of an activation-induced upregulation of CD56 and concomitant cleavage of CD16 by the ADAM17 protease (19, 20). Finally, we cannot exclude the possibility that tumor-infiltrating NK cells can originate from NK cell progenitors migrated to the tumor site (**Figure 1**). This hypothesis is supported by the recent discovery, in various human adult peripheral tissues, of an NK cell-restricted precursor (defined as Lin^{neg} CD34^{+} CD38^{+} CD123^{neg} CD45RA^{+} CD7^{+} CD10^{+} CD127^{neg}) able to give rise to both CD56^{dim} and CD56^{bright} NK cells *in vitro* and *in vivo* and expressing high levels of CD62L on their surface (21).

FUNCTIONS OF NK CELLS IN NSCLC MICROENVIRONMENT

Collectively, we know relatively little about the factors that regulate NK cell response under different physio-pathological conditions; therefore, the potential activity of NK cells in lung cancer microenvironment can just be speculative. The notion that CD56^{bright} NK cells are a subset endowed with immunoregulatory properties is now in general accepted: this is related to their ability to secrete cytokines able to influence innate immunity as well as to their ability to assist in the polarization of primary adaptive immune responses in secondary lymphoid organs. However, in the context of human cancers, they may even exert an immune suppressive function. As a matter of fact, the CD56^{bright} CD16^{low}KIR$^{+/-}$ phenotype found in lung cancers is also reminiscent of NK cells present in decidua tissue and it might be induced upon NK cell entry into a tumor microenvironment expressing substantial TGFβ (22). Also, results from a recent investigation in melanoma are in line with a potential regulatory role by CD56^{bright} NK cells (23). According to this study, blood CD56^{bright}CD16^{neg} NK cells express higher levels of ectoenzymes than CD56^{dim} NK cells; hence, they may act as "regulatory cells" by inhibiting autologous CD4^{+} T cell proliferation. This activity is mediated by production of adenosine, with a pivotal role played by CD38. Moreover, upon proper activation, $CD56^{bright}$ NK cells are equally, if not more, cytotoxic than $CD56^{dim}$ NK cells against autologous activated $CD4^{+}$ T cells (24). This observation may be of interest in NSCLC, where NK cells were mainly found in the tumor stroma, in close contact with other leukocytes (8). It should be noted that also NK cells from PE, upon IL-2 activation, gain a strong cytolitic activity against tumor cells (12), even stronger than PB-NK cells undergoing the same activation. These findings raise new questions about a suitable therapeutic exploitation of NK cells: more specifically, about the possibility to boost cytotoxic function of $CD56^{bright}$ NK cells against cancer cells while avoiding elimination of activated autologous $CD4^{+}$ T cells.

As a whole, however, few *in vivo* data have so far been collected in support of an immunosuppressive role of NK cells within tumor tissues and in lung cancer in particular. Nevertheless, this remains an interesting and poorly explored field of investigation.

On the other hand, NK cells have been implicated in the response to tumors also because of their abundant release of IFNγ. Experimental animal studies highlighted the importance of NK cell-mediated secretion of IFNγ in tumor rejection. In particular, the initial expression of IFNγ by NK cells may induce expression of CXCR3 ligands and, in turn, contribute to further accumulation of NK cells and tumor rejection (25). Tu et al. showed a pivotal role for CD160 and IFNγ secretion in controlling tumor growth by NK cells (26). An early release of IFNγ may be of relevance also for upregulating co-stimulatory molecules on dendritic cell (DC) surface, endowing DCs with a more potent triggering potential for tumor-specific $CD8^{+}$ T cells (27), but also, as highlighted below, for the expression of ligands of immune inhibitory receptors.

ROLE OF NK CELLS IN CONVENTIONAL AND UNCONVENTIONAL ANTICANCER THERAPY

The role of NK cells in the conventional therapy of lung cancer is only partially known; in addition, the possible modifications of NK cell activities and homeostasis upon administration of therapies currently used for lung cancers, ranging from conventional chemotherapy to anti-PD1 blocking mAbs, have been thus far scarcely investigated. Nevertheless, various evidences support the notion that beneficial effects of conventional therapies may also be mediated as "off-target" mechanisms by this innate immune cell population. For example, both cisplatin (28) and gemcitabine treatments (29) induce upregulation of NKG2D ligands on cancer cells. Similarly, also inhibitors of the MEK/ERK signaling pathway can upregulate the expression of ULBP2, one of the ligand of NKG2D, by inducing demethylation of the ULBP2 promoter (30). Moreover, NK cells may contribute to the efficacy of targeted therapies, such as cancer-associated fibroblast-targeted therapy used in combination with DC-based vaccines (31), or agents targeting EGFR, such as erlotinib (32) and necitumumab. The latter has been approved as a first-line treatment for advanced squamous NSCLC, the less immunogenic type of NSCLC tumor. Interestingly, the necitumumab–EGFR complex can induce ADCC by various types of immune system cells, such as NK lymphocytes (33).

In light of the therapeutic benefit obtained with immune checkpoint inhibitors, such as mAbs blocking programed cell death protein 1 (PD1) pathway, it would be interesting to further investigate the effects of these classes of drugs on NK cell compartment. Anti-PD1 (nivolumab and pembrolizumab), anti-KIR (lirilumab), and anti-NKG2A (monalizumab) may act by removing a brake on the effector functions of NK cells. Anti-PD1 agents have been already approved for the treatment of advanced NSCLC and, very recently, have shown promising results when used in the neoadjuvant setting (34). Anti-PD1 therapy might unleash the cytotoxic potential of activated NK cells expressing PD1 (35), thus contributing to the anticancer activities of this drug. On the other hand, NK cells might also be detrimental for avoiding PD1 blocking since IFNγ derived from tumor-infiltrating NK cells may further increase the expression of the ligand PD-L1 in the tumor microenvironment.

While we are now accomplishing groundbreaking results by anti-PD1 agents, anti-KIR agents are currently in phase II; nonetheless, they may be important in various settings, in neoadjuvant and advanced settings but also as adjuvant therapy in order to reduce risks of recurrence. Along with KIRs, targeting of the inhibitory receptor NKG2A (monalizumab) is currently under investigation as single agent or in combination with other agents in different types of solid tumors (36). The rationale for the use of this agent in lung cancer is also supported by the higher expression levels of NKG2A in intratumoral NK cells if compared to NK cells in normal lung tissues (6, 8). Another therapy targeting the inhibitory receptor LAG3 (BMS-986016), expressed on lymphocytes and NK cells, is also currently investigated (NCT01968109) alone and in combination with nivolumab in subjects with select advanced (metastatic and/or unresectable) solid tumors. Further increasing the number of targetable checkpoints on NK cells, recent preclinical data demonstrated the utility of blocking CD96 for enhancing NK cell activity against experimental and spontaneous metastases (37). Finally, targeting costimulatory receptors such as CD137 (4-1BB), expressed on activated NK cells and cytotoxic T cells, may represent another option, especially in combination with mAbs targeting TAA, as shown for head and neck cancer patients (38).

CONCLUSION AND PERSPECTIVE

Despite knowledge on NK cell biology significantly increased in these last years, we surprisingly still know relatively little about their functions in TM. This is particularly astonishing considering that NK cell discovery was related to their strong ability to recognize and kill cancer cells without prior sensitiza-

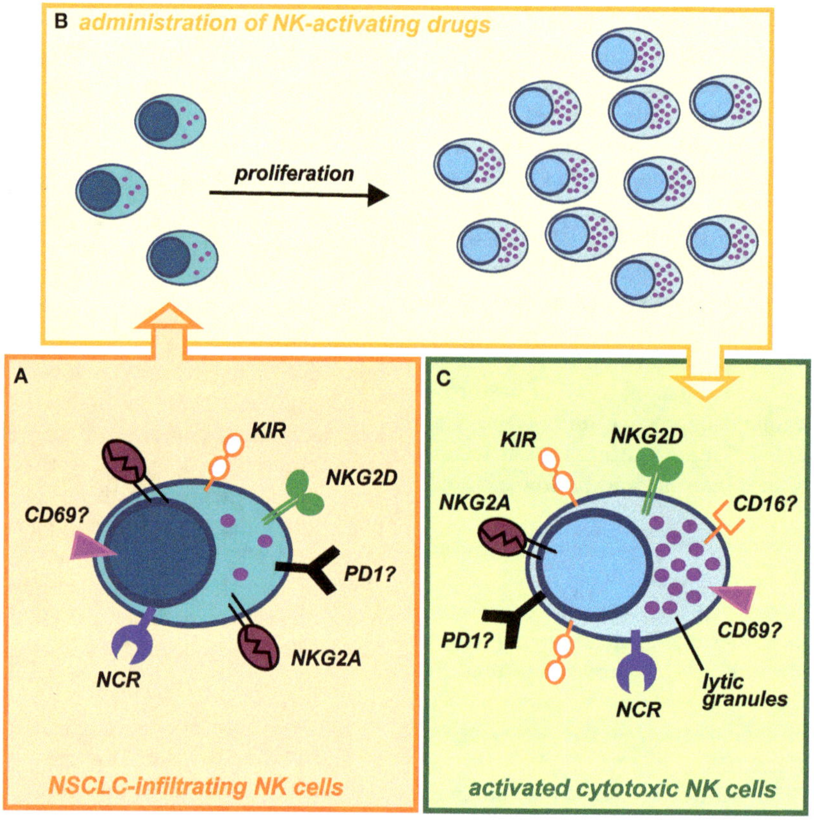

FIGURE 2 | Possible conversion of lung cancer-infiltrating natural killer (NK) cells into highly cytotoxic killer cells upon administration of NK-activating drugs. (A) NK cells isolated from NSCLC tissues display phenotypic features similar to those of peripheral blood (PB) CD56^bright NK cells, with the exception of KIRs, that can be detectable on these cells. Similarly to PB CD56^bright NK cells, they are endowed with high proliferative potential but impaired in their cytotoxic capability. In addition, it is not clear whether NK cells infiltrating human lung tumors express other immune checkpoints (such as PD-1), which might further contribute to their poor cytolytic activity. **(B)** Upon activation with stimulating cytokines (recombinant IL-15, IL-2) or mAbs-blocking inhibitory receptors (e.g., anti-NKG2A, which is highly expressed on these cells, or anti-PD1), intratumoral NK cells may proliferate *in situ* and modify their phenotype acquiring strong cytolytic capabilities against cancer cells. **(C)** Administration of cytokines or other NK-activating drugs, currently available, can upregulate the expression of cytolytic granules on expanded tumor-infiltrating NK cells but also induce higher levels of KIR inhibitory receptors: this might be circumvented by administration of anti-KIRs mAbs. It remains to be investigated whether *in vivo* stimulation of lung cancer-infiltrating NK cells can upregulate on these cells the expression of activating receptors (e.g., CD16), immune checkpoints, or markers of tissue residency.

tion. Clearly, further knowledge, particularly in human setting, is required to better exploit this abundant innate lymphocyte population in lung cancer treatment. However, recent evidences indicate that CD56^{bright} NK cell population infiltrating lung cancers, endowed with low ability to kill cancer cells, can spontaneously acquire KIR receptors and thus further differentiate, suggesting that they might convert in cytotoxic killer cells. Remarkably, CD56^{bright} NK cells, the main NK cell subset infiltrating lung cancer, can undergo robust proliferation upon appropriate cytokine stimulation. At the same time, factors able to efficiently activate NK cells, induce their proliferation and to terminally differentiate them into highly cytotoxic NK cells are currently available as clinical-grade agents (e.g., IL-15 and/or mAbs blocking NK inhibitory receptors) (**Figure 2**). Finally, although few data are currently available, we recently started to shed some light on the mechanisms ruling he recruitment of NK cells within lung cancer tissues. The use of factors able to modify lung cancer expression of chemokines involved in NK cell recruitment could represent an interesting therapeutic avenue.

While further basic investigation on the biology of lung cancer NK cells is clearly required, we recently gained a better insight in this field. Novel immunotherapeutic strategies for lung cancers, conceivably in combination with established anticancer therapies, should now aim at increasing the amount of cytotoxic NK cells in lung cancers: that could be achieved not only recruiting them from PB but also driving their proliferation and differentiation within TM.

AUTHOR CONTRIBUTIONS

PC and GF equally contributed in writing of this manuscript.

REFERENCES

1. Takanami I, Takeuchi K, Giga M. The prognostic value of natural killer cell infiltration in resected pulmonary adenocarcinoma. *J Thorac Cardiovasc Surg* (2001) 121(6):1058–63. doi:10.1067/mtc.2001.113026
2. Villegas FR, Coca S, Villarrubia VG, Jimenez R, Chillon MJ, Jareno J, et al. Prognostic significance of tumor infiltrating natural killer cells subset CD57 in patients with squamous cell lung cancer. *Lung Cancer* (2002) 35(1):23–8. doi:10.1016/S0169-5002(01)00292-6
3. Platonova S, Cherfils-Vicini J, Damotte D, Crozet L, Vieillard V, Validire P, et al. Profound coordinated alterations of intratumoral NK cell phenotype and function in lung carcinoma. *Cancer Res* (2011) 71(16):5412–22. doi:10.1158/0008-5472.CAN-10-4179
4. Jin S, Deng Y, Hao JW, Li Y, Liu B, Yu Y, et al. NK cell phenotypic modulation in lung cancer environment. *PLoS One* (2014) 9(10):e109976. doi:10.1371/journal.pone.0109976
5. Tomasello E, Yessaad N, Gregoire E, Hudspeth K, Luci C, Mavilio D, et al. Mapping of NKp46(+) cells in healthy human lymphoid and non-lymphoid tissues. *Front Immunol* (2012) 3:344. doi:10.3389/fimmu.2012.00344
6. Carrega P, Bonaccorsi I, Di Carlo E, Morandi B, Paul P, Rizzello V, et al. CD56brightPerforinlow noncytotoxic human NK cells are abundant in both healthy and neoplastic solid tissues and recirculate to secondary lymphoid organs via afferent lymph. *J Immunol* (2014) 192(8):3805–15. doi:10.4049/jimmunol.1301889
7. Marquardt N, Kekalainen E, Chen P, Kvedaraite E, Wilson JN, Ivarsson MA, et al. Human lung natural killer cells are predominantly comprised of highly differentiated hypofunctional CD69-CD56dim cells. *J Allergy Clin Immunol* (2016). doi:10.1016/j.jaci.2016.07.043
8. Carrega P, Morandi B, Costa R, Frumento G, Forte G, Altavilla G, et al. Natural killer cells infiltrating human nonsmall-cell lung cancer are enriched in CD56 bright CD16(-) cells and display an impaired capability to kill tumor cells. *Cancer* (2008) 112(4):863–75. doi:10.1002/cncr.23239
9. Bruno A, Focaccetti C, Pagani A, Imperatori AS, Spagnoletti M, Rotolo N, et al. The proangiogenic phenotype of natural killer cells in patients with non-small cell lung cancer. *Neoplasia* (2013) 15(2):133–42. doi:10.1593/neo.121758
10. Bjorkstrom NK, Ljunggren HG, Michaelsson J. Emerging insights into natural killer cells in human peripheral tissues. *Nat Rev Immunol* (2016) 16(5):310–20. doi:10.1038/nri.2016.34
11. Carrega P, Loiacono F, Di Carlo E, Scaramuccia A, Mora M, Conte R, et al. NCR(+)ILC3 concentrate in human lung cancer and associate with intratumoral lymphoid structures. *Nat Commun* (2015) 6:8280. doi:10.1038/ncomms9280
12. Vacca P, Martini S, Garelli V, Passalacqua G, Moretta L, Mingari MC. NK cells from malignant pleural effusions are not anergic but produce cytokines and display strong antitumor activity on short-term IL-2 activation. *Eur J Immunol* (2013) 43(2):550–61. doi:10.1002/eji.201242783
13. Donatelli SS, Zhou JM, Gilvary DL, Eksioglu EA, Chen X, Cress WD, et al. TGF-beta-inducible microRNA-183 silences tumor-associated natural killer cells. *Proc Natl Acad Sci U S A* (2014) 111(11):4203–8. doi:10.1073/pnas.1319269111
14. Cantoni C, Grauwet K, Pietra G, Parodi M, Mingari MC, Maria AD, et al. Role of NK cells in immunotherapy and virotherapy of solid tumors. *Immunotherapy* (2015) 7(8):861–82. doi:10.2217/imt.15.53
15. Dieu-Nosjean MC, Giraldo NA, Kaplon H, Germain C, Fridman WH, Sautes-Fridman C. Tertiary lymphoid structures, drivers of the anti-tumor responses in human cancers. *Immunol Rev* (2016) 271(1):260–75. doi:10.1111/imr.12405
16. Mamessier E, Sylvain A, Thibult ML, Houvenaeghel G, Jacquemier J, Castellano R, et al. Human breast cancer cells enhance self tolerance by promoting evasion from NK cell antitumor immunity. *J Clin Invest* (2011) 121(9):3609–22. doi:10.1172/JCI45816
17. Romagnani C, Juelke K, Falco M, Morandi B, D'Agostino A, Costa R, et al. CD56brightCD16- killer Ig-like receptor- NK cells display longer telomeres and acquire features of CD56dim NK cells upon activation. *J Immunol* (2007) 178(8):4947–55. doi:10.4049/jimmunol.178.8.4947
18. Yu J, Mao HC, Wei M, Hughes T, Zhang J, Park IK, et al. CD94 surface density identifies a functional intermediary between the CD56bright and CD56dim human NK-cell subsets. *Blood* (2010) 115(2):274–81. doi:10.1182/blood-2009-04-215491
19. Jing Y, Ni Z, Wu J, Higgins L, Markowski TW, Kaufman DS, et al. Identification of an ADAM17 cleavage region in human CD16 (FcgammaRIII) and the engineering of a non-cleavable version of the receptor in NK cells. *PLoS One* (2015) 10(3):e0121788. doi:10.1371/journal.pone.0121788
20. Hoffmann U, Neudorfl C, Daemen K, Keil J, Stevanovic-Meyer M, Lehner F, et al. NK cells of kidney transplant recipients display an activated phenotype that is influenced by immunosuppression and pathological staging. *PLoS One* (2015) 10(7):e0132484. doi:10.1371/journal.pone.0132484
21. Renoux VM, Zriwil A, Peitzsch C, Michaelsson J, Friberg D, Soneji S, et al. Identification of a human natural killer cell lineage-restricted progenitor in fetal and adult tissues. *Immunity* (2015) 43(2):394–407. doi:10.1016/j.immuni.2015.07.011
22. Keskin DB, Allan DS, Rybalov B, Andzelm MM, Stern JN, Kopcow HD, et al. TGFbeta promotes conversion of CD16+ peripheral blood NK cells into CD16- NK cells with similarities to decidual NK cells. *Proc Natl Acad Sci U S A* (2007) 104(9):3378–83. doi:10.1073/pnas.0611098104
23. Morandi F, Horenstein AL, Chillemi A, Quarona V, Chiesa S, Imperatori A, et al. CD56brightCD16- NK cells produce adenosine through a CD38-mediated pathway and act as regulatory cells inhibiting autologous CD4+ T cell proliferation. *J Immunol* (2015) 195(3):965–72. doi:10.4049/jimmunol.1500591
24. Nielsen N, Odum N, Urso B, Lanier LL, Spee P. Cytotoxicity of CD56(bright) NK cells towards autologous activated CD4+ T cells is mediated through NKG2D, LFA-1 and TRAIL and dampened via CD94/NKG2A. *PLoS One* (2012) 7(2):e31959. doi:10.1371/journal.pone.0031959
25. Wendel M, Galani IE, Suri-Payer E, Cerwenka A. Natural killer cell accumulation in tumors is dependent on IFN-gamma and CXCR3 ligands. *Cancer Res* (2008) 68(20):8437–45. doi:10.1158/0008-5472.CAN-08-1440
26. Tu TC, Brown NK, Kim TJ, Wroblewska J, Yang X, Guo X, et al. CD160 is essential for NK-mediated IFN-gamma production. *J Exp Med* (2015) 212(3):415–29. doi:10.1084/jem.20131601
27. Viaud S, Ploix S, Lapierre V, Thery C, Commere PH, Tramalloni D, et al. Updated technology to produce highly immunogenic dendritic cell-derived exosomes of clinical grade: a critical role of interferon-gamma. *J Immunother* (2011) 34(1):65–75. doi:10.1097/CJI.0b013e3181fe535b
28. Okita R, Yukawa T, Nojima Y, Maeda A, Saisho S, Shimizu K, et al. MHC class I chain-related molecule A and B expression is upregulated by cisplatin and associated with good prognosis in patients with non-small cell lung cancer. *Cancer Immunol Immunother* (2016) 65(5):499–509. doi:10.1007/s00262-016-1814-9
29. Okita R, Wolf D, Yasuda K, Maeda A, Yukawa T, Saisho S, et al. Contrasting effects of the cytotoxic anticancer drug gemcitabine and the EGFR tyrosine kinase inhibitor gefitinib on NK cell-mediated cytotoxicity via regulation of NKG2D ligand in non-small-cell lung cancer cells. *PLoS One* (2015) 10(10):e0139809. doi:10.1371/journal.pone.0139809
30. Sers C, Kuner R, Falk CS, Lund P, Sueltmann H, Braun M, et al. Down-regulation of HLA Class I and NKG2D ligands through a concerted action of MAPK and DNA methyltransferases in colorectal cancer cells. *Int J Cancer* (2009) 125(7):1626–39. doi:10.1002/ijc.24557
31. Ohshio Y, Teramoto K, Hanaoka J, Tezuka N, Itoh Y, Asai T, et al. Cancer-associated fibroblast-targeted strategy enhances antitumor immune responses in dendritic cell-based vaccine. *Cancer Sci* (2015) 106(2):134–42. doi:10.1111/cas.12584
32. Venugopalan A, Lee MJ, Niu G, Medina-Echeverz J, Tomita Y, Lizak MJ, et al. EGFR-targeted therapy results in dramatic early lung tumor regression accompanied by imaging response and immune infiltration in EGFR mutant transgenic mouse models. *Oncotarget* (2016) 7(34):54137–56. doi:10.18632/oncotarget.11021
33. Genova C, Hirsch FR. Clinical potential of necitumumab in non-small cell lung carcinoma. *Onco Targets Ther* (2016) 9:5427–37. doi:10.2147/OTT.S114039
34. Forde PM, Smith KN, Chaft JE, Hellmann M, Merghoub T, Wolchok JD, et al. NSCLC, early stageNeoadjuvant anti-PD1, nivolumab, in early stage resectable non-small-cell lung cancer. *Ann Oncol* (2016) 27(Suppl 6):vi552–87. doi:10.1093/annonc/mdw435.38
35. Pesce S, Greppi M, Tabellini G, Rampinelli F, Parolini S, Olive D, et al. Identification of a subset of human natural killer cells expressing high levels of programmed death 1: a phenotypic and functional characterization. *J Allergy Clin Immunol* (2016) 139(1):335–46.e3. doi:10.1016/j.jaci.2016.04.025

36. Carotta S. Targeting NK cells for anticancer immunotherapy: clinical and preclinical approaches. *Front Immunol* (2016) 7:152. doi:10.3389/fimmu.2016.00152
37. Blake SJ, Stannard K, Liu J, Allen S, Yong MC, Mittal D, et al. Suppression of metastases using a new lymphocyte checkpoint target for cancer immunotherapy. *Cancer Discov* (2016) 6(4):446–59. doi:10.1158/2159-8290.CD-15-0944
38. Srivastava RM, Trivedi S, Concha-Benavente F, Gibson SP, Reeder C, Ferrone S, et al. CD137 stimulation enhances cetuximab-induced natural killer: dendritic cell priming of antitumor T-cell immunity in patients with head and neck cancer. *Clin Cancer Res* (2016) 23(3):707–16. doi:10.1158/1078-0432.CCR-16-0879

16

Genetic Causes of Human NK Cell Deficiency and their Effect on NK Cell Subsets

Emily M. Mace and Jordan S. Orange**

Center for Human Immunobiology, Baylor College of Medicine and Texas Children's Hospital, Houston, TX, USA

Edited by:
Laurent Brossay, Brown University, USA

Reviewed by:
Michael G. Brown, University of Virginia, USA
Stephen Noel Waggoner, Cincinnati Children's Hospital Medical Center, USA
Aimee Beaulieu, Rutgers New Jersey Medical School, USA

***Correspondence:**
Emily M. Mace
mace@bcm.edu;
Jordan S. Orange
orange@bcm.edu

Human NK cells play critical roles in human host defense, particularly the control of viral infection and malignancy, and patients with congenital immunodeficiency affecting NK cell function or number can suffer from severe illness. The importance of NK cell function is particularly underscored in patients with primary immunodeficiency in which NK cells are the primary or sole affected population (NK cell deficiency, NKD). While NKD may lead to the absence of NK cells, we are also gaining an increasing appreciation of the effect that NKD may have on the generation of specific NK cell subsets. In turn, this leads to improved insights into the requirements for human NK cell subset generation, as well as their importance in immune homeostasis. The presence of inherently abnormally developed or functionally impaired NK cells, in particular, appears to be problematic in the way of interfering with normal human host defense and may be more impactful than low numbers of NK cells alone. Here, we review the known genetic causes of NKD and the insight that is derived by these into the requirements for human subset generation and, by extension, for NK cell-mediated immunity.

Keywords: NK cell deficiency, NK cell development, NK cells, primary immunodeficiency, immune homeostasis

INTRODUCTION

While human NK cells and NK cell subsets have been well described, understanding the true physiologic role that they serve in humans has been difficult. The heterogeneity of NK cells as a result of both genetic and environmental diversity, as well as the divergence in phenotypic markers between mouse and man has made advances in the field incremental. An important tool in the understanding of human NK cells and NK cell subsets has been the discovery of human inborn genetic immunodeficiency diseases (primary immunodeficiencies) that affect the generation or homeostasis of NK cells or specific NK cell subsets as well as those that affect NK cell function. There are over 300 genetic deficiencies of human immunity, and nearly 50 are known to have at least some impact on NK cells. This latter category has been reviewed by a number of authors in recent years (1–4). Five of these primary immunodeficiencies, however, may have their primary impact upon NK cells and are known as NK cell deficiency (NKD) (these are summarized in **Table 1**). These have provided insight into the requirements for human NK cell maturation, the importance of the successful execution of NK cell development, and the overall value of certain NK cell functions. Here, we will describe recent advances in our understanding of the importance of human NK cell subsets and NK cell function as informed by NKD.

TABLE 1 | Human NK cell deficiencies and their effect on NK cell subsets, function, and proliferation.

NKD type	Gene	Inheritance	NK cell number	NK cell lytic function	NK cell proliferation	CD56^{dim}	CD56^{bright}	Associated immune phenotype	Reference
cNKD	*GATA2*	AD or spontaneous	Normal or ↓	↓	ND	Normal or ↓	Absent/ severely ↓	DC cytopenia, B cell lymphopenia, monocytopenia, CD4 lymphocytopenia, neutropenia (all variable across reported cases)	(33, 37–39)
cNKD	*IRF8*	AR	↓	↓	ND	↓	Normal/↑	Mild DC phenotype	(66)
cNKD	*MCM4*	AR	↓	↓	↓	↓	Normal/↑	None	(53, 54)
cNKD	*RTEL1*	AR	↓	↓	Normal	ND	ND	None in NKD patient reported	(60, 61)
fNKD	*FCGR3A*	AR	Normal	↓	ND	Normal	Normal	None	(71)

AR, autosomal recessive; AD, autosomal dominant; ND, not done.

HUMAN NK CELL SUBSETS

Human NK cells in peripheral blood are broadly classified as being CD56^{+}CD3^{-} and generally comprise 1–17% of mononuclear cells within peripheral blood of healthy adults (5). They are found primarily in two broadly defined phenotypic subsets, namely the CD56^{bright} and CD56^{dim} subsets, with the CD56^{dim} subset being the predominant one in peripheral blood. While the basis for this stratification is the surface density of CD56 itself, each is a distinct functional subset with its own repertoire of markers that represent its capabilities. CD56^{bright} NK cells are broadly defined as being the potent producers of cytokines, whereas the CD56^{dim} subset is the classical mediator of contact-dependent cytotoxicity, both natural and antibody-dependent. However, exceptions to these rules exist, and there is evidence that both subsets can perform the tasks commonly attributed to the other (6, 7). In addition, there are several well-defined functional and phenotypic intermediaries between the two, as well as several subsets that are specifically expanded after infection or other immune stimulation. These include the CD56^{neg} subset, which is commonly found expanded in patients with HIV and HCV (8, 9). It should be noted that these cells are not truly CD56 negative, although the expression of CD56 on them is significantly lower than that of the CD56^{dim} subset. They retain expression of CD16, killer cell immunoglobulin receptors (KIR, CD158 family), and CD57 and are generally thought to be terminally differentiated or exhausted cells (10). An additional important subset expanded in disease is the adaptive NKG2C^{+} population expanded following certain viral infections, particularly CMV (11–13), but also in cases of reinfection of CMV seropositive individuals by hantavirus (14), HIV (15, 16), Epstein–Barr Virus (17), Hepatitis C Virus (18), and Chikungunya virus (19). Further studies have revealed distinct subsets associated with the generation of adaptive NK cells following CMV infection. These cells are high affinity IgE receptor (FcεRIγ)-deficient and express low amounts of Syk and/or EWS/FLI1-activated transcript 2 (EAT2) (20, 21). Further adaptation of the NK cell repertoire occurs with epigenetic silencing of promyelocytic leukemia zinc finger (PLZF) (22). There is also remarkable diversity generated by both environmental and genetic factors within the human population as a whole with regards to NK cell receptor expression and functional potential (23, 24).

The relationship between CD56^{bright} and CD56^{dim} subsets is not entirely clear, although it is thought that they represent the two terminally mature populations of NK cells derived from less mature NK cell precursors. There are oft-cited lines of evidence suggesting that CD56^{bright} NK cells are the direct precursors of the CD56^{dim} subset, namely the longer telomeres in CD56^{bright} NK cells (25), their earlier regeneration after HSCT (26), and the generation of CD56^{dim} NK cells from CD56^{bright} cells in certain experimental systems (27–29). The expression of CD56 on NK cells is unique to human and non-human primate systems, which has made the study of this specific aspect of NK cell development difficult to model prior to the advent of humanized mouse models. The use of such humanized mice, in which human NK cells develop in a rodent system, showed that CD56^{bright} NK cells give rise to CD56^{dim} cells (29); however, a recent study using genetic bar coding of NK cell lineages in rhesus macaques was indicative of the two subsets having distinct ontologies (30). Ultimately, answering this important question for NK cell biology will require a dissection of the mechanism that drives NK cell development using a satisfying human model. The study of primary human NKDs that affect NK cell subset generation are an informative tool in this endeavor, specifically for identifying novel requirements for the generation and homeostasis of human NK cell subsets (discussed further below). While complex, these disorders also provide insight into the functions of particular NK cell subsets in humans, as well as the overall value of NK cells in human host defense.

CLASSICAL NKD

Classical (c)NKD refers to NKD as a result of absent or profoundly decreased CD56^{+} CD3^{-} NK cell number, specifically defined as ≤1% of peripheral lymphocytes (3, 31). As a result, these are disorders in which overall NK cell development and/or homeostasis is impaired. Recent studies enabled by next generation sequencing techniques have identified genetic causes for previously described but unidentified cNKD. In combination with these genetic advances, we have also seen an increase in our understanding of NK cell subsets and the cell biological tools used to evaluate them. As a result, we can now appreciate that cNKD also includes cases in which human peripheral blood NK cell subsets are selectively underrepresented or absent. Thus, while cNKD has been historically used to define individuals with very low percentages of NK cells among peripheral blood lymphocytes, we would propose that this definition be expanded to include any example of impaired development or survival leading to the gross underrepresentation of a major NK cell subset. We would also argue that abnormally developed NK cells or the absence of particular NK cell subsets in the presence of others is far more relevant to host defense than simply having low overall numbers of NK cells.

To date, there are four genetically defined reported causes of cNKD, and of these at least three (*MCM4, GATA2, IRF8*) contain an aspect of subset aberration, in addition to NK cell numbers ranging from the low end of normal to undetectable. The fourth (*RTEL1*) presents information too limited to make a definitive statement other than there being underrepresented NK cells seemingly in isolation from other immune defects. In some cases, namely GATA2 deficiency, a single gene defect may account for a widely variable clinical phenotype with regards to NK cell number. This likely is related to the effect of specific mutations but may also be reflective of innate plasticity of NK cell numbers and subsets over time, in even healthy individuals, and the infectious history of the patient (5, 23, 24).

GATA2

GATA2 deficiency is the genetic lesion behind the most frequently cited case of classic NKD in the literature. A 13-year-old girl, originally presenting with varicella pneumonia, was found to have decreased NK cell number and function in the presence of T cell and B cell function (32). In addition to VZV, CMV and HSV infections followed, and she ultimately developed aplastic anemia and died during the course of hematopoietic stem cell transplantation (32, 33). Posthumous sequencing from a cryopreserved T cell line identified a pathogenic frameshift GATA mutation (c.1025_1026insGCCG; p.A342GfsX41), confirming GATA2 haploinsufficiency as the cause of her disease (33). GATA2 haploinsufficiency was also described as a cause of DC and monocyte deficiency (MonoMAC, DCML) by independent groups in 2009, and NK cell cytopenia was noted as a feature of disease (34–36). More detailed phenotyping has revealed that there is variation in the NK cell number in these patients, with many of them having less than 1% NK cells within their lymphocyte population. However, there is a substantive range in NK cell number, with some patients having NK cells within the normal range of healthy donors. Despite the presence of NK cells in some patients, a consistent and notable feature of GATA2 deficiency is the absolute loss of the $CD56^{bright}$ NK cell subset (33). Originally described in a cohort of eight patients, this has since been reported by other groups (37, 38) and is a consistent feature of the >30 patients that we have studied with GATA2 deficiency and accompanying infectious history or hematologic disease.

While it is difficult to estimate the prevalence of GATA2 mutation within the general population, the report of a cohort of 57 patients, of which 82% have clinically detected NK cell lymphopenia and 70% have severe and early onset viral infection suggestive of NK cell dysfunction, is the largest group of NKD patients that we are aware of (39). This suggests it is perhaps the most common NKD, as this cohort alone significantly outnumbers those reported with *RTEL1, MCM4, IRF8*, and *FCGR3A* mutations. Furthermore, *GATA2* is also the only gene associated with NKD that is autosomal dominant (*via* haploinsufficiency) and the only for which spontaneous cases have been reported (35, 36), thus supporting the statement of commonality from a genetic standpoint as well.

GATA2 deficiency is complex as it can be a multi-syndromic disease affecting multiple organs and presenting in multiple different ways. Patients can be susceptible to atypical mycobacterial infections, fungal infections, and severe and recalcitrant viral infections. Interestingly, there is a seemingly progressive nature of the disease, with many patients presenting in young adulthood and some even later in life. The range of clinical presentations and natural history has been well described elsewhere (39). Interestingly, one of the earliest clinical features to appear is a susceptibility to HPV disease, which could point to a role for NK cell-mediated defenses. In addition to immune deficiency, GATA2 deficiency is a cause of familial bone marrow failure (40), and a recent study of over 400 children and adolescents revealed GATA2 mutation to be the most common germline mutation leading to myelodysplastic syndrome in children and young adults (41). While this group did not specifically examine NKD in their cohort, they report half of their patients to have immunodeficiency, suggesting that the high rates of NK cell cytopenias reported by Spinner et al. (39) are likely present in this group.

However, there are some notable aspects of GATA2 deficiency with regards to NK cell biology. While other NK cell deficiencies have been reported as affecting the frequency of subset distribution (particularly MCM4, described below), the loss of the $CD56^{bright}$ NK cell subset is as near to absolute as has been described. The mechanism by which GATA2 regulates the development or maintenance of the $CD56^{bright}$ NK cell pool is not understood. GATA2 is a zinc finger transcription factor that is required for embryonic hematopoiesis, and maintenance of the stem cell pool in adults (42, 43) and GATA2 haploinsufficiency can also lead to loss of dendritic cell subsets and B cell cytopenias. Therefore, given its important role in stem cell maintenance and hematopoiesis, GATA2 deficiency may affect multiple immune cell lineages, which are again highly variable from patient to patient. *In vitro* differentiation of NK cells from patient hematopoietic stem cells leads to aberrant NK cell development, suggesting that NKD is cell intrinsic in these patients (33). However, given the interdependence on particularly NK and DC cross talk, it is likely that loss of other subsets affects NK cell numbers and functions in these patients, although this has not been explicitly studied. Immune manifestations include susceptibility to mycobacterial disease, frequently *M. avium* and *M. kansasii*, susceptibility to fungal infection, and severe herpesviral infections (39). An increased rate of malignancy may be in part due to viral infection as this includes increased rates of cervical cancer in young women potentially attributable to human papillomavirus infection. Less common clinical symptoms include miscarriage, solid organ tumors, and lymphedema (Emberger's syndrome) (39).

Due to the requirement for GATA2 in renewal of the adult stem cells, it is thought that depletion of the stem cell pool leads to subsequent lymphopenia, and *in vitro* differentiation experiments demonstrate an NK cell intrinsic role for GATA2 through the phenocopying of the $CD56^{bright}$ NK cell subset loss in NK cells derived from patient $CD34^{+}$ HSC (33). Whether GATA2 is required for the generation or homeostasis of $CD56^{bright}$ NK cells, or whether the defect arises earlier in lineage commitment, remains to be determined. GATA2 is highly expressed not only in $CD34^+$ HSC, but also across a range of lineages including monocytes, monocyte-derived dendritic cells, B cells, and mature NK cells, as well as the common myeloid precursor (44–46). Conditional deletion of *Gata2* in mice has recently demonstrated its requirement for DC differentiation from lineage negative precursors (47). Interestingly, this is at least partially through repression of genes that control T cell and ILC lineages, including *Tcf7*, *Eomes*, and *Gata3*. Direct binding of the Gata3 promoter by Gata2 in common myeloid progenitor cells suggests that these two transcription factors may play a role in tuning fate decisions similar to the well described GATA switch that occurs between GATA1 and GATA2 and is independent of transcription factor expression (47, 48).

This particular deficiency also allows for one to hypothesize regarding potential contribution of NK cells to human host defense. Given that some GATA2 deficiency patients lack only the $CD56^{bright}$ subset, it could speak to an important role for human NK cell-derived cytokine production specifically in the defense against HPV and/or herpesviruses. As there is strong animal model support for this role (49–51), the connection is at least plausible. It is not possible, however, to draw a direct connection as there is the variable impact upon other elements of innate immunity in these patients as discussed above. Furthermore, there are also those GATA2 deficiency patients that essentially lack NK cells altogether as well as those who have $CD56^{dim}$ NK cells. As deficient natural cytotoxicity is a conserved clinical hallmark of GATA2 deficiency (32, 33), this leaves open the contribution of contact-dependent cytotoxicity as the critical NK cell-mediated component of viral control. Thus, while GATA2 deficiency provides both insights and compelling leads, further work needs be done to formally attach a specific role to NK cell defenses through this disorder.

MCM4

Mutations in minichromosomal maintenance complex member 4 (MCM4) were described in 2012 as a cause of NKD accompanied by adrenal insufficiency, developmental delay, and short stature in a population of endogamous Irish travelers (52–55). Notably, patients in this cohort had increased susceptibility to viral infection, including cytomegalovirus, and malignancy (52–54). An initial description selective NKD (52) was followed by a genetic diagnosis of splice-site mutations in MCM4 and more detailed analysis of the NK cell phenotype (53, 54). MCM4 patients were the first to have demonstrated heritability to NKD and were the first listed in the Online Mendelian Inheritance in Man database as such (OMIM# 609981). Patients with MCM4 mutation have decreased frequency of $CD56^{dim}$ NK cells and, as a result, a relative overrepresentation of the $CD56^{bright}$ subset with an overall decrease in absolute numbers of NK cells (53). Careful analysis of both NK cells and fibroblasts from six patients in the Irish traveler cohort yielded novel insight into both NK cell biology and the role of the MCM complex in NK cell homeostasis and human disease. Specifically, $CD56^{bright}$ NK cells from patients had decreased rates of proliferation and impaired terminal maturation. This likely accounts for the severe viral susceptibility in these patients as they are lacking the more mature subset of NK cells and have at least some instability within the $CD56^{bright}$ population (53).

Why does MCM4 mutation exert such a specific, profound effect on NK cell maturation and function? MCM4, along with MCM2-7, MCM10, and GINS1 form a highly conserved complex that plays a critical role in the initiation and elongation of eukaryotic DNA replication (56, 57). Embryos from the Mcm4 knockout mouse do not survive implantation, and mice carrying the hypomorphic *Chaos3* mutation affecting Mcm4 have genomic instability reminiscent of that found in MCM4 patients (58). Careful analysis of the patients' fibroblasts showed normal MCM2–7 complex formation and DNA-binding but impaired DNA replication and, subsequently, cell cycle arrest (53). Genomic instability frequently accompanies impaired replication, and this is also the case in MCM4 patient fibroblasts, which had increased rates of chromosomal breakage (53). Interestingly, accompanying the decrease in NK cell number and specific decrease in frequency of $CD56^{dim}$ NK cells, patient NK cells had impaired proliferation in response to cytokine stimulation and increased rates of apoptosis (53).

The apparent role of MCM4 in the generation of $CD56^{dim}$ NK cells is not fully understood; however, it may speak to a previously hypothesized requirement for homeostatic proliferation, specifically in the $CD56^{bright}$ subset, leading to generation of the $CD56^{dim}$ subset. This model was originally proposed based on differential rates of apoptosis and proliferation detected in $CD56^{bright}$ and $CD56^{dim}$ NK cells in peripheral blood (59). The rapid proliferation of peripheral blood $CD56^{bright}$ cells with little apoptosis suggests that this subset either becomes recruited to tissue or generates the $CD56^{dim}$ subset, as substantiated by mouse xenograft studies (29). Should the latter be the case, the loss of $CD56^{dim}$ NK cells in MCM4 patients could be explained by the inability of the $CD56^{bright}$ NK cells to successfully undergo this proliferative burst and effectively generate $CD56^{dim}$ NK cells. Clearly, MCM4 deficiency raises many important new questions in NK cell biology, and answers will undoubtedly provide leads into both how NK cells develop as well as how developing NK cells figure into human host defense.

RTEL1

The third case of classic NKD, similarly to GATA2, is a case of a young girl who died of varicella infection at the age of 2 years. Originally reported in 2005 (60), her genetic lesion was recently solved by whole exome sequencing and found to be due to homozygous mutation in *RTEL1* (61). Consistent with RTEL1 being a cause of specific NKD, the girl had seemingly normal immunoglobulin levels and specific antibody titers, normal T and B lymphocyte counts and subsets, and absence of bone marrow failure. Notably, however, she had severely decreased NK cell number and function that was not rescued by IL-2. As common γ chain and JAK3 mutations lead to SCID, her IL-15 levels and signaling pathways were tested, yet were found to be intact (61). *RTEL1* deficiency is a reported cause of Hoyeraal–Hreidarsson Syndrome, an X-linked telomere deficiency prevalent among the Ashkenazi Jewish population that causes dyskeratosis congenita, bone marrow failure syndrome, and immunodeficiency (62). The mutation reported in this (2-year-old female) patient is a founder mutation causing Hoyeraal–Hreidarsson syndrome and lymphopenia, including progressive NK cell immunodeficiency (63, 64). Interestingly, the patient had no clinical features of Hoyeraal–Hreidarsson syndrome, and her growth and neurological development were normal for her age (personal communication, A. Etzioni, 2016). She did have an inverted CD4/CD8 T cell ratio but was virally infected (61). The patient also did not have recognizable platelet abnormalities or anemia (personal communication, A. Etzioni, 2016). Since the discovery of this patient, there have not been other reports of biallelic *RTEL1* mutations causing selective NKD. However, given the molecular mechanism of MCM4 deficiency, the link between RTEL1, which also plays a role in DNA repair, and NK cell development is interesting. As the patient was deceased at the time of genetic study, unfortunately in-depth studies of NK cell subsets or developmental processes were not tenable. The effect of RTEL1 deficiency on NK cell development underscores the specific sensitivity of human NK cells to DNA damage and the requirement for the DNA damage response in normal human NK cell development and homeostasis.

IRF8

The most recent description of cNKD identifies the cause of NKD in a family first described in 1982 with severe EBV susceptibility and absent NK cell function in affected individuals (65). The longitudinal study of a surviving affected sibling shows a distinctive, stable NK cell phenotype, with decreased NK cell function and increased CD56^{bright} NK cells relative to the CD56^{dim} subset (66). This phenotype is conserved in other patients with biallelic IRF8 mutation and is NK cell intrinsic. Gene expression analysis of NK cells from the proband shows deregulation of genes involved in NK cell maturation and effector function, suggesting that IRF8 is playing a role in the regulation of genes that include *NFIL3*, *PRDM1*, *TBX21*, *GRZB*, *STAT5a*, *STAT5b*, and *PRF*. Strikingly, a similar block in terminal maturation is also observed in mice with homozygous, but not heterozygous *Irf8* mutations. IRF8 deficiency as a result of both homozygous and heterozygous mutation can lead to DC deficiency (67, 68), although DC subsets in the proband reported here were only minimally affected (66). While mutation and gene dosage effects may be at play, the NK cell phenotype was conserved between all patients with biallelic IRF8 mutations studied but was not identified in any individuals with heterozygous IRF8 mutation. These findings demonstrate a requirement for IRF8 in terminal NK cell maturation and human antiviral defense and identify biallelic mutations in IRF8 as a newly described cause of cNKD.

FUNCTIONAL NKD

In contrast to cNKD, functional NKD (fNKD) represents a scenario in which NK cells are present in normal numbers but have some impaired functional capacity. In fNKD, a patient's NK cells appear to have gone through appropriate developmental progression and are not unduly susceptible to cell death (i.e., impaired survival). This distinction makes this category particularly difficult to diagnose clinically as the most common screening tests of quantitative flow cytometry will most likely be normal. Thus, a suspicion given a patient's particular clinical susceptibility is necessary to prompt a more substantive assessment of NK cell function. While presently fewer in number, it is possible that the fNKD category may ultimately outweigh the cNKD category in number of mechanistically defined diagnoses. Presently, there is only one known fNKD that is caused by biallelic specific mutations in *FCGR3A*.

FCGR3A

The only cause of fNKD reported to date is a rare mutation in *FCGR3A* (OMIM #615705), the gene encoding the low affinity IgG Fc receptor found on NK cells and macrophages (FcγRIIIA, CD16). Originally reported as the first monogenic cause of isolated NKD (69, 70), the c.230T-A transversion leads to L66H substitution in the first extracellular Ig domain of CD16 and is disease-causing when homozygous (69–71). Notably, while protein is expressed, the L66H substitution leads to loss of recognition of CD16 by mAb B73.1 while retaining recognition by the more commonly used 3G8 mAb (71). Therefore, dual detection with these antibodies is a rapid screen for this rare mutation as in certain individuals the B73.1 epitope will not be detected, although genotyping is required to confirm this molecular diagnosis (as individuals have been identified lacking the epitope for other reasons). Aside from the apparent loss of CD16^{+} NK cells if using mAb B73.1, NK cells from these patients appear to have appropriate development and subset generation (71).

This cause of fNKD was first identified in two patients with recurrent frequent upper respiratory infections and recurrent HSV (as well as herpes whitlow in one patient and recurrent varicella zoster in the second) (69, 70). Our group subsequently identified homozygous L66H substitution in a 14-year-old male with recurrent EBV-driven Castleman's disease and recalcitrant cutaneous warts (71). While CD16 function is required to mediate ADCC, strikingly the patients with CD16 L66H mutation have normal ADCC-mediated cellular cytotoxicity, yet impaired natural cytotoxicity. This seemingly paradoxical effect is explained by the location of the mutation in the membrane distal Ig-like domain, which does not affect binding to IgG Fc mediated by the membrane proximal domain. Instead, molecular studies of the effect of L66H mutation revealed a role for the distal Ig-like domain of CD16 in binding to and stabilizing the NK cell coactivating receptor CD2, thus delineating a crucial role for CD16 in the co-stimulation of natural cytotoxicity in addition to its well-defined role in ADCC (71). Recently, the biology of this interaction has been further advanced through the study of individuals who do not possess the NKG2C locus in which a compensatory co-stimulatory relationship between CD2 and CD16 were demonstrated to be of value (72). Thus while, only one example of fNKD, this disorder has provided valuable insights into how NK cells react to challenges and focus upon patients with abnormal NK cell function, but grossly normal appearing NK cells should lead to important new leads both biologically and clinically.

PUTTING IT ALL IN CONTEXT

While originally classified as classical or functional, the lines between the types of NK cell deficiencies has been blurred by increased resolution of NK cell subsets as well as the normative ranges and diversity of human NK cells (5, 23, 24). This is exemplified by the example of GATA2 mutation, which was initially reported as leading to absence of circulating NK cells but which, upon careful analysis was shown to lead to decreased NK cell number, specifically loss of the CD56^{bright} NK cell subset, accompanied by decreased function (33). Thus, it is possible to have a scenario in which NK cells may be present by clinical enumeration in a patient with classical NKD. The use of multiparametric flow cytometry to evaluate even the most rudimentary of NK cell subsets provides invaluable insight into the potential source of NKD. This is particularly true of GATA2 deficiency, which has a distinctive and seemingly immutable phenotype with regards to the absence of CD56^{bright} NK cells (33, 37, 38).

A fascinating biological question has emerged from these discoveries of NKD, namely that of the interdependence of subsets upon each other for function, as well as the relative importance of NK cell number versus functional potential. Particularly in the case of GATA2 and MCM4, the skewing of CD56^{bright} and CD56^{dim} subset generation has a seemingly profound effect on the function of the other subset (33, 53). Namely, in cases where CD56^{dim} NK cells are present in GATA2 patients, these cells remain unable to mediate the contact-dependent target cell lysis that is a functional hallmark of the CD56^{dim} subset. In the case of MCM4 deficiency, although it seems that the CD56^{bright} subset is impacted more so than the CD56^{dim} subset, profoundly decreased NK cell function by the CD56^{dim} subset is a likely a contributing cause of the patients' disease (54).

These observations add to an interesting but complicated literature on the ontology of human NK cells and ultimately speak to the need for a better understanding of the mechanism of, and requirement for human NK cell development. Finally, the question of the relationship between NK cell number, subset distribution, function, and disease state remains incompletely understood. Whether it is more advantageous to have fewer NK cells that are appropriately developed, as opposed to a greater number of inappropriately mature cells, remains to be seen. The evidence points to the former, begging the analogy that it is more important to have a small, functional team than a large one with badly behaved players. By extension, a small team may be a result of a disruptive individual, so determining the relative contribution of the deregulation of function and/or phenotype will be increasingly insightful into the pathogenesis of NKD. Emerging evidence in transplantation for primary immunodeficiency emphasizes this point, as having potentially very low numbers of NK cells that have the potential to work effectively may suffice at the right time and under the right circumstance. Along these lines, it is even possible that having a few defective NK cells could be more damaging than having none at all. Answering these questions will require further studies that carefully and thoughtfully dissect human NK cell development in health and disease. Furthermore, there are likely to be context-, time-, and exposure-specific components of NKD as there are likely to be critical windows in which an absence of NK cell subsets or functions may be particularly relevant for human host defense.

It should also be noted that all these studies in NKD will benefit from a greater understanding of NK cells within unique microenvironments. By necessity, most of this work has used peripheral blood as the source of NK cells, yet the tissue-specific NK cell phenotypes in liver, gravid uterus, and secondary lymphoid tissue speak to their shaping by these sites. Understanding how these seldom-visualized cells are affected by disease will be an important component of a greater understanding of the non-redundant function of human NK cells in health and disease. The present state of knowledge, however, suggests that there are patients with deficiencies of immunity reflected primarily within the NK cell compartment who are in need of mechanistic and therapeutic answers. Further study into individuals with suggestive phenotypes will likely produce new causative genes for NKD and insights into NK cell biology that can ultimately lead to new and improved therapeutic directions.

AUTHOR CONTRIBUTIONS

Both the authors contributed equally to the writing and editing of this manuscript.

ACKNOWLEDGMENTS

The authors wish to thank their clinical and research collaborators, as well as patients and their families. They also acknowledge funding support to JO by NIH-NIAID Grants R01AI067946 and R01AI120989 and the Jeffrey Modell Foundation.

REFERENCES

1. Ham H, Billadeau DD. Human immunodeficiency syndromes affecting human natural killer cell cytolytic activity. *Front Immunol* (2014) 5:2. doi:10.3389/fimmu.2014.00002
2. Orange JS. Human natural killer cell deficiencies. *Curr Opin Allergy Clin Immunol* (2006) 6(6):399–409. doi:10.1097/ACI.0b013e3280106b65
3. Orange JS. Natural killer cell deficiency. *J Allergy Clin Immunol* (2013) 132(3):515–25; quiz 526. doi:10.1016/j.jaci.2013.07.020
4. Voss M, Bryceson YT. Natural killer cell biology illuminated by primary immunodeficiency syndromes in humans. *Clin Immunol* (2015). doi:10.1016/j.clim.2015.11.004
5. Angelo LS, Banerjee PP, Monaco-Shawver L, Rosen JB, Makedonas G, Forbes LR, et al. Practical NK cell phenotyping and variability in healthy adults. *Immunol Res* (2015) 62(3):341–56. doi:10.1007/s12026-015-8664-y
6. De Maria A, Bozzano F, Cantoni C, Moretta L. Revisiting human natural killer cell subset function revealed cytolytic CD56(dim)CD16+ NK cells as rapid producers of abundant IFN-gamma on activation. *Proc Natl Acad Sci U S A* (2011) 108(2):728–32. doi:10.1073/pnas.1012356108
7. Nielsen N, Ødum N, Ursø B, Lanier LL, Spee P. Cytotoxicity of CD56(bright) NK cells towards autologous activated CD4+ T cells is mediated through NKG2D, LFA-1 and TRAIL and dampened via CD94/NKG2A. *PLoS One* (2012) 7(2):e31959. doi:10.1371/journal.pone.0031959

8. Gonzalez VD, Falconer K, Björkström NK, Blom KG, Weiland O, Ljunggren HG, et al. Expansion of functionally skewed CD56-negative NK cells in chronic hepatitis C virus infection: correlation with outcome of pegylated IFN-alpha and ribavirin treatment. *J Immunol* (2009) 183(10):6612–8. doi:10.4049/jimmunol.0901437
9. Hu PF, Hultin LE, Hultin P, Hausner MA, Hirji K, Jewett A, et al. Natural killer cell immunodeficiency in HIV disease is manifest by profoundly decreased numbers of CD16+CD56+ cells and expansion of a population of CD16dimCD56- cells with low lytic activity. *J Acquir Immune Defic Syndr Hum Retrovirol* (1995) 10(3):331–40. doi:10.1097/00042560-199511000-00005
10. Bjorkstrom NK, Ljunggren HG, Sandberg JK. CD56 negative NK cells: origin, function, and role in chronic viral disease. *Trends Immunol* (2010) 31(11):401–6. doi:10.1016/j.it.2010.08.003
11. Foley B, Cooley S, Verneris MR, Pitt M, Curtsinger J, Luo X, et al. Cytomegalovirus reactivation after allogeneic transplantation promotes a lasting increase in educated NKG2C+ natural killer cells with potent function. *Blood* (2012) 119(11):2665–74. doi:10.1182/blood-2011-10-386995
12. Gumá M, Angulo A, Vilches C, Gómez-Lozano N, Malats N, López-Botet M. Imprint of human cytomegalovirus infection on the NK cell receptor repertoire. *Blood* (2004) 104(12):3664–71. doi:10.1182/blood-2004-05-2058
13. Gumá M, Budt M, Sáez A, Brckalo T, Hengel H, Angulo A, et al. Expansion of CD94/NKG2C+ NK cells in response to human cytomegalovirus-infected fibroblasts. *Blood* (2006) 107(9):3624–31. doi:10.1182/blood-2005-09-3682
14. Björkström NK, Lindgren T, Stoltz M, Fauriat C, Braun M, Evander M, et al. Rapid expansion and long-term persistence of elevated NK cell numbers in humans infected with hantavirus. *J Exp Med* (2011) 208(1):13–21. doi:10.1084/jem.20100762
15. Goodier MR, Mela CM, Steel A, Gazzard B, Bower M, Gotch F. NKG2C+ NK cells are enriched in AIDS patients with advanced-stage Kaposi's sarcoma. *J Virol* (2007) 81(1):430–3. doi:10.1128/JVI.01567-06
16. Brunetta E, Fogli M, Varchetta S, Bozzo L, Hudspeth KL, Marcenaro E, et al. Chronic HIV-1 viremia reverses NKG2A/NKG2C ratio on natural killer cells in patients with human cytomegalovirus co-infection. *AIDS* (2010) 24(1):27–34. doi:10.1097/QAD.0b013e3283328d1f
17. Sohlberg E, Saghafian-Hedengren S, Rasul E, Marchini G, Nilsson C, Klein E, et al. Cytomegalovirus-seropositive children show inhibition of in vitro EBV infection that is associated with CD8+CD57+ T cell enrichment and IFN-gamma. *J Immunol* (2013) 191(11):5669–76. doi:10.4049/jimmunol.1301343
18. Béziat V, Dalgard O, Asselah T, Halfon P, Bedossa P, Boudifa A, et al. CMV drives clonal expansion of NKG2C+ NK cells expressing self-specific KIRs in chronic hepatitis patients. *Eur J Immunol* (2012) 42(2):447–57. doi:10.1002/eji.201141826
19. Petitdemange C, Becquart P, Wauquier N, Béziat V, Debré P, Leroy EM, et al. Unconventional repertoire profile is imprinted during acute chikungunya infection for natural killer cells polarization toward cytotoxicity. *PLoS Pathog* (2011) 7(9):e1002268. doi:10.1371/journal.ppat.1002268
20. Lee J, Zhang T, Hwang I, Kim A, Nitschke L, Kim M, et al. Epigenetic modification and antibody-dependent expansion of memory-like NK cells in human cytomegalovirus-infected individuals. *Immunity* (2015) 42(3):431–42. doi:10.1016/j.immuni.2015.02.013
21. Zhang T, Scott JM, Hwang I, Kim S. Cutting edge: antibody-dependent memory-like NK cells distinguished by FcRγ deficiency. *J Immunol* (2013) 190(4):1402–6. doi:10.4049/jimmunol.1203034
22. Schlums H, Cichocki F, Tesi B, Theorell J, Beziat V, Holmes TD, et al. Cytomegalovirus infection drives adaptive epigenetic diversification of NK cells with altered signaling and effector function. *Immunity* (2015) 42(3):443–56. doi:10.1016/j.immuni.2015.02.008
23. Horowitz A, Strauss-Albee DM, Leipold M, Kubo J, Nemat-Gorgani N, Dogan OC, et al. Genetic and environmental determinants of human NK cell diversity revealed by mass cytometry. *Sci Transl Med* (2013) 5(208):208ra145. doi:10.1126/scitranslmed.3006702
24. Strauss-Albee DM, Fukuyama J, Liang EC, Yao Y, Jarrell JA, Drake AL, et al. Human NK cell repertoire diversity reflects immune experience and correlates with viral susceptibility. *Sci Transl Med* (2015) 7(297):297ra115. doi:10.1126/scitranslmed.aac5722
25. Romagnani C, Juelke K, Falco M, Morandi B, D'Agostino A, Costa R, et al. CD56brightCD16- killer Ig-like receptor- NK cells display longer telomeres and acquire features of CD56dim NK cells upon activation. *J Immunol* (2007) 178(8):4947–55. doi:10.4049/jimmunol.178.8.4947
26. Dulphy N, Haas P, Busson M, Belhadj S, Peffault de Latour R, Robin M, et al. An unusual CD56(bright) CD16(low) NK cell subset dominates the early posttransplant period following HLA-matched hematopoietic stem cell transplantation. *J Immunol* (2008) 181(3):2227–37. doi:10.4049/jimmunol.181.3.2227
27. Chan A, Hong DL, Atzberger A, Kollnberger S, Filer AD, Buckley CD, et al. CD56bright human NK cells differentiate into CD56dim cells: role of contact with peripheral fibroblasts. *J Immunol* (2007) 179(1):89–94. doi:10.4049/jimmunol.179.1.89
28. Cooley S, Xiao F, Pitt M, Gleason M, McCullar V, Bergemann TL, et al. A subpopulation of human peripheral blood NK cells that lacks inhibitory receptors for self-MHC is developmentally immature. *Blood* (2007) 110(2):578–86. doi:10.1182/blood-2006-07-036228
29. Huntington ND, Legrand N, Alves NL, Jaron B, Weijer K, Plet A, et al. IL-15 trans-presentation promotes human NK cell development and differentiation in vivo. *J Exp Med* (2009) 206(1):25–34. doi:10.1084/jem.20082013
30. Wu C, Li B, Lu R, Koelle SJ, Yang Y, Jares A, et al. Clonal tracking of rhesus macaque hematopoiesis highlights a distinct lineage origin for natural killer cells. *Cell Stem Cell* (2014) 14(4):486–99. doi:10.1016/j.stem.2014.01.020
31. Orange JS. Unraveling human natural killer cell deficiency. *J Clin Invest* (2012) 122(3):798–801. doi:10.1172/JCI62620
32. Biron CA, Byron KS, Sullivan JL. Severe herpesvirus infections in an adolescent without natural killer cells. *N Engl J Med* (1989) 320(26):1731–5. doi:10.1056/NEJM198906293202605
33. Mace EM, Hsu AP, Monaco-Shawver L, Makedonas G, Rosen JB, Dropulic L, et al. Mutations in GATA2 cause human NK cell deficiency with specific loss of the CD56(bright) subset. *Blood* (2013) 121(14):2669–77. doi:10.1182/blood-2012-09-453969
34. Dickinson RE, Griffin H, Bigley V, Reynard LN, Hussain R, Haniffa M, et al. Exome sequencing identifies GATA-2 mutation as the cause of dendritic cell, monocyte, B and NK lymphoid deficiency. *Blood* (2011) 118(10):2656–8. doi:10.1182/blood-2011-06-360313
35. Hsu AP, Sampaio EP, Khan J, Calvo KR, Lemieux JE, Patel SY, et al. Mutations in GATA2 are associated with the autosomal dominant and sporadic monocytopenia and mycobacterial infection (MonoMAC) syndrome. *Blood* (2011) 118(10):2653–5. doi:10.1182/blood-2011-05-356352
36. Vinh DC, Patel SY, Uzel G, Anderson VL, Freeman AF, Olivier KN, et al. Autosomal dominant and sporadic monocytopenia with susceptibility to mycobacteria, fungi, papillomaviruses, and myelodysplasia. *Blood* (2010) 115(8):1519–29. doi:10.1182/blood-2009-03-208629
37. Maciejewski-Duval A, Meuris F, Bignon A, Aknin ML, Balabanian K, Faivre L, et al. Altered chemotactic response to CXCL12 in patients carrying GATA2 mutations. *J Leukoc Biol* (2016) 99(6):1065–76. doi:10.1189/jlb.5MA0815-388R
38. Dickinson RE, Milne P, Jardine L, Zandi S, Swierczek SI, McGovern N, et al. The evolution of cellular deficiency in GATA2 mutation. *Blood* (2014) 123(6):863–74. doi:10.1182/blood-2013-07-517151
39. Spinner MA, Sanchez LA, Hsu AP, Shaw PA, Zerbe CS, Calvo KR, et al. GATA2 deficiency: a protean disorder of hematopoiesis, lymphatics, and immunity. *Blood* (2014) 123(6):809–21. doi:10.1182/blood-2013-07-515528
40. Hahn CN, Chong CE, Carmichael CL, Wilkins EJ, Brautigan PJ, Li XC, et al. Heritable GATA2 mutations associated with familial myelodysplastic syndrome and acute myeloid leukemia. *Nat Genet* (2011) 43(10):1012–7. doi:10.1038/ng.913
41. Wlodarski MW, Hirabayashi S, Pastor V, Starý J, Hasle H, Masetti R, et al. Prevalence, clinical characteristics, and prognosis of GATA2-related myelodysplastic syndromes in children and adolescents. *Blood* (2016) 127(11):1387–97. doi:10.1182/blood-2015-09-669937
42. Rodrigues NP, Janzen V, Forkert R, Dombkowski DM, Boyd AS, Orkin SH, et al. Haploinsufficiency of GATA-2 perturbs adult hematopoietic stem-cell homeostasis. *Blood* (2005) 106(2):477–84. doi:10.1182/blood-2004-08-2989
43. Tipping AJ, Pina C, Castor A, Hong D, Rodrigues NP, Lazzari L, et al. High GATA-2 expression inhibits human hematopoietic stem and progenitor cell function by effects on cell cycle. *Blood* (2009) 113(12):2661–72. doi:10.1182/blood-2008-06-161117

44. Su AI, Wiltshire T, Batalov S, Lapp H, Ching KA, Block D, et al. A gene atlas of the mouse and human protein-encoding transcriptomes. *Proc Natl Acad Sci U S A* (2004) 101(16):6062–7. doi:10.1073/pnas.0400782101
45. Mabbott NA, Baillie JK, Brown H, Freeman TC, Hume DA. An expression atlas of human primary cells: inference of gene function from coexpression networks. *BMC Genomics* (2013) 14:632. doi:10.1186/1471-2164-14-632
46. Wu C, Orozco C, Boyer J, Leglise M, Goodale J, Batalov S, et al. BioGPS: an extensible and customizable portal for querying and organizing gene annotation resources. *Genome Biol* (2009) 10(11):R130. doi:10.1186/gb-2009-10-11-r130
47. Onodera K, Fujiwara T, Onishi Y, Itoh-Nakadai A, Okitsu Y, Fukuhara N, et al. GATA2 regulates dendritic cell differentiation. *Blood* (2016) 128(4):508–18. doi:10.1182/blood-2016-02-698118
48. Bresnick EH, Lee HY, Fujiwara T, Johnson KD, Keles S. GATA switches as developmental drivers. *J Biol Chem* (2010) 285(41):31087–93. doi:10.1074/jbc.R110.159079
49. Bukowski JF, Woda BA, Welsh RM. Pathogenesis of murine cytomegalovirus infection in natural killer cell-depleted mice. *J Virol* (1984) 52(1):119–28.
50. Welsh RM, Brubaker JO, Vargas-Cortes M, O'Donnell CL. Natural killer (NK) cell response to virus infections in mice with severe combined immunodeficiency. The stimulation of NK cells and the NK cell-dependent control of virus infections occur independently of T and B cell function. *J Exp Med* (1991) 173(5):1053–63. doi:10.1084/jem.173.5.1053
51. Welsh RM, O'Donnell CL, Shultz LD. Antiviral activity of NK 1.1+ natural killer cells in C57BL/6 scid mice infected with murine cytomegalovirus. *Nat Immun* (1994) 13(5):239–45.
52. Eidenschenk C, Dunne J, Jouanguy E, Fourlinnie C, Gineau L, Bacq D, et al. A novel primary immunodeficiency with specific natural-killer cell deficiency maps to the centromeric region of chromosome 8. *Am J Hum Genet* (2006) 78(4):721–7. doi:10.1086/503269
53. Gineau L, Cognet C, Kara N, Lach FP, Dunne J, Veturi U, et al. Partial MCM4 deficiency in patients with growth retardation, adrenal insufficiency, and natural killer cell deficiency. *J Clin Invest* (2012) 122(3):821–32. doi:10.1172/JCI61014
54. Hughes CR, Guasti L, Meimaridou E, Chuang CH, Schimenti JC, King PJ, et al. MCM4 mutation causes adrenal failure, short stature, and natural killer cell deficiency in humans. *J Clin Invest* (2012) 122(3):814–20. doi:10.1172/JCI60224
55. O'Riordan SM, Lynch SA, Hindmarsh PC, Chan LF, Clark AJ, Costigan C. A novel variant of familial glucocorticoid deficiency prevalent among the Irish Traveler population. *J Clin Endocrinol Metab* (2008) 93(7):2896–9. doi:10.1210/jc.2008-0034
56. Aparicio T, Guillou E, Coloma J, Montoya G, Méndez J. The human GINS complex associates with Cdc45 and MCM and is essential for DNA replication. *Nucleic Acids Res* (2009) 37(7):2087–95. doi:10.1093/nar/gkp065
57. Costa A, Onesti S. Structural biology of MCM helicases. *Crit Rev Biochem Mol Biol* (2009) 44(5):326–42. doi:10.1080/10409230903186012
58. Shima N, Alcaraz A, Liachko I, Buske TR, Andrews CA, Munroe RJ, et al. A viable allele of Mcm4 causes chromosome instability and mammary adenocarcinomas in mice. *Nat Genet* (2007) 39(1):93–8. doi:10.1038/ng1936
59. Lutz CT, Karapetyan A, Al-Attar A, Shelton BJ, Holt KJ, Tucker JH, et al. Human NK cells proliferate and die in vivo more rapidly than T cells in healthy young and elderly adults. *J Immunol* (2011) 186(8):4590–8. doi:10.4049/jimmunol.1002732
60. Etzioni A, Eidenschenk C, Katz R, Beck R, Casanova JL, Pollack S. Fatal varicella associated with selective natural killer cell deficiency. *J Pediatr* (2005) 146(3):423–5. doi:10.1016/j.jpeds.2004.11.022
61. Hanna S, Béziat V, Jouanguy E, Casanova JL, Etzioni A. A homozygous mutation of RTEL1 in a child presenting with an apparently isolated natural killer cell deficiency. *J Allergy Clin Immunol* (2015) 136(4):1113–4. doi:10.1016/j.jaci.2015.04.021
62. Fedick AM, Shi L, Jalas C, Treff NR, Ekstein J, Kornreich R, et al. Carrier screening of RTEL1 mutations in the Ashkenazi Jewish population. *Clin Genet* (2015) 88(2):177–81. doi:10.1111/cge.12459
63. Ballew BJ, Joseph V, De S, Sarek G, Vannier JB, Stracker T, et al. A recessive founder mutation in regulator of telomere elongation helicase 1, RTEL1, underlies severe immunodeficiency and features of Hoyeraal Hreidarsson syndrome. *PLoS Genet* (2013) 9(8):e1003695. doi:10.1371/journal.pgen.1003695
64. Ballew BJ, Yeager M, Jacobs K, Giri N, Boland J, Burdett L, et al. Germline mutations of regulator of telomere elongation helicase 1, RTEL1, in Dyskeratosis congenita. *Hum Genet* (2013) 132(4):473–80. doi:10.1007/s00439-013-1265-8
65. Fleisher G, Starr S, Koven N, Kamiya H, Douglas SD, Henle W. A non-x-linked syndrome with susceptibility to severe Epstein-Barr virus infections. *J Pediatr* (1982) 100(5):727–30. doi:10.1016/S0022-3476(82)80572-6
66. Mace EM, Bigley V, Gunesch JT, Chinn IK, Cao TN, Keller MD, et al. *Biallelic Mutations in IRF8 Cause Human NK Cell Deficiency*. Barcelona, Spain: 17th Biennial Meeting of the European Society for Immunodeficiencies (2016).
67. Salem S, Langlais D, Lefebvre F, Bourque G, Bigley V, Haniffa M, et al. Functional characterization of the human dendritic cell immunodeficiency associated with the IRF8(K108E) mutation. *Blood* (2014) 124(12):1894–904. doi:10.1182/blood-2014-04-570879
68. Hambleton S, Salem S, Bustamante J, Bigley V, Boisson-Dupuis S, Azevedo J, et al. IRF8 mutations and human dendritic-cell immunodeficiency. *N Engl J Med* (2011) 365(2):127–38. doi:10.1056/NEJMoa1100066
69. de Vries E, Koene HR, Vossen JM, Gratama JW, von dem Borne AE, Waaijer JL, et al. Identification of an unusual Fc gamma receptor IIIa (CD16) on natural killer cells in a patient with recurrent infections. *Blood* (1996) 88(8):3022–7.
70. Jawahar S, Moody C, Chan M, Finberg R, Geha R, Chatila T. Natural killer (NK) cell deficiency associated with an epitope-deficient Fc receptor type IIIA (CD16-II). *Clin Exp Immunol* (1996) 103(3):408–13. doi:10.1111/j.1365-2249.1996.tb08295.x
71. Grier JT, Forbes LR, Monaco-Shawver L, Oshinsky J, Atkinson TP, Moody C, et al. Human immunodeficiency-causing mutation defines CD16 in spontaneous NK cell cytotoxicity. *J Clin Invest* (2012) 122(10):3769–80. doi:10.1172/JCI64837
72. Liu LL, Landskron J, Ask EH, Enqvist M, Sohlberg E, Traherne JA, et al. Critical role of CD2 co-stimulation in adaptive natural killer cell responses revealed in NKG2C-deficient humans. *Cell Rep* (2016) 15(5):1088–99. doi:10.1016/j.celrep.2016.04.005

17

Modeling Human Natural Killer Cell Development in the Era of Innate Lymphoid Cells

Steven D. Scoville[1,2], Aharon G. Freud[2,3] and Michael A. Caligiuri[2,4]**

[1] Biomedical Sciences Graduate Program, Medical Scientist Training Program, The James Cancer Hospital and Solove Research Institute, The Ohio State University, Columbus, OH, USA, [2] Comprehensive Cancer Center, The James Cancer Hospital and Solove Research Institute, The Ohio State University, Columbus, OH, USA, [3] Department of Pathology, The James Cancer Hospital and Solove Research Institute, The Ohio State University, Columbus, OH, USA, [4] Division of Hematology and Oncology, Department of Internal Medicine, The James Cancer Hospital and Solove Research Institute, The Ohio State University, Columbus, OH, USA

Edited by:
Luigi Daniele Notarangelo, Harvard Medical School, USA

Reviewed by:
Georges Leclercq, Ghent University, Belgium
Sylvain Latour, Centre national de la recherche scientifique (CNRS), France
Roberto Biassoni, Istituto Giannina Gaslini (IRCCS), Italy

***Correspondence:**
Aharon G. Freud
aharon.freud@osumc.edu;
Michael A. Caligiuri
michael.caligiuri@osumc.edu

Decades after the discovery of natural killer (NK) cells, their developmental pathways in mice and humans have not yet been completely deciphered. Accumulating evidence indicates that NK cells can develop in multiple tissues throughout the body. Moreover, detailed and comprehensive models of NK cell development were proposed soon after the turn of the century. However, with the recent identification and characterization of other subtypes of innate lymphoid cells (ILCs), which show some overlapping functional and phenotypic features with NK cell developmental intermediates, the distinct stages through which human NK cells develop from early hematopoietic progenitor cells remain unclear. Thus, there is a need to reassess and refine older models of NK cell development in the context of new data and in the era of ILCs. Our group has focused on elucidating the developmental pathway of human NK cells in secondary lymphoid tissues (SLTs), including tonsils and lymph nodes. Here, we provide an update of recent progress that has been made with regard to human NK cell development in SLTs, and we discuss these new findings in the context of contemporary models of ILC development.

Keywords: natural killer, innate lymphoid cells, development, secondary lymphoid tissues, human lymphopoiesis

INTRODUCTION

Natural killer (NK) cells belong to the family of innate lymphoid cells (ILCs) whose common features include reliance on the transcription factor ID2 for development and rapid elaboration of effector function in response to microbial products, cytokine stimulation, and contact with other leukocytes (1). ILCs share phenotypic and functional features with T cells, yet, ILCs lack expression of markers specific for other leukocytes, including antigen-specific T and B cell receptors (i.e., CD3/TCR and CD19/BCR). In 2013, Spits et al. published a proposal for ILC nomenclature and classification, including the designation of three major groups according to functional and phenotypic characteristics (2). Group 1 ILCs, which express the transcription factor T-BET and produce the T helper cell type 1 (Th1)-associated cytokine interferon gamma (IFN-γ), include NK cells as well as functionally distinct "ILC1s." NK cells are unique in their expression of the transcription factor EOMES and in their ability to recognize and destroy virally infected and malignantly transformed cells that have downregulated major histocompatibility (MHC) class I molecules and/or upregulated stress-induced molecules (3, 4). Group 2 ILCs or "ILC2s" produce Th2-associated cytokines, such as interleukin (IL)-5 and IL-13; they express the transcription factors BCL11B, GATA-3, and

RORα; and they are involved in many processes, including fat metabolism, allergy, and protection against parasites (5). Group 3 ILCs share features with Th17 cells, including expression of the transcription factors AHR and RORγt and production of IL-17 and IL-22. Group 3 ILCs consist of "ILC3s" as well as lymphoid tissue inducer cells whose roles include the formation and restoration of lymphoid tissues following infection (6). We refer the readers to other excellent comprehensive reviews exploring the development, transcriptional regulation, and diverse roles of ILCs in physiology and disease (1, 7–12).

As mentioned above, ILCs are identified as "lineage" (Lin) negative lymphocytes, lacking expression of surface markers more specifically expressed on T cells (CD3, CD5, TCR), B cells (CD19, CD20, BCR), myelomonocytic cells (CD14, CD15, CD36), and dendritic cells (DCs) (CD116, CD123, CD303). In humans, all non-NK ILCs express CD127 (IL-7Rα) and CD161 (NKRP1A) (13–16), and they may be further distinguished according to the expression of other subset-associated surface antigens including CXCR3, CD294 (CRTH2), and CD117 (c-Kit) for ILC1s, ILC2s, and ILC3s, respectively (10, 17). Human NK cells can also express the pan-ILC markers CD127 and CD161 (18–20), but NK cells are typically distinguished by their surface expression of CD16 (FcγRIIIA), CD94/NKG2 heterodimers, killer immunoglobulin-like receptors (KIRs), NKG2D, and NKp80 (21). Two subsets of human peripheral blood (PB) NK cells can be distinguished according to their relative expression of the pan-NK cell surface marker, CD56 (NCAM-1): "CD56^{bright}" and "CD56^{dim}" (22, 23). CD56^{bright} human NK cells express high levels of CD62L (L-selectin) and CD94 but absent or low levels of CD16 and KIRs; they predominate in secondary lymphoid tissues (SLTs) such as lymph nodes (LNs) and tonsils; they show low baseline perforin expression and cytotoxic activity *ex vivo*; and they rapidly produce cytokines, including IFN-γ, following stimulation by monocyte-derived cytokines ("monokines"), such as IL-12, IL-15, and IL-18 (22). In contrast, CD56^{dim} NK cells are CD16^{hi} and express more KIRs but less CD94 and CD62L; they predominate in PB; they show high baseline perforin expression and cytotoxicity against MHC class I negative target cells; and they preferentially produce cytokines in response to direct target cell interactions rather than *via* monokine stimulation (3). While the developmental relationship between these human NK cell subsets has not been definitively established, evidence suggests that CD56^{bright} NK cells represent immediate physiologic precursors of CD56^{dim} NK cells (19, 24–29). Alternative hypotheses include that CD56^{bright} NK cells represent activated NK cells *in vivo* and/or that PB NK cell subsets derive from distinct hematopoietic progenitor cells (HPCs) and developmental pathways (22, 30–33). Recent published data from Dunbar and colleagues suggest that the latter may be the case in rhesus macaques (34).

HUMAN NK CELL DEVELOPMENT IN SLTs

Human NK cells were originally thought to develop strictly within the bone marrow (BM) (3, 35). This notion was supported by the observation that Lin$^-$CD56$^+$ cytotoxic NK cells can be generated *in vitro* following culture of purified human BM CD34$^+$ HPCs with either BM-derived stroma or with IL-15, which can be produced by stroma (36, 37). Nonetheless, more recent extensive *ex vivo* characterization of HPCs and putative downstream NK cell developmental intermediates (NKDIs) reveals that the latter are naturally enriched in SLTs, including tonsils, spleen, and LNs, suggesting that in humans NK cells can also, if not preferentially, develop in SLTs (**Figure 1A**) (38–42). Similar NKDIs have also been identified in the thymus, liver, and uterus (43–45). Thus, human NK cell development is likely not restricted to SLTs (46).

In 2006, five putative stages of human SLT NK cell development were described according to the differential expression of CD34, CD117, CD94, and CD16 (41, 47, 48). Stage 1 cells (Lin$^-$CD34$^+$CD117$^-$CD94$^-$CD16$^-$) lack expression of the common IL-2/IL-15 receptor beta chain (IL-2/15Rβ, CD122) and are thus not responsive to exogenous soluble IL-2 or IL-15 *ex vivo*. However, they can generate NK cells when cultured in IL-15 plus other cytokines, such as Flt3 ligand and c-Kit ligand (KL) that likely induce CD122 expression and hence IL-15 responsiveness (49). In contrast, stage 2 cells (Lin$^-$CD34$^+$CD117$^+$CD94$^-$CD16$^-$) constitutively express CD122 (albeit below the level of detection by flow cytometry) and can generate functionally mature NK cells *in vitro* in the presence of exogenous soluble IL-15 in media without other cytokines or support cells (41). Stage 2 cells also constitutively express a functional high affinity IL-2 receptor, including the IL-2Rα subunit (CD25), and can differentiate in response to picomolar concentrations of IL-2 *in vitro* (39). The physiologic relevance of this cytokine receptor expression is not yet known and has not been tested *in vivo*. One likely possibility is that these HPCs, which are naturally enriched and reside within the parafollicular T cell-rich regions of SLTs, can respond to T cell-derived IL-2 and differentiate into NK cells following T cell activation *in vivo* (39).

When originally tested in bulk polyclonal cultures under supportive *in vitro* conditions, stage 1 and stage 2 cells were multipotent and could give rise to T cells and DCs as well as to NK cells, although they could not generate B cells or myeloid cells (41). In contrast, human stage 3 cells (Lin$^-$CD34$^-$CD117$^+$CD94$^-$CD16$^-$) lacked T cell and DC developmental potential. Stage 3 cells could, however, give rise to mature NK cells *in vitro* and *in vivo* and were thus originally proposed to represent committed NK cell precursors (41). Stage 3 cells are distinct from mature NK cells in that they lack high expression of T-BET and EOMES, cannot produce IFN-γ, and are incapable of mediating perforin-dependent cytotoxicity against MHC class I$^-$ target cells. In contrast, these features are detected within the stage 4 (CD34$^-$CD117$^{+/-}$CD94$^+$CD16$^-$) and stage 5 (CD34$^-$CD117$^{lo/-}$CD94$^{+/-}$CD16$^+$) populations in SLTs (40, 41). Stage 5 cells are further distinguished by the constitutive expression of CD16, a low affinity receptor for the Fc portion of immunoglobulin, which provides for antibody-mediated cellular cytotoxicity (50).

RECENT ADVANCES IN OUR UNDERSTANDING OF HUMAN NKDIs

Continued *ex vivo* phenotypic and functional characterization of the aforementioned five putative NKDI populations in

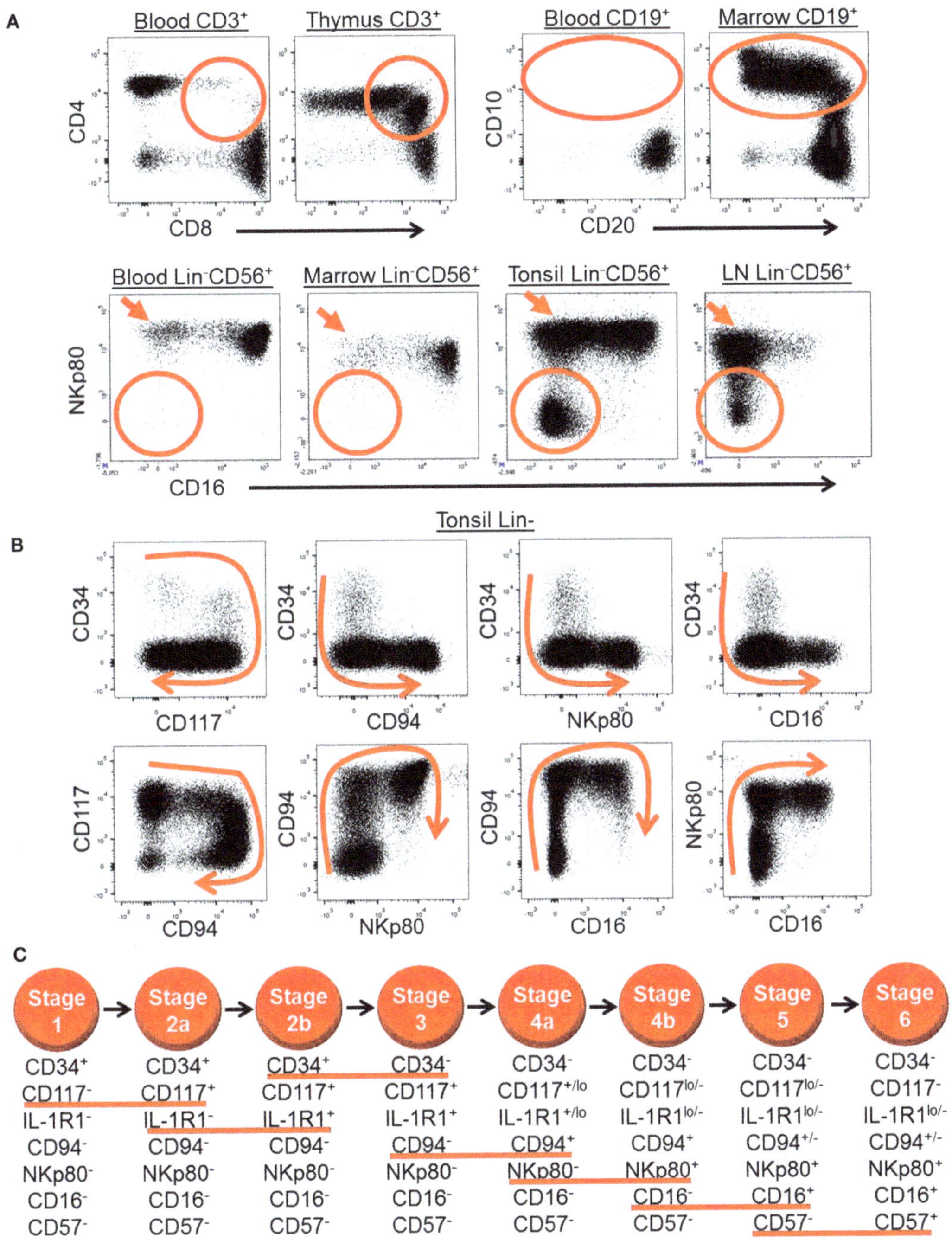

FIGURE 1 | ***Ex vivo* patterns of surface antigen expression support a model of human natural killer (NK) cell development in secondary lymphoid tissues (SLTs)**. **(A)** *Ex vivo* immunophenotypic analyses of CD3$^+$ cells (top row, left plots), CD19$^+$ cells (top row, right plots), and Lin$^-$CD56$^+$ cells (bottom row) in the indicated tissues demonstrate how immature T, B, and NK cell developmental intermediates (designated by the red circles and ovals) are naturally enriched in the thymus, bone marrow, and SLTs, respectively. Of note, the SLT populations designated by the red circles in the bottom row also likely contain some ILC3s, which can express CD56 (14). The red arrows in the bottom row highlight the relative enrichment of stage 4b CD56^{bright}NKp80$^+$CD16$^-$ NK cells in SLTs. **(B)** Immunophenotypic analysis of Lin$^-$ ILCs in human tonsil demonstrating the two-way patterns of CD34, CD117, CD94, NKp80, and CD16 expression as they relate to one another. The red arrows depict the putative directions of progressive NK cell development in SLTs. **(C)** Schematic representation of the proposed stages of human NK cell development in SLTs. The stages are defined according to the differential expression of CD34, CD117, interleukin (IL)-1R1, CD94, NKp80, CD16, and CD57, and the red lines underline the surface antigen changes that define each stage transition. Although not depicted, it is noted that CD56 expression is first detected at stage 2b (heterogeneous), peaks at stage 4b (CD56^{bright}), and then decreases to the level of most peripheral blood NK cells at stage 6 (CD56^{dim}). Also not depicted is killer immunoglobulin-like receptor expression, which is first detected within stage 4b in SLTs (40).

human SLTs has revealed a remarkable degree of heterogeneity within each stage. For example, in a study comparing tonsil- and thymus-derived HPCs, McClory et al. demonstrated that the human SLT stage 1 population, which expresses CD45RA and CD10, contains a minute subset of $CD1a^+CD11c^-$ cells that gives rise to T cells *ex vivo* and that shows substantial phenotypic overlap with $CD34^+CD45RA^+CD10^+CD1a^+$ T cell precursors in the thymus (51). McClory et al. were also able to trace a full putative pathway of tonsil T cell development branching directly from the tonsil $CD34^+CD45RA^+CD10^+CD1a^+CD11c^-$ stage 1 subset and closely paralleling T cell development in the thymus. These data suggest that the developmental pathways of other lymphoid subsets overlap/intersect with the originally characterized NK cell developmental pathway in SLTs. Consistent with this notion, Montaldo et al. recently showed that human SLT-derived stage 2 cells constitutively express RORγt and can give rise to $ROR\gamma t^+CD117^+$ ILC3s (52). In that study, Montaldo et al. observed relatively low NK cell production from stage 2 cells under the *in vitro* conditions tested. Given those findings, the investigators concluded that the stage 2 cells represent lineage-specified ILC3 progenitors. However, Montaldo et al. did not assess for the capacity of stage 2 cells to differentiate into other lineages such as DCs and T cells, and as mentioned earlier, Freud et al. demonstrated that stage 2 cells can differentiate into T cells, DCs, and NK cells under supportive conditions (41).

In a subsequent study, Scoville et al. characterized two functionally distinct subsets of SLT stage 2 cells according to surface expression of the IL-1β receptor, IL-1R1 (42). Both $IL\text{-}1R1^-$ (i.e., stage "2a") and $IL\text{-}1R1^+$ (i.e., stage "2b") subsets express CD45RA, integrin β7, and *ID2*, and they show low or undetectable expression of CD10. However, the stage 2b cells are unique in their near uniform expression of the pan-ILC marker, CD161, their lack of detectable expression of *RAG1* mRNA, which is expressed in stage 1 and stage 2a cells, and their natural restriction to SLTs. In a series of experiments in which stage 1, stage 2a, and stage 2b cells were freshly purified and then cultured *in vitro* or transplanted into non-obese diabetic (NOD)-scid $IL2Rgamma^{null}$ (NSG) immunodeficient mice treated with human IL-15, the stage 2b population was shown to be capable of giving rise to all four major subsets of ILCs (ILC1s, ILC2s, ILC3s, and NK cells), yet, they lacked T cell and DC developmental potential. In contrast, stage 1 and stage 2a cells could give rise to all ILC subsets as well as to T cells and DCs under the conditions tested (42). Thus, stage 2b cells appear to represent common ILC progenitors (CILPs) in humans.

Following the original discovery of SLT stage 3 cells (41), it was determined that cells within this population ($Lin^-CD34^-CD117^+CD94^-CD16^-$) express AHR, CD127, RORγt, IL-1R1, and IL-22 (14, 53–55). According to the 2013 ILC classification, these features denote Group 3 ILCs (2). Thus, it is not yet clear if stage 3 cells and ILC3s are entirely overlapping in their phenotypic characteristics, and this is a subject of ongoing investigation (see below) (56). Within the stage 3/ILC3 population in human SLTs, there is marked heterogeneity with regards to the expression of numerous surface markers, including CD7, CD56, CD62L, HLA-DR, and NKp44 (14, 41, 57, 58). The significance of this heterogeneity is largely unknown, although it was shown that NKp44 expression closely correlates with IL-22 production *ex vivo* (53, 58).

Last, in a recent study by Freud and colleagues, the SLT stage 4 population, defined as $Lin^-CD34^-CD117^{+/-}CD94^+CD16^-$, was shown to contain two functionally distinct subsets according to expression of the surface activating C-type lectin-like receptor, NKp80 (40), which is expressed on most if not all PB NK cells in healthy humans (59). Freud et al. described these SLT stage 4 subsets as stage "4a" ($NKp80^-$) and stage "4b" ($NKp80^+$), and they demonstrated that only the stage 4b subset is capable of IFN-γ production and perforin-dependent cellular cytotoxicity *ex vivo* (40). Consistent with these functional data, the surface expression of NKp80 among total SLT ILCs closely correlates with intracellular expression of T-BET, EOMES, and perforin. In contrast to stage 4b cells, stage 4a cells, which express CD94/NKG2A, lack the aforementioned functional and phenotypic features associated with mature NK cells and rather show a stage 3/ILC3-like profile including expression of CD117, CD127, IL-1R1, AHR, RORγt, and IL-22 (40). Following co-culture with allogeneic monocyte-derived DCs or transplantation into NSG mice treated with IL-15, purified SLT stage 4a cells gave rise to functional NK cells including some that were $NKp80^+CD16^+$ and coexpressed KIRs and CD57, the latter of which has been associated with terminal maturation and may represent a putative "stage 6" of human NK cell development (60, 61). Thus, stage 4a cells appear to comprise a naturally occurring and physiologic developmental intermediate population of cells emerging from stage 3 cells and giving rise to stage 4b cells. This hypothesis is supported by *ex vivo* patterns of surface antigen expression among total Lin^- ILCs in human SLTs (**Figure 1B**).

MODELING HUMAN NK CELL DEVELOPMENT IN THE CONTEXT AND ERA OF ILCs

Collectively, these recently published data described above support an updated linear model of human NK cell development (**Figure 1C**), which incorporates SLT NKDIs as well as the putative progressive development of human PB NK cells from $CD56^{bright}CD94^+NKp80^+CD16^-CD57^-$ (i.e., stage 4b) to $CD56^{dim}CD94^{+/-}NKp80^+CD16^+CD57^-$ (i.e., stage 5) to $CD56^{dim}CD94^{+/-}NKp80^+CD16^+CD57^+$ (i.e., stage 6). Although this is still somewhat controversial, a linear developmental relationship between PB $CD56^{bright}$ and $CD56^{dim}$ NK cell subsets is supported by the recent detection of PB NK cell populations that appear to represent naturally occurring NKDI spanning the developmental continuum between $CD56^{bright}$ and $CD56^{dim}$ NK cells (27, 29). In addition, other studies have shown that $CD56^{bright}$ NK cells can give rise to $CD56^{dim}$ NK cells *in vitro* and *in vivo* (19, 24, 26).

Despite the accumulating evidence in support of this proposed model, it is likely that other pathways of human NK cell development exist *in vivo* and potentially account for the marked diversity among tissue-resident NK cells (62–64). For example, Renoux et al. recently described a putative lineage committed NK cell progenitor (NKP) population in human

BM, blood, and SLTs characterized as Lin$^-$CD34$^+$CD38$^+$CD123$^-$ CD45RA$^+$CD7$^+$CD10$^+$CD127$^-$ and that could only give rise to NK cells *in vitro* and *in vivo* (65). Given that this NKP is only partially CD117$^+$ and reportedly lacks detectable expression of CD161 and CD127, it appears to be distinct from the aforementioned SLT CILP (i.e., stage 2b) population that is CD117$^+$CD161$^+$CD127$^+$ and that can generate all ILCs including NK cells (42). Rather, the immunophenotype of the NKP described in the study by Renoux et al. appears to overlap with that of the stage 1 and stage 2a populations described above. Indeed, it will be important in future studies to determine how these various HPC populations are related. Regardless, the study by Renoux et al. raises the intriguing possibility of other NK cell developmental pathways in humans.

It is also important to note that the human model of NK cell development depicted here differs from contemporary models of murine NK cell development (**Figure 2**) (1, 7, 12). Aside from inherent challenges in comparing these developmental processes due to differences in species-specific antigen expression, there may be fundamental differences between the species with regards to the developmental relationship(s) between NK cells and other ILCs and with regards to transcription factor expression patterns. For example, in mice, all ILC subsets are thought to derive from a CILP population that subsequently gives rise to committed NKPs as well as to common helper ILC progenitor cells (CHILPs) that can differentiate into ILC1s, ILC2s, and ILC3s, but not NK cells (66–70). While putative human CILPs (SLT stage 2b cells) and NKPs have been described as discussed above (42, 65), it is not yet clear if these two populations are related in the same way that the analogous populations in mice have been described. In addition, a human CHILP population has not been identified.

Another potential difference between mouse and human NK cell development relates to the relationship between NK cells and ILC3s. In mice, these two ILC subsets appear to be developmentally distinct (6), because in addition to these CHILP data described above, genetic fate-mapping (fm) studies for *Rorc2*, which encodes Rorγt, showed that mouse ILC3s are *Rorc2* fm$^+$ whereas mouse Eomes$^+$ NK cells are *Rorc2* fm$^-$ (68, 71, 72). In contrast, Scoville et al. recently demonstrated that the putative human CILP (SLT stage 2b) population expresses RORγt and that human PB CD56^{bright} NK cells, which are NKp80$^+$EOMES$^+$ (4, 40, 59, 73, 74), constitutively express detectable *RORC2* transcript (42). Thus, as opposed to mouse NK cells, at least some human NK cells would be fm$^+$ for *RORC2* if such an experiment could be performed. In addition, as mentioned earlier, there is substantial phenotypic overlap between cells originally described as "stage 3" NK cell precursors and cells denoted as "ILC3s" by other groups, raising the question as to whether or not these cell types are different or comprise the same population in humans (56). To address this issue, Crellin et al. previously evaluated the

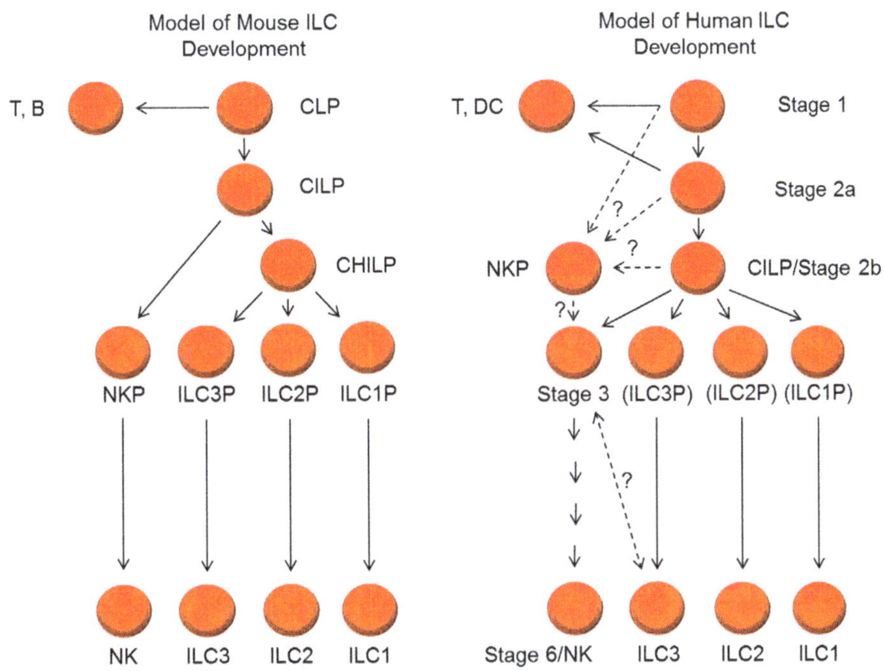

FIGURE 2 | **Models of mouse and human innate lymphoid cell (ILC) development**. Shown on the left and right are schematic representations of the cellular intermediates and developmental pathways of mouse and human ILCs, respectively. The black lines represent progressive steps of differentiation and maturation that are currently supported by published data. The dashed lines represent possible steps and relationships of differentiation and maturation, but these have not yet been tested or definitively established. In particular, the relationships between stage 1, 2a, 2b, and 3 cells with NK cell progenitors (NKPs) described by Renoux et al. (65) are not yet known. In addition, the relationship and possible distinction between human stage 3 natural killer (NK) cell precursors and ILC3s is also not yet clear. In the diagram to the right depicting human ILC development, the labels ILC3P (for ILC3 progenitor), ILC2P (for ILC2 progenitor), and ILC1P (for ILC1 progenitor) are shown in parentheses to convey that these are theoretical populations that have not been identified. CILP, common innate lymphoid progenitor; CHILP, common helper innate lymphoid progenitor.

in vitro differentiation potentials of Lin⁻CD34⁻CD161⁺CD117⁺ ILCs according to the differential expression of CD127 (75). They demonstrated that a minute CD127⁻ fraction preferentially gave rise to NK cells whereas the predominant CD127⁺ fraction mostly retained ILC3 features. Thus CD127 has since been touted as a critical marker distinguishing human stage 3 NK cell precursors from ILC3s (1, 6). However, in the study by Crellin et al., CD94⁺ NK cells, which can express CD117 and CD161 (18, 76), were not excluded in the pre-culture sorting preparations (75). Moreover, other reports suggest that the *in vitro* conditions used by Crellin et al. may be better optimized for supporting NK cell differentiation; conditions such as co-culture with OP9-DL1 stroma or DCs may be needed (40, 42). As such, the results provided in the Crellin et al. study could be due at least in part to the culture conditions employed and/or the preferential expansion of CD94⁺ NK cells that were present in the sorted Lin⁻CD34⁻CD161⁺CD117⁺CD127⁻ fractions. Last, we note that CD127 is constitutively expressed on PB CD56^{bright} NK cells (19, 20), and so excluding any CD127⁺ cells as potential NK cell precursors may be incorrect. Thus, to the best of our knowledge, there is as yet no reliable immunophenotypic strategy to distinguish human stage 3 NK cell precursors from ILC3s. Further investigation is needed.

CONCLUDING REMARKS AND FUTURE DIRECTIONS

The goals of investigating human NK cell development are to understand malignant counterparts and to facilitate the design and implementation of optimal immunotherapies for patients with cancer and potentially other diseases. Much progress has been made; however, the recent discovery of non-NK ILCs has required a reassessment of models of NK cell development in both mice and humans. More work is also needed in order to understand how all ILCs develop and to determine if the putative pathway of human NK cell development in SLTs is representative of NK cell development in other tissues.

AUTHOR CONTRIBUTIONS

SS, AF, and MC conceived the idea and wrote the manuscript.

ACKNOWLEDGMENTS

The authors thank Karen A. Young and Dr. Bethany L. Mundy-Bosse for insightful discussions and assistance in generating the figures. We apologize to those authors whose works were not cited due to space constraints.

REFERENCES

1. Artis D, Spits H. The biology of innate lymphoid cells. *Nature* (2015) 517(7534):293–301. doi:10.1038/nature14189
2. Spits H, Artis D, Colonna M, Diefenbach A, Di Santo JP, Eberl G, et al. Innate lymphoid cells – a proposal for uniform nomenclature. *Nat Rev Immunol* (2013) 13(2):145–9. doi:10.1038/nri3365
3. Caligiuri MA. Human natural killer cells. *Blood* (2008) 112(3):461–9. doi:10.1182/blood-2007-09-077438
4. Spits H, Bernink JH, Lanier L. NK cells and type 1 innate lymphoid cells: partners in host defense. *Nat Immunol* (2016) 17(7):758–64. doi:10.1038/ni.3482
5. Kim BS, Artis D. Group 2 innate lymphoid cells in health and disease. *Cold Spring Harb Perspect Biol* (2015) 7(5):a016337. doi:10.1101/cshperspect.a016337
6. Montaldo E, Juelke K, Romagnani C. Group 3 innate lymphoid cells (ILC3s): origin, differentiation, and plasticity in humans and mice. *Eur J Immunol* (2015) 45(8):2171–82. doi:10.1002/eji.201545598
7. Diefenbach A, Colonna M, Koyasu S. Development, differentiation, and diversity of innate lymphoid cells. *Immunity* (2014) 41(3):354–65. doi:10.1016/j.immuni.2014.09.005
8. Eberl G, Colonna M, Di Santo JP, McKenzie AN. Innate lymphoid cells. Innate lymphoid cells: a new paradigm in immunology. *Science* (2015) 348(6237):aaa6566. doi:10.1126/science.aaa6566
9. Eberl G, Di Santo JP, Vivier E. The brave new world of innate lymphoid cells. *Nat Immunol* (2015) 16(1):1–5. doi:10.1038/ni.3059
10. Hazenberg MD, Spits H. Human innate lymphoid cells. *Blood* (2014) 124(5):700–9. doi:10.1182/blood-2013-11-427781
11. Klose CS, Artis D. Innate lymphoid cells as regulators of immunity, inflammation and tissue homeostasis. *Nat Immunol* (2016) 17(7):765–74. doi:10.1038/ni.3489
12. Zook EC, Kee BL. Development of innate lymphoid cells. *Nat Immunol* (2016) 17(7):775–82. doi:10.1038/ni.3481
13. Bernink JH, Peters CP, Munneke M, te Velde AA, Meijer SL, Weijer K, et al. Human type 1 innate lymphoid cells accumulate in inflamed mucosal tissues. *Nat Immunol* (2013) 14(3):221–9. doi:10.1038/ni.2534
14. Cupedo T, Crellin NK, Papazian N, Rombouts EJ, Weijer K, Grogan JL, et al. Human fetal lymphoid tissue-inducer cells are interleukin 17-producing precursors to RORC+ CD127+ natural killer-like cells. *Nat Immunol* (2009) 10(1):66–74. doi:10.1038/ni.1668
15. Fuchs A, Vermi W, Lee JS, Lonardi S, Gilfillan S, Newberry RD, et al. Intraepithelial type 1 innate lymphoid cells are a unique subset of IL-12- and IL-15-responsive IFN-γ-producing cells. *Immunity* (2013) 38(4):769–81. doi:10.1016/j.immuni.2013.02.010
16. Mjösberg JM, Trifari S, Crellin NK, Peters CP, van Drunen CM, Piet B, et al. Human IL-25- and IL-33-responsive type 2 innate lymphoid cells are defined by expression of CRTH2 and CD161. *Nat Immunol* (2011) 12(11):1055–62. doi:10.1038/ni.2104
17. Cortez VS, Robinette ML, Colonna M. Innate lymphoid cells: new insights into function and development. *Curr Opin Immunol* (2015) 32:71–7. doi:10.1016/j.coi.2015.01.004
18. Azzoni L, Zatsepina O, Abebe B, Bennett IM, Kanakaraj P, Perussia B. Differential transcriptional regulation of CD161 and a novel gene, 197/15a, by IL-2, IL-15, and IL-12 in NK and T cells. *J Immunol* (1998) 161(7):3493–500.
19. Romagnani C, Juelke K, Falco M, Morandi B, D'Agostino A, Costa R, et al. CD56brightCD16- killer Ig-like receptor- NK cells display longer telomeres and acquire features of CD56dim NK cells upon activation. *J Immunol* (2007) 178(8):4947–55. doi:10.4049/jimmunol.178.8.4947
20. Vosshenrich CA, García-Ojeda ME, Samson-Villéger SI, Pasqualetto V, Enault L, Richard-Le Goff O, et al. A thymic pathway of mouse natural killer cell development characterized by expression of GATA-3 and CD127. *Nat Immunol* (2006) 7(11):1217–24. doi:10.1038/ni1395
21. Moretta L, Montaldo E, Vacca P, Del Zotto G, Moretta F, Merli P, et al. Human natural killer cells: origin, receptors, function, and clinical applications. *Int Arch Allergy Immunol* (2014) 164(4):253–64. doi:10.1159/000365632

22. Cooper MA, Fehniger TA, Caligiuri MA. The biology of human natural killer-cell subsets. *Trends Immunol* (2001) 22(11):633–40. doi:10.1016/S1471-4906(01)02060-9
23. Lanier LL, Le AM, Civin CI, Loken MR, Phillips JH. The relationship of CD16 (Leu-11) and Leu-19 (NKH-1) antigen expression on human peripheral blood NK cells and cytotoxic T lymphocytes. *J Immunol* (1986) 136(12):4480–6.
24. Chan A, Hong DL, Atzberger A, Kollnberger S, Filer AD, Buckley CD, et al. CD56bright human NK cells differentiate into CD56dim cells: role of contact with peripheral fibroblasts. *J Immunol* (2007) 179(1):89–94. doi:10.4049/jimmunol.179.1.89
25. Dulphy N, Haas P, Busson M, Belhadj S, Peffault de Latour R, Robin M, et al. An unusual CD56(bright) CD16(low) NK cell subset dominates the early posttransplant period following HLA-matched hematopoietic stem cell transplantation. *J Immunol* (2008) 181(3):2227–37. doi:10.4049/jimmunol.181.3.2227
26. Huntington ND, Legrand N, Alves NL, Jaron B, Weijer K, Plet A, et al. IL-15 trans-presentation promotes human NK cell development and differentiation in vivo. *J Exp Med* (2009) 206(1):25–34. doi:10.1084/jem.20082013
27. Juelke K, Killig M, Luetke-Eversloh M, Parente E, Gruen J, Morandi B, et al. CD62L expression identifies a unique subset of polyfunctional CD56dim NK cells. *Blood* (2010) 116(8):1299–307. doi:10.1182/blood-2009-11-253286
28. Ouyang Q, Baerlocher G, Vulto I, Lansdorp PM. Telomere length in human natural killer cell subsets. *Ann N Y Acad Sci* (2007) 1106:240–52. doi:10.1196/annals.1392.001
29. Yu J, Mao HC, Wei M, Hughes T, Zhang J, Park IK, et al. CD94 surface density identifies a functional intermediary between the CD56bright and CD56dim human NK-cell subsets. *Blood* (2010) 115(2):274–81. doi:10.1182/blood-2009-04-215491
30. Grzywacz B, Kataria N, Kataria N, Blazar BR, Miller JS, Verneris MR. Natural killer-cell differentiation by myeloid progenitors. *Blood* (2011) 117(13):3548–58. doi:10.1182/blood-2010-04-281394
31. Mailliard RB, Alber SM, Shen H, Watkins SC, Kirkwood JM, Herberman RB, et al. IL-18-induced CD83+CCR7+ NK helper cells. *J Exp Med* (2005) 202(7):941–53. doi:10.1084/jem.20050128
32. Perussia B, Chen Y, Loza MJ. Peripheral NK cell phenotypes: multiple changing of faces of an adapting, developing cell. *Mol Immunol* (2005) 42(4):385–95. doi:10.1016/j.molimm.2004.07.017
33. Vukicevic M, Chalandon Y, Helg C, Matthes T, Dantin C, Huard B, et al. CD56bright NK cells after hematopoietic stem cell transplantation are activated mature NK cells that expand in patients with low numbers of T cells. *Eur J Immunol* (2010) 40(11):3246–54. doi:10.1002/eji.200940016
34. Wu C, Li B, Lu R, Koelle SJ, Yang Y, Jares A, et al. Clonal tracking of rhesus macaque hematopoiesis highlights a distinct lineage origin for natural killer cells. *Cell Stem Cell* (2014) 14(4):486–99. doi:10.1016/j.stem.2014.01.020
35. Colucci F, Caligiuri MA, Di Santo JP. What does it take to make a natural killer? *Nat Rev Immunol* (2003) 3(5):413–25. doi:10.1038/nri1088
36. Miller JS, Alley KA, McGlave P. Differentiation of natural killer (NK) cells from human primitive marrow progenitors in a stroma-based long-term culture system: identification of a CD34+7+ NK progenitor. *Blood* (1994) 83(9):2594–601.
37. Mrozek E, Anderson P, Caligiuri MA. Role of interleukin-15 in the development of human CD56+ natural killer cells from CD34+ hematopoietic progenitor cells. *Blood* (1996) 87(7):2632–40.
38. Eissens DN, Spanholtz J, van der Meer A, van Cranenbroek B, Dolstra H, Kwekkeboom J, et al. Defining early human NK cell developmental stages in primary and secondary lymphoid tissues. *PLoS One* (2012) 7(2):e30930. doi:10.1371/journal.pone.0030930
39. Freud AG, Becknell B, Roychowdhury S, Mao HC, Ferketich AK, Nuovo GJ, et al. A human CD34(+) subset resides in lymph nodes and differentiates into CD56bright natural killer cells. *Immunity* (2005) 22(3):295–304. doi:10.1016/j.immuni.2005.01.013
40. Freud AG, Keller KA, Scoville SD, Mundy-Bosse BL, Cheng S, Youssef Y, et al. NKp80 defines a critical step during human natural killer cell development. *Cell Rep* (2016) 16(2):379–91. doi:10.1016/j.celrep.2016.05.095
41. Freud AG, Yokohama A, Becknell B, Lee MT, Mao HC, Ferketich AK, et al. Evidence for discrete stages of human natural killer cell differentiation in vivo. *J Exp Med* (2006) 203(4):1033–43. doi:10.1084/jem.20052507
42. Scoville SD, Mundy-Bosse BL, Zhang MH, Chen L, Zhang X, Keller KA, et al. A progenitor cell expressing transcription factor RORgammat generates all human innate lymphoid cell subsets. *Immunity* (2016) 44(5):1140–50. doi:10.1016/j.immuni.2016.04.007
43. Hidalgo L, Martinez VG, Valencia J, Hernandez-Lopez C, Vazquez MN, Nunez JR, et al. Expression of BMPRIA on human thymic NK cell precursors: role of BMP signaling in intrathymic NK cell development. *Blood* (2012) 119(8):1861–71. doi:10.1182/blood-2011-07-370650
44. Male V, Hughes T, McClory S, Colucci F, Caligiuri MA, Moffett A. Immature NK cells, capable of producing IL-22, are present in human uterine mucosa. *J Immunol* (2010) 185(7):3913–8. doi:10.4049/jimmunol.1001637
45. Moroso V, Famili F, Papazian N, Cupedo T, van der Laan LJ, Kazemier G, et al. NK cells can generate from precursors in the adult human liver. *Eur J Immunol* (2011) 41(11):3340–50. doi:10.1002/eji.201141760
46. Yu J, Freud AG, Caligiuri MA. Location and cellular stages of natural killer cell development. *Trends Immunol* (2013) 34(12):573–82. doi:10.1016/j.it.2013.07.005
47. Freud AG, Caligiuri MA. Human natural killer cell development. *Immunol Rev* (2006) 214:56–72. doi:10.1111/j.1600-065X.2006.00451.x
48. Grzywacz B, Kataria N, Sikora M, Oostendorp RA, Dzierzak EA, Blazar BR, et al. Coordinated acquisition of inhibitory and activating receptors and functional properties by developing human natural killer cells. *Blood* (2006) 108(12):3824–33. doi:10.1182/blood-2006-04-020198
49. Yu H, Fehniger TA, Fuchshuber P, Thiel KS, Vivier E, Carson WE, et al. Flt3 ligand promotes the generation of a distinct CD34(+) human natural killer cell progenitor that responds to interleukin-15. *Blood* (1998) 92(10):3647–57.
50. Lanier LL, Le AM, Phillips JH, Warner NL, Babcock GF. Subpopulations of human natural killer cells defined by expression of the Leu-7 (HNK-1) and Leu-11 (NK-15) antigens. *J Immunol* (1983) 131(4):1789–96.
51. McClory S, Hughes T, Freud AG, Briercheck EL, Martin C, Trimboli AJ, et al. Evidence for a stepwise program of extrathymic T cell development within the human tonsil. *J Clin Invest* (2012) 122(4):1403–15. doi:10.1172/JCI46125
52. Montaldo E, Teixeira-Alves LG, Glatzer T, Durek P, Stervbo U, Hamann W, et al. Human RORγt(+)CD34(+) cells are lineage-specified progenitors of group 3 RORγt(+) innate lymphoid cells. *Immunity* (2014) 41(6):988–1000. doi:10.1016/j.immuni.2014.11.010
53. Cella M, Fuchs A, Vermi W, Facchetti F, Otero K, Lennerz JK, et al. A human natural killer cell subset provides an innate source of IL-22 for mucosal immunity. *Nature* (2009) 457(7230):722–5. doi:10.1038/nature07537
54. Hughes T, Becknell B, Freud AG, McClory S, Briercheck E, Yu J, et al. Interleukin-1beta selectively expands and sustains interleukin-22+ immature human natural killer cells in secondary lymphoid tissue. *Immunity* (2010) 32(6):803–14. doi:10.1016/j.immuni.2010.06.007
55. Hughes T, Becknell B, McClory S, Briercheck E, Freud AG, Zhang XL, et al. Stage 3 immature human natural killer cells found in secondary lymphoid tissue constitutively and selectively express the T(H)17 cytokine interleukin-22. *Blood* (2009) 113(17):4008–10. doi:10.1182/blood-2008-12-192443
56. Freud AG, Yu J, Caligiuri MA. Human natural killer cell development in secondary lymphoid tissues. *Semin Immunol* (2014) 26(2):132–7. doi:10.1016/j.smim.2014.02.008
57. Bjorklund AK, Forkel M, Picelli S, Konya V, Theorell J, Friberg D, et al. The heterogeneity of human CD127(+) innate lymphoid cells revealed by single-cell RNA sequencing. *Nat Immunol* (2016) 17(4):451–60. doi:10.1038/ni.3368
58. Hoorweg K, Peters CP, Cornelissen F, Aparicio-Domingo P, Papazian N, Kazemier G, et al. Functional differences between human NKp44(-) and NKp44(+) RORC(+) innate lymphoid cells. *Front Immunol* (2012) 3:72. doi:10.3389/fimmu.2012.00072
59. Vitale M, Falco M, Castriconi R, Parolini S, Zambello R, Semenzato G, et al. Identification of NKp80, a novel triggering molecule expressed by human NK cells. *Eur J Immunol* (2001) 31(1):233–42. doi:10.1002/1521-4141(200101)31:1<233::AID-IMMU233>3.0.CO;2-4
60. Bjorkstrom NK, Riese P, Heuts F, Andersson S, Fauriat C, Ivarsson MA, et al. Expression patterns of NKG2A, KIR, and CD57 define a process of CD56dim NK-cell differentiation uncoupled from NK-cell education. *Blood* (2010) 116(19):3853–64. doi:10.1182/blood-2010-04-281675
61. Lopez-Verges S, Milush JM, Pandey S, York VA, Arakawa-Hoyt J, Pircher H, et al. CD57 defines a functionally distinct population of mature NK cells in

the human CD56dimCD16+ NK-cell subset. *Blood* (2010) 116(19):3865–74. doi:10.1182/blood-2010-04-282301
62. Lugthart G, Melsen JE, Vervat C, van Ostaijen-Ten Dam MM, Corver WE, Roelen DL, et al. Human lymphoid tissues harbor a distinct CD69+CXCR6+ NK cell population. *J Immunol* (2016) 197(1):78–84. doi:10.4049/jimmunol.1502603
63. Melsen JE, Lugthart G, Lankester AC, Schilham MW. Human circulating and tissue-resident CD56(bright) natural killer cell populations. *Front Immunol* (2016) 7:262. doi:10.3389/fimmu.2016.00262
64. Sojka DK, Plougastel-Douglas B, Yang L, Pak-Wittel MA, Artyomov MN, Ivanova Y, et al. Tissue-resident natural killer (NK) cells are cell lineages distinct from thymic and conventional splenic NK cells. *Elife* (2014) 3:e01659. doi:10.7554/eLife.01659
65. Renoux VM, Zriwil A, Peitzsch C, Michaelsson J, Friberg D, Soneji S, et al. Identification of a human natural killer cell lineage-restricted progenitor in fetal and adult tissues. *Immunity* (2015) 43(2):394–407. doi:10.1016/j.immuni.2015.07.011
66. Constantinides MG, Gudjonson H, McDonald BD, Ishizuka IE, Verhoef PA, Dinner AR, et al. PLZF expression maps the early stages of ILC1 lineage development. *Proc Natl Acad Sci U S A* (2015) 112(16):5123–8. doi:10.1073/pnas.1423244112
67. Constantinides MG, McDonald BD, Verhoef PA, Bendelac A. A committed precursor to innate lymphoid cells. *Nature* (2014) 508(7496):397–401. doi:10.1038/nature13047
68. Klose CS, Flach M, Möhle L, Rogell L, Hoyler T, Ebert K, et al. Differentiation of type 1 ILCs from a common progenitor to all helper-like innate lymphoid cell lineages. *Cell* (2014) 157(2):340–56. doi:10.1016/j.cell.2014.03.030
69. Yang Q, Li F, Harly C, Xing S, Ye L, Xia X, et al. TCF-1 upregulation identifies early innate lymphoid progenitors in the bone marrow. *Nat Immunol* (2015) 16(10):1044–50. doi:10.1038/ni.3248
70. Yu X, Wang Y, Deng M, Li Y, Ruhn KA, Zhang CC, et al. The basic leucine zipper transcription factor NFIL3 directs the development of a common innate lymphoid cell precursor. *Elife* (2014) 3:e04406. doi:10.7554/eLife.04406
71. Satoh-Takayama N, Lesjean-Pottier S, Vieira P, Sawa S, Eberl G, Vosshenrich CA, et al. IL-7 and IL-15 independently program the differentiation of intestinal CD3-NKp46+ cell subsets from Id2-dependent precursors. *J Exp Med* (2010) 207(2):273–80. doi:10.1084/jem.20092029
72. Vonarbourg C, Mortha A, Bui VL, Hernandez PP, Kiss EA, Hoyler T, et al. Regulated expression of nuclear receptor RORγt confers distinct functional fates to NK cell receptor-expressing RORγt(+) innate lymphocytes. *Immunity* (2010) 33(5):736–51. doi:10.1016/j.immuni.2010.10.017
73. Knox JJ, Cosma GL, Betts MR, McLane LM. Characterization of T-bet and eomes in peripheral human immune cells. *Front Immunol* (2014) 5:217. doi:10.3389/fimmu.2014.00217
74. Simonetta F, Pradier A, Roosnek E. T-bet and eomesodermin in NK cell development, maturation, and function. *Front Immunol* (2016) 7:241. doi:10.3389/fimmu.2016.00241
75. Crellin NK, Trifari S, Kaplan CD, Cupedo T, Spits H. Human NKp44+IL-22+ cells and LTi-like cells constitute a stable RORC+ lineage distinct from conventional natural killer cells. *J Exp Med* (2010) 207(2):281–90. doi:10.1084/jem.20091509
76. Matos ME, Schnier GS, Beecher MS, Ashman LK, William DE, Caligiuri MA. Expression of a functional c-kit receptor on a subset of natural killer cells. *J Exp Med* (1993) 178(3):1079–84. doi:10.1084/jem.178.3.1079

18

Dual Role of Natural Killer Cells on Graft Rejection and Control of Cytomegalovirus Infection in Renal Transplantation

Miguel López-Botet[1,2,3], Carlos Vilches[4], Dolores Redondo-Pachón[1,5], Aura Muntasell[1], Aldi Pupuleku[3], José Yélamos[1,2], Julio Pascual[1,5] and Marta Crespo[1,5]*

[1] Hospital del Mar Medical Research Institute (IMIM), Barcelona, Spain, [2] Department of Immunology, Hospital del Mar, Barcelona, Spain, [3] Univ. Pompeu Fabra, Barcelona, Spain, [4] Immunogenetics-Histocompatibility, Instituto de Investigación Sanitaria Puerta de Hierro, Majadahonda, Spain, [5] Department of Nephrology, Hospital del Mar, Barcelona, Spain

Edited by:
Emanuela Marcenaro,
University of Genoa, Italy

Reviewed by:
Michael G. Brown,
University of Virginia, USA
Ennio Carbone,
Magna Græcia University, Italy

***Correspondence:**
Miguel López-Botet
lbotet@imim.es,
miguel.lopez-botet@upf.edu

Allograft rejection constitutes a major complication of solid organ transplantation requiring prophylactic/therapeutic immunosuppression, which increases susceptibility of patients to infections and cancer. Beyond the pivotal role of alloantigen-specific T cells and antibodies in the pathogenesis of rejection, natural killer (NK) cells may display alloreactive potential in case of mismatch between recipient inhibitory killer-cell immunoglobulin-like receptors (KIRs) and graft HLA class I molecules. Several studies have addressed the impact of this variable in kidney transplant with conflicting conclusions; yet, increasing evidence supports that alloantibody-mediated NK cell activation *via* FcγRIIIA (CD16) contributes to rejection. On the other hand, human cytomegalovirus (HCMV) infection constitutes a risk factor directly associated with the rate of graft loss and reduced host survival. The levels of HCMV-specific $CD8^+$ T cells have been reported to predict the risk of posttransplant infection, and KIR-B haplotypes containing activating KIR genes have been related with protection. HCMV infection promotes to a variable extent an adaptive differentiation and expansion of a subset of mature NK cells, which display the CD94/NKG2C-activating receptor. Evidence supporting that adaptive $NKG2C^+$ NK cells may contribute to control the viral infection in kidney transplant recipients has been recently obtained. The dual role of NK cells in the interrelation of HCMV infection with rejection deserves attention. Further phenotypic, functional, and genetic analyses of NK cells may provide additional insights on the pathogenesis of solid organ transplant complications, leading to the development of biomarkers with potential clinical value.

Keywords: human, natural killer, cytomegalovirus, renal, transplantation, rejection

INTRODUCTION

Kidney transplantation is a widely used therapeutic intervention for chronic renal failure. Graft rejection remains a major complication, requiring prophylactic/therapeutic administration of immunosuppressive drugs. Consequently, kidney transplant recipients (KTR) are exposed to an increased susceptibility to infections, particularly by herpesviruses (e.g., cytomegalovirus and Epstein–Barr virus). Besides the pivotal role played by alloantigen-specific T cells and antibodies in the pathogenesis of graft rejection, natural killer (NK) cells alloreactivity and their contribution to antiviral defense receive increasing attention.

Diversity of the Human NK Cell Receptor Repertoire and NK Cell Subsets Distribution

Natural killer cells constitute an innate lymphoid lineage involved in early defense against certain intracellular pathogens and tumors, which mediate cytotoxicity and pro-inflammatory cytokine production upon interaction with pathological cells (1–3). NK cells are controlled by an array of germ line-encoded inhibitory and activating/co-stimulatory receptors (NKR), as well as by different cytokines (e.g., IL-2, IL-12, IL-15, IL-18, and type I interferons), which regulate their differentiation, proliferation, and effector functions. Inhibitory killer-cell immunoglobulin-like receptors (KIRs) and CD94/NKG2A complement each other, scanning potential target cells for altered surface expression of HLA class I (HLA-I) molecules.

The combinatorial distribution of these NKR along differentiation determines the existence of a variety of NK cell subsets capable of responding against pathological cells, which have downregulated HLA-I expression, as predicted by the "missing-self" hypothesis (4). In the context of transplantation, NK cell subsets may also react against normal allogeneic cells lacking specific HLA-I ligands for their inhibitory KIR (iKIR).

Killer-cell immunoglobulin-like receptor and NKG2 NK cell receptor families include other members with activating function whose physiological role is being investigated. At late differentiation stages, cytolytic T lymphocytes (TCRαβ CD8$^+$, CD4$^+$, and TCRγδ) may also display HLA-specific NKR (i.e., KIR, CD94/NKG2A, CD94/NKG2C, and LILRB1) (5, 6).

KIRs for HLA-A, -B, and -C

The human KIR family comprises (i) six receptors (four KIR2DL and two KIR3DL) with cytoplasmic "immunoreceptor tyrosine-based inhibition motifs" (ITIMs), which recruit the SHP-1/2 tyrosine phosphatases preventing NK cell activation; (ii) six KIR with short cytoplasmic tails lacking ITIMs (i.e., KIR2DS and KIR3DS), which interact with DAP12; this adaptor molecule contains "immunoreceptor tyrosine-based activation motifs" (ITAM) linked to protein tyrosine kinase (PTK) activation pathways; and (iii) two KIR (2DL4 and 3DL3) displaying ambiguous signaling motifs (7, 8).

Most iKIRs specifically recognize sets of HLA class Ia (i.e., HLA-A, -B, and -C) allotypes sharing structural polymorphisms at the α1 domain; yet, the ligands for some of them (e.g., KIR2DL5) and most activating KIR (aKIR) remain elusive. In an example of convergent evolution, the physiological role of KIR is undertaken in mice by members of the Ly49 lectin-like family; the Ly49H receptor triggers NK cell functions upon interaction with the m157 viral protein, contributing to defense against murine CMV (9–11). The low affinity interaction of some aKIR with HLA-I molecules suggests that they might specifically recognize pathogen-derived HLA–peptide complexes or other as yet unknown molecules.

At the population level, KIR repertoires are quite diverse due to the fact that not all KIR loci are found in the genome of every individual, and to the existence of a variety of alleles. Each KIR is encoded by a different gene in chromosome 19q13.4, and multiple KIR haplotypes/genotypes have been described worldwide (8). Moreover, iKIR–ligand interactions modulate functional NK cell maturation through an education process termed "licensing," ill-defined at the molecular level, which dictates that most mature NK cells display at least an inhibitory NKR specific for self HLA-I molecules (12, 13).

CD94/NKG2 Killer Lectin-Like Receptors for HLA-E

CD94 and members of the NKG2 family are lectin-like membrane glycoproteins encoded at the NK gene complex on human chromosome 12. Similar to KIRs, the CD94/NKG2A heterodimer constitutes an inhibitory receptor linked to the SHP-1 tyrosine phosphatase, and CD94/NKG2C is coupled through DAP12 to a PTK activation pathway (14). The specific ligand for both CD94/NKG2 receptors is constituted by the HLA-E class Ib molecule, which binds to leader sequence peptides from other HLA-I molecules, including alleles not recognized by iKIRs (15–17). Thus, CD94/NKG2A prevents the response against cells with a normal expression of HLA-I molecules, complementing the function of KIRs. HLA-E may present pathogen-derived peptides [e.g., human cytomegalovirus (HCMV), HIV-1, and HCV] altering CD94/NKG2A recognition (18–20). On the other hand, CD94/NKG2C binds to HLA-E with lower affinity than its inhibitory counterpart (21, 22) and has been reported to be involved in the response to human HCMV (see Adaptive NK Cell Response to HCMV).

Additional Activating and Inhibitory NKR

The CD16A (FcγRIIIA) receptor is coupled through CD3ζ or FcεRIγ chain adapters to a PTK activation pathway, triggering cytotoxicity and cytokine production upon interaction with IgG-opsonized cells (23). A CD16A allelic dimorphism (158V or F) influences the affinity of its interaction with IgG, modulating receptor-mediated signaling and activation of effector functions (24). Surface CD16 expression is downregulated in activated NK cells through a shedding process mediated by ADAM-17 metalloprotease (25, 26).

The human NKG2D C-type lectin triggers phosphatidyl inositol-3 kinase signaling through the DAP10 adaptor (27). NKG2D functions as an activating/co-stimulatory receptor specific for a set of ligands (MICA, MICB, and "UL16-binding proteins") displayed by pathological cells, which are also inducible by cellular stress in normal tissues (6). Several immune evasion mechanisms that prevent NKG2D ligand (NKG2D-L) expression in HCMV-infected cells have been identified (28).

Natural cytotoxicity receptors, i.e., NCR1 (NKp46), NCR2 (NKp44), and NCR3 (NKp30), are connected to PTK signaling pathways through different ITAM-bearing adapters (29). In addition to their putative role in recognition of pathogen-derived molecules, there is evidence supporting the expression of ligands in normal cells that may trigger NK cell functions when control by inhibitory receptors is reduced (30). NKp46 is coupled to the CD3ζ or FcεRIγ chain, triggering cytotoxicity and cytokine production upon recognition of an ill-defined cellular ligand(s). NKp46 has been shown to be involved in the NK cell response to HCMV-infected dendritic cells and macrophages (31, 32). The nature of cellular ligands for NKp44 also remains open, and several ligands have been reported for the CD3ζ-linked NKp30 (30, 33).

In addition to the pivotal role played by adhesion molecules (i.e., LFA-1 and CD2) in the NK cell interaction with target cells, engagement of DNAM1, a co-stimulatory receptor specific for Nectin-2 (CD112) and PVR (CD155), contributes to the response against tumor and virus-infected cells (32, 34). NK cells may acquire additional inhibitory NKR upon activation or at late differentiation stages. Among these checkpoints, LILRB1 (ILT2, LIR-1, or CD85j) interacts with a wide spectrum of HLA-I molecules and binds with a higher affinity to the UL18 HCMV glycoprotein (35, 36); similarly, TIGIT (T cell Ig and ITIM domain) binds to CD155 competing with DNAM1 (37).

Peripheral Blood NK Cell Subsets

The human peripheral blood NK cell compartment includes a variety of cell subsets, which represent distinct maturation stages and display different combinations of HLA-I-specific NKR. Similar to T and B lymphocytes, NK cells may undergo clonal expansion and late differentiation events, skewing the NKR repertoire and further diversifying their phenotypic/functional profile (**Figure 1**).

Two NK cell populations are identified in peripheral blood according to their surface expression levels of the CD56 neural-cell adhesion molecule isoform (i.e., CD56^{bright} and CD56^{dim}) (38). CD56^{bright} NK cells constitute a minor fraction (~10%) of the normal circulating NK cell compartment. They display a low cytotoxic potential but secrete pro-inflammatory cytokines and are conventionally considered to represent an early maturation stage (39). Most CD56^{bright} NK cells express CD94/NKG2A, NKG2D, and NCR, but lack KIR and CD16. The predominant (~90%) CD16^{+}NKG2D^{+}CD56^{dim} NK cell population comprises distinct subsets, defined according to KIR, NKG2A, and NKG2C expression (e.g., NKG2A^{+}KIR^{+}NKG2C$^{+/-}$ and NKG2A^{-}KIR^{+} NKG2C$^{+/-}$). Evidences have been obtained indirectly supporting a linear differentiation model in which CD56^{bright} NK cells sequentially give rise to the other NK cell subsets (38, 40). Yet, the possibility that alternative differentiation pathways branching from NK cell precursors may independently generate CD56^{bright} and CD56^{dim} subsets cannot be formally ruled out.

Further levels of NK cell phenotypic/functional heterogeneity are determined by (i) the diversity of human NKR repertoires, conditioned by the existence of hundreds of different KIR haplotypes diverging in gene and allotype content; (ii) the clonal distribution of KIR combinations among CD56^{dim} NK cells, modulated by the influence of KIR–ligand interactions on NK cell maturation; (iii) the oligoclonal adaptive expansion of NK cell subsets in response to HCMV infection (see Adaptive NK Cell Response to HCMV); and (iv) the incidence of late differentiation events, which determine additional phenotypic and functional changes (e.g., expression of CD57 and LILRB1) (**Figure 1**).

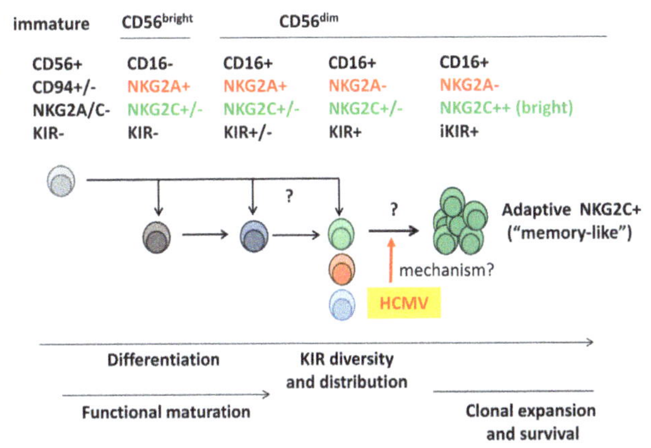

FIGURE 1 | Diversity of the human peripheral blood natural killer (NK) cell compartment. Two main NK cell populations are identified according to expression levels of the CD56 marker. CD56^{bright} NK cells secrete pro-inflammatory cytokines but display a low cytotoxic potential and are often considered to represent an early maturation stage. According to such view, this subset is believed to differentiate into the major cytotoxic CD56^{dim} NK cell population, which includes a variety of subsets differing in NKR expression [e.g., killer-cell immunoglobulin-like receptors (KIRs), CD94/NKG2A, and CD94/NKG2C]. Whether some CD56^{dim} subsets (e.g., NKG2A^{-} KIR^{+}) might directly derive from immature NK cell precursors rather than from NKG2A^{+} KIR^{-} CD56^{bright} NK cells is not formally ruled out. Human cytomegalovirus (HCMV) infection promotes the differentiation and stable expansion of an NK cell subset, which displays high levels of the CD94/NKG2C receptor and an oligoclonal inhibitory KIR (iKIR) expression pattern, associated with other phenotypic and functional characteristics (see details in Section "Adaptive NK Cell Response to HCMV"). The nature of their precursors and the mechanism(s) underlying such adaptive NK cell response to HCMV are investigated.

NK CELLS AND HCMV INFECTION IN KTR

Human cytomegalovirus is a member of the herpesviridae family which causes highly prevalent lifelong infections in all human populations, generally asymptomatic in immunocompetent hosts. The virus establishes latency, undergoing occasional reactivation which allows its efficient transmission through secretions (41, 42). HCMV may cause severe congenital disorders (43) and increases the morbidity/mortality rate in immunocompromised individuals (44, 45), being associated with some chronic inflammatory disorders (i.e., atherosclerosis) and immune senescence (46). As a consequence of immunosuppression to prevent graft rejection, KTR are exposed to HCMV reactivation/reinfection, leading to potentially severe complications (47, 48).

Together with specific T lymphocytes and antibodies, commonly analyzed to assess the adaptive immune response to HCMV, NK cells contribute to defense against this pathogen (49, 50). To escape from CD8^{+} T cells, HCMV downregulates surface expression of HLA-I molecules in infected cells, interfering with antigen presentation (51, 52). Consequently, engagement of inhibitory NKR is impaired promoting NK cell activation, which is counteracted by a variety of viral immune evasion strategies (53–55).

Adaptive NK Cell Response to HCMV

In 2004, we discovered that healthy HCMV-seropositive (HCMV$^+$) individuals display increased proportions of NK and T cells hallmarked by high surface levels of CD94/NKG2C (NKG2C^{bright}) (56). The imprint of HCMV in the NK cell compartment is perceived to a variable extent only in some HCMV$^+$ subjects, persisting under steady state conditions. A number of reports have extended these observations in different settings, and the terms "adaptive" or "memory-like" are currently employed to designate the human differentiated NKG2C^{bright} NK cell population (55). For the sake of precision, we have strictly used this original definition along the text. Yet, it is of note that these terms have been used by some authors to define other NK cell populations (e.g., *in vitro* cytokine-differentiated NK cells) (57).

Expansions of NKG2C^{bright} cells are not induced by other herpesviruses (i.e., EBV and HSV-1) but have been reported in the course of different viral infections, yet associated with HCMV coinfection (58–61). As compared to other NK cell subsets, including the low proportions of NKG2C^{dim} cells detected in HCMV(−) and some HCMV(+) individuals, adaptive NKG2C$^+$ NK cells display a phenotype characterized by an oligoclonal pattern of iKIR specific for self HLA-I molecules (preferentially HLA-C). Moreover, they express reduced levels of NCR (i.e., NKp30 and NKp46), Siglec7, and CD161 (56, 62–64), acquire late differentiation markers (e.g., CD57 and LILRB1) (65, 66), maintain surface expression of NKG2D and CD16, and display increased levels of CD2 involved in their activation (67, 68). Epigenetic downregulation of signaling molecules (e.g., FcεRIγ chain and Syk) and certain transcription factors have been associated with adaptive NK cell differentiation (69, 70). From a functional standpoint, they contain greater levels of Granzyme B and efficiently secrete TNF-α and IFN-γ (62, 63), mediating antibody-dependent cytotoxicity (ADCC) and cytokine production against HCMV-infected cells (71–73).

Expansions of NKG2C$^+$ cells following HCMV infection were reported in immunosuppressed transplant recipients (65, 66, 74), in a severe T cell primary immunodeficiency (75), as well as in children and newborns with congenital or postnatal HCMV infection (76, 77), independently of aging (78–80). Altogether, these observations suggest that the magnitude of the HCMV imprint on the NK cell compartment in healthy individuals is likely fixed at the time of primary infection, presumably depending on host/virus genetics and other circumstantial factors (e.g., age at infection, viral load, etc.) (81).

By analogy with the role of Ly49H$^+$ cells in the response to murine CMV (82), we hypothesized that CD94/NKG2C-mediated specific recognition of virus-infected cells drives the adaptive differentiation, proliferation, and survival of this lymphocyte subset (55). Indirectly supporting this view, *in vitro* stimulation of PBMC from HCMV$^+$ donors with virus-infected cells elicited a preferential expansion of CD94/NKG2C$^+$ NK cells (83, 84). Yet, at variance with Ly49H, the nature of a hypothetical viral ligand remains uncertain, and there is no experimental evidence supporting that the CD94/NKG2C receptor may trigger NK cell effector functions against HCMV-infected cells (32, 55, 83, 85). By contrast, NKG2C$^+$ adaptive NK cells have been shown to efficiently mediate antibody-dependent effector functions, particularly pro-inflammatory cytokine production, against HCMV and HSV-1 infected cells (24, 71). It is of note that CD16 remains functionally coupled to the CD3ζ adapter (73) following downregulation of FcεRIγ. The molecular mechanisms driving this pattern of response to HCMV and the existence of a putative CD94/NKG2C viral ligand are investigated (**Figure 2**).

A deletion of the *NKG2C* gene (officially designated *KLRC2*) is frequently detected in different human populations, with some variation depending on their ethnic/geographic origin (86–89). *NKG2C* gene copy number is directly related with surface expression levels and the activating function of CD94/NKG2C (62). Moreover, the *NKG2C* genotype is as well associated with steady state numbers of circulating NKG2C$^+$ NK cells, which appear reduced in *NKG2C*$^{+/del}$ as compared to *NKG2C*$^{+/+}$ individuals, further supporting a role of the NKR in driving the generation of adaptive NK cells (62, 76, 88). The identification of ~5% HCMV(+) healthy *NKG2C*$^{del/del}$ blood donors illustrates that the receptor is dispensable for controlling the viral infection under normal conditions, being redundant with other cell types (i.e., T lymphocytes). Moreover, NKG2C$^-$ NK cell subsets sharing some phenotypic features with canonical adaptive NKG2C$^+$ NK cells have been reported in HCMV(+) *NKG2C*$^{del/del}$ blood donors (68, 90) and HCMV-infected hematopoietic stem cell transplantation (HSCT) recipients (91). On the other hand, the lack of NKG2C$^+$ NK cells has been suggested to alter the control of primary HCMV infection in childhood (88); a putative relevance of the *NKG2C* deletion in immunosuppressed patients is discussed in the next section.

NK Cell Response to HCMV Infection in KTR

Posttransplant HCMV infection constitutes a risk factor for cardiac and renal allograft vasculopathy associated with chronic graft dysfunction and is directly associated with the rate of graft

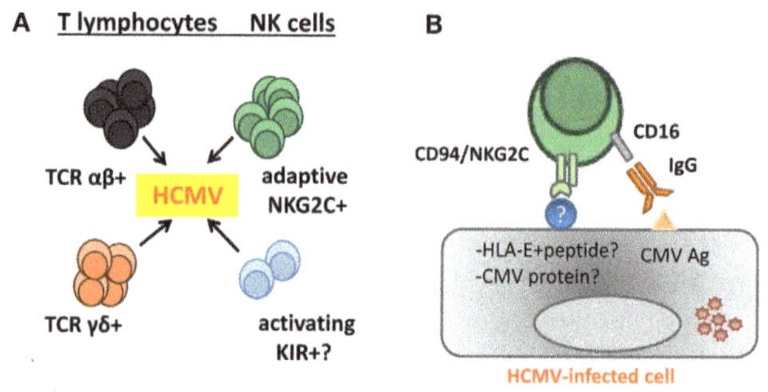

FIGURE 2 | Contribution of adaptive natural killer (NK) cells to human cytomegalovirus (HCMV) control. (A) Evidences supporting a contribution of different T and NK cell subsets in the control of HCMV infection in kidney transplant recipients have been reported. **(B)** Adaptive NKG2C^{bright} NK cells generated in response to HCMV infection efficiently mediate antibody-dependent cytotoxicity and cytokine production (e.g., TNF-α and IFN-γ) in response to HCMV-infected cells. Yet, there is no consistent evidence supporting an involvement of CD94/NKG2C in triggering NK cell effector functions against infected cells, and the nature of a hypothetical viral ligand remains elusive.

loss and reduced host survival (47, 48, 92). Antiviral prophylaxis is commonly administered to HCMV(−) KTR transplanted from an HCMV(+) donor or treated with intensive immunosuppression; patients developing HCMV viremia receive antiviral therapy, not free of adverse effects. Identification of biomarkers predicting the risk of posttransplant HCMV infection is warranted to improve its clinical management. Regular immunosuppressive therapy in KTR is aimed to prevent rejection, impairing the development of alloreactive T cells and production of alloantibodies, but has been proposed to be less effective on differentiated CTL and mature NK cells (93). Yet, alterations of the phenotypic and functional profile of circulating NK cells following immunosuppression were detected in other studies (94, 95). After low-dose therapy with anti-thymocyte globulin (ATG) NK cells recovered faster than T cells (96). In this regard, following induction with ATG functionally competent NK cells were reported to display for several months an NKG2A$^+$ KIR$^-$ phenotype (97). Thus, it is plausible that NK cells may contribute to antiviral defense in KTR, partially compensating their impaired T cell response.

The putative influence of KIR and HLA-I genotypes in the control of HCMV infection in KTR has been addressed. A relation of the KIR repertoire with viral load was reported in primary HCMV infection (98), even though the risk of HCMV disease was not influenced by KIR–ligand matching (99). De Rham et al. detected increased numbers of KIR3DL1$^+$ NK cells in KTR during the acute phase of HCMV reactivation (100). In both KTR and healthy blood donors, this NK cell subset efficiently killed *in vitro* infected fibroblasts; different interpretations for this observation were proposed. On the other hand, KIR-B haplotypes encoding aKIR were related with a lower rate of HCMV infection (101). In cases receiving thymoglobulin and intensive immunosuppression, KIR-associated control of HCMV was limited to seropositive KTR (102). A role of activating NKR in the control of other viral infections (e.g., BK and varicella zoster) has been also proposed (103, 104).

We recently explored the relationship of adaptive NKG2C$^+$ NK cells with the outcome of HCMV infection in KTR, monitoring pre- and posttransplant the NK cell immunophenotype and the incidence of viremia (105). NKG2C$^+$ NK cell expansions did not systematically follow detection of HCMV viremia in KTR, thus suggesting that a prompt control of the infection by antiviral therapy and preexisting differentiated CTL may hamper the adaptive NK cell response development. Conversely, late NKG2C$^+$ NK cell expansions might reflect clinically unnoticed HCMV replication after withdrawal of antiviral therapy. In this regard, symptom-free HCMV reactivations in KTR have been associated with altered phenotypic and functional profiles of NK cells, which expressed LILRB1 and downregulated FcϵRIγ (106). In the same line, increased proportions of LILRB1$^+$ (LIR-1$^+$) NK cells were originally associated with HCMV infection in lung transplant recipients (107).

Regular immunosuppressive protocols did not modify the levels of adaptive NK cells in KTR without detectable viremia along the follow-up, nor did they impair their expansion in some cases undergoing HCMV infection (105). Nevertheless, the possibility that immunosuppression may interfere with *de novo* adaptive NK cell differentiation, as it does with alloreactive T cell development, is not ruled out. Further studies are warranted to precisely assess the impact of different drugs on the development and effector functions of adaptive NK cells.

Of note, high pretransplant levels of NKG2C$^+$ NK cells were associated with a reduced incidence of posttransplant HCMV viremia, independently of other related variables (e.g., thymoglobulin induction, antiviral prophylaxis, and age), suggesting that adaptive NK cells might confer some protection against viral reactivation/reinfection (105). In this regard, a low NK cell count post-liver transplantation has been reported to be an independent risk factor for HCMV disease (108). Despite their limited direct *in vitro* response against HCMV-infected cells, adaptive NKG2C$^+$ NK cells may contribute to antiviral defense. In particular, they efficiently mediate antibody-dependent effector functions and likely participate in the response to HCMV reactivation in KTR, in combination with specific IgG (70, 71, 73) (**Figure 2B**). In this context, the influence of CD16A dimorphism and IgG allotypes on the magnitude of ADCC deserves attention (24). The possibility that aKIR may be involved in the putative antiviral effect of adaptive NKG2C$^+$ NK cells appears unlikely, considering that they do express iKIR (63, 64, 90) and that their expansion is independent of KIR-A/B haplotypes (56). Nevertheless, NK cell subsets expressing CD94/NKG2C or aKIR might play complementary roles in the response to HCMV.

The frequencies of TcR$\alpha\beta$ T cells specific for HCMV antigens (e.g., IE-1 and pp65) have been reported to predict the risk of posttransplant infection (109, 110); moreover, TcR$\gamma\delta$ T cells were associated with control of posttransplant HCMV viremia (111). Adaptive NKG2C$^+$ NK cells and CTL have been proposed to be independent (78–80). Thus the possibility that the association of adaptive NKG2C$^+$ NK cells with a lower risk of HCMV infection might indirectly reflect a central role of HCMV-specific TcR$\alpha\beta$ T cells (**Figure 3**) appears unlikely; further studies are warranted to precisely address this issue.

The distributions of the *NKG2C* genotypes in two different KTR cohorts, studied pre- and posttransplant, appeared comparable to the frequencies detected in blood donors; as reported, the magnitude of the NKG2C$^+$ NK cell expansion was greater in *NKG2C*$^{+/+}$ than in *NKG2C*$^{+/del}$ subjects (105). Remarkably, somewhat increased frequencies of the *NKG2C*$^{+/del}$ genotype and a reciprocal reduction of *NKG2C*$^{+/+}$ cases were detected among KTR suffering symptomatic HCMV infection; unexpectedly, an opposite reduction of the NKG2C$^{del/del}$ frequency was observed among this KTR group. Despite that differences did not reach statistical significance, the coincident trends in both cohorts suggested a relation of *NKG2C* copy number with the outcome of HCMV infection and its impact in KTR; larger studies are warranted to confirm these observations.

Altogether these results indirectly support that adaptive NKG2C$^+$ NK cells may play an active role in defense against HCMV, partially compensating in KTR the effect of immunosuppression on T cells. High pretransplant levels of NKG2C$^+$ cells may predict a lower risk of posttransplant HCMV replication/disease in KTR receiving regular immunosuppression, particularly in *NKG2C*$^{+/+}$ HCMV(+) patients (**Figure 3**). On the other hand, posttransplant expansions of differentiated adaptive NKG2C$^+$ NK cells reflect the incidence of viral replication and, once established, might contribute to its control. It is plausible that antibody-mediated response to other viral infections may as well contribute to the expansion of adaptive NKG2C⁺ cells (70, 112). It is uncertain whether adaptive NK cells may comparably respond to HCMV reactivation or reinfection, reported to have a different clinical impact (113). From a practical standpoint, monitoring basal and posttransplant levels of adaptive NK cells may provide biomarkers to evaluate the control of HCMV, with practical implications in the clinical management of the viral infection. Assessing the relation with other phenotypic features displayed at late stages of adaptive NKG2C⁺ NK cells differentiation (e.g., CD57 expression and FcεRIγ chain loss) deserves attention. Furthermore, studies in larger cohorts are required to assess the relation of the adaptive NK cell response in KTR with the incidence of other viral infections, as well as with the risk of chronic graft rejection, cardiovascular disease, and cancer (48, 114). In this regard, the possibility that antibody-dependent activation of adaptive NK cells may participate in donor-specific alloantibodies (DSA)-mediated rejection is addressed in the next section.

NK CELLS AND ALLOGRAFT REJECTION

NK Cell Alloreactivity

Mature NK cells whose iKIR fail to recognize HLA-I alleles on an allograft are predicted to mediate cytotoxicity and pro-inflammatory cytokine production, as long as activating NKR are engaged by ligands displayed on target cells. Some degree of KIR–ligand mismatching between donors and recipients is estimated to occur in 50–75% of HLA non-identical transplants, and several studies have addressed the impact of this variable in kidney transplant outcome (**Figure 4A**). On one hand, KIR–ligand mismatches were suggested to influence short-term outcome in KTR (115) and were associated with a reduced long-term graft survival in HLA-incompatible KTR (116), proposing a beneficial effect of NK cell-targeted immunosuppression. Conversely, KIR–ligand mismatch was reported by others to be irrelevant for predicting long-term allograft survival (117) and, in the same line, no effect on the risk of rejection was perceived after reduction of immunosuppressive therapy (118).

These apparently conflicting observations might be reconciled considering the implications of inhibitory NKR–MHC class I mismatch in other experimental and clinical transplant settings. Classical animal models of "F1-hybrid resistance" revealed a role of NK cells in rejection of allogeneic hematopoietic transplants, but not of other tissues grafts (119). In HSCT, donor NK cell-mediated alloreactivity has been shown to potentially exert an antileukemic effect without promoting graft-versus-host disease (120). In the same line, adoptive immunotherapy with allogeneic NK cells in HSCT recipients has been proven a safe procedure (121, 122). Altogether these observations support that the NK cell alloreactive potential, determined by KIR–ligand mismatch, may have a negligible pathogenic impact in solid organ transplantation, unless engagement of activating NKR triggers NK cell effector functions. Accordingly, NK cell alloreactivity would be favored by stimuli promoting graft expression of activating NKR ligands (e.g., NKG2D-L). This situation may take place in the context of infections (e.g., HCMV) or T cell/DSA-mediated rejection reactions, enhancing the pathogenic impact of these adverse events (**Figure 4A**). From a methodological standpoint, the genotypic prediction of KIR–ligand mismatching should be complemented

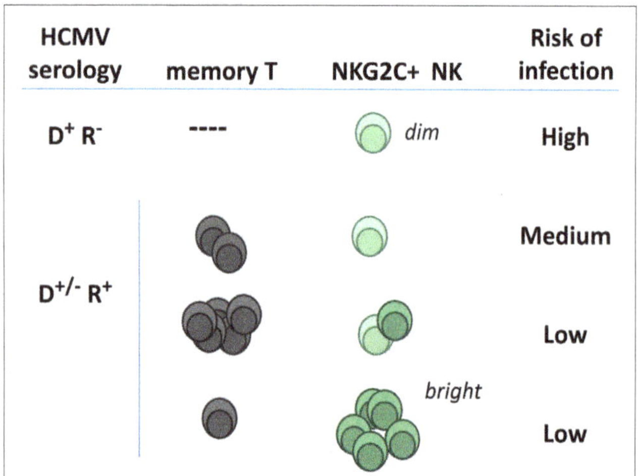

FIGURE 3 | Hypothetical relation of adaptive natural killer (NK) cells and specific T lymphocytes with posttransplant infection risk in kidney transplant recipients. NKG2C^{dim} NK cells detected in human cytomegalovirus (HCMV)(–) individuals express lower surface levels of the receptor and differ phenotypically and functionally from adaptive NK cells expanding in response to HCMV infection (NKG2C^{bright}) (see details in Section "Adaptive NK Cell Response to HCMV"). Pretransplant frequencies of virus-specific T lymphocytes and of NKG2C⁺ NK cells in seropositive recipients have been independently related with a reduced incidence of HCMV infection (D, donor; R, recipient).

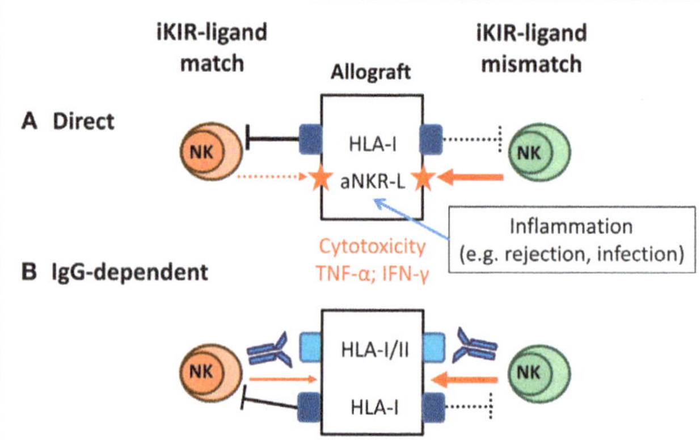

FIGURE 4 | Natural killer (NK) cell-mediated alloreactivity in solid organ transplantation. (A) NK cells lacking inhibitory KIR (iKIR) specific for donor HLA class I ligands ("mismatched," right) may potentially mediate cytotoxicity and cytokine production against the allograft. NK cell alloreactivity is favored by cellular stress conditions (e.g., pro-inflammatory stimuli) inducing graft expression of ligands for activating NKR (e.g., NKG2D ligand). **(B)** Donor-specific alloantibodies (DSA) trigger CD16⁺ NK cell-mediated cytotoxicity and cytokine production against the allograft, overcoming the control by inhibitory NKR (e.g., iKIR) (left); iKIR–ligand mismatch might synergize with a DSA-CD16-mediated response (right).

by a direct assessment of the frequencies of potential alloreactive NK cells, using specific mAbs to discriminate homologous activating and iKIR as reported for HSCT (123).

Alloantibody-Dependent NK Cell Activation

Posttransplant donor-specific anti-HLA antibodies (DSA) are a major risk factor in kidney transplant, causing microvascular damage associated with humoral rejection. In addition to complement activation, HLA-specific alloantibodies may trigger NK cells through CD16 to mediate ADCC and cytokine production (**Figure 4B**). Indications that NK cells contribute to chronic antibody-mediated rejection (ABMR) have been obtained in experimental models and analyzing kidney biopsies (124, 125). Consistent with a pathogenic role of NK cells, increased CD56^{+} cells have been observed in graft lesions from patients suffering ABMR. NK cell-associated gene expression has been associated with microvascular inflammation (126, 127), providing biomarkers with potential diagnostic/prognostic value (128–130). CD16A is also expressed by TCRγδ and some TCRαβ T lymphocyte subsets (131, 132). CD16^{+} TCRγδ T cells have been related with the response to posttransplant HCMV infection in KTR, and evidences supporting their involvement in ABMR have been reported (111, 133).

CD16 downregulation and expression of activation markers have been observed in circulating NK cells from KTR, likely reflecting IgG-dependent NK cell activation triggered by infectious pathogens (e.g., HCMV) or DSA (134). In the same line, altered distributions of circulating NK cells have been associated with the presence of alloantibodies in KTR. DSA^{+} patients were reported to display lower proportions of the major CD56^{dim} NK cell subset as compared with cases without anti-HLA antibodies (95). Increased proportions of CD56^{bright} and CD56^{dim} NKG2A^{+} cells, but not their absolute numbers, were observed in DSA^{+} KTR (135). The data suggest that alloantibody-mediated activation of NK cells *via* CD16 may promote their turnover, accounting for the imbalanced NK cell subset distribution.

This hypothesis predicts that CD56^{bright} NKG2A^{+} CD16^{-} NK cells should be spared from the effect of alloantibodies, consistent with their increased proportions in DSA^{+} KTR. On the other hand, the association of DSA with increased proportions of CD56^{dim}NKG2A^{+} NK cells suggests that engagement of CD94/NKG2A by HLA-E, conserved in all individuals, might also dampen the alloantibody effect on this subpopulation. Conversely, KIR–ligand mismatch would add to alloantibody activation of CD56^{dim}NKG2A^{-} KIR^{+} NK cells, synergizing with the pathogenic effects of DSA and accelerating their turnover. Given the oligoclonal expression by adaptive NKG2C^{+} NK cells of self-reactive KIR, preferentially specific for HLA-C molecules (63, 64), and their ability to mediate antibody-dependent effector functions (71, 73), it is likely that they may play a relevant pathogenic role in DSA-mediated graft rejection of KIR–ligand-I mismatched transplants.

In summary, consistent evidence has been obtained supporting a functional duality of NK cells in the context of kidney transplantation, reflected by their positive involvement in the response to HCMV infection as opposed to their participation in graft rejection. Further studies integrating phenotypic, functional, and genetic analysis of NK cells should provide valuable insights on the pathogenesis of solid organ transplant complications, leading to the potential development of clinically useful biomarkers.

AUTHOR CONTRIBUTIONS

All authors have actively contributed to build up the conceptual framework developed in this review and revised the draft written by ML-B.

ACKNOWLEDGMENTS

The authors thank the technical help of Dulce Soto, Judit Rigol, Andrea Vera, Sara Alvarez, Anna Faura, and Rosa Causadias.

REFERENCES

1. Moretta A, Bottino C, Mingari MC, Biassoni R, Moretta L. What is a natural killer cell? *Nat Immunol* (2002) 3(1):6–8.
2. Lanier LL. NK cell recognition. *Annu Rev Immunol* (2005) 23:225–74.
3. Vivier E, Tomasello E, Baratin M, Walzer T, Ugolini S. Functions of natural killer cells. *Nat Immunol* (2008) 9(5):503–10. doi:10.1038/ni1582
4. Ljunggren HG, Karre K. In search of the 'missing self': MHC molecules and NK cell recognition. *Immunol Today* (1990) 11:237–44.
5. Vivier E, Anfossi N. Inhibitory NK-cell receptors on T cells: witness of the past, actors of the future. *Nat Rev Immunol* (2004) 4(3):190–8.
6. Raulet DH, Gasser S, Gowen BG, Deng W, Jung H. Regulation of ligands for the NKG2D activating receptor. *Annu Rev Immunol* (2013) 31:413–41. doi:10.1146/annurev-immunol-032712-095951
7. Vilches C, Parham P. KIR: diverse, rapidly evolving receptors of innate and adaptive immunity. *Annu Rev Immunol* (2002) 20:217–51.
8. Parham P, Moffett A. Variable NK cell receptors and their MHC class I ligands in immunity, reproduction and human evolution. *Nat Rev Immunol* (2013) 13(2):133–44. doi:10.1038/nri3370
9. Arase H, Mocarski ES, Campbell AE, Hill AB, Lanier LL. Direct recognition of cytomegalovirus by activating and inhibitory NK cell receptors. *Science* (2002) 296(5571):1323–6.
10. Smith HR, Heusel JW, Mehta IK, Kim S, Dorner BG, Naidenko OV, et al. Recognition of a virus-encoded ligand by a natural killer cell activation receptor. *Proc Natl Acad Sci U S A* (2002) 99(13):8826–31.
11. Sun JC, Beilke JN, Lanier LL. Adaptive immune features of natural killer cells. *Nature* (2009) 457(7229):557–61. doi:10.1038/nature07665

12. Orr MT, Lanier LL. Natural killer cell education and tolerance. *Cell* (2010) 142(6):847–56. doi:10.1016/j.cell.2010.08.031
13. Horowitz A, Djaoud Z, Nemat-Gorgani N, Blokhuis J, Hilton HG, Béziat V, et al. Class I HLA haplotypes form two schools that educate NK cells in different ways. *Sci Immunol* (2016) 1(3):eaag1672.
14. López-Botet M, Bellón T. Natural killer cell activation and inhibition by receptors for MHC class I. *Curr Opin Immunol* (1999) 11(3):301–7.
15. Braud VM, Allan DS, O'Callaghan CA, Soderstrom K, D'Andrea A, Ogg GS, et al. HLA-E binds to natural killer cell receptors CD94/NKG2A, B and C. *Nature* (1998) 391(6669):795–9.
16. Lee N, Llano M, Carretero M, Ishitani A, Navarro F, López-Botet M, et al. HLA-E is a major ligand for the natural killer inhibitory receptor CD94/NKG2A. *Proc Natl Acad Sci U S A* (1998) 95:5199–204.
17. Borrego F, Ulbrecht M, Weiss EH, Coligan JE, Brooks AG. Recognition of human histocompatibility leukocyte antigen (HLA)-E complexed with HLA class I signal sequence-derived peptides by CD94/NKG2 confers protection from natural killer cell-mediated lysis. *J Exp Med* (1998) 187(5):813–8.
18. Tomasec P, Braud VM, Rickards C, Powell MB, McSharry BP, Gadola S, et al. Surface expression of HLA-E, an inhibitor of natural killer cells, enhanced by human cytomegalovirus gpUL40. *Science* (2000) 287(5455):1031–33.
19. Cohen GB, Gandhi RT, Davis DM, Mandelboim O, Chen BK, Strominger JL, et al. The selective downregulation of class I major histocompatibility complex proteins by HIV-1 protects HIV-infected cells from NK cells. *Immunity* (1999) 10(6):661–71.
20. Nattermann J, Nischalke HD, Hofmeister V, Ahlenstiel G, Zimmermann H, Leifeld L, et al. The HLA-A2 restricted T cell epitope HCV core 35-44 stabilizes HLA-E expression and inhibits cytolysis mediated by natural killer cells. *Am J Pathol* (2005) 166(2):443–53.
21. Valés-Gómez M, Reyburn HT, Erskine RA, López-Botet M, Strominger JL. Kinetics and peptide dependency of the binding of the inhibitory NK receptor CD94/NKG2-A and the activating receptor CD94/NKG2-C to HLA-E. *EMBO J* (1999) 18(15):4250–60.
22. Heatley SL, Pietra G, Lin J, Widjaja JM, Harpur CM, Lester S, et al. Polymorphism in human cytomegalovirus UL40 impacts on recognition of human leukocyte antigen-E (HLA-E) by natural killer cells. *J Biol Chem* (2013) 288(12):8679–90. doi:10.1074/jbc.M112.409672
23. Bryceson YT, March ME, Ljunggren HG, Long EO. Synergy among receptors on resting NK cells for the activation of natural cytotoxicity and cytokine secretion. *Blood* (2006) 107(1):159–66.
24. Moraru M, Black LE, Muntasell A, Portero F, López-Botet M, Reyburn HT, et al. NK cell and Ig interplay in defense against herpes simplex virus type 1: epistatic interaction of CD16A and IgG1 allotypes of variable affinities modulates antibody-dependent cellular cytotoxicity and susceptibility to clinical reactivation. *J Immunol* (2015) 195(4):1676–84. doi:10.4049/jimmunol.1500872
25. Romee R, Schneider SE, Leong JW, Chase JM, Keppel CR, Sullivan RP, et al. Cytokine activation induces human memory-like NK cells. *Blood* (2012) 120(24):4751–60. doi:10.1182/blood-2012-04-419283
26. Sun JC, Lanier LL. NK cell development, homeostasis and function: parallels with CD8(+) T cells. *Nat Rev Immunol* (2011) 11(10):645–57. doi:10.1038/nri3044
27. Raulet DH. Roles of the NKG2D immunoreceptor and its ligands. *Nat Rev Immunol* (2003) 3(10):781–90.
28. Jonjic S, Babic M, Polic B, Krmpotic A. Immune evasion of natural killer cells by viruses. *Curr Opin Immunol* (2008) 20(1):30–8. doi:10.1016/j.coi.2007.11.002
29. Moretta A, Bottino C, Vitale M, Pende D, Cantoni C, Mingari MC, et al. Activating receptors and coreceptors involved in human natural killer cell-mediated cytolysis. *Annu Rev Immunol* (2001) 19:197–223.
30. Koch J, Steinle A, Watzl C, Mandelboim O. Activating natural cytotoxicity receptors of natural killer cells in cancer and infection. *Trends Immunol* (2013) 34(4):182–91. doi:10.1016/j.it.2013.01.003
31. Romo N, Magri G, Muntasell A, Heredia G, Baia D, Angulo A, et al. Natural killer cell-mediated response to human cytomegalovirus-infected macrophages is modulated by their functional polarization. *J Leukoc Biol* (2011) 90(4):717–26. doi:10.1189/jlb.0311171
32. Magri G, Muntasell A, Romo N, Sáez-Borderias A, Pende D, Geraghty DE, et al. NKp46 and DNAM-1 NK-cell receptors drive the response to human cytomegalovirus-infected myeloid dendritic cells overcoming viral immune evasion strategies. *Blood* (2011) 117(3):848–56. doi:10.1182/blood-2010-08-301374
33. Kruse PH, Matta J, Ugolini S, Vivier E. Natural cytotoxicity receptors and their ligands. *Immunol Cell Biol* (2014) 92(3):221–9. doi:10.1038/icb.2013.98
34. Pende D, Spaggiari GM, Marcenaro S, Martini S, Rivera P, Capobianco A, et al. Analysis of the receptor-ligand interactions in the natural killer-mediated lysis of freshly isolated myeloid or lymphoblastic leukemias: evidence for the involvement of the poliovirus receptor (CD155) and Nectin-2 (CD112). *Blood* (2005) 105(5):2066–73.
35. Colonna M, Navarro F, Bellón T, Llano M, García P, Samaridis J, et al. A common inhibitory receptor for major histocompatibility complex class I molecules on human lymphoid and myelomonocytic cells. *J Exp Med* (1997) 186(11):1809–18.
36. Cosman D, Fanger N, Borges L, Kubin M, Chin W, Peterson L, et al. A novel immunoglobulin superfamily receptor for cellular and viral MHC class I molecules. *Immunity* (1997) 7(2):273–82.
37. Stanietsky N, Simic H, Arapovic J, Toporik A, Levy O, Novik A, et al. The interaction of TIGIT with PVR and PVRL2 inhibits human NK cell cytotoxicity. *Proc Natl Acad Sci U S A* (2009) 106(42):17858–63. doi:10.1073/pnas.0903474106
38. Caligiuri MA. Human natural killer cells. *Blood* (2008) 112(3):461–9. doi:10.1182/blood-2007-09-077438
39. Michel T, Poli A, Cuapio A, Briquemont B, Iserentant G, Ollert M, et al. Human CD56bright NK cells: an update. *J Immunol* (2016) 196(7):2923–31. doi:10.4049/jimmunol.1502570
40. Bjorkström NK, Riese P, Heuts F, Andersson S, Fauriat C, Ivarsson MA, et al. Expression patterns of NKG2A, KIR, and CD57 define a process of CD56dim NK-cell differentiation uncoupled from NK-cell education. *Blood* (2010) 116(19):3853–64. doi:10.1182/blood-2010-04-281675
41. Mocarski ES, Shenk T, Griffiths PD, Pass RF. Cytomegaloviruses. 6th ed. In: Knipe DM, Howley PM, editors. *Fields Virology*. Philadelphia: Lippincott Williams & Wilkins (2013). p. 1960–2015.
42. Crumpacker CS. Cytomegalovirus. 5th ed. In: Mandell GL, Bennett JE, Dolin R, editors. *Mandell, Douglas, and Bennett's Principles and Practice of Infectious Diseases*. Philadelphia: Churchill Livingston (2000). p. 1586–99.
43. Pass RF, Fowler KB, Boppana SB, Britt WJ, Stagno S. Congenital cytomegalovirus infection following first trimester maternal infection: symptoms at birth and outcome. *J Clin Virol* (2006) 35(2):216–20.
44. Fishman JA. Infection in solid-organ transplant recipients. *N Engl J Med* (2007) 357(25):2601–14.
45. Boeckh M, Ljungman P. How we treat cytomegalovirus in hematopoietic cell transplant recipients. *Blood* (2009) 113(23):5711–9. doi:10.1182/blood-2008-10-143560
46. Pawelec G, Goldeck D, Derhovanessian E. Inflammation, ageing and chronic disease. *Curr Opin Immunol* (2014) 29:23–8. doi:10.1016/j.coi.2014.03.007
47. Sagedal S, Hartmann A, Nordal KP, Osnes K, Leivestad T, Foss A, et al. Impact of early cytomegalovirus infection and disease on long-term recipient and kidney graft survival. *Kidney Int* (2004) 66(1):329–37.
48. Tong CY, Bakran A, Peiris JS, Muir P, Herrington CS. The association of viral infection and chronic allograft nephropathy with graft dysfunction after renal transplantation. *Transplantation* (2002) 74(4):576–8.
49. Orange JS. Natural killer cell deficiency. *J Allergy Clin Immunol* (2013) 132(3):515–25. doi:10.1016/j.jaci.2013.07.020
50. Biron CA, Nguyen KB, Pien GC, Cousens LP, Salazar-Mather TP. Natural killer cells in antiviral defense: function and regulation by innate cytokines. *Annu Rev Immunol* (1999) 17:189–220.
51. Halenius A, Gerke C, Hengel H. Classical and non-classical MHC I molecule manipulation by human cytomegalovirus: so many targets-but how many arrows in the quiver? *Cell Mol Immunol* (2015) 12(2):139–53. doi:10.1038/cmi.2014.105
52. Tortorella D, Gewurz BE, Furman MH, Schust DJ, Ploegh HL. Viral subversion of the immune system. *Annu Rev Immunol* (2000) 18:861–926.
53. Lisnic B, Lisnic VJ, Jonjic S. NK cell interplay with cytomegaloviruses. *Curr Opin Virol* (2015) 15:9–18. doi:10.1016/j.coviro.2015.07.001
54. Wilkinson GW, Tomasec P, Stanton RJ, Armstrong M, Prod'homme V, Aicheler R, et al. Modulation of natural killer cells by human cytomegalovirus. *J Clin Virol* (2008) 41(3):206–12.
55. López-Botet M, Muntasell A, Vilches C. The CD94/NKG2C+ NK-cell subset on the edge of innate and adaptive immunity to human cytomegalovirus

infection. *Semin Immunol* (2014) 26(2):145–51. doi:10.1016/j.smim.2014.03.002

56. Gumá M, Angulo A, Vilches C, Gómez-Lozano N, Malats N, López-Botet M. Imprint of human cytomegalovirus infection on the NK cell receptor repertoire. *Blood* (2004) 104(12):3664–71.
57. Wagner JA, Fehniger TA. Human adaptive natural killer cells: beyond NKG2C. *Trends Immunol* (2016) 37(6):351–3. doi:10.1016/j.it.2016.05.001
58. Gumá M, Cabrera C, Erkizia I, Bofill M, Clotet B, Ruiz L, et al. Human cytomegalovirus infection is associated with increased proportions of NK cells that express the CD94/NKG2C receptor in aviremic HIV-1-positive patients. *J Infect Dis* (2006) 194(1):38–41.
59. Petitdemange C, Becquart P, Wauquier N, Béziat V, Debré P, Leroy EM, et al. Unconventional repertoire profile is imprinted during acute chikungunya infection for natural killer cells polarization toward cytotoxicity. *PLoS Pathog* (2011) 7(9):e1002268. doi:10.1371/journal.ppat.1002268
60. Béziat V, Dalgard O, Asselah T, Halfon P, Bedossa P, Boudifa A, et al. CMV drives clonal expansion of NKG2C+ NK cells expressing self-specific KIRs in chronic hepatitis patients. *Eur J Immunol* (2012) 42(2):447–57. doi:10.1002/eji.201141826
61. Bjorkström NK, Lindgren T, Stoltz M, Fauriat C, Braun M, Evander M, et al. Rapid expansion and long-term persistence of elevated NK cell numbers in humans infected with hantavirus. *J Exp Med* (2011) 208(1):13–21. doi:10.1084/jem.20100762
62. Muntasell A, López-Montañés M, Vera A, Heredia G, Romo N, Peñafiel J, et al. NKG2C zygosity influences CD94/NKG2C receptor function and the NK-cell compartment redistribution in response to human cytomegalovirus. *Eur J Immunol* (2013) 43:3268–78. doi:10.1002/eji.201343773
63. Béziat V, Liu L, Malmberg JA, Ivarsson MA, Sohlberg E, Bjorklund AT, et al. NK cell responses to cytomegalovirus infection lead to stable imprints in the human KIR repertoire and involve activating KIRs. *Blood* (2013) 121:2678–88. doi:10.1182/blood-2012-10-459545
64. Djaoud Z, David G, Bressollette C, Willem C, Rettman P, Gagne K, et al. Amplified NKG2C+ NK cells in cytomegalovirus (CMV) infection preferentially express killer cell Ig-like receptor 2DL: functional impact in controlling CMV-infected dendritic cells. *J Immunol* (2013) 191(5):2708–16. doi:10.4049/jimmunol.1301138
65. López-Vergès S, Milush JM, Schwartz BS, Pando MJ, Jarjoura J, York VA, et al. Expansion of a unique CD57+NKG2Chi natural killer cell subset during acute human cytomegalovirus infection. *Proc Natl Acad Sci U S A* (2011) 108(36):14725–32. doi:10.1073/pnas.1110900108
66. Della Chiesa M, Falco M, Podesta M, Locatelli F, Moretta L, Frassoni F, et al. Phenotypic and functional heterogeneity of human NK cells developing after umbilical cord blood transplantation: a role for human cytomegalovirus? *Blood* (2012) 119(2):399–410. doi:10.1182/blood-2011-08-372003
67. Rölle A, Halenius A, Ewen EM, Cerwenka A, Hengel H, Momburg F. CD2-CD58 interactions are pivotal for the activation and function of adaptive natural killer cells in human cytomegalovirus infection. *Eur J Immunol* (2016) 46(10):2420–5. doi:10.1002/eji.201646492
68. Liu LL, Landskron J, Ask EH, Enqvist M, Sohlberg E, Traherne JA, et al. Critical role of CD2 co-stimulation in adaptive natural killer cell responses revealed in NKG2C-deficient humans. *Cell Rep* (2016) 15(5):1088–99. doi:10.1016/j.celrep.2016.04.005
69. Schlums H, Cichocki F, Tesi B, Theorell J, Béziat V, Holmes TD, et al. Cytomegalovirus infection drives adaptive epigenetic diversification of NK cells with altered signaling and effector function. *Immunity* (2015) 42(3):443–56. doi:10.1016/j.immuni.2015.02.008
70. Lee J, Zhang T, Hwang I, Kim A, Nitschke L, Kim M, et al. Epigenetic modification and antibody-dependent expansion of memory-like NK cells in human cytomegalovirus-infected individuals. *Immunity* (2015) 42(3):431–42. doi:10.1016/j.immuni.2015.02.013
71. Costa-García M, Vera A, Moraru M, Vilches C, López-Botet M, Muntasell A. Antibody-mediated response of NKG2Cbright NK cells against human cytomegalovirus. *J Immunol* (2015) 194(6):2715–24. doi:10.4049/jimmunol.1402281
72. Wu Z, Sinzger C, Frascaroli G, Reichel J, Bayer C, Wang L, et al. Human cytomegalovirus-induced NKG2C(hi) CD57(hi) natural killer cells are effectors dependent on humoral antiviral immunity. *J Virol* (2013) 87(13):7717–25. doi:10.1128/JVI.01096-13
73. Zhang T, Scott JM, Hwang I, Kim S. Cutting edge: antibody-dependent memory-like NK cells distinguished by FcRγ deficiency. *J Immunol* (2013) 190(4):1402–6. doi:10.4049/jimmunol.1203034
74. Foley B, Cooley S, Verneris MR, Pitt M, Curtsinger J, Luo X, et al. Cytomegalovirus reactivation after allogeneic transplantation promotes a lasting increase in educated NKG2C+ natural killer cells with potent function. *Blood* (2012) 119(11):2665–74. doi:10.1182/blood-2011-10-386995
75. Kuijpers TW, Baars PA, Dantin C, van den Burg M, van Lier RA, Roosnek E. Human NK cells can control CMV infection in the absence of T cells. *Blood* (2008) 112(3):914–5. doi:10.1182/blood-2008-05-157354
76. Noyola DE, Fortuny C, Muntasell A, Noguera-Julián A, Muñoz-Almagro C, Alarcón A, et al. Influence of congenital human cytomegalovirus infection and the NKG2C genotype on NK-cell subset distribution in children. *Eur J Immunol* (2012) 42:3256–66. doi:10.1002/eji.201242752
77. Noyola DE, Alarcón A, Noguera-Julian A, Muntasell A, Muñoz-Almagro C, García J, et al. Dynamics of the NK-cell subset redistribution induced by cytomegalovirus infection in preterm infants. *Hum Immunol* (2015) 76(2–3):118–23. doi:10.1016/j.humimm.2015.01.017
78. Bengner M, Béziat V, Ernerudh J, Nilsson BO, Lofgren S, Wikby A, et al. Independent skewing of the T cell and NK cell compartments associated with cytomegalovirus infection suggests division of labor between innate and adaptive immunity. *Age (Dordr)* (2014) 336(2):571–82. doi:10.1007/s11357-013-9587-y
79. Bigley AB, Spielmann G, Agha N, O'Connor DP, Simpson RJ. Dichotomous effects of latent CMV infection on the phenotype and functional properties of CD8+ T-cells and NK-cells. *Cell Immunol* (2016) 300:26–32. doi:10.1016/j.cellimm.2015.11.005
80. López-Botet M, Muntasell A, Martínez-Rodríguez JE, López-Montañés M, Costa-García M, Pupuleku A. Development of the adaptive NK cell response to human cytomegalovirus in the context of aging. *Mech Ageing Dev* (2016) 158:23–6. doi:10.1016/j.mad.2016.06.010
81. Muntasell A, Vilches C, Angulo A, López-Botet M. Adaptive reconfiguration of the human NK-cell compartment in response to cytomegalovirus: a different perspective of the host-pathogen interaction. *Eur J Immunol* (2013) 43(5):1133–41. doi:10.1002/eji.201243117
82. O'Sullivan TE, Sun JC, Lanier LL. Natural killer cell memory. *Immunity* (2015) 43(4):634–45. doi:10.1016/j.immuni.2015.09.013
83. Gumá M, Budt M, Sáez A, Brckalo T, Hengel H, Angulo A, et al. Expansion of CD94/NKG2C+ NK cells in response to human cytomegalovirus-infected fibroblasts. *Blood* (2006) 107(9):3624–31.
84. Rölle A, Pollmann J, Ewen EM, Le VT, Halenius A, Hengel H, et al. IL-12-producing monocytes and HLA-E control HCMV-driven NKG2C+ NK cell expansion. *J Clin Invest* (2014) 124(12):5305–16. doi:10.1172/JCI77440
85. Djaoud Z, Riou R, Gavlovsky PJ, Mehlal S, Bressollette C, Gérard N, et al. Cytomegalovirus-infected primary endothelial cells trigger NKG2C+ natural killer cells. *J Innate Immun* (2016) 8(4):374–85. doi:10.1159/000445320
86. Miyashita R, Tsuchiya N, Hikami K, Kuroki K, Fukazawa T, Bijl M, et al. Molecular genetic analyses of human NKG2C (KLRC2) gene deletion. *Int Immunol* (2004) 16(1):163–8.
87. Moraru M, Cañiizares M, Muntasell A, de Pablo R, López-Botet M, Vilches C. Assessment of copy-number variation in the NKG2C receptor gene in a single-tube and characterization of a reference cell panel, using standard polymerase chain reaction. *Tissue Antigens* (2012) 80(2):184–7. doi:10.1111/j.1399-0039.2012.01911.x
88. Goodier MR, White MJ, Darboe A, Nielsen CM, Goncalves A, Bottomley C, et al. Rapid NK cell differentiation in a population with near-universal human cytomegalovirus infection is attenuated by NKG2C deletions. *Blood* (2014) 124(14):2213–22. doi:10.1182/blood-2014-05-576124
89. Rangel-Ramírez VV, García-Sepúlveda CA, Escalante-Padrón F, Pérez-González LF, Rangel-Castilla A, Aranda-Romo S, et al. NKG2C gene deletion in the Mexican population and lack of association to respiratory viral infections. *Int J Immunogenet* (2014) 41(2):126–30. doi:10.1111/iji.12104
90. Muntasell A, Pupuleku A, Cisneros E, Vera A, Moraru M, Vilches C, et al. Relationship of NKG2C copy number with the distribution of distinct cytomegalovirus-induced adaptive NK cell subsets. *J Immunol* (2016) 196:3818–27. doi:10.4049/jimmunol.1502438
91. Della Chiesa M, Falco M, Bertaina A, Muccio L, Alicata C, Frassoni F, et al. Human cytomegalovirus infection promotes rapid maturation of NK cells expressing activating killer Ig-like receptor in patients transplanted with

NKG2C-/- umbilical cord blood. *J Immunol* (2014) 192:1471–9. doi:10.4049/jimmunol.1302053
92. Razonable RR, Humar A. Cytomegalovirus in solid organ transplantation. *Am J Transplant* (2013) 13(Suppl 4):93–106. doi:10.1111/ajt.12103
93. Demmers MW, Korevaar SS, Betjes MG, Weimar W, Rowshani AT, Baan CC. Limited efficacy of immunosuppressive drugs on CD8+ T cell-mediated and natural killer cell-mediated lysis of human renal tubular epithelial cells. *Transplantation* (2014) 97(11):1110–8. doi:10.1097/TP.0000000000000108
94. Morteau O, Blundell S, Chakera A, Bennett S, Christou CM, Mason PD, et al. Renal transplant immunosuppression impairs natural killer cell function in vitro and in vivo. *PLoS One* (2010) 5(10):e13294. doi:10.1371/journal.pone.0013294
95. Neudoerfl C, Mueller BJ, Blume C, Daemen K, Stevanovic-Meyer M, Keil J, et al. The peripheral NK cell repertoire after kidney transplantation is modulated by different immunosuppressive drugs. *Front Immunol* (2013) 4:46. doi:10.3389/fimmu
96. Kho MM, Bouvy AP, Cadogan M, Kraaijeveld R, Baan CC, Weimar W. The effect of low and ultra-low dosages thymoglobulin on peripheral T, B and NK cells in kidney transplant recipients. *Transpl Immunol* (2012) 26(4): 186–90. doi:10.1016/j.trim.2012.02.003
97. Hadaya K, Avila Y, Valloton L, de Rham, Bandelier C, Ferrari-Lacraz S, et al. Natural killer cell receptor – repertoire and functions after induction therapy by polyclonal rabbit anti-thymocyte globulin in unsensitized kidney transplant recipients. *Clin Immunol* (2010) 137(2):250–60. doi:10.1016/j.clim.2010.07.004
98. Jones DC, Peacock S, Hughes D, Traherne JA, Allen RL, Barnardo MC, et al. Killer immunoglobulin-like receptor gene repertoire influences viral load of primary human cytomegalovirus infection in renal transplant patients. *Genes Immun* (2014) 15(8):562–8. doi:10.1038/gene.2014.53
99. Michelo CM, van der Meer A, Tijssen HJ, Zomer R, Stelma F, Hilbrands LB, et al. KIR and human leukocyte antigen genotype associated risk of cytomegalovirus disease in renal transplant patients. *Transplantation* (2015) 99(7):1506–13. doi:10.1097/TP.0000000000000497
100. De Rham C, Hadaya K, Bandelier C, Ferrari-Lacraz S, Villard J. Expression of killer cell immunoglobulin-like receptors (KIRs) by natural killer cells during acute CMV infection after kidney transplantation. *Transpl Immunol* (2014) 31(3):157–64. doi:10.1016/j.trim.2014.08.002
101. Stern M, Hadaya K, Honger G, Martin PY, Steiger J, Hess C, et al. Telomeric rather than centromeric activating KIR genes protect from cytomegalovirus infection after kidney transplantation. *Am J Transplant* (2011) 11(6):1302–7. doi:10.1111/j.1600-6143.2011.03516.x
102. González A, Schmitter K, Hirsch HH, Garzoni C, van Delden C, Boggian K, et al. KIR-associated protection from CMV replication requires pre-existing immunity: a prospective study in solid organ transplant recipients. *Genes Immun* (2014) 15(7):495–9. doi:10.1038/gene.2014.39
103. Trydzenskaya H, Juerchott K, Lachmann N, Kotsch K, Kunert K, Weist B, et al. The genetic predisposition of natural killer cell to BK virus-associated nephropathy in renal transplant patients. *Kidney Int* (2013) 84(2):359–65. doi:10.1038/ki.2013.59
104. Schmied L, Terszowski G, González A, Schmitter K, Hirsch HH, Garzoni C, et al. Protection from varicella zoster in solid organ transplant recipients carrying killer cell immunoglobulin-like receptor B haplotypes. *Transplantation* (2015) 99(12):2651–5. doi:10.1097/TP.0000000000000778
105. Redondo-Pachón D, Crespo M, Yélamos J, Muntasell A, Pérez-Sáez MJ, Pérez-Fernández S, et al. Adaptive NKG2C+ NK cell response and the risk of cytomegalovirus infection in kidney transplant recipients. *J Immunol* (2017) 198(1):94–101.
106. Makwana NB, Foley B, Lee S, Fernandez S, Irish AB, Price P. Asymptomatic CMV infections in long-term renal transplant recipients are associated with the loss of FcRgamma from LIR-1+ NK cells. *Eur J Immunol* (2016) 46(11):2597–608. doi:10.1002/eji.201646422
107. Berg L, Riise GC, Cosman D, Bergström T, Olofsson S, Kärre K, et al. LIR-1 expression on lymphocytes, and cytomegalovirus disease in lung-transplant recipients. *Lancet* (2003) 361(9363):1099–101.
108. Fernández-Ruiz M, Silva JT, López-Medrano F, Allende LM, San Juan R, Cambra F, et al. Post-transplant monitoring of NK cell counts as a simple approach to predict the occurrence of opportunistic infection in liver transplant recipients. *Transpl Infect Dis* (2016) 18(4):552–65. doi:10.1111/tid.12564
109. Bestard O, Lucia M, Crespo E, Van LB, Palacio D, Melilli E, et al. Pretransplant immediately early-1-specific T cell responses provide protection for CMV infection after kidney transplantation. *Am J Transplant* (2013) 13(7):1793–805. doi:10.1111/ajt.12256
110. Lucia M, Crespo E, Cruzado JM, Grinyó JM, Bestard O. Human CMV-specific T-cell responses in kidney transplantation; toward changing current risk-stratification paradigm. *Transpl Int* (2014) 27(7):643–56. doi:10.1111/tri.12318
111. Kaminski H, Garrigue I, Couzi L, Taton B, Bachelet T, Moreau JF, et al. Surveillance of gammadelta T cells predicts cytomegalovirus infection resolution in kidney transplants. *J Am Soc Nephrol* (2016) 27(2):637–45. doi:10.1681/ASN.2014100985
112. Saghafian-Hedengren S, Sohlberg E, Theorell J, Carvalho-Queiroz C, Nagy N, Persson JO, et al. Epstein-Barr virus coinfection in children boosts cytomegalovirus-induced differentiation of natural killer cells. *J Virol* (2013) 87(24):13446–55. doi:10.1128/JVI.02382-13
113. Ishibashi K, Yamaguchi O, Suzutani T. Reinfection of cytomegalovirus in renal transplantation. *Fukushima J Med Sci* (2011) 57(1):1–10.
114. Achour A, Baychelier F, Besson C, Arnoux A, Marty M, Hannoun L, et al. Expansion of CMV-mediated NKG2C+ NK cells associates with the development of specific de novo malignancies in liver-transplanted patients. *J Immunol* (2014) 192(1):503–11. doi:10.4049/jimmunol.1301951
115. Kunert K, Seiler M, Mashreghi MF, Klippert K, Schonemann C, Neumann K, et al. KIR/HLA ligand incompatibility in kidney transplantation. *Transplantation* (2007) 84(11):1527–33.
116. Van Bergen J, Thompson A, Haasnoot GW, Roodnat JI, de Fijter JW, Claas FH, et al. KIR-ligand mismatches are associated with reduced long-term graft survival in HLA-compatible kidney transplantation. *Am J Transplant* (2011) 11(9):1959–64. doi:10.1111/j.1600-6143.2011.03621.x
117. Tran TH, Unterrainer C, Fiedler G, Dohler B, Scherer S, Ruhenstroth A, et al. No impact of KIR-ligand mismatch on allograft outcome in HLA-compatible kidney transplantation. *Am J Transplant* (2013) 13(4):1063–8. doi:10.1111/ajt.12134
118. Kreijveld E, van der Meer A, Tijssen HJ, Hilbrands LB, Joosten I. KIR gene and KIR ligand analysis to predict graft rejection after renal transplantation. *Transplantation* (2007) 84(8):1045–51.
119. Kumar V, George T, Yu YYL, Liu J, Bennett M. Role of murine NK cells and their receptors in hybrid resistance. *Curr Opin Immunol* (1997) 9:52–6.
120. Velardi A, Ruggeri L, Mancusi A. Killer-cell immunoglobulin-like receptors reactivity and outcome of stem cell transplant. *Curr Opin Hematol* (2012) 19(4):319–23. doi:10.1097/MOH.0b013e32835423c3
121. Locatelli F, Moretta F, Brescia L, Merli P. Natural killer cells in the treatment of high-risk acute leukaemia. *Semin Immunol* (2014) 26(2):173–9. doi:10.1016/j.smim.2014.02.004
122. Knorr DA, Bachanova V, Verneris MR, Miller JS. Clinical utility of natural killer cells in cancer therapy and transplantation. *Semin Immunol* (2014) 26(2):161–72. doi:10.1016/j.smim.2014.02.002
123. Locatelli F, Pende D, Mingari MC, Bertaina A, Falco M, Moretta A, et al. Cellular and molecular basis of haploidentical hematopoietic stem cell transplantation in the successful treatment of high-risk leukemias: role of alloreactive NK cells. *Front Immunol* (2013) 4:15. doi:10.3389/fimmu.2013.00015
124. Zhang ZX, Huang X, Jiang J, Lau A, Yin Z, Liu W, et al. Natural killer cells mediate long-term kidney allograft injury. *Transplantation* (2015) 99(5):916–24. doi:10.1097/TP.0000000000000665
125. Baldwin WM III, Valujskikh A, Fairchild RL. Mechanisms of antibody-mediated acute and chronic rejection of kidney allografts. *Curr Opin Organ Transplant* (2016) 21(1):7–14. doi:10.1097/MOT.0000000000000262
126. Hidalgo LG, Sellares J, Sis B, Mengel M, Chang J, Halloran PF. Interpreting NK cell transcripts versus T cell transcripts in renal transplant biopsies. *Am J Transplant* (2012) 12(5):1180–91. doi:10.1111/j.1600-6143.2011.03970.x
127. Venner JM, Hidalgo LG, Famulski KS, Chang J, Halloran PF. The molecular landscape of antibody-mediated kidney transplant rejection: evidence for NK involvement through CD16a Fc receptors. *Am J Transplant* (2015) 15(5):1336–48. doi:10.1111/ajt.13115
128. Sellares J, de Freitas DG, Mengel M, Reeve J, Einecke G, Sis B, et al. Understanding the causes of kidney transplant failure: the dominant role

of antibody-mediated rejection and nonadherence. *Am J Transplant* (2012) 12(2):388–99. doi:10.1111/j.1600-6143.2011.03840.x

129. Gupta A, Broin PO, Bao Y, Pullman J, Kamal L, Ajaimy M, et al. Clinical and molecular significance of microvascular inflammation in transplant kidney biopsies. *Kidney Int* (2016) 89(1):217–25. doi:10.1038/ki.2015.276
130. Halloran PF, Famulski K, Reeve J. The molecular phenotypes of rejection in kidney transplant biopsies. *Curr Opin Organ Transplant* (2015) 20(3):359–67. doi:10.1097/MOT.0000000000000193
131. Clemenceau B, Vivien R, Berthome M, Robillard N, Garand R, Gallot G, et al. Effector memory alphabeta T lymphocytes can express FcgammaRIIIa and mediate antibody-dependent cellular cytotoxicity. *J Immunol* (2008) 180(8):5327–34.
132. Couzi L, Pitard V, Sicard X, Garrigue I, Hawchar O, Merville P, et al. Antibody-dependent anti-cytomegalovirus activity of human gammadelta T cells expressing CD16 (FcgammaRIIIa). *Blood* (2012) 119(6):1418–27. doi:10.1182/blood-2011-06-363655
133. Bachelet T, Couzi L, Pitard V, Sicard X, Rigothier C, Lepreux S, et al. Cytomegalovirus-responsive gammadelta T cells: novel effector cells in antibody-mediated kidney allograft microcirculation lesions. *J Am Soc Nephrol* (2014) 25(11):2471–82. doi:10.1681/ASN.2013101052
134. Hoffmann U, Neudorfl C, Daemen K, Keil J, Stevanovic-Meyer M, Lehner F, et al. NK cells of kidney transplant recipients display an activated phenotype that is influenced by immunosuppression and pathological staging. *PLoS One* (2015) 10(7):e0132484. doi:10.1371/journal.pone.0132484
135. Crespo M, Yélamos J, Redondo D, Muntasell A, Pérez-Sáez MJ, López-Montañés M, et al. Circulating NK-cell subsets in renal allograft recipients with anti-HLA donor-specific antibodies. *Am J Transplant* (2015) 15(3):806–14. doi:10.1111/ajt.13010

19

Manufacturing Natural Killer Cells as Medicinal Products

Christian Chabannon[1,2], Bechara Mfarrej[1,2], Sophie Guia[3], Sophie Ugolini[3], Raynier Devillier[2], Didier Blaise[2], Eric Vivier[3,4] and Boris Calmels[1,2]*

[1]CBT-1409: INSERM, Aix Marseille Univ, Institut Paoli-Calmettes, AP-HM, Marseille, France, [2]CRCM: INSERM, CNRS, Aix Marseille Univ, Institut Paoli-Calmettes, CRCM, Marseille, France, [3]UM2, INSERM, Centre d'Immunologie de Marseille-Luminy, U1104, CNRS UMR7280, Aix-Marseille University, Marseille, France, [4]Laboratoire d'Immunologie, Hôpital de la Conception, Assistance Publique – Hôpitaux de Marseille, Aix-Marseille University, Marseille, France

Edited by:
Lutz Walter, Leibniz-Institute for Primate Research, Germany

Reviewed by:
Markus Uhrberg, University of Düsseldorf, Germany
Ulrike Koehl, Hannover Medical School, Germany

***Correspondence:**
Christian Chabannon
chabannonc@ipc.unicancer.fr, christian.chabannon@univ-amu.fr, thercell@ipc.unicancer.fr

Natural Killer (NK) cells are innate lymphoid cells (ILC) with cytotoxic and regulatory properties. Their functions are tightly regulated by an array of inhibitory and activating receptors, and their mechanisms of activation strongly differ from antigen recognition in the context of human leukocyte antigen presentation as needed for T-cell activation. NK cells thus offer unique opportunities for new and improved therapeutic manipulation, either *in vivo* or *in vitro*, in a variety of human diseases, including cancers. NK cell activity can possibly be modulated *in vivo* through direct or indirect actions exerted by small molecules or monoclonal antibodies. NK cells can also be adoptively transferred following more or less substantial modifications through cell and gene manufacturing, in order to empower them with new or improved functions and ensure their controlled persistence and activity in the recipient. In the present review, we will focus on the technological and regulatory challenges of NK cell manufacturing and discuss conditions in which these innovative cellular therapies can be brought to the clinic.

Keywords: natural killer cells, innate lymphoid cells, immunotherapy, immuno-oncology, cellular therapy, cell transplant, manufacturing

INTRODUCTION

Cellular therapies are, nowadays, increasing in numbers and in diversity. Manufacturing of various types of immune cells is likely to provide additional therapeutic resources and complete the portfolio of immunotherapies, along with chemical molecules and engineered monoclonal antibodies. It is envisioned that combination of these different medicinal products, tailored to disease characteristics as well as to the host (immune) environment (1), will contribute to precision medicine and expectedly to higher rates of success in the cure of a variety of health disorders, including, but not restricted to, cancers. Among immune effectors amenable to cell and genetic manipulation prior to adoptive transfer, natural killer (NK) cells present with appealing biological characteristics. NK cells are innate

---

Abbreviations: ADCC, antibody-dependent cytotoxicity; ATMP, advanced therapy medicinal product; CAR, chimeric antigen receptor; CB, cord blood, placental blood; CBU, cord blood unit; CLL, chronic lymphocytic leukemia; DLI, donor lymphocyte infusion; EMA, European Medicines Agency; ES, embryonic stem (cells); FDA, Food and Drug Agency; Flt3-L, flt3 ligand; G-CSF, granulocyte colony-stimulating factor; GM-CSF, granulocyte-macrophage colony-stimulating factor; GMP, good manufacturing practices; GvHD, graft-versus-host disease; HLA, human leukocyte antigens; HSCT, hematopoietic stem cell transplantation; IL-x, interleukin x; ILC, innate lymphoid cells; iPS, induced pluripotent stem (cells); KAR, killer activating receptors; KIR, killer cell Immunoglobulin-like receptors or killer inhibitory receptors; LIF, leukemia inhibitory factor; LMWH, low molecular weight heparin; MIP1α, Macrophage Inflammatory Protein-1α; NK, natural killer; PBMC, peripheral blood mononuclear cells; SCF, stem cell factor; TPO, thrombopoietin; TRAIL, TNF-related apoptosis-inducing ligand.

lymphoid cells (ILC) and contribute to innate immunity (2). Their activities are regulated through the biological modulation of a large array of both inhibitory and activating receptors, including killer cell immunoglobulin-like receptors (KIR), NKp44, and NKp46. These receptors do not bind specific antigens on target cells as do T cells, but rather molecules induced by cellular stress that provide an activating signal, or human leukocyte antigen (HLA) molecules that predominantly provide inhibitory signals; already published material provides in-depth description of these pathways. Here, we review the existing literature that describes the rationale for various technological approaches to NK cell manufacturing, either autologous or allogeneic, as a prerequisite to adoptive transfer and clinical evaluation of these peculiar populations of immune effectors.

MEDICAL APPLICATIONS OF NK CELL MANUFACTURING AND ADOPTIVE TRANSFER

Autologous NK Cells

Adoptive transfer of NK cells engineered to express new or augmented functions represent an interesting avenue for the treatment of various high-risk malignancies in which conventional options have failed (**Figure 1**). Examples include B-chronic lymphocytic leukemia (B-CLL) (3), multiple myeloma (4), and also tumors of non-hematopoietic origin such as breast cancer (5), melanoma, or renal cell carcinoma (6).

Early studies reporting the adoptive transfer of $CD56^+$ bead-selected autologous NK cells into patients with metastatic cancers, in combination with high doses of IL-2 demonstrated feasibility yet poor clinical efficacy (7). Patients experienced severe toxic side effects due to the high doses of IL-2. In addition to activation-induced NK cell death, NK cell function may have been inhibited due to regulatory T cell expansion in response to high IL-2 doses (8, 9). Reducing the daily IL-2 doses after NK cell transfer resulted in limited clinical success (10). Several strategies are under investigation to overcome this hurdle [reviewed in Ref. (11)]. Another mechanism by which autologous NK cells are inhibited is by self-HLA molecules. Therefore "releasing the brakes" with anti-KIR antibodies such as Lirilumab® that targets the inhibitory KIR receptors on NK cells could be one approach (12–14).

Allogeneic NK Cells

A number of arguments support an important role for donor-derived NK cells in the context of allogeneic hematopoietic stem cell transplantation (allo-HSCT). Following the administration of either myelo-ablative (15) or reduced-intensity (15–17) conditioning regimen and allo–HSCT, the rapid reconstitution of high numbers of circulating and phenotypically defined NK cells is associated with better clinical outcome. The recovery of various functions for donor-derived cells may be further modulated and improved *in vivo* with additional intervention (18).

Transplantation of high doses of immune-selected $CD34^+$ cells collected from haploidentical donors after myelo-ablative conditioning regimen has provided a setting which demonstrates that "KIR-incompatibility" was associated with lower incidence of disease relapses, at least for AML (19). Transplantation of T-replete marrow or blood cell grafts obtained from haploidentical donors, using modified immune-suppressive conditioning regimen such as those including posttransplant cyclophosphamide, represent a more widely applicable procedure, in which to further explore the potential contribution of alloreactive NK cells in posttransplant clinical events. Unexpectedly, a recently published report suggests that, in this context, the presence of recipient class I ligands to donor KIR receptors confers some protection to the recipient against leukemia relapse, an observation that needs further confirmation and would imply a role for killer activating receptors (KAR) as much as for KIR (20). The role of alloreactive NK cells remains more elusive in the context of HSCT performed from other categories of donors. Expression of specific KIR receptors in HLA-matched unrelated donors was demonstrated to produce superior or inferior clinical outcomes in recipients, depending on donor–recipient combinations (21–23).

Adoptive transfer of allogeneic NK cells either with a stem cell graft *ex vivo* depleted of immune effectors or as a substitute to posttransplant "donor lymphocyte infusions" (DLIs) is thus appealing as a way to improve engraftment, immune reconstitution, and antitumor activity with reduced chances of triggering graft-versus-host disease (GVHD) (24). Results of a small number of clinical trials have been reported so far, demonstrating the feasibility of manufacturing allogeneic NK cells from matched related, matched unrelated, or mostly from haploidentical donors (25–29). Although allogeneic NK cell infusions were generally reported as safe, a recent publication describes the clinical outcome of a small cohort of pediatric patients treated for

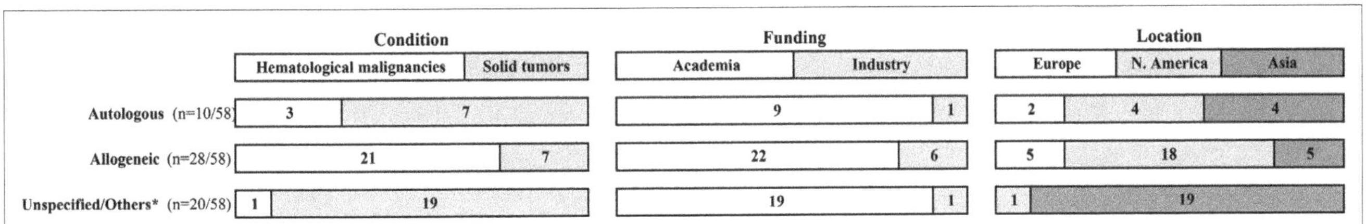

FIGURE 1 | Ongoing clinical trials that evaluate NK cell-based cellular therapies. Data were collected at https://clinicaltrials.gov/ in September 2016, using the search terms: "recruiting studies," "NK," or "natural killer" in the "Title" field. Raw data are available in Data Sheet S1 in Supplementary Material. *Unspecified = 18 studies with no designation whether NK cells are autologous or allogeneic; others = 1 study tests NK cell line and 1 study compares autologous to allogeneic NK cells.

non-hematological high-risk malignancies and a high proportion of aGVHD triggered by HLA-matched donor-derived NK cells (30). Mostly, these limited clinical results suggest that additional improvements are needed either during the *ex-vivo* manufacturing process (31) or after infusion of manufactured NK cells (25) to improve long-term persistence and activity *in vivo*.

FACTORS AFFECTING NK CELL PRODUCT MANUFACTURING

Many variables contribute to an efficient NK cell generation protocol (**Table 1**). Donor–recipient combinations, the source of starting material and culture conditions are factors that must be carefully selected to optimize the manufacturing process and potentially the clinical efficacy of the resulting medicinal product upon administration to the recipient.

Donor Selection

In the setting of allogeneic NK-DLI, donor selection can affect the clinical outcome of NK cell therapy, since certain KIR, HLA, and FcγR polymorphisms influence NK cell function (32, 33). KIR typing can be genotypic, classifying donors on the basis of gene expression of activating and inhibitory KIR (34), thereby assigning them scores to select "preferable donors" (35–37). Additionally, KIR typing can be phenotypic, assessing surface protein expression of KIRs (38), adding another layer of complexity to the selection of "preferable donors." KIR "allele-typing" is a recent addition to the donor selection algorithm, whereby alleles that possess better functional properties (stronger licensing capability and more durable surface expression upon ligand binding) are favored (22, 39). Typing FcγR polymorphism is relevant in NK cell therapy settings that use monoclonal antibodies to enhance NK cell activation and consequently empower their antibody-dependent cytotoxicity (ADCC) properties (40, 41). All these strategies have helped define "preferable donor" profiles.

TABLE 1 | Factors affecting the ouctome of the manufacturing process of NK cell-based medicinal products.

Cell source
Bone marrow
Umbilical cord blood
Embryonic stem cells
Induced pluripotent stem cells
NK cell lines
Culture conditions
Cytokines (IL-2, IL-15, IL-12, IL-18)
Feeder cells (autologous PBMC, EBV-TCT-LCL, K562-mb15-41BBL)
Antibodies (anti-CD3, anti-CD52)
Genetic manipulation (retro- or lentiviral-based transduction, mRNA transfection)
Culture containers
Standard culture flasks
Culture bags
Gas-permeable static cell culture flasks
Bioreactors
Final product evaluation
Viability (live/dead)
Identity and contamination (CD56, CD16, CD3, CD14, CD45, CD19)
Yield
Phenotype (KIR, NKp44, NKp46, NKG2A, NKG2C)
Functionality (degranulation, cytokine release, target cell lysis, activation)

Source of Cells

Natural killer cell therapies can be manufactured from a variety of sources: these include peripheral blood, either steady-state or taking advantage of apheresis performed to collect hematopoietic stem and progenitor cells mobilized with growth factors such as granulocyte colony-stimulating factor (G-CSF), bone marrow, and cord blood.

Peripheral Blood Mononuclear Cells

Peripheral blood mononuclear cells (PBMC) can be collected in large numbers using apheresis. Nowadays, it is the preferred source for allo-HSCT; donor apheresis is collected after receiving a mobilization treatment that increases the percentage and number of circulating progenitor and stem cells (as evaluated by the number of circulating $CD34^+$ cells); G-CSF is the only marketed agent for $CD34^+$ cell mobilization that can be used in donors; the use of other mobilizing agents such as acutely myelo-suppressive drugs or plerixafor is restricted to patients undergoing autologous collection. Since $CD34^+$ cells represent only a small proportion of collected PBMC, these collected cell products may also represent a source of immune effectors, and thus either an alternative to PBMC collected in homeostatic conditions for standard DLI or a starting material for further immune cell manufacturing, including NK cell manufacturing. One caveat to this approach is that studies looking at the effects of G-CSF on NK cell function have produced controversial results; some studies suggest minimal consequences (42, 43), while others suggest significant changes (44).

Non-mobilized apheresis products contain 5–15% NK cells. To isolate NK cells, the strategy commonly used is $CD3^+$ cell depletion of PBMC, followed by $CD56^+$ cell enrichment using immune-magnetic bead separation with medical devices and clinical-grade reagents.

Bone Marrow

Since bone marrow is nowadays used as the source of stem cells for a minority of recipients – in part due to the increased resource needed for the logistics of BM collection compared to aphereses – there is little preclinical or clinical experience in the manufacturing of NK cells from this starting material.

Cord Blood Cells

The use of cord blood (CB) as a source of stem cells has raised great hope in the field 30 years ago, when the first clinical transplants were reported (45). Cord blood unit (CBU) can be used even when not fully matched to the recipient, offering the opportunity to identify a "donor" even for patients who had no HLA-matched related or unrelated donor. CB transplantation is nowadays facing tough competition from the rapidly emerging field of related haploidentical transplantation, and is further hampered by the lengthy immune reconstitution and the lack of possibility to use pre-emptive or curative DLI posttransplant.

However, more than 600,000 CBU that are stored in public banks worldwide – not to mention the unknown cumulative number of CBU preserved in private CB banks – represent a unique source of human material to start with the manufacturing process. Indeed, preclinical validation studies reported the production of significant numbers of functional NK cells either from a complete CBU (46) or even from a minute sample of a CBU (47), the latter opening the way to posttransplant immune cellular therapies in recipients of CBU transplantation while the former offers the promise of "off-the-shelf" allogeneic NK cellular therapies.

Initial attempts to use CD56$^+$ bead-selected NK cells from CB followed by culture on mesenchymal stromal cells in the presence of cytokines resulted in modest yields incompatible with therapeutic needs (48). Starting from immune-selected CB CD34$^+$ cells and using refined protocols produced more interesting results (49–51). Such efforts culminated in a novel good manufacturing practice (GMP)-compliant technique that mimics the extracellular bone marrow environment: stromal cell-free/serum-free medium, heparin, and cytokine supplements (52, 53). Consequently, Glycostem Therapeutics (Oss, the Netherlands) and Radboud University Medical Center (Nijmegen, the Netherlands) are currently conducting a Phase I/II clinical trial in elderly AML patients using NK cell products generated using this method (CCMO nr. NL31699 and Dutch Trial Register nr. 2818) (see **Figure 1** and Data Sheet S1 in Supplementary Material for a list of ongoing clinical trials).

ES and iPS Cells

Manufacturing of clinical-grade NK cells from either embryonic (ES) or induced pluripotent stem (iPS) cells nowadays appear as a futuristic option, although preclinical demonstrations that NK cells can be differentiated from these sources of pluripotent stem cells were already published (54–56). For iPS cells, several factors affect the pluripotency and differentiation abilities of reprogramed cells: the choice of target donor somatic cell type and the reprograming protocol, including the nature and combination of genes as well as the method used to deliver transcription factors into somatic cells (57).

An important step in the specific hematopoietic lineage-differentiating protocols starting from ES or iPS cells is the generation of CD34$^+$ hematopoietic precursors, particularly CD34$^+$CD45$^+$ cells, preferred for their high content in hematopoietic progenitors (58). A 30-day culture protocol of sorted ES cells-derived CD34$^+$ cells together with feeder cells (murine fetal-liver-derived stromal cell line) and cytokines generates NK cells with typical maturation markers and target cell lysis capabilities (58, 59). Another expansion method has been described for ES or iPS cells, using an embryonic body assay followed by culture with feeder cells and cytokines (60, 61).

Similar to what has been mentioned for CBU, progress in the development of safe, efficient, and standardized clinical-grade manufacturing protocols will offer an opportunity to develop off-the-shelf personalized and non-immunogenic cellular therapies.

NK Cell Lines

Ex vivo-expanded primary NK cells persist *in vivo* for short periods of time after adoptive transfer. In an attempt to take advantage of the long lifetime of established cell lines, several groups have evaluated their therapeutic potential. Although other cell lines exist (NKG, YT, NK-YS, YTS cells, HANK-1, and NKL cells), the NK-92 cell line (NantKWest Inc., Culver City, CA, USA) characterized by good cytotoxicity and expansion kinetics (62, 63) has been predominantly evaluated in preclinical investigations and clinical trials (NCT00900809 and NCT00990717) (64). It has been tested in a small number of clinical contexts, yet with minimal efficacy (65–67). Recently, chimeric antigen receptor (CAR) modification by gene transfer for NK cells has opened a new avenue to explore (68, 69). NK cell lines represent a more homogeneous population for CAR modification, compared to peripheral blood NK cells; however, this advantage is largely offset by the need to additionally transfect CD16 to gain ADCC function and the necessary irradiation before infusion for safety reasons, rendering them unable to expand *in vivo*. Choice of the CAR construct adds another layer of complexity (69).

Culture Conditions: Medium, Cytokines, and Cell Culture Systems

As already described, NK cells are generally isolated through immune-selection techniques, using the canonical CD3$^-$/CD56$^+$ phenotype (42, 70), then cultured for functional activation and possibly expansion. Furthermore, NK cells can be genetically engineered to express natural or chimeric molecules empowering them with improved immune functions (5, 64).

Expansion and activation of potent cytotoxic NK cells require several signals for survival, proliferation, and activation. Culture conditions, thus, incorporate media and serum supplements, together with clinical-grade cytokines, monoclonal antibodies, or other soluble molecules, and possibly native or engineered cell feeders. Culture conditions can further be improved through the substitution of bioreactors to static conditions (**Table 1**).

As already mentioned, and since most protocols that use only cytokines result in limited NK cell expansion, the introduction of feeder cells in the culture protocol has been extensively tested. Feeder cells provide additional stimulatory signals necessary for NK cell proliferation. Monocytes provide humoral signals and cell-to-cell contacts hence can serve as feeder cells (71); irradiated autologous PBMC have been used as feeder cells to produce sufficient numbers of NK cells with acceptable purity (6, 72, 73). Alternatively, irradiated allogeneic cells have been evaluated: Epstein–Barr virus-transformed lymphoblastoid B cell lines (EBV-TM-LCL), K562 cells (leukemic cell line) engineered to express a membrane-bound form of IL-15 fused to the T-cell receptor CD8α and the 41BB ligand (74–78), or K562 cells transduced with IL-21 (79). Such feeder cells proved effective in preclinical validation of the production of clinically relevant numbers of NK cells, however, raise regulatory issues when it comes to manufacture medicinal products.

Since the presence of residual feeder cells in the final product is of major concern for clinical applications (80), alternative approaches are evaluated as substitutes. Anti-CD3 (OKT3) antibodies in addition to IL-2, with or without IL-15, produced substantial although lower fold expansion of $CD3^-/CD56^+$-enriched cells (81–86) than protocols that use feeder cells.

Expansion strategies of clinical-grade NK cells usually require 7–21 days of culture; up to 28 days of culture have been reported (87). There is an incentive to substitute animal or human serum-replete medium with animal and human component-free medium. The most commonly used media for CB-derived NK cells is glycostem basal growth medium (Clear Cell Technologies, Beernem, Belgium), preferred for being free of animal-derived components (52, 53). For PBMC-derived NK cells, preferred media are: X-Vivo™ serum-free media (Biowhittaker, Verviers, Belgium), AIM V® serum-free medium (Thermo Fisher Scientific, Grand Island, NY, USA), Stem Cell Growth medium (Cell Genix, Freiburg, Germany), or complete Roswell Park Memorial Institute 1640 (Biowhittaker, Verviers, Belgium).

Media supplements still being used by some groups include GMP-grade human AB serum, pooled human AB plasma, or fetal bovine serum (FBS). GMP-grade cytokines (recombinant human IL-2 and IL-15), antibodies (anti-CD3-OKT3), and other ancillary reagents (nicotinamide-NAM) commonly serve as medium supplements for PBMC-derived NK cell generation. Additional growth factors and cytokines are necessary for CB-derived NK cells since the starting material is commonly $CD34^+$ stem cells [stem cell factor (SCF), IL-7, IL-15, IL-2, IL-6, flt3 ligand (Flt3-L), thrombopoietin (TPO), G-CSF, low molecular weight heparin (LMWH), granulocyte-macrophage colony-stimulating factor (GM-CSF), leukemia inhibitory factor (LIF), MIP1α].

In addition to using T75 cell culture flasks, several groups have used culture bags (Baxter LifeCell® or VueLife®) (76, 83). On larger scales, gas-permeable static cell culture (G-Rex®) flasks (Wilson Wolf Manufacturing, New Brighton, MN, USA) (78) or WAVE Bioreactor™ (GE Healthcare Life Sciences, Chicago, IL, USA) (83) served as expansion platforms.

AMPLITUDE OF NK CELLS EXPANSION AND DEFINITION OF AN OPTIMAL CELL DOSE

Numbers of infused NK cells in clinical trials typically range from 5 to 50 × 10^6 NK cells/kg, but infusion of as many as 10^8 NK cells/kg has been reported (88). Based on percentages of NK cells in the starting materials, manufacturing the higher doses implies significant expansion during *in vitro* cultures. This raises a practical issue, since, in the absence of feeder cells, NK cells expansion is modest if any. Using autologous irradiated PBMC as feeder cells, up to 2,500-fold expansion of functionally active NK cells at day 17 has been reported (89). The use of genetically modified cell lines as feeder leads to a 30,000-fold expansion of NK cells after 21 days of culture (79).

A recent study took advantage of the introduction of anti-CD3 and anti-CD52 monoclonal antibodies over a period of 14 days and reports a median 1500-fold increase in NK cell numbers; however, it must be emphasized that T cells represent up to 40% of the final cell product and that NK cells were not obtained through a cGMP protocol (90).

QUALITY CONTROLS AND RELEASE CRITERIA FOR ENGINEERED NK CELL CELLS

Tools for assessing the efficacy of NK cell generation protocols are necessary for comparing technical results from different NK cell therapy studies. Furthermore, European Medicine Agency (EMA), Food and Drug Administration (FDA), and several guidelines require the characterization of the final product to define release criteria in order to ensure safety and efficacy.

Basic, yet essential, criteria are generally used to characterize the final product: these include purity and viability of the target cell population, contamination with undesirable cells such as residual T and B cells, and sterility. These are commonly used as release criteria although their relevance may vary for different clinical conditions: T cell contamination for instance is most important in an allogeneic, but not so much in an autologous setting. More sophisticated testing may provide additional information: a reduction in telomere length indicates cell senescence due to extensive long-term culturing.

Phenotype and function (tumor cytotoxicity) are additional characteristics that should help identify the most effective NK cell products. When expanded NK cells were compared with freshly isolated and IL-2-activated NK cells, a higher expression of NKG2D, TNF-related apoptosis-inducing ligand (TRAIL), and natural cytotoxicity receptors NKp30, NKp44, and NKp46 was reported (91), in addition to higher cytotoxicity to K562 cells. Efforts are, however, much needed to harmonize technical protocols and identify a panel of phenotypic and functional biomarkers that would allow comparisons between protocols that evaluate adoptive transfer of NK cells (92). It is essential to mention that such a panel needs to be run within a reasonably short time to release the product in time for a "fresh infusion."

Cryopreservation and conservation of cytolytic activity of thawed NK cells would render multiple rounds of adoptive NK cell infusions feasible. Lapteva et al. and Berg et al. reported that an overnight activation with IL-2 would rescue the reduced cytolytic activity of thawed NK cells, yet at the cost of a diminished recovery (74, 78). Efforts to optimize cryopreservation and thawing methods are in progress.

REGULATORY STATUS OF ENGINEERED NK CELLS AND COMMERCIAL PERSPECTIVES

Autologous and allogeneic NK cells engineered from primary human cells are individually produced for a unique and designated individual, rather than manufactured as batches. Since manufacturing incorporates *ex-vivo* culture and activation of immune-selected cells from the primary material, these will be considered as substantially manipulated or more-than minimally manipulated cell products, and thus will qualify as "advanced therapy medicinal products" (ATMPs) and somatic cell therapy products as defined in EC regulation 1394/2007. Before the regulation was released, such cell therapies were engineered as part of clinical research protocols by cell processing facilities, usually supported and operated by academia, no differently from minimally manipulated cell transplants. Since the regulation has been published, the view is that somatic cell therapy products and gene therapy products will be manufactured in compliance with *good manufacturing practices* (85, 86, 93), and eventually marketed by industry, when a marketing authorization is granted by competent authorities at a European level, i.e., by the EMA or by the FDA in the USA.

Beyond academic investigations, NK cells have now aroused the interest of a significant number of pharma companies (94) (see **Figure 1**), although no NK-based cellular therapy has so far been authorized as an ATMP in Europe, nor a somewhat comparable status in the USA. However, in December 2014, orphan designation (EU/3/14/1395) was granted by the European Commission to Glycostem, for the GCT-NK cell product, made of allogeneic *ex vivo*-generated natural killer (NK) cells from $CD34^+$ CB progenitor cells for the treatment of acute myeloid leukemia (EMA/COMP/730059/2014 Committee for Orphan Medicinal Products).

CONCLUSION

Over the last 30 years, enormous progress has been made in our understanding of the biology of NK cells. New agents targeting their activity *in vivo* have been evaluated, and technological improvements in NK cell manufacturing have been introduced in the clinic. However, the demonstration that modulation of NK cell activity by any of these means can achieve therapeutic activity over a wide range of diseases is still awaited. The ability to follow and image *in vivo* adoptively transferred autologous or allogeneic NK cells would represent a major advantage to understand the "pharmacokinetics" and mechanisms of action of these immune effectors, as illustrated in preclinical (95) as well as in clinical (96) studies. It would help to understand the consequences of culture conditions on *in vivo* persistence and activity. Ongoing developments for innovative cellular therapies in the academic sector as well as in the commercial sector suggest that such progress may result in broader clinical applications in the near future.

AUTHOR CONTRIBUTIONS

All authors are members of a Marseille-based consortium that aims at translation of scientific discoveries to clinically applicable innovations, with a focus on innovative cellular therapies for patients affected with hematological diseases. All authors have contributed to the writing of this manuscript and had the opportunity to review the draft and final versions.

REFERENCES

1. Blank CU, Haanen JB, Ribas A, Schumacher TN. CANCER IMMUNOLOGY. The "cancer immunogram". *Science* (2016) 352(6286):658–60. doi:10.1126/science.aaf2834
2. Spits H, Artis D, Colonna M, Diefenbach A, Di Santo JP, Eberl G, et al. Innate lymphoid cells – a proposal for uniform nomenclature. *Nat Rev Immunol* (2013) 13(2):145–9. doi:10.1038/nri3365
3. Guven H, Gilljam M, Chambers BJ, Ljunggren HG, Christensson B, Kimby E, et al. Expansion of natural killer (NK) and natural killer-like T (NKT)-cell populations derived from patients with B-chronic lymphocytic leukemia (B-CLL): a potential source for cellular immunotherapy. *Leukemia* (2003) 17(10):1973–80. doi:10.1038/sj.leu.2403083
4. Garg TK, Szmania SM, Khan JA, Hoering A, Malbrough PA, Moreno-Bost A, et al. Highly activated and expanded natural killer cells for multiple myeloma immunotherapy. *Haematologica* (2012) 97(9):1348–56. doi:10.3324/haematol.2011.056747
5. Schonfeld K, Sahm C, Zhang C, Naundorf S, Brendel C, Odendahl M, et al. Selective inhibition of tumor growth by clonal NK cells expressing an ErbB2/HER2-specific chimeric antigen receptor. *Mol Ther* (2015) 23(2):330–8. doi:10.1038/mt.2014.219
6. Parkhurst MR, Riley JP, Dudley ME, Rosenberg SA. Adoptive transfer of autologous natural killer cells leads to high levels of circulating natural killer cells but does not mediate tumor regression. *Clin Cancer Res* (2011) 17(19):6287–97. doi:10.1158/1078-0432.CCR-11-1347
7. Rosenberg SA, Lotze MT, Muul LM, Leitman S, Chang AE, Ettinghausen SE, et al. Observations on the systemic administration of autologous lymphokine-activated killer cells and recombinant interleukin-2 to patients with metastatic cancer. *N Engl J Med* (1985) 313(23):1485–92. doi:10.1056/NEJM198512053132327
8. Ghiringhelli F, Menard C, Terme M, Flament C, Taieb J, Chaput N, et al. CD4+CD25+ regulatory T cells inhibit natural killer cell functions in a transforming growth factor-beta-dependent manner. *J Exp Med* (2005) 202(8):1075–85. doi:10.1084/jem.20051511
9. Rodella L, Zamai L, Rezzani R, Artico M, Peri G, Falconi M, et al. Interleukin 2 and interleukin 15 differentially predispose natural killer cells to apoptosis mediated by endothelial and tumour cells. *Br J Haematol* (2001) 115(2):442–50. doi:10.1046/j.1365-2141.2001.03055.x
10. Hayakawa M, Hatano T, Ogawa Y, Gakiya M, Ogura H, Osawa A. Treatment of advanced renal cell carcinoma using regional arterial administration of lymphokine-activated killer cells in combination with low doses of rIL-2. *Urol Int* (1994) 53(3):117–24. doi:10.1159/000282651
11. Abdel-Azim H, Heisterkamp N. Potential of autologous NK cell therapy to eradicate leukemia: "Education is [not] the best provision for old age" – Aristotle. *Oncoimmunology* (2015) 4(2):e984549. doi:10.4161/2162402X.2014.984549
12. Kohrt HE, Thielens A, Marabelle A, Sagiv-Barfi I, Sola C, Chanuc F, et al. Anti-KIR antibody enhancement of anti-lymphoma activity of natural killer cells as monotherapy and in combination with anti-CD20 antibodies. *Blood* (2014) 123(5):678–86. doi:10.1182/blood-2013-08-519199
13. Terszowski G, Klein C, Schmied L, Stern M. How to outsmart NK cell tolerance. *Oncoimmunology* (2015) 4(9):e1016708. doi:10.1080/2162402X.2015.1016708
14. Vey N, Bourhis JH, Boissel N, Bordessoule D, Prebet T, Charbonnier A, et al. A phase 1 trial of the anti-inhibitory KIR mAb IPH2101 for AML in complete remission. *Blood* (2012) 120(22):4317–23. doi:10.1182/blood-2012-06-437558
15. Kheav VD, Busson M, Scieux C, Peffault de Latour R, Maki G, Haas P, et al. Favorable impact of natural killer cell reconstitution on chronic graft-versus-host disease and cytomegalovirus reactivation after allogeneic hematopoietic stem cell transplantation. *Haematologica* (2014) 99(12):1860–7. doi:10.3324/haematol.2014.108407
16. Dunbar EM, Buzzeo MP, Levine JB, Schold JD, Meier-Kriesche HU, Reddy V. The relationship between circulating natural killer cells after reduced intensity conditioning hematopoietic stem cell transplantation and relapse-free survival and graft-versus-host disease. *Haematologica* (2008) 93(12):1852–8. doi:10.3324/haematol.13033

17. Pical-Izard C, Crocchiolo R, Granjeaud S, Kochbati E, Just-Landi S, Chabannon C, et al. Reconstitution of natural killer cells in HLA-matched HSCT after reduced-intensity conditioning: impact on clinical outcome. *Biol Blood Marrow Transplant* (2015) 21(3):429–39. doi:10.1016/j.bbmt.2014.11.681
18. Foley B, Cooley S, Verneris MR, Curtsinger J, Luo X, Waller EK, et al. NK cell education after allogeneic transplantation: dissociation between recovery of cytokine-producing and cytotoxic functions. *Blood* (2011) 118(10):2784–92. doi:10.1182/blood-2011-04-347070
19. Ruggeri L, Capanni M, Urbani E, Perruccio K, Shlomchik WD, Tosti A, et al. Effectiveness of donor natural killer cell alloreactivity in mismatched hematopoietic transplants. *Science* (2002) 295(5562):2097–100. doi:10.1126/science.1068440
20. Zhao XY, Chang YJ, Xu LP, Zhang XH, Liu KY, Li D, et al. HLA and KIR genotyping correlates with relapse after T-cell-replete haploidentical transplantation in chronic myeloid leukaemia patients. *Br J Cancer* (2014) 111(6):1080–8. doi:10.1038/bjc.2014.423
21. Venstrom JM, Pittari G, Gooley TA, Chewning JH, Spellman S, Haagenson M, et al. HLA-C-dependent prevention of leukemia relapse by donor activating KIR2DS1. *N Engl J Med* (2012) 367(9):805–16. doi:10.1056/NEJMoa1200503
22. Bari R, Rujkijyanont P, Sullivan E, Kang G, Turner V, Gan K, et al. Effect of donor KIR2DL1 allelic polymorphism on the outcome of pediatric allogeneic hematopoietic stem-cell transplantation. *J Clin Oncol* (2013) 31(30):3782–90. doi:10.1200/JCO.2012.47.4007
23. Gagne K, Busson M, Bignon JD, Balere-Appert ML, Loiseau P, Dormoy A, et al. Donor KIR3DL1/3DS1 gene and recipient Bw4 KIR ligand as prognostic markers for outcome in unrelated hematopoietic stem cell transplantation. *Biol Blood Marrow Transplant* (2009) 15(11):1366–75. doi:10.1016/j.bbmt.2009.06.015
24. Vivier E, Ugolini S, Blaise D, Chabannon C, Brossay L. Targeting natural killer cells and natural killer T cells in cancer. *Nat Rev Immunol* (2012) 12(4):239–52. doi:10.1038/nri3174
25. Bachanova V, Cooley S, Defor TE, Verneris MR, Zhang B, McKenna DH, et al. Clearance of acute myeloid leukemia by haploidentical natural killer cells is improved using IL-2 diphtheria toxin fusion protein. *Blood* (2014) 123(25):3855–63. doi:10.1182/blood-2013-10-532531
26. Choi I, Yoon SR, Park SY, Kim H, Jung SJ, Jang YJ, et al. Donor-derived natural killer cells infused after human leukocyte antigen-haploidentical hematopoietic cell transplantation: a dose-escalation study. *Biol Blood Marrow Transplant* (2014) 20(5):696–704. doi:10.1016/j.bbmt.2014.01.031
27. Stern M, Passweg JR, Meyer-Monard S, Esser R, Tonn T, Soerensen J, et al. Pre-emptive immunotherapy with purified natural killer cells after haploidentical SCT: a prospective phase II study in two centers. *Bone Marrow Transplant* (2013) 48(3):433–8. doi:10.1038/bmt.2012.162
28. Curti A, Ruggeri L, D'Addio A, Bontadini A, Dan E, Motta MR, et al. Successful transfer of alloreactive haploidentical KIR ligand-mismatched natural killer cells after infusion in elderly high risk acute myeloid leukemia patients. *Blood* (2011) 118(12):3273–9. doi:10.1182/blood-2011-01-329508
29. Curti A, Ruggeri L, Parisi S, Bontadini A, Dan E, Motta MR, et al. Larger size of donor alloreactive NK cell repertoire correlates with better response to NK cell immunotherapy in elderly acute myeloid leukemia patients. *Clin Cancer Res* (2016) 22(8):1914–21. doi:10.1158/1078-0432.CCR-15-1604
30. Shah NN, Baird K, Delbrook CP, Fleisher TA, Kohler ME, Rampertaap S, et al. Acute GVHD in patients receiving IL-15/4-1BBL activated NK cells following T-cell-depleted stem cell transplantation. *Blood* (2015) 125(5):784–92. doi:10.1182/blood-2014-07-592881
31. Miller JS, Rooney CM, Curtsinger J, McElmurry R, McCullar V, Verneris MR, et al. Expansion and homing of adoptively transferred human natural killer cells in immunodeficient mice varies with product preparation and in vivo cytokine administration: implications for clinical therapy. *Biol Blood Marrow Transplant* (2014) 20(8):1252–7. doi:10.1016/j.bbmt.2014.05.004
32. Leung W. Infusions of allogeneic natural killer cells as cancer therapy. *Clin Cancer Res* (2014) 20(13):3390–400. doi:10.1158/1078-0432.CCR-13-1766
33. Lim O, Jung MY, Hwang YK, Shin EC. Present and future of allogeneic natural killer cell therapy. *Front Immunol* (2015) 6:286. doi:10.3389/fimmu.2015.00286
34. Leung W. Use of NK cell activity in cure by transplant. *Br J Haematol* (2011) 155(1):14–29. doi:10.1111/j.1365-2141.2011.08823.x
35. Cooley S, Weisdorf DJ, Guethlein LA, Klein JP, Wang T, Le CT, et al. Donor selection for natural killer cell receptor genes leads to superior survival after unrelated transplantation for acute myelogenous leukemia. *Blood* (2010) 116(14):2411–9. doi:10.1182/blood-2010-05-283051
36. Cooley S, Weisdorf DJ, Guethlein LA, Klein JP, Wang T, Marsh SG, et al. Donor killer cell Ig-like receptor B haplotypes, recipient HLA-C1, and HLA-C mismatch enhance the clinical benefit of unrelated transplantation for acute myelogenous leukemia. *J Immunol* (2014) 192(10):4592–600. doi:10.4049/jimmunol.1302517
37. Leung W, Iyengar R, Turner V, Lang P, Bader P, Conn P, et al. Determinants of antileukemia effects of allogeneic NK cells. *J Immunol* (2004) 172(1):644–50. doi:10.4049/jimmunol.172.1.644
38. Leung W, Iyengar R, Triplett B, Turner V, Behm FG, Holladay MS, et al. Comparison of killer Ig-like receptor genotyping and phenotyping for selection of allogeneic blood stem cell donors. *J Immunol* (2005) 174(10):6540–5. doi:10.4049/jimmunol.174.10.6540
39. Bari R, Bell T, Leung WH, Vong QP, Chan WK, Das Gupta N, et al. Significant functional heterogeneity among KIR2DL1 alleles and a pivotal role of arginine 245. *Blood* (2009) 114(25):5182–90. doi:10.1182/blood-2009-07-231977
40. Cartron G, Dacheux L, Salles G, Solal-Celigny P, Bardos P, Colombat P, et al. Therapeutic activity of humanized anti-CD20 monoclonal antibody and polymorphism in IgG Fc receptor FcgammaRIIIa gene. *Blood* (2002) 99(3):754–8. doi:10.1182/blood.V99.3.754
41. Koene HR, Kleijer M, Algra J, Roos D, von dem Borne AE, de Haas M. Fc gammaRIIIa-158V/F polymorphism influences the binding of IgG by natural killer cell Fc gammaRIIIa, independently of the Fc gammaRIIIa-48L/R/H phenotype. *Blood* (1997) 90(3):1109–14.
42. Leung W, Iyengar R, Leimig T, Holladay MS, Houston J, Handgretinger R. Phenotype and function of human natural killer cells purified by using a clinical-scale immunomagnetic method. *Cancer Immunol Immunother* (2005) 54(4):389–94. doi:10.1007/s00262-004-0609-6
43. Pascal V, Brunet C, Pradel V, Thirion X, Andre P, Faucher C, et al. Analysis of donor NK and T cells infused in patients undergoing MHC-matched allogeneic hematopoietic transplantation. *Leukemia* (2002) 16(11):2259–66. doi:10.1038/sj.leu.2402670
44. Su YC, Li SC, Hsu CK, Yu CC, Lin TJ, Lee CY, et al. G-CSF downregulates natural killer cell-mediated cytotoxicity in donors for hematopoietic SCT. *Bone Marrow Transplant* (2012) 47(1):73–81. doi:10.1038/bmt.2011.22
45. Ballen KK, Gluckman E, Broxmeyer HE. Umbilical cord blood transplantation: the first 25 years and beyond. *Blood* (2013) 122(4):491–8. doi:10.1182/blood-2013-02-453175
46. Shah N, Martin-Antonio B, Yang H, Ku S, Lee DA, Cooper LJ, et al. Antigen presenting cell-mediated expansion of human umbilical cord blood yields log-scale expansion of natural killer cells with anti-myeloma activity. *PLoS One* (2013) 8(10):e76781. doi:10.1371/journal.pone.0076781
47. Vasu S, Berg M, Davidson-Moncada J, Tian X, Cullis H, Childs RW. A novel method to expand large numbers of CD56(+) natural killer cells from a minute fraction of selectively accessed cryopreserved cord blood for immunotherapy after transplantation. *Cytotherapy* (2015) 17(11):1582–93. doi:10.1016/j.jcyt.2015.07.020
48. Boissel L, Tuncer HH, Betancur M, Wolfberg A, Klingemann H. Umbilical cord mesenchymal stem cells increase expansion of cord blood natural killer cells. *Biol Blood Marrow Transplant* (2008) 14(9):1031–8. doi:10.1016/j.bbmt.2008.06.016
49. Beck RC, Padival M, Yeh D, Ralston J, Cooke KR, Lowe JB. The Notch ligands Jagged2, Delta1, and Delta4 induce differentiation and expansion of functional human NK cells from CD34+ cord blood hematopoietic progenitor cells. *Biol Blood Marrow Transplant* (2009) 15(9):1026–37. doi:10.1016/j.bbmt.2009.06.002
50. Vitale C, Cottalasso F, Montaldo E, Moretta L, Mingari MC. Methylprednisolone induces preferential and rapid differentiation of CD34+ cord blood precursors toward NK cells. *Int Immunol* (2008) 20(4):565–75. doi:10.1093/intimm/dxn014
51. Kao IT, Yao CL, Kong ZL, Wu ML, Chuang TL, Hwang SM. Generation of natural killer cells from serum-free, expanded human umbilical cord blood CD34+ cells. *Stem Cells Dev* (2007) 16(6):1043–51. doi:10.1089/scd.2007.0033
52. Spanholtz J, Preijers F, Tordoir M, Trilsbeek C, Paardekooper J, de Witte T, et al. Clinical-grade generation of active NK cells from cord blood hematopoietic

progenitor cells for immunotherapy using a closed-system culture process. *PLoS One* (2011) 6(6):e20740. doi:10.1371/journal.pone.0020740
53. Spanholtz J, Tordoir M, Eissens D, Preijers F, van der Meer A, Joosten I, et al. High log-scale expansion of functional human natural killer cells from umbilical cord blood CD34-positive cells for adoptive cancer immunotherapy. *PLoS One* (2010) 5(2):e9221. doi:10.1371/journal.pone.0009221
54. Lan T, Wang L, Xu L, Jin N, Yan G, Xia J, et al. Induced pluripotent stem cells can effectively differentiate into multiple functional lymphocyte lineages in vivo with negligible bias. *Stem Cells Dev* (2016) 25(6):462–71. doi:10.1089/scd.2015.0248
55. Bock AM, Knorr D, Kaufman DS. Development, expansion, and in vivo monitoring of human NK cells from human embryonic stem cells (hESCs) and and induced pluripotent stem cells (iPSCs). *J Vis Exp* (2013) 74:e50337. doi:10.3791/50337
56. Woll PS, Grzywacz B, Tian X, Marcus RK, Knorr DA, Verneris MR, et al. Human embryonic stem cells differentiate into a homogeneous population of natural killer cells with potent in vivo antitumor activity. *Blood* (2009) 113(24):6094–101. doi:10.1182/blood-2008-06-165225
57. Eguizabal C, Zenarruzabeitia O, Monge J, Santos S, Vesga MA, Maruri N, et al. Natural killer cells for cancer immunotherapy: pluripotent stem cells-derived NK cells as an immunotherapeutic perspective. *Front Immunol* (2014) 5:439. doi:10.3389/fimmu.2014.00439
58. Woll PS, Martin CH, Miller JS, Kaufman DS. Human embryonic stem cell-derived NK cells acquire functional receptors and cytolytic activity. *J Immunol* (2005) 175(8):5095–103. doi:10.4049/jimmunol.175.8.5095
59. Knorr DA, Bock A, Brentjens RJ, Kaufman DS. Engineered human embryonic stem cell-derived lymphocytes to study in vivo trafficking and immunotherapy. *Stem Cells Dev* (2013) 22(13):1861–9. doi:10.1089/scd.2012.0608
60. Tabatabaei-Zavareh N, Vlasova A, Greenwood CP, Takei F. Characterization of developmental pathway of natural killer cells from embryonic stem cells in vitro. *PLoS One* (2007) 2(2):e232. doi:10.1371/journal.pone.0000232
61. Knorr DA, Ni Z, Hermanson D, Hexum MK, Bendzick L, Cooper LJ, et al. Clinical-scale derivation of natural killer cells from human pluripotent stem cells for cancer therapy. *Stem Cells Transl Med* (2013) 2(4):274–83. doi:10.5966/sctm.2012-0084
62. Gong JH, Maki G, Klingemann HG. Characterization of a human cell line (NK-92) with phenotypical and functional characteristics of activated natural killer cells. *Leukemia* (1994) 8(4):652–8.
63. Tam YK, Martinson JA, Doligosa K, Klingemann HG. Ex vivo expansion of the highly cytotoxic human natural killer-92 cell-line under current good manufacturing practice conditions for clinical adoptive cellular immunotherapy. *Cytotherapy* (2003) 5(3):259–72. doi:10.1080/14653240310001523
64. Glienke W, Esser R, Priesner C, Suerth JD, Schambach A, Wels WS, et al. Advantages and applications of CAR-expressing natural killer cells. *Front Pharmacol* (2015) 6:21. doi:10.3389/fphar.2015.00021
65. Tonn T, Becker S, Esser R, Schwabe D, Seifried E. Cellular immunotherapy of malignancies using the clonal natural killer cell line NK-92. *J Hematother Stem Cell Res* (2001) 10(4):535–44. doi:10.1089/15258160152509145
66. Tonn T, Schwabe D, Klingemann HG, Becker S, Esser R, Koehl U, et al. Treatment of patients with advanced cancer with the natural killer cell line NK-92. *Cytotherapy* (2013) 15(12):1563–70. doi:10.1016/j.jcyt.2013.06.017
67. Arai S, Meagher R, Swearingen M, Myint H, Rich E, Martinson J, et al. Infusion of the allogeneic cell line NK-92 in patients with advanced renal cell cancer or melanoma: a phase I trial. *Cytotherapy* (2008) 10(6):625–32. doi:10.1080/14653240802301872
68. Boissel L, Betancur-Boissel M, Lu W, Krause DS, Van Etten RA, Wels WS, et al. Retargeting NK-92 cells by means of CD19- and CD20-specific chimeric antigen receptors compares favorably with antibody-dependent cellular cytotoxicity. *Oncoimmunology* (2013) 2(10):e26527. doi:10.4161/onci.26527
69. Hermanson DL, Kaufman DS. Utilizing chimeric antigen receptors to direct natural killer cell activity. *Front Immunol* (2015) 6:195. doi:10.3389/fimmu.2015.00195
70. Carlens S, Gilljam M, Chambers BJ, Aschan J, Guven H, Ljunggren HG, et al. A new method for in vitro expansion of cytotoxic human CD3-CD56+ natural killer cells. *Hum Immunol* (2001) 62(10):1092–8. doi:10.1016/S0198-8859(01)00313-5
71. Miller JS, Oelkers S, Verfaillie C, McGlave P. Role of monocytes in the expansion of human activated natural killer cells. *Blood* (1992) 80(9):2221–9.
72. Luhm J, Brand JM, Koritke P, Hoppner M, Kirchner H, Frohn C. Large-scale generation of natural killer lymphocytes for clinical application. *J Hematother Stem Cell Res* (2002) 11(4):651–7. doi:10.1089/15258160260194794
73. Siegler U, Meyer-Monard S, Jorger S, Stern M, Tichelli A, Gratwohl A, et al. Good manufacturing practice-compliant cell sorting and large-scale expansion of single KIR-positive alloreactive human natural killer cells for multiple infusions to leukemia patients. *Cytotherapy* (2010) 12(6):750–63. doi:10.3109/14653241003786155
74. Berg M, Lundqvist A, McCoy P Jr, Samsel L, Fan Y, Tawab A, et al. Clinical-grade ex vivo-expanded human natural killer cells up-regulate activating receptors and death receptor ligands and have enhanced cytolytic activity against tumor cells. *Cytotherapy* (2009) 11(3):341–55. doi:10.1080/14653240902807034
75. Lundqvist A, Berg M, Smith A, Childs RW. Bortezomib treatment to potentiate the anti-tumor immunity of ex-vivo expanded adoptively infused autologous natural killer cells. *J Cancer* (2011) 2:383–5. doi:10.7150/jca.2.383
76. Fujisaki H, Kakuda H, Shimasaki N, Imai C, Ma J, Lockey T, et al. Expansion of highly cytotoxic human natural killer cells for cancer cell therapy. *Cancer Res* (2009) 69(9):4010–7. doi:10.1158/0008-5472.CAN-08-3712
77. Granzin M, Soltenborn S, Muller S, Kollet J, Berg M, Cerwenka A, et al. Fully automated expansion and activation of clinical-grade natural killer cells for adoptive immunotherapy. *Cytotherapy* (2015) 17(5):621–32. doi:10.1016/j.jcyt.2015.03.611
78. Lapteva N, Durett AG, Sun J, Rollins LA, Huye LL, Fang J, et al. Large-scale ex vivo expansion and characterization of natural killer cells for clinical applications. *Cytotherapy* (2012) 14(9):1131–43. doi:10.3109/14653249.2012.700767
79. Denman CJ, Senyukov VV, Somanchi SS, Phatarpekar PV, Kopp LM, Johnson JL, et al. Membrane-bound IL-21 promotes sustained ex vivo proliferation of human natural killer cells. *PLoS One* (2012) 7(1):e30264. doi:10.1371/journal.pone.0030264
80. Geraghty RJ, Capes-Davis A, Davis JM, Downward J, Freshney RI, Knezevic I, et al. Guidelines for the use of cell lines in biomedical research. *Br J Cancer* (2014) 111(6):1021–46. doi:10.1038/bjc.2014.166
81. Alici E, Sutlu T, Bjorkstrand B, Gilljam M, Stellan B, Nahi H, et al. Autologous antitumor activity by NK cells expanded from myeloma patients using GMP-compliant components. *Blood* (2008) 111(6):3155–62. doi:10.1182/blood-2007-09-110312
82. Klingemann HG, Martinson J. Ex vivo expansion of natural killer cells for clinical applications. *Cytotherapy* (2004) 6(1):15–22. doi:10.1080/14653240310004548
83. Sutlu T, Stellan B, Gilljam M, Quezada HC, Nahi H, Gahrton G, et al. Clinical-grade, large-scale, feeder-free expansion of highly active human natural killer cells for adoptive immunotherapy using an automated bioreactor. *Cytotherapy* (2010) 12(8):1044–55. doi:10.3109/14653249.2010.504770
84. Iyengar R, Handgretinger R, Babarin-Dorner A, Leimig T, Otto M, Geiger TL, et al. Purification of human natural killer cells using a clinical-scale immunomagnetic method. *Cytotherapy* (2003) 5(6):479–84. doi:10.1080/14653240310003558
85. Lim O, Lee Y, Chung H, Her JH, Kang SM, Jung MY, et al. GMP-compliant, large-scale expanded allogeneic natural killer cells have potent cytolytic activity against cancer cells in vitro and in vivo. *PLoS One* (2013) 8(1):e53611. doi:10.1371/journal.pone.0053611
86. Koehl U, Brehm C, Huenecke S, Zimmermann SY, Kloess S, Bremm M, et al. Clinical grade purification and expansion of NK cell products for an optimized manufacturing protocol. *Front Oncol* (2013) 3:118. doi:10.3389/fonc.2013.00118
87. Suck G, Oei VY, Linn YC, Ho SH, Chu S, Choong A, et al. Interleukin-15 supports generation of highly potent clinical-grade natural killer cells in long-term cultures for targeting hematological malignancies. *Exp Hematol* (2011) 39(9):904–14. doi:10.1016/j.exphem.2011.06.003
88. Szmania S, Lapteva N, Garg T, Greenway A, Lingo J, Nair B, et al. Ex vivo-expanded natural killer cells demonstrate robust proliferation in vivo in high-risk relapsed multiple myeloma patients. *J Immunother* (2015) 38(1):24–36. doi:10.1097/CJI.0000000000000059
89. Ahn YO, Kim S, Kim TM, Song EY, Park MH, Heo DS. Irradiated and activated autologous PBMCs induce expansion of highly cytotoxic human NK cells in vitro. *J Immunother* (2013) 36(7):373–81. doi:10.1097/CJI.0b013e3182a3430f
90. Masuyama J, Murakami T, Iwamoto S, Fujita S. Ex vivo expansion of natural killer cells from human peripheral blood mononuclear cells co-stimulated

with anti-CD3 and anti-CD52 monoclonal antibodies. *Cytotherapy* (2016) 18(1):80–90. doi:10.1016/j.jcyt.2015.09.011

91. Childs RW, Berg M. Bringing natural killer cells to the clinic: ex vivo manipulation. *Hematology Am Soc Hematol Educ Program* (2013) 2013:234–46. doi:10.1182/asheducation-2013.1.234
92. Koehl U, Kalberer C, Spanholtz J, Lee DA, Miller JS, Cooley S, et al. Advances in clinical NK cell studies: donor selection, manufacturing and quality control. *Oncoimmunology* (2016) 5(4):e1115178. doi:10.1080/2162402X.2015.1115178
93. McKenna DH Jr, Sumstad D, Bostrom N, Kadidlo DM, Fautsch S, McNearney S, et al. Good manufacturing practices production of natural killer cells for immunotherapy: a six-year single-institution experience. *Transfusion* (2007) 47(3):520–8. doi:10.1111/j.1537-2995.2006.01145.x
94. Garber K. Natural killer cells blaze into immuno-oncology. *Nat Biotech* (2016) 34(3):219–20. doi:10.1038/nbt0316-219
95. Mallett CL, McFadden C, Chen Y, Foster PJ. Migration of iron-labeled KHYG-1 natural killer cells to subcutaneous tumors in nude mice, as detected by magnetic resonance imaging. *Cytotherapy* (2012) 14(6):743–51. doi:10.3109/14653249.2012.667874
96. Brand JM, Meller B, Von Hof K, Luhm J, Bahre M, Kirchner H, et al. Kinetics and organ distribution of allogeneic natural killer lymphocytes transfused into patients suffering from renal cell carcinoma. *Stem Cells Dev* (2004) 13(3):307–14. doi:10.1089/154732804323099235

20

NK Cells and Other Innate Lymphoid Cells in Hematopoietic Stem Cell Transplantation

Paola Vacca[1†], Elisa Montaldo[2†], Daniele Croxatto[1], Francesca Moretta[3,4], Alice Bertaina[5], Chiara Vitale[1,6], Franco Locatelli[5,7], Maria Cristina Mingari[1,6,8] and Lorenzo Moretta[9*]*

[1]Department of Experimental Medicine, University of Genova, Genova, Italy, [2]Division of Regenerative Medicine, Stem Cells and Gene Therapy, San Raffaele Telethon Institute for Gene Therapy, IRCCS San Raffaele Scientific Institute, Milano, Italy, [3]Department of Internal Medicine, University of Verona, Verona, Italy, [4]Ospedale Sacro Cuore Negrar, Verona, Italy, [5]Department of Pediatric Hematology and Oncology, IRCCS Bambino Gesù Children's Hospital, Rome, Italy, [6]U.O. Immunology IRCCS AOU San Martino-IST, Genova, Italy, [7]Department of Pediatrics, University of Pavia, Pavia, Italy, [8]Centro di Eccellenza per la Ricerca Biomedica – CEBR, Genova, Italy, [9]Department of Immunology, IRCCS Bambino Gesù Children's Hospital, Rome, Italy

Edited by:
Luigi Daniele Notarangelo, Harvard Medical School, USA

Reviewed by:
Andrew Gennery, Newcastle University, UK
Mette D. Hazenberg, Academic Medical Centre, Netherlands

***Correspondence:**
Paola Vacca
paola.vacca@unige.it;
Lorenzo Moretta
lorenzo.moretta@opbg.net

†Paola Vacca and Elisa Montaldo contributed equally.

Natural killer (NK) cells play a major role in the T-cell depleted haploidentical hematopoietic stem cell transplantation (haplo-HSCT) to cure high-risk leukemias. NK cells belong to the expanding family of innate lymphoid cells (ILCs). At variance with NK cells, the other ILC populations (ILC1/2/3) are non-cytolytic, while they secrete different patterns of cytokines. ILCs provide host defenses against viruses, bacteria, and parasites, drive lymphoid organogenesis, and contribute to tissue remodeling. In haplo-HSCT patients, the extensive T-cell depletion is required to prevent graft-versus-host disease (GvHD) but increases risks of developing a wide range of life-threatening infections. However, these patients may rely on innate defenses that are reconstituted more rapidly than the adaptive ones. In this context, ILCs may represent important players in the early phases following transplantation. They may contribute to tissue homeostasis/remodeling and lymphoid tissue reconstitution. While the reconstitution of NK cell repertoire and its role in haplo-HSCT have been largely investigated, little information is available on ILCs. Of note, $CD34^+$ cells isolated from different sources of HSC may differentiate *in vitro* toward various ILC subsets. Moreover, cytokines released from leukemia blasts (e.g., IL-1β) may alter the proportions of NK cells and ILC3, suggesting the possibility that leukemia may skew the ILC repertoire. Further studies are required to define the timing of ILC development and their potential protective role after HSCT.

Keywords: hematopoietic stem cell transplantation, innate lymphoid cells, NK cells, GVHD

INTRODUCTION

Allogeneic hematopoietic stem cell transplantation (allo-HSCT) still represents a major therapeutic option for severe hematological and immunological disorders (1). However, success of allo-HSCT can be impaired by graft-versus-host disease (GvHD) and, in the case of high-risk hematological malignancies, also by disease relapse. The clinical outcome can also be hampered by infections favored by the delayed immune reconstitution in transplanted patients (1). Moreover, only 60% of patients may find a related or unrelated HLA-matched donor. For the remaining patients, umbilical cord blood (UCB) may represent an alternative source of HSC (2). UCB transplant allows a less stringent HLA-matching between donor and recipient, but it is frequently associated with delayed neutrophil engraftment and delayed T-cell reconstitution. Moreover, UCB transplant recipient cannot benefit from the adoptive transfer of antigen-experienced donor T-cells (2). Another important therapeutic option is represented by the haploidentical (haplo)-HSCT. In this setting, the donor (in most instances, a parent) is identical for one HLA haplotype and mismatched for the other one (3). Given the high degree of HLA disparity, haplo-HSCT requires an extensive T-cell depletion of the graft (3) or heavily posttransplantation immune-suppressive therapy to prevent severe GvHD (4). In both UCB-HSCT and haplo-HSCT settings, the immune-compromised hosts are highly susceptible to a wide range of opportunistic infections. Thus, cells of the innate immunity are the only players exerting a major defensive role for several months before the restoring of adaptive immune responses. In particular, natural killer (NK) cells can provide protection against viral reactivation and/or primary infections. Perhaps, more importantly, the presence of alloreactive NK cells provides a potent graft-versus-leukemia (GvL) effect that contributes to tumor eradication (4, 5). It is now clear that NK cells are one of the components of a broad family of innate lymphoid cells (ILCs). However, so far, little is known on the possible role of the other ILC subsets in haplo-HSCT. Here, we will summarize our current knowledge on ILCs both in murine models and in human studies, since they could result crucial in host defenses after HSCT.

ILC Subsets

Different from T-cells and B-cells, ILCs are a group of lymphocytes that do not express rearranged antigen-specific receptors (6). ILCs represent a heterogeneous family of cells classified on the basis of their transcriptional and functional profile. Similar to T-cells, ILCs have been grouped into cytotoxic-ILC and helper-ILC (6). NK cells represent the cytotoxic-ILC population (7). They express eomesodermin (Eomes) and T-box transcription factor (T-bet), display cytolytic activity, and produce pro-inflammatory cytokines, primarily IFNγ and TNF. Helper-ILC population is further subdivided into three groups, namely: ILC1, ILC2, and ILC3 (6). ILC1 depend on the expression of the T-bet transcription factor for their development and secrete IFNγ, but, different from NK cells, they neither express Eomes nor exert cytolytic activity (7). ILC2 express GATA-binding protein 3 (GATA3) and produce type-2 cytokines, including IL-13 and IL-5 (8). Finally, ILC3 are a heterogeneous cell population, including fetal lymphoid tissue-inducer (LTi) cells and adult ILC3 that are further subdivided into natural cytotoxicity receptors$^-$ (NCR$^-$) and NCR$^+$ subsets. Collectively, ILC3 are defined by the expression of the retinoic acid receptor-related orphan receptor (RORγt) and produce mainly IL-17 and IL-22 (9). Studies in mice revealed that ILC, similar to T-cells and B-cells, derive from the common lymphoid progenitors (CLPs). The expression of the Id2 transcription factor determines further commitment toward a precursor common to all ILC subsets. While the NK cell precursor diverges early from the other ILC lineages, all helper-ILCs share a common helper-ILC progenitor (CHILP). Subsequently, upon exposure to different cytokines and/or to environmental cues, the CHILP differentiate toward ILC1, ILC2, or ILC3 (9). In humans, the developmental pathways are less characterized (10). However, NK and ILC3-committed precursors have recently been identified. Indeed, Renoux and coworkers identified, in several fetal and adult tissues, Lin$^-$CD34$^+$CD38$^+$CD123$^-$CD45RA$^+$CD7$^+$CD10$^+$CD127$^-$ cells able to differentiate exclusively toward cytotoxic NK cells both *in vitro* and *in vivo* (11). The ILC3 precursors, identified according to the Lin$^-$CD34$^+$RORγt$^+$ phenotype, have been detected selectively in tonsils and intestinal lamina propria (12).

ILC in Host Defenses against Pathogens and in Tissue Remodeling

In view of the heterogeneous cytokine profile and function of different ILC subsets, it is conceivable that ILCs may contribute to host defenses against a broad variety of pathogens (13, 14). Our knowledge on human ILC1 and their functional profile are still rather limited (15–17). Taking advantage of murine models, it has been shown that ILC1, thanks to the production of IFN-γ and TNF, contribute to immune responses against intracellular pathogens, such as *Toxoplasma gondii* (18). Also, NK cells are an important source of IFN-γ and TNF and, in addition, display very important effector functions, such as natural cytotoxicity and antibody-dependent cell-mediated cytotoxicity (ADCC). In the context of antimicrobial defenses, NK cells are primarily involved in the control of different viral infections, primarily herpes-viruses, but may also exert a protective role against bacterial and parasitic infections (19, 20). Of note, NK cells, thanks to their potent cytolytic activity, play also an important role against tumors (**Figure 1**) (21).

Host protection against parasites requires type-2 responses. A number of findings indicate that, during helminthic infections, epithelial cell-derived IL-25, IL-33, and thymic stromal lymphopoietin (TSLP) induce IL-13 release from ILC2 (22–24). In turn, IL-13 increases mucus production and smooth muscle contractility, thus contributing to the control of parasitic infections (25). In addition, dendritic cell (DC)-derived IL-23 and IL-1β cytokines induce the release of IL-22 and IL-17 by ILC3 (26–29). In turn, these ILC3-derived cytokines promote the production of antimicrobial molecules and neutrophil recruitment, enhancing the response against extracellular bacteria and fungi (14, 27, 30, 31) (**Figure 1**). In humans, the role played by ILC2 and ILC3 in defenses against pathogens is still undefined (10). Notably, patients experiencing helminthic infections show increases in ILC2 proportions; however, the role of ILC2 in anti-parasitic responses needs further investigation to be precisely clarified (8). Besides their antimicrobial function, ILCs are also involved in processes of tissues remodeling/repair. In particular, ILC2 appear to be involved in resolution of damages caused by viral or parasitic infections in lung tissues. Indeed, in response to IL-33, ILC2 also produce amphiregulin that promotes airway epithelial cell repair (32). Fetal LTi cells were the first ILC3 population to be described. LTi cells coordinate lymphoid organogenesis through their interaction with stromal cells by means of the LTαβ/LTβ receptor, leading to the upregulation of ICAM-1 and VCAM adhesion molecules on stromal cells (33). More recently,

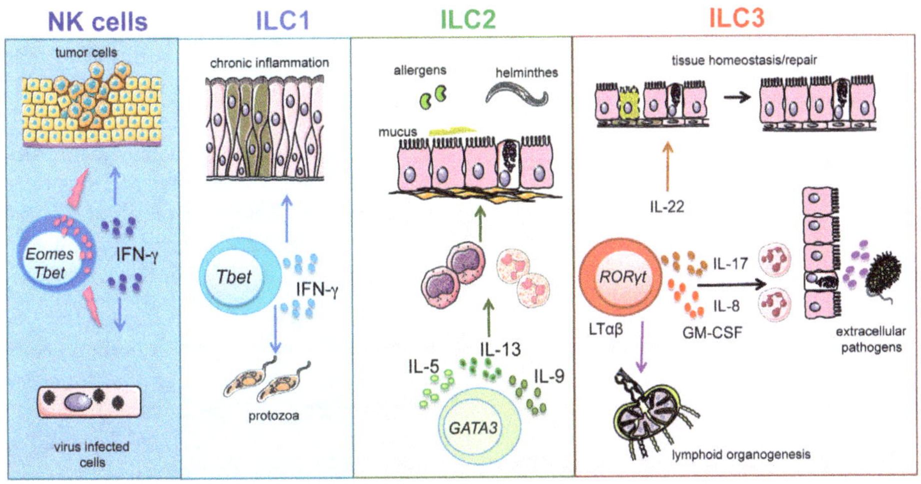

FIGURE 1 | ILC subsets and function. Graphic representation of the role played by ILC subsets in host defenses and tissue homeostasis/repair.

TABLE 1 | This table summarizes the main function exerted by distinct ILC subsets and the possible role exerted by these cells in the context of HSCT.

Cell type	Function	Role in HSCT
NK cells	Anti-tumor activity (21)	GvL (21, 55, 56)
	Defense against virus-infected cells (19, 20, 53)	Control of viral reactivation and/or primary infections (55, 56)
ILC1	Defense against protozoa (14, 18)	*Control of posttransplant opportunistic infections?*
ILC2	Defense against helminthic infection (14, 22–24)	*Control of posttransplant opportunistic infections?*
	Wound healing (32)	*Contribute to tissue repair?*
ILC3	Lymphoid organogenesis (33)	*Regeneration of secondary lymphoid organs?*
	Lymphoid tissue remodeling (34)	Thymic epithelial cell recovery (41)
	Epithelial homeostasis (35, 36)	Protection against therapy-induced epithelial damage and mucositis and promotion of tissue regeneration (35, 37)
		Reduction of GvHD occurrence (37)
	Defense against extracellular bacteria and fungi (14, 27–30)	*Control of posttransplant opportunistic infections?*

Only for some of the ILC populations, a role in the context of HSCT has been demonstrated. The possible roles exerted by other cell subsets are indicated in italics.

postnatal ILC3 have been shown to promote both survival and proliferation of stromal cells, following lymphoid tissue damage caused by viral infection and/or irradiation (34) (**Figure 1**). In addition, ILC3-derived IL-22 exerts a protective role on intestinal epithelial stem cells, particularly in the context of tissue damage caused by irradiation and/or acute GvHD (35, 36).

ILC and HSCT

So far, only a limited number of studies addressed the role of helper-ILC in the context of HSCT (35–37).

Reconstitution

Chemotherapy and radiotherapy treatment before HSCT induces extensive tissue damages in the host, including severe intestinal mucositis (38). Such damages can be even worse after allo-HSCT, if donor T lymphocytes attack the recipient intestinal epithelium (GvH reaction) (39). In a murine model of acute GvHD, Hanash and coworkers showed that host-derived IL-22 could substantially limit the development of GvHD (35). They could identify intestinal ILC3 subset as a main producers of IL-22 after total body irradiation treatment. In particular, IL-22 seemed to play a crucial role in the protection against epithelial cell damage and in preserving intestinal stem cells. These data are further supported by the finding that treatment with IL-22, in mice receiving bone marrow transplantation, resulted in increased intestinal stem cell recovery, in enhanced epithelial cell regeneration, and in reduction of intestinal GvHD (36). Given the role of ILC3 in lymphoid organogenesis and in lymphoid tissue remodeling, a role for these cells could also be envisaged in the regeneration of lymphoid tissues damaged by radiations (38, 40). Of note, ILC3-derived IL-22 can also favor the recovery of thymic epithelial cells, thus allowing a more efficient and rapid reconstitution of T-cell compartment (**Table 1**) (41). Conversely, it remains to be determined whether ILC3 also contribute to the regeneration of secondary lymphoid organs. In this context, it is recently shown that gamma irradiation used in conditioning regimen before HSCT may exert a long-lasting effect on secondary lymphoid organ structure and function (40). Also, ILC2 appear to be involved in epithelial tissue repair, particularly in lung tissues; however, no data are available to support an actual protection exerted by these cells in GvHD-induced tissue damages (42).

Graft-versus-Host Disease

In the context of human HSCT, only a single study investigated the possible role of ILCs in the protection from GvHD. It was suggested that both host and donor ILCs might exert a protective role (37). The expression of activation markers and of gut and skin homing receptors on host ILCs, detected prior to HSCT, correlated with a lower incidence of both mucositis and GvHD. Notably, after HSCT, ILCs detectable in peripheral blood (PB) are of donor origin. An early appearance of activated NCR$^+$ILC3 correlated with reduced risk of developing GvHD. In light of these finding, it is conceivable that the induction of a rapid ILC3 expansion/generation after HSCT may protect from GvHD. In this context, we have recently shown that granulocyte-colony-stimulating factor (G-CSF) could affect ILC3 and NK cell differentiation (43). Of note, G-CSF is used in UCB transplantation to accelerate engraftment and neutrophil recovery and is also used as a potent HSC mobilizing agent, before collection of HSC from donor PB (2, 44). Accordingly, we observed that HSC recovered after G-CSF-induced mobilization display a delayed and lower ILC3 and NK cell differentiation *in vitro* as compared to HSC isolated from bone marrow or UCB (43). These findings suggest that pre- and posttransplant treatment with G-CSF may affect ILC3 generation. Further studies should confirm these results *in vivo* and establish possible correlations with the occurrence of GvHD. Of note, it has been shown that ILC development may be impaired in patients with acute myeloid leukemia (AML) (45). Thus, after HSCT, ILC development might be affected by the presence of high residual leukemia burden or leukemia relapse. Indeed, it has been shown that HSC, when cultured in the presence of IL-1β-releasing AML blasts, display an impaired ability to differentiate toward ILC3 (46). Although in these culture setting the generation of NK cells seemed to be favored over ILC3, the final number of NK cells recovered was dramatically lower than those recovered in control cultures. Thus, if this inhibitory effect occurs also *in vivo*, it could have a negative impact on the NK-mediated GvL in haplo-HSCT. Of note, NK cell generation and differentiation after HSCT may be affected by immune-suppressor drugs, such as calcineurin inhibitors, used for treatment of GvHD (47, 48). On the other hand, helper-ILC reconstitution does not seem to be affected by cyclosporine or corticosteroids (37).

Opportunistic Infections

Studies in mice revealed that ILC might contribute to host defenses against different pathogens. In particular, while they are crucial in the control of infections in immune-compromised mice (18, 28, 49, 50), their actual role in the presence of a functional T-cell compartment seems to be marginal [as in the case of ILC3 during *Citrobacter rodentium* infection (51)]. However, as discussed above, patients transplanted with UCB cells or recipients of T-cell depleted haploidentical allograft experience a delayed recovery of both T-cell and B-cell adaptive responses, thus suggesting a possible relevant role of ILC in these transplantation settings. Accordingly, a rapid ILC differentiation after HSCT could guarantee an efficient host defense against opportunistic infections. Whether ILC1, ILC2, and ILC3 may indeed play a role in the control of infections in immune-compromised host, such as HSCT patients, has not been addressed yet. In contrast, clear evidence exists that patients with NK cell deficiencies and patients with functional NK cell defects display a higher susceptibility to viral infections [reviewed in Ref. (52)]. Moreover, in humanized mouse models, NK cells are required to effectively control Epstein–Barr virus (EBV) reactivation even in the presence of CD8$^+$ T-cells (53). NK cells also contribute to host protection against cytomegalovirus (CMV) (20, 54) also in the context of HSCT (55). In particular, NK cell involvement in CMV control is suggested by the finding that certain KIR haplotypes correlate with decreased CMV reactivation after transplantation (56). On the other hand, CMV, similar to other viral infections, can dramatically shape the NK cell repertoire (57–68). In humans, CMV infection is accompanied by a rapid NK cell maturation, the acquisition of KIR and CD57, and a selective expansion of a NKG2C$^+$ NK subset (62, 63, 67).

CONCLUDING REMARKS

Information available on ILC development and function derives primarily from studies performed in mice. Although these studies could provide reliable models of ILC differentiation, further analyses are required to address the dynamics of helper-ILC reconstitution after HSCT, the influence of HSC source, and the possible interference of cytokines produced by leukemia cells with ILC development. In addition, it will be crucial to clarify the role of specific ILC subsets in response to infections. Key information is still lacking in humans, not only on the role of ILC during infections but also in lymphoid tissue homeostasis. The possible exploitation of ILC in the context of HSCT requires a deeper knowledge of the mechanisms regulating their function and of the stimuli that drive their development.

AUTHOR CONTRIBUTIONS

All the authors provided data reported in this review. PV, EM, CV, and LM wrote and revised the manuscript.

REFERENCES

1. Copelan EA. Hematopoietic stem-cell transplantation. *N Engl J Med* (2006) **354**:1813–26. doi:10.1056/NEJMra052638
2. Danby R, Rocha V. Improving engraftment and immune reconstitution in umbilical cord blood transplantation. *Front Immunol* (2014) **5**:68. doi:10.3389/fimmu.2014.00068
3. Locatelli F, Pende D, Mingari MC, Bertaina A, Falco M, Moretta A, et al. Cellular and molecular basis of haploidentical hematopoietic stem cell transplantation in the successful treatment of high-risk leukemias: role of alloreactive NK cells. *Front Immunol* (2013) **4**:15. doi:10.3389/fimmu.2013.00015
4. Ruggeri L, Capanni M, Urbani E, Perruccio K, Shlomchik WD, Tosti A, et al. Effectiveness of donor natural killer cell alloreactivity in mismatched hematopoietic transplants. *Science* (2002) **295**:2097–100. doi:10.1126/science.1068440
5. Moretta L, Locatelli F, Pende D, Marcenaro E, Mingari MC, Moretta A. Killer Ig-like receptor-mediated control of natural killer cell alloreactivity in haploidentical hematopoietic stem cell transplantation. *Blood* (2011) **117**:764–71. doi:10.1182/blood-2010-08-264085
6. Diefenbach A, Colonna M, Koyasu S. Development, differentiation, and diversity of innate lymphoid cells. *Immunity* (2014) **41**:354–65. doi:10.1016/j.immuni.2014.09.005
7. Seillet C, Belz GT, Huntington ND. Development, homeostasis, and heterogeneity of NK cells and ILC1. *Curr Top Microbiol Immunol* (2015) **395**:37–61. doi:10.1007/82_2015_474
8. Halim TY. Group 2 innate lymphoid cells in disease. *Int Immunol* (2016) **28**:13–22. doi:10.1093/intimm/dxv050
9. Montaldo E, Juelke K, Romagnani C. Group 3 innate lymphoid cells (ILC3s): origin, differentiation, and plasticity in humans and mice. *Eur J Immunol* (2015) **45**:2171–82. doi:10.1002/eji.201545598
10. Montaldo E, Vacca P, Vitale C, Moretta F, Locatelli F, Mingari MC, et al. Human innate lymphoid cells. *Immunol Lett* (2016). doi:10.1016/j.imlet.2016.01.007
11. Renoux VM, Zriwil A, Peitzsch C, Michaelsson J, Friberg D, Soneji S, et al. Identification of a human natural killer cell lineage-restricted progenitor in fetal and adult tissues. *Immunity* (2015) **43**:394–407. doi:10.1016/j.immuni.2015.07.011
12. Montaldo E, Teixeira-Alves LG, Glatzer T, Durek P, Stervbo U, Hamann W, et al. Human RORgammat(+)CD34(+) cells are lineage-specified progenitors of group 3 RORgammat(+) innate lymphoid cells. *Immunity* (2014) **41**:988–1000. doi:10.1016/j.immuni.2014.11.010
13. McKenzie AN, Spits H, Eberl G. Innate lymphoid cells in inflammation and immunity. *Immunity* (2014) **41**:366–74. doi:10.1016/j.immuni.2014.09.006
14. Sonnenberg GF, Artis D. Innate lymphoid cells in the initiation, regulation and resolution of inflammation. *Nat Med* (2015) **21**:698–708. doi:10.1038/nm.3892
15. Bernink JH, Peters CP, Munneke M, te Velde AA, Meijer SL, Weijer K, et al. Human type 1 innate lymphoid cells accumulate in inflamed mucosal tissues. *Nat Immunol* (2013) **14**:221–9. doi:10.1038/ni.2534
16. Fuchs A, Vermi W, Lee JS, Lonardi S, Gilfillan S, Newberry RD, et al. Intraepithelial type 1 innate lymphoid cells are a unique subset of IL-12- and IL-15-responsive IFN-gamma-producing cells. *Immunity* (2013) **38**:769–81. doi:10.1016/j.immuni.2013.02.010
17. Vacca P, Montaldo E, Croxatto D, Loiacono F, Canegallo F, Venturini PL, et al. Identification of diverse innate lymphoid cells in human decidua. *Mucosal Immunol* (2015) **8**:254–64. doi:10.1038/mi.2014.63
18. Klose CS, Flach M, Mohle L, Rogell L, Hoyler T, Ebert K, et al. Differentiation of type 1 ILCs from a common progenitor to all helper-like innate lymphoid cell lineages. *Cell* (2014) **157**:340–56. doi:10.1016/j.cell.2014.03.030
19. Horowitz A, Stegmann KA, Riley EM. Activation of natural killer cells during microbial infections. *Front Immunol* (2012) **2**:88. doi:10.3389/fimmu.2011.00088
20. Lanier LL. Evolutionary struggles between NK cells and viruses. *Nat Rev Immunol* (2008) **8**:259–68. doi:10.1038/nri2276
21. Moretta L, Pietra G, Montaldo E, Vacca P, Pende D, Falco M, et al. Human NK cells: from surface receptors to the therapy of leukemias and solid tumors. *Front Immunol* (2014) **5**:87. doi:10.3389/fimmu.2014.00087
22. Moro K, Yamada T, Tanabe M, Takeuchi T, Ikawa T, Kawamoto H, et al. Innate production of T(H)2 cytokines by adipose tissue-associated c-Kit(+)Sca-1(+) lymphoid cells. *Nature* (2010) **463**:540–4. doi:10.1038/nature08636
23. Neill DR, Wong SH, Bellosi A, Flynn RJ, Daly M, Langford TK, et al. Nuocytes represent a new innate effector leukocyte that mediates type-2 immunity. *Nature* (2010) **464**:1367–70. doi:10.1038/nature08900
24. Price AE, Liang HE, Sullivan BM, Reinhardt RL, Eisley CJ, Erle DJ, et al. Systemically dispersed innate IL-13-expressing cells in type 2 immunity. *Proc Natl Acad Sci U S A* (2010) **107**:11489–94. doi:10.1073/pnas.1003988107
25. Maizels RM, Hewitson JP, Smith KA. Susceptibility and immunity to helminth parasites. *Curr Opin Immunol* (2012) **24**:459–66. doi:10.1016/j.coi.2012.06.003
26. Cella M, Fuchs A, Vermi W, Facchetti F, Otero K, Lennerz JK, et al. A human natural killer cell subset provides an innate source of IL-22 for mucosal immunity. *Nature* (2009) **457**:722–5. doi:10.1038/nature07537
27. Gladiator A, Wangler N, Trautwein-Weidner K, LeibundGut-Landmann S. Cutting edge: IL-17-secreting innate lymphoid cells are essential for host defense against fungal infection. *J Immunol* (2013) **190**:521–5. doi:10.4049/jimmunol.1202924
28. Satoh-Takayama N, Vosshenrich CA, Lesjean-Pottier S, Sawa S, Lochner M, Rattis F, et al. Microbial flora drives interleukin 22 production in intestinal NKp46+ cells that provide innate mucosal immune defense. *Immunity* (2008) **29**:958–70. doi:10.1016/j.immuni.2008.11.001
29. Sonnenberg GF, Monticelli LA, Elloso MM, Fouser LA, Artis D. CD4(+) lymphoid tissue-inducer cells promote innate immunity in the gut. *Immunity* (2011) **34**:122–34. doi:10.1016/j.immuni.2010.12.009
30. Zheng Y, Valdez PA, Danilenko DM, Hu Y, Sa SM, Gong Q, et al. Interleukin-22 mediates early host defense against attaching and effacing bacterial pathogens. *Nat Med* (2008) **14**:282–9. doi:10.1038/nm1720
31. Croxatto D, Micheletti A, Montaldo E, Orecchia P, Loiacono F, Canegallo F, et al. Group 3 innate lymphoid cells regulate neutrophil migration and function in human decidua. *Mucosal Immunol* (2016). doi:10.1038/mi.2016.10
32. Monticelli LA, Sonnenberg GF, Abt MC, Alenghat T, Ziegler CG, Doering TA, et al. Innate lymphoid cells promote lung-tissue homeostasis after infection with influenza virus. *Nat Immunol* (2011) **12**:1045–54. doi:10.1031/ni.2131
33. Finke D. Fate and function of lymphoid tissue inducer cells. *Curr Opin Immunol* (2005) **17**:144–50. doi:10.1016/j.coi.2005.01.006
34. Scandella E, Bolinger B, Lattmann E, Miller S, Favre S, Littman DR, et al. Restoration of lymphoid organ integrity through the interaction of lymphoid tissue-inducer cells with stroma of the T cell zone. *Nat Immunol* (2008) **9**:667–75. doi:10.1038/ni.1605
35. Hanash AM, Dudakov JA, Hua G, O'Connor MH, Young LF, Singer NV, et al. Interleukin-22 protects intestinal stem cells from immune-mediated tissue damage and regulates sensitivity to graft versus host disease. *Immunity* (2012) **37**:339–50. doi:10.1016/j.immuni.2012.05.028
36. Lindemans CA, Calafiore M, Mertelsmann AM, O'Connor MH, Dudakov JA, Jenq RR, et al. Interleukin-22 promotes intestinal-stem-cell-mediated epithelial regeneration. *Nature* (2015) **528**:560–4. doi:10.1038/nature16460
37. Munneke JM, Bjorklund AT, Mjosberg JM, Garming-Legert K, Bernink JH, Blom B, et al. Activated innate lymphoid cells are associated with a reduced susceptibility to graft-versus-host disease. *Blood* (2014) **124**:812–21. doi:10.1182/blood-2013-11-536888
38. Mohty M, Malard F, Savani BN. High-dose total body irradiation and myeloablative conditioning before allogeneic hematopoietic cell transplantation: time to rethink? *Biol Blood Marrow Transplant* (2015) **21**:620–4. doi:10.1016/j.bbmt.2014.09.010
39. Shlomchik WD. Graft-versus-host disease. *Nat Rev Immunol* (2007) **7**:340–52. doi:10.1038/nri2000
40. Rangel-Moreno J, de la Luz Garcia-Hernandez M, Ramos-Payan R, Biear J, Hernady E, Sangster MY, et al. Long-lasting impact of neonatal exposure to total body gamma radiation on secondary lymphoid organ structure and function. *Radiat Res* (2015) **184**:352–66. doi:10.1667/RR14047.1
41. Dudakov JA, Hanash AM, Jenq RR, Young LF, Ghosh A, Singer NV, et al. Interleukin-22 drives endogenous thymic regeneration in mice. *Science* (2012) **336**:91–5. doi:10.1126/science.1218004
42. Konya V, Mjosberg J. Innate lymphoid cells in graft-versus-host disease. *Am J Transplant* (2015) **15**:2795–801. doi:10.1111/ajt.13394

43. Moretta F, Petronelli F, Lucarelli B, Pitisci A, Bertaina A, Locatelli F, et al. The generation of human innate lymphoid cells is influenced by the source of hematopoietic stem cells and by the use of G-CSF. *Eur J Immunol* (2016) **46**:1271–8. doi:10.1002/eji.201546079
44. Bendall LJ, Bradstock KF. G-CSF: from granulopoietic stimulant to bone marrow stem cell mobilizing agent. *Cytokine Growth Factor Rev* (2014) **25**:355–67. doi:10.1016/j.cytogfr.2014.07.011
45. Trabanelli S, Curti A, Lecciso M, Salome B, Riether C, Ochsenbein A, et al. CD127+ innate lymphoid cells are dysregulated in treatment naive acute myeloid leukemia patients at diagnosis. *Haematologica* (2015) **100**:e257–60. doi:10.3324/haematol.2014.119602
46. Vitale C, Ambrosini P, Montaldo E, Ballerini F, Moretta L, Mingari MC. IL-1beta-releasing human acute myeloid leukemia blasts modulate natural killer cell differentiation from CD34+ precursors. *Haematologica* (2015) **100**:e42–5. doi:10.3324/haematol.2014.110494
47. Vitale C, Chiossone L, Morreale G, Lanino E, Cottalasso F, Moretti S, et al. Analysis of the activating receptors and cytolytic function of human natural killer cells undergoing in vivo differentiation after allogeneic bone marrow transplantation. *Eur J Immunol* (2004) **34**:455–60. doi:10.1002/eji.200324668
48. Wang H, Grzywacz B, Sukovich D, McCullar V, Cao Q, Lee AB, et al. The unexpected effect of cyclosporin A on CD56+CD16- and CD56+CD16+ natural killer cell subpopulations. *Blood* (2007) **110**:1530–9. doi:10.1182/blood-2006-10-048173
49. Rankin LC, Groom JR, Chopin M, Herold MJ, Walker JA, Mielke LA, et al. The transcription factor T-bet is essential for the development of NKp46+ innate lymphocytes via the Notch pathway. *Nat Immunol* (2013) **14**:389–95. doi:10.1038/ni.2545
50. Satoh-Takayama N, Serafini N, Verrier T, Rekiki A, Renauld JC, Frankel G, et al. The chemokine receptor CXCR6 controls the functional topography of interleukin-22 producing intestinal innate lymphoid cells. *Immunity* (2014) **41**:776–88. doi:10.1016/j.immuni.2014.10.007
51. Rankin LC, Girard-Madoux MJ, Seillet C, Mielke LA, Kerdiles Y, Fenis A, et al. Complementarity and redundancy of IL-22-producing innate lymphoid cells. *Nat Immunol* (2016) **17**:179–86. doi:10.1038/ni.3332
52. Orange JS. Natural killer cell deficiency. *J Allergy Clin Immunol* (2013) **132**:515–25. doi:10.1016/j.jaci.2013.07.020 quiz 526
53. Chijioke O, Muller A, Feederle R, Barros MH, Krieg C, Emmel V, et al. Human natural killer cells prevent infectious mononucleosis features by targeting lytic Epstein-Barr virus infection. *Cell Rep* (2013) **5**:1489–98. doi:10.1016/j.celrep.2013.11.041
54. Della Chiesa M, Sivori S, Carlomagno S, Moretta L, Moretta A. Activating KIRs and NKG2C in viral infections: toward NK cell memory? *Front Immunol* (2015) **6**:573. doi:10.3389/fimmu.2015.00573
55. Foley B, Felices M, Cichocki F, Cooley S, Verneris MR, Miller JS. The biology of NK cells and their receptors affects clinical outcomes after hematopoietic cell transplantation (HCT). *Immunol Rev* (2014) **258**:45–63. doi:10.1111/imr.12157
56. Palmer JM, Rajasekaran K, Thakar MS, Malarkannan S. Clinical relevance of natural killer cells following hematopoietic stem cell transplantation. *J Cancer* (2013) **4**:25–35. doi:10.7150/jca.5049
57. Beck JC, Wagner JE, DeFor TE, Brunstein CG, Schleiss MR, Young JA, et al. Impact of cytomegalovirus (CMV) reactivation after umbilical cord blood transplantation. *Biol Blood Marrow Transplant* (2010) **16**:215–22. doi:10.1016/j.bbmt.2009.09.019
58. Beziat V, Dalgard O, Asselah T, Halfon P, Bedossa P, Boudifa A, et al. CMV drives clonal expansion of NKG2C+ NK cells expressing self-specific KIRs in chronic hepatitis patients. *Eur J Immunol* (2012) **42**:447–57. doi:10.1002/eji.201141826
59. Beziat V, Liu LL, Malmberg JA, Ivarsson MA, Sohlberg E, Bjorklund AT, et al. NK cell responses to cytomegalovirus infection lead to stable imprints in the human KIR repertoire and involve activating KIRs. *Blood* (2013) **121**:2678–88. doi:10.1182/blood-2012-10-459545
60. Bjorkstrom NK, Lindgren T, Stoltz M, Fauriat C, Braun M, Evander M, et al. Rapid expansion and long-term persistence of elevated NK cell numbers in humans infected with hantavirus. *J Exp Med* (2011) **208**:13–21. doi:10.1084/jem.20100762
61. Bjorkstrom NK, Svensson A, Malmberg KJ, Eriksson K, Ljunggren HG. Characterization of natural killer cell phenotype and function during recurrent human HSV-2 infection. *PLoS One* (2011) **6**:e27664. doi:10.1371/journal.pone.0027664
62. Della Chiesa M, Falco M, Bertaina A, Muccio L, Alicata C, Frassoni F, et al. Human cytomegalovirus infection promotes rapid maturation of NK cells expressing activating killer Ig-like receptor in patients transplanted with NKG2C-/- umbilical cord blood. *J Immunol* (2014) **192**:1471–9. doi:10.4049/jimmunol.1302053
63. Della Chiesa M, Falco M, Podesta M, Locatelli F, Moretta L, Frassoni F, et al. Phenotypic and functional heterogeneity of human NK cells developing after umbilical cord blood transplantation: a role for human cytomegalovirus? *Blood* (2012) **119**:399–410. doi:10.1182/blood-2011-08-372003
64. Foley B, Cooley S, Verneris MR, Curtsinger J, Luo X, Waller EK, et al. Human cytomegalovirus (CMV)-induced memory-like NKG2C(+) NK cells are transplantable and expand in vivo in response to recipient CMV antigen. *J Immunol* (2012) **189**:5082–8. doi:10.4049/jimmunol.1201964
65. Foley B, Cooley S, Verneris MR, Pitt M, Curtsinger J, Luo X, et al. Cytomegalovirus reactivation after allogeneic transplantation promotes a lasting increase in educated NKG2C+ natural killer cells with potent function. *Blood* (2012) **119**:2665–74. doi:10.1182/blood-2011-10-386995
66. Guma M, Angulo A, Vilches C, Gomez-Lozano N, Malats N, Lopez-Botet M. Imprint of human cytomegalovirus infection on the NK cell receptor repertoire. *Blood* (2004) **104**:3664–71. doi:10.1182/blood-2004-05-2058
67. Muccio L, Bertaina A, Falco M, Pende D, Meazza R, Lopez-Botet M, et al. Analysis of memory-like natural killer cells in human cytomegalovirus-infected children undergoing αβ+T- and B-cell depleted HSCT for hematological malignancies. *Haematologica* (2016) **101**:371–81. doi:10.3324/haematol.2015.134155
68. Petitdemange C, Becquart P, Wauquier N, Beziat V, Debre P, Leroy EM, et al. Unconventional repertoire profile is imprinted during acute chikungunya infection for natural killer cells polarization toward cytotoxicity. *PLoS Pathog* (2011) **7**:e1002268. doi:10.1371/journal.ppat.1002268

21

T-bet and Eomesodermin in NK Cell Development, Maturation, and Function

Federico Simonetta, Amandine Pradier and Eddy Roosnek*

Department of Medical Specialties, Division of Hematology, Geneva University Hospitals, University of Geneva, Geneva, Switzerland

Edited by:
Emanuela Marcenaro, University of Genova, Italy

Reviewed by:
Guido Ferlazzo, University of Messina, Italy
Elisa Montaldo, San Raffaele Telethon Institute for Gene Therapy (SR-TIGET), Italy

***Correspondence:**
Federico Simonetta
federico.simonetta@hcuge.ch

Recent reports give insights into the role of the T-box transcription factors, T-bet and Eomesodermin (Eomes), in NK cell biology. In this mini-review, we recapitulate the initial reports that delineate T-bet and Eomes as master regulators of NK cell development, maturation, and function. We discuss how T-bet and Eomes expression is regulated during NK cell development and peripheral maturation. Furthermore, we summarize the current literature on the role of T-bet and Eomes in the transcriptional regulation of NK cell function and review possible effects of T-box transcription factor anomalies during aging, infection, cancer, and after hematopoietic stem cell transplantation. We discuss how the current data argue in favor of a model of T-bet and Eomes synergy in transcriptional regulation of NK cell function and identify T-box transcription factors as potential targets for therapeutic interventions.

Keywords: natural killer cells, T-box transcription factors, T-bet, Eomes

INTRODUCTION

The phylogenetically conserved family of T-box transcription factors, which share T-box DNA-binding domains, is critically involved in developmental processes in vertebrates. The T-box protein in T cells (T-bet) is a tyrosine- and serine-phosphorylated protein encoded by the *Tbx21* gene that is expressed only in cells of hematopoietic origin. T-bet was originally identified in T lymphocytes as the key transcription factor involved in interferon-gamma (IFN-γ) production that commits CD4 T cells to the Th1 lineage (1). Eomesodermin (Eomes), another T-box transcription factor sharing homology with T-bet, was originally described as a key player in vertebrate embryogenesis (2). More recently, Eomes and T-bet have been reported to coordinate the differentiation of CD8 T cells into effector cells (3–5) as well as their transition to the memory cell pool (6, 7). T-bet and Eomes are therefore considered as master regulators of T cell function. The bulk of mature murine (6, 8, 9) and human (10–12) NK cells express high levels of T-bet and Eomes, but until recently, their impact on NK cell function was not known. In the present work, we summarize the current knowledge about the role of T-bet and Eomes in NK cell development, peripheral maturation, and function.

T-BET AND EOMES IN NK CELL DEVELOPMENT

The first evidence for a role of T-bet in NK cell biology came from the observation that T-bet deficient (T-bet$^{-/-}$) mice have slightly higher NK cell numbers in the bone marrow but reduced numbers of NK cells in spleen, liver, and peripheral blood (13). Because many NK cells in T-bet$^{-/-}$ mice expressed an immature CD27^{pos}CD11b^{pos} phenotype, it was suggested that T-bet played a role in NK cell maturation without being essential for the early stages of NK cell development. Because Eomes$^{-/-}$ mice die in an early embryonic stage, the role of Eomes in NK cell development has initially been assessed only in compound mutant Eomes$^{+/-}$ Tbx21$^{-/-}$ mice (6). Interestingly, Eomes$^{+/-}$ Tbx21$^{-/-}$ mice displayed a severely exacerbated defect in the NK cell compartment compared to mice only lacking T-bet, suggesting a distinct, but complimentary function for Eomes in NK cell development. Importantly, the loss of one allele of *Eomes* results in a severe downregulation of CD122 (6), the beta-chain of IL-2R and IL-15R, which is essential for IL-15 signaling and NK cell development. Chromatin Immunoprecipitation (ChIP) assays showed that Eomes regulated CD122 transcription (6) for which T-bet appeared to be unnecessary (13). More recently, the role of Eomes has been studied in mice harboring floxed alleles of Eomes and expressing hematopoietic-restricted Cre recombinase under control of Vav regulatory elements (Eomes$^{flox/flox}$Vav–Cre^{+} mice), which restricts the Eomes-inactivation to cells of the hematopoietic lineage (8). Deletion of Eomes resulted in a severe reduction of NK cells in spleen and blood whereas only a modest reduction in NK cell numbers was observed in liver, lymph node, and bone marrow. Deletion of Eomes and T-bet in Tbx21$^{-/-}$Eomes$^{flox/flox}$Vav–Cre^{+} mice resulted in complete absence of NK cells in all organs (8). Hence, T-bet and Eomes are essential for normal NK cell development, but in the absence of either T-bet or Eomes, an incomplete development may still occur suggesting that the two T-box transcription factors share several functions.

The analysis of the contribution of Eomes and T-bet to NK cell development also led to the identification of an ontologically distinct subset of innate lymphocyte (ILC) cells residing in the liver. Lymphocytes expressing NK cell markers in murine liver contain up to 40% of Eomes-negative cells that express high levels of T-bet (8, 9, 14). Hepatic Eomes^{neg}T-bet^{high} NK cells display an immature phenotype characterized by the expression of Trail and lack of expression of DX5 (Trail^{pos}DX5^{neg}) (8). Initial experiments suggested that Eomes^{neg}T-bet^{high}Trail^{pos}DX5^{neg} cells represented an intermediate developmental stage that could differentiate into mature Eomes^{pos}Trail^{neg}DX5^{pos} cells (8). However, experiments performed with Eomes-negative cells isolated from Eomes-GFP reporter mice demonstrated that Eomes^{neg}T-bet^{high}Trail^{pos}DX5^{neg} cells are in fact an ontologically and functionally different subset of ILCs differentiating in the liver (14).

According to the current model, type 1 ILCs differentiate into two developmentally distinct lineages, type 1 helper innate lymphoid cells (hILC1s) and conventional NK cells (cNKs), which can be discriminated by the T-box transcription factors expressed (**Figure 1**). hILC1 differentiate in the liver when T-bet is upregulated and Eomes transcription is suppressed (14). Conversely, Eomes expression directs ILC1 development toward bone marrow-derived conventional NK cells that express relatively low levels of T-bet (15). Whether similar developmental pathways exist in human NK cells is still unknown (**Figure 2**). A recent study identified a T-bet^{pos}Eomes^{neg} CD49a^{pos} NK cell subset in human liver, absent from hepatic venous or peripheral blood, with a CD56^{bright}CD16^{neg}CD57^{neg}perforin^{neg} phenotype that may represent the human equivalent of murine T-bet^{pos} intrahepatic hILC1 (16).

In recent years, other ontologically distinct tissue residing NK cell subsets have been identified in salivary glands, skin, and

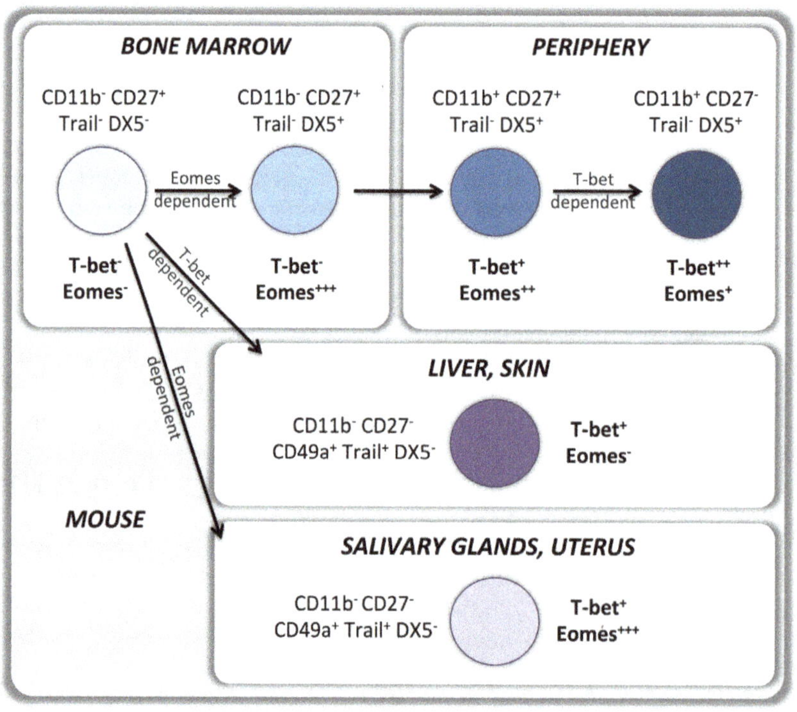

FIGURE 1 | **T-bet and Eomes in murine NK cell development and peripheral maturation.**

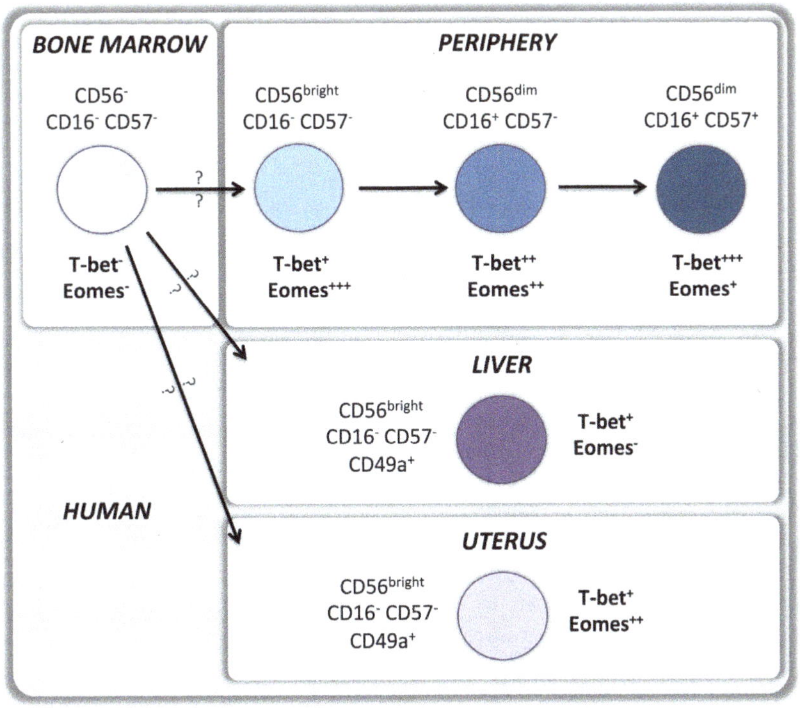

FIGURE 2 | **T-bet and Eomes in human NK cell development and peripheral maturation.**

uterus. Similarly to liver residing cells, NK cells isolated from these tissues display an immature CD49a^{pos}DX5^{neg} phenotype associated with the expression of markers of tissue residency. However T-box transcription factors expression varies with the tissues the NK cells reside in, which may point at distinct developmental pathways. Skin residing NK cells do not express Eomes and strictly depend on T-bet for their development (17), suggesting a developmental relationship with liver residing NK cells. Conversely, salivary gland (18) and uterine (17, 19–21) NK cells express high levels of Eomes and are T-bet-independent for their development and appear therefore to be a more distinct NK cell lineage (**Figure 1**).

Little is known on the mechanisms that induce cells to upregulate or repress T-box transcription factors in different organs. By contrast, several cell intrinsic mechanisms regulating Eomes and T-bet expression have been elucidated to date. First, T-bet and Eomes regulate each other's expression during NK cell development, and levels of Eomes expression correlate inversely with levels of T-bet in developing NK cells suggesting that active repressive mechanisms regulate the balance of T-bet/Eomes expression (14). This hypothesis has been confirmed by showing that T-bet$^{-/-}$ NK cells express high levels of Eomes whereas transgenic NK cells overexpressing T-bet display low levels of Eomes (14). In addition, expression of T-bet and Eomes has been shown to be strictly related with expression of other transcription factors crucial for NK cell development. T-bet expression is induced by the transcription factors ETS1 (V-Ets Avian Erythroblastosis Virus E26 Oncogene Homolog 1) (22), TOX1 (thymocyte selection-associated HMG box protein TOX-1) (23), and TOX2 (24). Conversely, Eomes expression depends on the bZIP transcription factor Nfil3 (Nuclear factor, interleukin 3 regulated, also known as E4BP4), which binds to the regulatory regions of the Eomes gene to promote its transcription (25). Nfil3 deficiency interferes with the development of the Eomes^{pos}TRAIL^{neg}DX5^{pos} bone marrow-derived NK cells while hepatic Tbet^{pos}Eomes^{neg}TRAIL^{pos}DX5^{neg} cells are unaffected (17, 26, 27). Interestingly, tissue-resident Eomes^{pos} NK cells localized in salivary glands and in the uterus can develop in the absence of Nfil3 (17, 18, 20, 27), suggesting the existence of alternative molecular mechanisms for the induction of Eomes transcription.

The only direct available information about the impact of T-box transcription factors deficiency on human NK cell development comes from a study describing a patient with a rare autosomal recessive microcephaly syndrome related to a translocation between chromosomes 3p and 10q leading to the silencing of the Eomes transcript (28). The fact that the infant displayed a normal distribution of T, B, and NK cells suggests that human NK cell development is possible in the absence of Eomes, although no information is available about the patient's NK cell functionality.

T-BET AND EOMES IN NK CELL PERIPHERAL MATURATION

Upregulation of T-bet and Eomes expression during development is maintained by most peripheral NK cells in mice (6) as well as in humans (10). Indeed, sustained expression of T-bet and Eomes in the periphery is necessary to maintain the NK cell maturation status while deletion of both T-box transcription factors results in reversion into an immature phenotype (8). T-bet expression is upregulated, and Eomes expression is downregulated during maturation of $CD11b^{pos}CD27^{pos}$ murine NK cells to the $CD11b^{pos}CD27^{neg}$ stage (14) (**Figure 1**). Importantly, T-bet appears to be essential for completion of this final maturation step as it controls the repression of CD27 and c-kit expression as well as the upregulation of S1P5 and KLRG1 (8, 13, 29–31). Therefore, bone marrow-derived $Eomes^{pos}TRAIL^{neg}DX5^{pos}$ NK cells can develop in the absence of T-bet, but are unable to undergo the terminal stages of maturation. Part of this terminal maturation process seems to be mediated by T-bet induction of the PR domain zinc finger protein 1 (Blimp-1) (30) and the zinc finger E-box-binding homeobox 2 (ZEB2) (32) transcription factors. Conversely, Forkhead box protein O1 (FOXO1) inhibits NK terminal maturation through repression of T-bet (33). Similar patterns of T-bet and Eomes expression exist in human NK cells that also upregulate T-bet and downregulate Eomes during peripheral maturation (**Figure 2**). Cytokine-producing $CD56^{bright}$ NK cells express higher levels of Eomes and lower levels of T-bet than cytotoxic $CD56^{dim}$ NK cells (10–12, 34, 35). Moreover, terminally differentiated $CD57^{pos}CD56^{dim}$ NK cells express the highest levels of T-bet and the lowest levels of Eomes (11, 12). Accordingly, upregulation of Killer-cell immunoglobulin-like receptors (KIRs) during maturation is associated with a decrease of Eomes and an increase of T-bet levels (11), which appears to be independent of the fact whether KIRs are licensing or not (Pradier et al., unpublished observations submitted for the present Frontiers Immunology Research Topic).

T-BET AND EOMES IN NK CELL FUNCTION

Chromatin Immunoprecipitation assays combined with the analysis of T-bet- and Eomes-deficient mice have partially uncovered the role of T-box transcription factors in NK cell biology. Similarly to what previously reported in $CD4^+$ Th1 cells, ChIP experiments identified IFN-γ as a target gene of T-bet in NK cells (13). By contrast, no evidence of Eomes binding to the IFN-γ promoter has been reported. Conditions that induce IFN-γ production, such as stimulation with IL-12 plus IL-15 or with IL-12 plus IL-18, also induce upregulation of T-bet (8, 13) and Eomes (12). Murine NK cells are still able to produce IFN-γ *in vivo* in the absence of T-bet, Eomes, or both (8, 13), but the maintenance of IFN-γ production is impaired in the absence of T-bet (13). In addition, T-bet and Eomes expression correlates positively with IFN-γ production *in vitro* in mice (36) as well as in humans (12, 37, 38). Furthermore, NK cells are less cytotoxic in the absence of T-bet (13, 39), which is possibly caused by a decreased production of perforin and granzyme B (8, 13). Murine studies suggest that T-bet but not Eomes is directly involved in the production of cytotoxic molecules (8). Accordingly, we found a positive correlation between T-bet levels and perforin production in human NK cells and no relationship between Eomes levels and expression of cytotoxic molecules (12). Collectively, these findings support a model in which T-bet and Eomes cooperatively regulate IFN-γ production in NK cells while T-bet seems to be the crucial regulator of their cytotoxic activity.

T-BET AND EOMES IN NK CELL BIOLOGY IN HEALTH AND DISEASE

Given their impact in NK cell development, peripheral maturation and function, alterations in T-bet and Eomes expression could account for NK cell abnormalities in pathological conditions in which NK cells exert an essential role, such as infections and cancer. Reduction in T-bet and Eomes levels in NK cells occurs during aging and is associated with an impaired NK cell cytotoxicity (40). Interestingly, T-bet and Eomes downregulation in aged mice is not related to a cell intrinsic defect but is induced by the aged environment pointing to a cell extrinsic induction of a senescent phenotype.

The role of T-bet and Eomes expression in NK cells has been investigated in several disease models. NK cells activated during murine cytomegalovirus or vaccinia virus infection do not undergo terminal maturation in the absence of T-bet (13, 41, 42), which concords with the typical role of T-bet in NK cell differentiation (13). However, despite the fact that this led to a significant reduction of NK cell virus-specific cytotoxicity early after infection (13, 42), the viral load remained unchanged suggesting that the NK cell activity in $T\text{-}bet^{-/-}$ mice is still sufficient to control viral replication. Similar sufficient *in vivo* NK cell responses have been reported after infection of $T\text{-}bet^{-/-}$ mice with *Listeria monocytogenes* (43) or with *Toxoplasma Gondii* (44).

Murine models of cancer have illustrated the impact of T-box transcription factors in NK cell antitumor responses. Peng and coworkers used a transgenic prostate adenocarcinoma mouse model to demonstrate that although T-bet deficiency only had a very limited impact on primary tumor development, it significantly affected the ability to control tumor spread (45). This observation, subsequently confirmed in a murine model of metastatic cancer (46), led to the conclusion that *in vivo* NK activity in metastasized cancer strongly depends on T-bet expression. Although NK cells are still capable to infiltrate metastatic tumors in the absence of T-bet, their survival and capacity to terminally differentiate into fully competent, cytotoxic $CD27^{neg}KLRG1^{pos}$ NK cells appears to be diminished (31, 46). It has been show that *in vivo* IL-15 administration overcomes the defect of $T\text{-}bet^{-/-}$ NK cells by inducing differentiation of $Eomes^{high}KLRG1^{pos}$ NK cells that are able to efficiently control metastatic pulmonary colorectal cancer, suggesting that IL-15 induced Eomes upregulation may compensate for the lack of T-bet inducing expansion of phenotypically and functionally mature NK cells.

Important insights into the relationship between T-bet and Eomes expression in NK cells and cancer come from the work of Gill and coworkers who identified the downregulation of T-bet and Eomes as the molecular signature of NK cell exhaustion in a murine NK adoptive transfer model of lymphoma (36). Importantly, downregulation of T-box transcription factors appeared to be not only the consequence of the NK cells' exposure to tumor cells but also of their homeostatic proliferation induced by the treatment-induced lymphopenic environment. Lymphopenia occurs frequently after cancer chemotherapy as well as after conditioning regimens for hematopoietic stem cell transplantation (HSCT). Indeed, we found the same exhausted phenotype in human NK cells isolated from patients undergoing HSCT (12). Similar to what had been observed in mice, T-bet and Eomes downregulation after HSCT was associated with impaired NK function and lower levels of T-bet in NK cells were associated with reduced patient overall survival (12). Surprisingly, improved survival associated with higher levels of T-bet in NK cells was not the consequence of improved cancer control but the result of a reduced non-relapse mortality, which suggests that sustained T-bet and Eomes expression in NK cell could participate to prevent the development of transplant related complications after HSCT. This hypothesis is supported by a recent study showing that adoptively transferred IL-12/15/18-preactivated NK cells, which do not undergo exhaustion and maintain high levels of Eomes and T-bet expression, suppressed acute Graft-versus-Host-Disease in a murine model of HSCT (47). These results suggest that T-bet and Eomes expression may also modulate NK cell function in immunopathological settings, similarly to what recently shown in multiple sclerosis patients (35).

CONCLUDING REMARKS

Recent findings clarifying the role of the two T-box transcription factors T-bet and Eomes in NK cells have considerably increased our knowledge of NK cell biology. Notably, they led to the characterization of previously unknown NK cells developmental pathways. Furthermore, they led to the identification of a molecular signature of NK cell exhaustion, which may represent a future target for immunomodulatory therapies.

AUTHOR CONTRIBUTIONS

FS wrote the manuscript and designed the figures. AP and ER critically discussed the work and edited the manuscript.

REFERENCES

1. Szabo SJ, Kim ST, Costa GL, Zhang X, Fathman CG, Glimcher LH. A novel transcription factor, T-bet, directs Th1 lineage commitment. *Cell* (2000) 100(6):655–69. doi:10.1016/S0092-8674(00)80702-3
2. Ryan K, Garrett N, Mitchell A, Gurdon J. Eomesodermin, a key early gene in *Xenopus* mesoderm differentiation. *Cell* (1996) 87(6):989–1000. doi:10.1016/S0092-8674(00)81794-8
3. Pearce EL, Mullen AC, Martins GA, Krawczyk CM, Hutchins AS, Zediak VP, et al. Control of effector CD8+ T cell function by the transcription factor Eomesodermin. *Science* (2003) 302(5647):1041–3. doi:10.1126/science.1090148
4. Intlekofer AM, Banerjee A, Takemoto N, Gordon SM, DeJong CS, Shin H, et al. Anomalous type 17 response to viral infection by CD8+ T cells lacking T-bet and eomesodermin. *Science* (2008) 321(5887):408–11. doi:10.1126/science.1159806
5. Cruz-Guilloty F, Pipkin ME, Djuretic IM, Levanon D, Lotem J, Lichtenheld MG, et al. Runx3 and T-box proteins cooperate to establish the transcriptional program of effector CTLs. *J Exp Med* (2009) 206(1):51–9. doi:10.1084/jem.20081242
6. Intlekofer AM, Takemoto N, Wherry EJ, Longworth SA, Northrup JT, Palanivel VR, et al. Effector and memory CD8+ T cell fate coupled by T-bet and eomesodermin. *Nat Immunol* (2005) 6(12):1236–44. doi:10.1038/ni1268
7. Banerjee A, Gordon SM, Intlekofer AM, Paley MA, Mooney EC, Lindsten T, et al. Cutting edge: the transcription factor eomesodermin enables CD8+ T cells to compete for the memory cell niche. *J Immunol* (2010) 185(9):4988–92. doi:10.4049/jimmunol.1002042
8. Gordon SM, Chaix J, Rupp LJ, Wu J, Madera S, Sun JC, et al. The transcription factors T-bet and Eomes control key checkpoints of natural killer cell maturation. *Immunity* (2012) 36(1):55–67. doi:10.1016/j.immuni.2011.11.016
9. Paley MA, Gordon SM, Bikoff EK, Robertson EJ, Wherry EJ, Reiner SL. Technical advance: fluorescent reporter reveals insights into eomesodermin biology in cytotoxic lymphocytes. *J Leukoc Biol* (2013) 93(2):307–15. doi:10.1189/jlb.0812400
10. Knox JJ, Cosma GL, Betts MR, McLane LM. Characterization of T-bet and eomes in peripheral human immune cells. *Front Immunol* (2014) 5:217. doi:10.3389/fimmu.2014.00217
11. Luetke-Eversloh M, Cicek BB, Siracusa F, Thom JT, Hamann A, Frischbutter S, et al. NK cells gain higher IFN-gamma competence during terminal differentiation. *Eur J Immunol* (2014) 44(7):2074–84. doi:10.1002/eji.201344072
12. Simonetta F, Pradier A, Bosshard C, Masouridi-Levrat S, Chalandon Y, Roosnek ENK. Cell functional impairment after allogeneic hematopoietic stem cell transplantation is associated with reduced levels of T-bet and eomesodermin. *J Immunol* (2015) 195(10):4712–20. doi:10.4049/jimmunol.1501522
13. Townsend MJ, Weinmann AS, Matsuda JL, Salomon R, Farnham PJ, Biron CA, et al. T-bet regulates the terminal maturation and homeostasis of NK and Vα14i NKT cells. *Immunity* (2004) 20(4):477–94. doi:10.1016/S1074-7613(04)00076-7
14. Daussy C, Faure F, Mayol K, Viel S, Gasteiger G, Charrier E, et al. T-bet and Eomes instruct the development of two distinct natural killer cell lineages in the liver and in the bone marrow. *J Exp Med* (2014) 211(3):563–77. doi:10.1084/jem.20131560
15. Pikovskaya O, Chaix J, Rothman NJ, Collins A, Chen Y-H, Scipioni AM, et al. Cutting edge: eomesodermin is sufficient to direct type 1 innate lymphocyte development into the conventional NK lineage. *J Immunol* (2016):1502396. doi:10.4049/jimmunol.1502396
16. Marquardt N, Beziat V, Nystrom S, Hengst J, Ivarsson MA, Kekalainen E, et al. Cutting edge: identification and characterization of human intrahepatic CD49a+ NK cells. *J Immunol* (2015) 194(6):2467–71. doi:10.4049/jimmunol.1402756
17. Sojka DK, Plougastel-Douglas B, Yang L, Pak-Wittel MA, Artyomov MN, Ivanova Y, et al. Tissue-resident natural killer (NK) cells are cell lineages distinct from thymic and conventional splenic NK cells. *Elife* (2014) 3:e01659. doi:10.7554/eLife.01659
18. Cortez VS, Fuchs A, Cella M, Gilfillan S, Colonna M. Cutting edge: salivary gland NK cells develop independently of Nfil3 in steady-state. *J Immunol* (2014) 192(10):4487–91. doi:10.4049/jimmunol.1303469
19. Tayade C, Fang Y, Black GP, Paffaro VA Jr, Erlebacher A, Croy BA. Differential transcription of Eomes and T-bet during maturation of mouse uterine natural killer cells. *J Leukoc Biol* (2005) 78(6):1347–55. doi:10.1189/jlb.0305142
20. Boulenouar S, Doisne J-M, Sferruzzi-Perri A, Gaynor LM, Kieckbusch J, Balmas E, et al. The residual innate lymphoid cells in nFil3-deficient mice support suboptimal maternal adaptations to pregnancy. *Front Immunol* (2016) 7:43. doi:10.3389/fimmu.2016.00043
21. Montaldo E, Vacca P, Chiossone L, Croxatto D, Loiacono F, Martini S, et al. Unique Eomes+ innate lymphoid cell subsets are present in uterus and decidua during early pregnancy. *Front Immunol* (2016) 6:646. doi:10.3389/fimmu.2015.00646
22. Ramirez K, Chandler KJ, Spaulding C, Zandi S, Sigvardsson M, Graves BJ, et al. Gene deregulation and chronic activation in natural killer cells deficient

in the transcription factor ETS1. *Immunity* (2012) 36(6):921–32. doi:10.1016/j.immuni.2012.04.006
23. Yun S, Lee SH, Yoon SR, Kim MS, Piao ZH, Myung PK, et al. TOX regulates the differentiation of human natural killer cells from hematopoietic stem cells in vitro. *Immunol Lett* (2011) 136(1):29–36. doi:10.1016/j.imlet.2010.11.008
24. Vong QP, Leung WH, Houston J, Li Y, Rooney B, Holladay M, et al. TOX2 regulates human natural killer cell development by controlling T-BET expression. *Blood* (2014) 124(26):3905–13. doi:10.1182/blood-2014-06-582965
25. Male V, Nisoli I, Kostrzewski T, Allan DS, Carlyle JR, Lord GM, et al. The transcription factor E4bp4/Nfil3 controls commitment to the NK lineage and directly regulates Eomes and Id2 expression. *J Exp Med* (2014) 211(4):635–42. doi:10.1084/jem.20132398
26. Crotta S, Gkioka A, Male V, Duarte JH, Davidson S, Nisoli I, et al. The transcription factor E4BP4 is not required for extramedullary pathways of NK cell development. *J Immunol* (2014) 192(6):2677–88. doi:10.4049/jimmunol.1302765
27. Seillet C, Huntington ND, Gangatirkar P, Axelsson E, Minnich M, Brady HJ, et al. Differential requirement for Nfil3 during NK cell development. *J Immunol* (2014) 192(6):2667–76. doi:10.4049/jimmunol.1302605
28. Baala L, Briault S, Etchevers HC, Laumonnier F, Natiq A, Amiel J, et al. Homozygous silencing of T-box transcription factor EOMES leads to microcephaly with polymicrogyria and corpus callosum agenesis. *Nat Genet* (2007) 39(4):454–6. doi:10.1038/ng1993
29. Jenne CN, Enders A, Rivera R, Watson SR, Bankovich AJ, Pereira JP, et al. T-bet-dependent S1P5 expression in NK cells promotes egress from lymph nodes and bone marrow. *J Exp Med* (2009) 206(11):2469–81. doi:10.1084/jem.20090525
30. Kallies A, Carotta S, Huntington ND, Bernard NJ, Tarlinton DM, Smyth MJ, et al. A role for Blimp1 in the transcriptional network controlling natural killer cell maturation. *Blood* (2011) 117(6):1869–79. doi:10.1182/blood-2010-08-303123
31. Malaise M, Rovira J, Renner P, Eggenhofer E, Sabet-Baktach M, Lantow M, et al. KLRG1+ NK cells protect T-bet-deficient mice from pulmonary metastatic colorectal carcinoma. *J Immunol* (2014) 192(4):1954–61. doi:10.4049/jimmunol.1300876
32. van Helden MJ, Goossens S, Daussy C, Mathieu AL, Faure F, Marcais A, et al. Terminal NK cell maturation is controlled by concerted actions of T-bet and Zeb2 and is essential for melanoma rejection. *J Exp Med* (2015) 212(12):2015–25. doi:10.1084/jem.20150809
33. Deng Y, Kerdiles Y, Chu J, Yuan S, Wang Y, Chen X, et al. Transcription factor Foxo1 is a negative regulator of natural killer cell maturation and function. *Immunity* (2015) 42(3):457–70. doi:10.1016/j.immuni.2015.02.006
34. Killig M, Friedrichs B, Meisig J, Gentilini C, Bluthgen N, Loddenkemper C, et al. Tracking in vivo dynamics of NK cells transferred in patients undergoing stem cell transplantation. *Eur J Immunol* (2014) 44(9):2822–34. doi:10.1002/eji.201444586
35. McKay FC, Gatt PN, Fewings N, Parnell GP, Schibeci SD, Basuki MA, et al. The low EOMES/TBX21 molecular phenotype in multiple sclerosis reflects CD56+ cell dysregulation and is affected by immunomodulatory therapies. *Clin Immunol* (2016) 163:96–107. doi:10.1016/j.clim.2015.12.015
36. Gill S, Vasey AE, De Souza A, Baker J, Smith AT, Kohrt HE, et al. Rapid development of exhaustion and down-regulation of eomesodermin limit the antitumor activity of adoptively transferred murine natural killer cells. *Blood* (2012) 119(24):5758–68. doi:10.1182/blood-2012-03-415364
37. Yu J, Wei M, Becknell B, Trotta R, Liu S, Boyd Z, et al. Pro- and antiinflammatory cytokine signaling: reciprocal antagonism regulates interferon-gamma production by human natural killer cells. *Immunity* (2006) 24(5):575–90. doi:10.1016/j.immuni.2006.03.016
38. Luetke-Eversloh M, Hammer Q, Durek P, Nordstrom K, Gasparoni G, Pink M, et al. Human cytomegalovirus drives epigenetic imprinting of the IFNG locus in NKG2Chi natural killer cells. *PLoS Pathog* (2014) 10(10):e1004441. doi:10.1371/journal.ppat.1004441
39. Svensson A, Nordstrom I, Sun JB, Eriksson K. Protective Immunity to genital herpes simpex virus type 2 infection is mediated by T-bet. *J Immunol* (2005) 174(10):6266–73. doi:10.4049/jimmunol.174.10.6266
40. Shehata HM, Hoebe K, Chougnet CA. The aged nonhematopoietic environment impairs natural killer cell maturation and function. *Aging Cell* (2015) 14(2):191–9. doi:10.1111/acel.12303
41. Robbins SH, Tessmer MS, Van Kaer L, Brossay L. Direct effects of T-bet and MHC class I expression, but not STAT1, on peripheral NK cell maturation. *Eur J Immunol* (2005) 35(3):757–65. doi:10.1002/eji.200425797
42. Matsui M, Moriya O, Yoshimoto T, Akatsuka T. T-bet is required for protection against vaccinia virus infection. *J Virol* (2005) 79(20):12798–806. doi:10.1128/JVI.79.20.12798-12806.2005
43. Way SS, Wilson CB. Cutting edge: immunity and IFN-production during *Listeria monocytogenes* infection in the absence of T-bet. *J Immunol* (2004) 173(10):5918–22. doi:10.4049/jimmunol.173.10.5918
44. Harms Pritchard G, Hall AO, Christian DA, Wagage S, Fang Q, Muallem G, et al. Diverse roles for T-bet in the effector responses required for resistance to infection. *J Immunol* (2015) 194(3):1131–40. doi:10.4049/jimmunol.1401617
45. Peng SL, Townsend MJ, Hecht JL, White IA, Glimcher LH. T-bet regulates metastasis rate in a murine model of primary prostate cancer. *Cancer Res* (2004) 64(2):452–5. doi:10.1158/0008-5472.CAN-03-3401
46. Werneck MBF, Lugo-Villarino G, Hwang ES, Cantor H, Glimcher LH. T-Bet plays a key role in NK-mediated control of melanoma metastatic disease. *J Immunol* (2008) 180(12):8004–10. doi:10.4049/jimmunol.180.12.8004
47. Huber CM, Doisne JM, Colucci F. IL-12/15/18-preactivated NK cells suppress GvHD in a mouse model of mismatched hematopoietic cell transplantation. *Eur J Immunol* (2015) 45(6):1727–35. doi:10.1002/eji.201445200

22

Effect of CMV and Aging on the Differential Expression of CD300a, CD161, T-bet, and Eomes on NK Cell Subsets

Nelson Lopez-Sejas[1†], *Carmen Campos*[1†], *Fakhri Hassouneh*[1], *Beatriz Sanchez-Correa*[2], *Raquel Tarazona*[2], *Alejandra Pera*[1*‡] *and Rafael Solana*[1*]

[1] *Maimonides Biomedicine Institute of Cordoba (IMIBIC), Reina Sofia Hospital, University of Cordoba, Cordoba, Spain,*
[2] *Immunology Unit, Department of Physiology, University of Extremadura, Cáceres, Spain*

Edited by:
Emanuela Marcenaro,
University of Genoa, Italy

Reviewed by:
Emily Mace,
Baylor College of Medicine, USA
Federico Simonetta,
Stanford University, Switzerland

***Correspondence:**
Alejandra Pera
alejandrapera@gmail.com;
Rafael Solana
rsolana@uco.es

[†]*Nelson Lopez-Sejas and Carmen Campos have contributed equally to the manuscript.*

Natural killer (NK) cells are innate lymphoid cells involved in the defense against virus-infected cells and tumor cells. NK cell phenotype and function is affected with age and cytomegalovirus (CMV) latent infection. Aging affects the frequency and phenotype of NK cells, and CMV infection also contributes to these alterations. Thus, a reduction of $CD56^{bright}$ NK cell subpopulation associated with age and an expansion of memory-like NK cells $CD56^{dim}CD57^{+}NKG2C^{+}$ probably related to CMV seropositivity have been described. NK cells express T-bet and Eomes transcription factors that are necessary for the development of NK cells. Here, we analyze the effect of age and CMV seropositivity on the expression of CD300a and CD161 inhibitory receptors, and T-bet and Eomes transcription factors in NK cell subsets defined by the expression of CD56 and CD57. CD300a is expressed by the majority of NK cells. $CD56^{bright}$ NK cells express higher levels of CD300a than $CD56^{dim}$ NK cells. An increase in the expression of CD300a was associated with age, whereas a decreased expression of CD161 in $CD56^{dim}$ NK cells was associated with CMV seropositivity. In $CD56^{dim}$ NK cells, an increased percentage of $CD57^{+}CD300a^{+}$ and a reduction in the percentage of $CD161^{+}CD300a^{+}$ cells were found to be associated with CMV seropositivity. Regarding T-bet and Eomes transcription factors, CMV seropositivity was associated with a decrease of $T\text{-}bet^{hi}$ in $CD56^{dim}CD57^{+}$ NK cells from young individuals, whereas Eomes expression was increased with CMV seropositivity in both $CD56^{bright}$ and $CD56^{dim}CD57^{+/-}$ (from middle age and young individuals, respectively) and was decreased with aging in all NK subsets from the three group of age. In conclusion, CMV infection and age induce significant changes in the expression of CD300a and CD161 in NK cell subsets defined by the expression of CD56 and CD57. T-bet and Eomes are differentially expressed on NK cell subsets, and their expression is affected by CMV latent infection and aging.

Keywords: aging, CMV, CD57, CD161, CD300a, Eomes, NK cell subsets, T-bet

INTRODUCTION

Natural killer (NK) cells are lymphocytes of innate immune response responsible for killing virus-infected cells and tumor cells. Frequency, phenotype, and function of NK cell subsets change in aging, and these changes are considered part of a general process of age-associated immune dysfunctions defined as immunosenescence (1–4). Immunosenescence affects adaptive and innate immunity, and it is associated with increased incidence and severity of infections and decreased response to vaccination (5–11).

It has been shown that percentage of CD56^{dim} NK cell subset (the main subset of NK cells and most cytotoxic) is increased (12, 13) or maintained (14) by age, whereas percentage of immature CD56^{bright} NK cells is decreased (12–15). On the other hand, the expression of different receptors on NK cells is also altered in aging (16–18). It has been found a decrease in the expression of natural cytotoxicity receptors (such as NKp30 and NKp46) and activating receptor DNAM-1 on CD56^{dim} NK subset (3, 19–21) an increased expression of CD57 (18, 20, 22), as well as reciprocal changes in NKG2A and killer immunoglobulin-like (KIR) inhibitory receptors, associated with age (18, 20, 22).

Human cytomegalovirus (CMV) chronic infection is related to a deterioration of the immune system that affect adaptive and innate immunity, and it has been postulated that CMV infection is a major driving force contributing to immunosenescence (10, 23, 24). CMV is a human herpesvirus type 5 (HHV5), which replicate in different cell types. Although CMV seropositivity is influenced by geographic, ethnic, and socio-economical factors, it has been shown that it increases with age in all populations studied (25). CMV prevalence is very high in Spain and more than 80% of individuals over the age of 40 years are CMV-seropositive (26).

Cytomegalovirus seropositivity can reshape the repertoire of NK cells (2, 27), particularly with an expansion of NKG2C^{+} NK cells, which also coexpress CD57 marker (14, 28, 29). Recent studies stratifying donors according to CMV serology have shown that increase of CD57 expression on CD56^{dim} NK cells is related to CMV seropositivity rather than aging (14) as well as the decreased expression of other surface receptors of NK cells, such as NKp30 (30) or CD161 (31). Thus, aging is associated with a loss of CD56^{bright} NK cell subpopulation (probably due to a decrease in the production of new NK cells in the bone marrow) and an expansion of memory-like NK cells CD56^{dim}CD57^{+}NKG2C^{+} that is mainly related to CMV seropositivity [reviewed in Ref. (2)].

Although it is well established that aging and CMV infection are associated with changes in NK cells, including alterations in the expression of activating (e.g., NCRs, NKG2C, and DNAM-1) and inhibitory receptors (e.g., KIR and NKG2A), little is known on the effect of aging and CMV on the expression of other inhibitory receptors such as CD300a and CD161 in NK cell subpopulations. CD300a is an inhibitory receptor expressed by NK cells that belongs to the CD300 family of molecules. These receptors are broadly expressed on immune cells and modulate their function *via* paired activating and inhibitory receptors that recognize lipids exposed on the plasma membrane of dead and activated cells including aminophospholipids such as phosphatidylserine (PS) and phosphatidylethanolamine (PE) (32). The analysis of its expression can be used in diagnosis and therapy in several pathological situations including infectious diseases, allergy, or cancer [for review, see Ref. (33)]. The human CD161 inhibitory receptor (also termed NKR-P1A, KLRB1, and CLEC5B) was originally described as a disulfide-linked homodimer of the C-type lectin superfamily expressed on subsets of NK cells and T lymphocytes (34) that binds the lectin-like transcript 1 (LLT1, also named CLEC2D, OCIL, and CLAX) (35, 36). The binding of CD161 on NK cells with its ligand on target cells results in inhibition of NK cell cytotoxicity by a mechanism involving the activation of acid sphingomyelinase (37). CD161 can also be expressed on subsets of other cells of the immune system, and different functional capacities have been shown after the interaction with its ligand, which can be upregulated during the immune response and during pathological circumstances. The current knowledge of NKRP1 receptors and their genetically linked CLEC2 ligand in human and other species has been recently reviewed (38, 39).

Natural killer cells are included in group 1 of the innate lymphoid cell (ILC), characterized by the release of interferon-gamma (IFN-γ) upon stimulation, and by the expression of T-bet and eomesodermin (Eomes) transcription factors (40–42). Both T-bet and Eomes are constitutively expressed by murine (43) and human (44, 45) NK cells and are necessary for the proper development of NK cells (46), sharing several functions. It has been observed that the frequency of T-bet^{+} cells and the level of T-bet expression per cell is significantly greater in the CD56^{dim} population compared to the CD56^{bright} population from peripheral human immune cells, contrary to Eomes expression pattern, suggesting the existence of a relationship among the expression levels of both transcription factors and the functionality of these cells (45). Thus, T-bet is related to terminal stages of maturation, while Eomes is downregulated during peripheral maturation (47).

Considering that aging affects the frequency and phenotype of NK cells and that CMV infection contributes to age-associated changes in NK cells; in this work, we have analyzed the effect of age and CMV seropositivity on inhibitory receptors CD300a and CD161 in NK cell subpopulations. Additionally, we have investigated the effect of age and CMV infection on T-bet and Eomes transcription factors expression in the CD56^{bright}CD57^{-}, CD56^{dim}CD57^{-}, and CD56^{dim}CD57^{+} NK cell subsets.

MATERIALS AND METHODS

Study Subjects

A total of 72 healthy adults voluntarily participated in the study, stratified according to age: 18–35 years (young), 40–65 years (middle age), and >70 years (old). Young and middle age donors were further divided according to CMV serology (CMV-seropositive and CMV-seronegative). However, all elderly donors included in the study were CMV-seropositive, given that the prevalence of CMV seropositivity in Spain in individuals over the age of 40 years is 80% (26) and, in our geographic area (Andalusia, Southern Spain), about 99% of individuals over

65 years are CMV-seropositive. Thus, the absence of a group of CMV-seronegative old donors represents a limitation of the study, making difficult to isolate age-related effect from the effect of chronic CMV infection in elderly individuals.

All donors were informed and signed informed consent to participate in the study and were included according to following inclusion criteria: no infection at the time of extraction, not suffer or have suffered cancer or autoimmune diseases, and not be under immunosuppressive drugs or calcium channel blockers. The study was approved by the Ethics Committee of Hospital Universitario Reina Sofia of Córdoba (Spain).

Sample Collection and Processing

Peripheral blood mononuclear cells (PBMCs) were obtained from blood samples (collected in lithium heparin tubes) and isolated by density gradient centrifugation using Ficoll Histopage-1077 (Sigma-Aldrich, St. Louis, MO, USA). Aliquots of cells were cryopreserved in FBS (Sigma-Aldrich, St Louis, MO, USA) with 10% DMSO (Panreac Chemistry SAU, Barcelona, Spain) in cryotubes at concentrations 5–6 × 10^6 cells/mL until further use.

A sample of plasma or serum was retrieved from all donors to analyze CMV-specific IgG and IgM. CMV serology was determined by using automated enzyme-linked immunosorbent assay (ELISA) (DRG International, Mountainside, NY, USA).

Analysis of NK Cell Receptors by Flow Cytometry

For surface marker analysis and computation of frequency of cells expressing CD57, CD161, or CD300a, cryopreserved PBMCs were used. Cell thawing was carried out in RPMI 1640 (Sigma-Aldrich) with 20% FBS (Gibco, Life Technologies California, USA). For flow cytometry staining, the following antibodies (mAbs) were used: anti-CD3 PerCP (clone: BW 264/56, Miltenyi Biotec), anti-CD56 PE-Cy7 (clone: B159, BD Pharmingen), anti-CD57 VioBlue (clone: TB03, Miltenyi Biotec), anti-CD300a PE (clone: E59.126, Beckman Coulter), and anti-CD161 APC (clone: DX12, BD Pharmingen). Cells were acquired on a MACSQuant cytometer (Miltenyi Biotec, Bergisch Gladbach, Germany). Compensation for flow cytometry was performed using single cell staining.

Data were analyzed using FlowJo v10 (Tree Star, Inc.). $CD3^{-}CD56^{+}$ NK cells were gated from total peripheral blood lymphocytes (PBLs), after singlets gating (Figure S1 in Supplementary Material). Two NK cell subsets were defined ($CD56^{bright}$ and $CD56^{dim}$) according to the level of CD56 marker expression. Subsequently, cells were gated according to the coexpression of CD57, CD161, and CD300a markers. Fluorescence minus one controls (FMO control) were used to identify and gate cells. FMO controls contain all the fluorochromes of the panel, except the one that was being measured. Analysis of coexpression of the three receptors was performed using FlowJo's Boolean gating options.

Transcription Factors Expression Analysis

The expression of T-bet and Eomes transcription factors was analyzed using cryopreserved PBMCs, thawed as indicated above. Surface staining was performed using anti-CD7 APC (clone: M-T701, BD Pharmingen), anti-CD56 BV421 (clone: NCAM16.2, BD Horizon), anti-CD16 PE- Vio770 (clone: VEP13, Miltenyi Biotec), anti-CD57 Biotin-Anti-Biotin-Viogreen (Miltenyi Biotec), and anti-CD3/anti-CD14/anti-CD19 conjugated with APC-Vio770 (clones: BW264/56, TÜK4, LT19 Miltenyi Biotec). After cell fixation and permeabilization using the Kit FoxP3 Staining Buffer Set (Miltenyi Biotec), following the manufacturer's instructions, intracellular staining was realized with anti-T-bet PerCP Cy5.5 (clone: 04-46, BD Pharmingen) and anti-Eomes FITC (clone: WD1928, eBioscience) antibodies. Cells were then acquired on a 10 parameter MACSQuant cytometer (Miltenyi Biotec, Bergisch Gladbach, Germany).

Data were analyzed using FlowJo v10 (Tree Star, Inc.). Once selected the $CD7^{+}CD3^{-}CD19^{-}CD14^{-}$ cells from PBLs singlets, three populations of NK cells were described: $CD56^{bright}CD57^{-}$, $CD56^{dim}CD57^{-}$, and $CD56^{dim}CD57^{+}$. Then, the intracellular expression of T-bet and Eomes was measured in each of these three subpopulations (Figure S2 in Supplementary Material). FMO controls and flow cytometry compensation were performed as indicated above.

Processing of Data and Statistical Analysis

For statistical analysis, Shapiro–Wilk normality test was performed for different groups. Since the distribution of measured values was not normal, the groups were evaluated by the non-parametric Kruskal–Wallis test (for comparison multiple) and Mann–Whitney test (for comparison sample pairs). Non-parametric Friedman test (for comparison multiple) and Wilcoxon test were used to test for differences between groups when the samples are related. The results are shown in graphs with interquartile medium range using GraphPad Prism (version 5.0), and analysis of Boolean gating data was performed by SPICE 5.35 software (Mario Roederer, ImmunoTechnology Section, Vaccine Research Centre, NIH, Bethesda, MD, USA; http://www.niaid.nih.gov) (48). To compare the pie charts, we used SPICE's permutation analysis, which asks how often the samples that comprise two different pies charts, would be different simply by chance (10,000 permutations). All statistical analysis was performed with SPSS 18.0 (SPSS Inc., Chicago, IL, USA). p-Values <0.05 were considered significant.

RESULTS

Increase of CD300a and Decrease of CD161 Expression from CMV-Seropositive Old Individuals

We analyzed by flow cytometry the expression of CD300a and CD161 inhibitory receptors on NK cells ($CD56^{bright}$ and $CD56^{dim}$) from healthy individuals stratified by age and CMV latent infection. Our data showed that both receptors exhibit a differential expression pattern. Particularly, the majority of $CD56^{dim}$ and $CD56^{bright}$ NK cells expressed CD300a, whereas CD161 was expressed on a subpopulation of NK cells regardless of the NK cell subset analyzed (Figure S1 in Supplementary Material). Moreover, when we gated the NK cell subsets according to CD56 and CD300a (referred to CD3⁻ cells), we observed that the majority of CD56^{bright} NK cells were CD300a^{hi}, whereas a high percentage of CD56^{dim} NK cells were CD300a^{lo} (**Figure 1A**). Analysis of the effect of age and CMV seropositivity on the expression of these receptors by NK cell subsets (gating strategy in Figure S1 in Supplementary Material) showed an increase of CD300a on CD56^{bright} and CD56^{dim} subsets from CMV-seropositive old individuals compared with middle-aged and young donors (independently of CMV seropositivity) (**Figure 1B**).

The analysis of CD161 expression on CD56^{dim} NK cells showed that young CMV-seronegative donors expressed higher levels of CD161 than young CMV-seropositive or middle age CMV-seronegative individuals, supporting that CMV and age independently associated with decreased expression of CD161 in

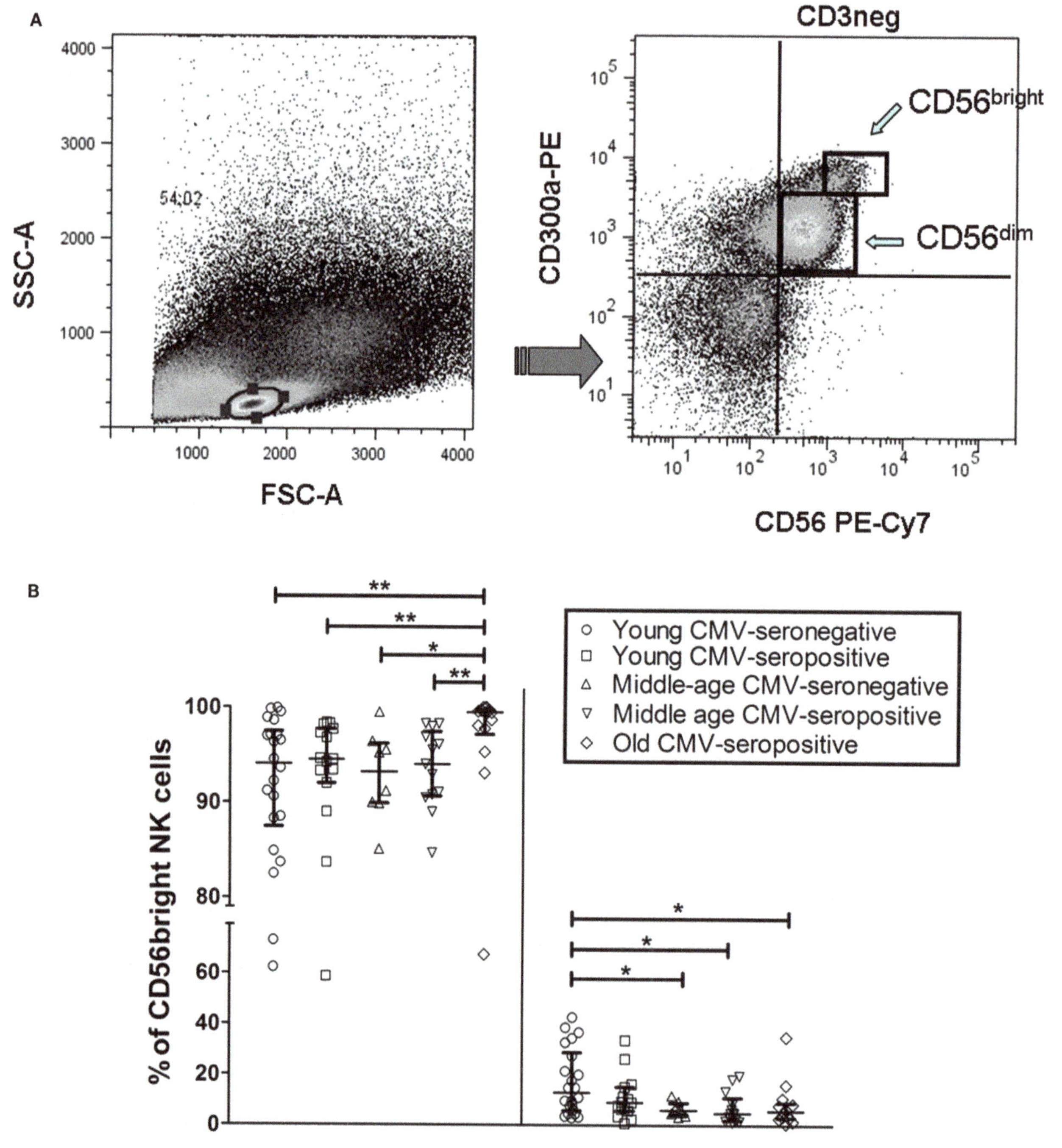

FIGURE 1 | CD300a and CD161 expression on CD56^{dim} and CD56^{bright} NK cells from young, middle age, and old individuals, according to CMV seropositivity.

(Continued)

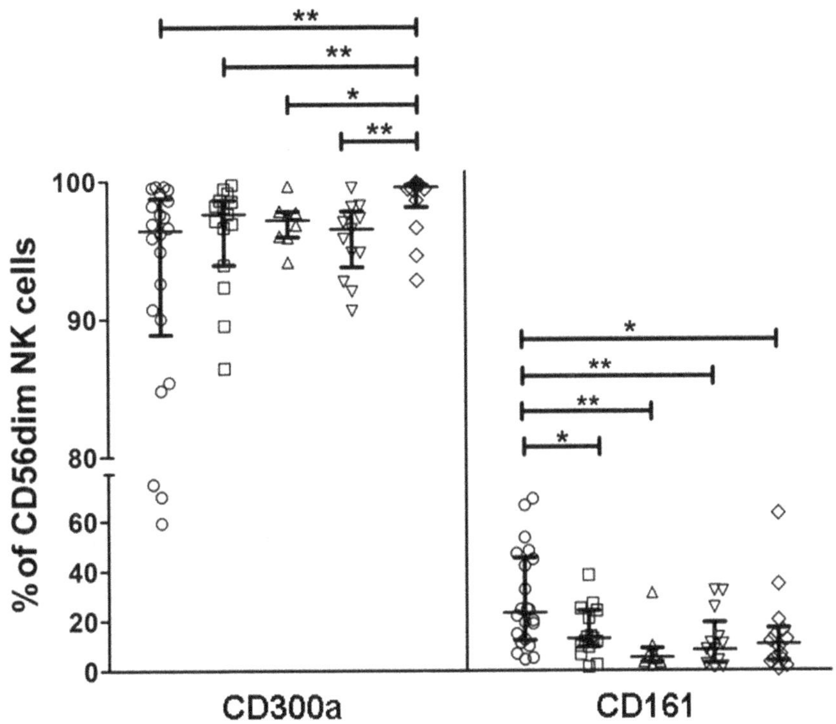

FIGURE 1 | **Continued**
(A) Differential expression of CD300a on NK cell subsets. NK cells were defined as CD3$^-$CD56$^+$ and characterized by CD300a expression as CD56^{bright}CD300a^{hi} and CD56^{dim}CD300a^{lo}. **(B)** Effect of age and CMV seropositivity on the expression of CD300a and CD161. Expression (percentage) of CD300a and CD161 markers was determined on NK cell subsets from young CMV-seronegative ($n = 22$), young CMV-seropositive ($n = 15$), middle age CMV-seronegative ($n = 8$), middle age CMV-seropositive ($n = 13$), and old CMV-seropositive donors ($n = 14$). CD56$^+$CD3$^-$ NK cells were gated from singlets peripheral blood lymphocytes (PBLs). Then, two NK cell subsets were defined (CD56^{bright} and CD56^{dim}) according to the level of CD56 marker expression, and cells were gated according to the expression of CD57, CD161, and CD300a markers. Non-parametric Kruskal–Wallis test (for multiple comparisons) and Mann–Whitney test (for paired comparisons) was used. Graphs showed the median with interquartile range, and results were considered significant at $^*p < 0.05$, $^{**}p < 0.01$, and $^{***}p < 0.001$.

this NK cell subset. On the contrary, the expression of CD161 on CD56^{bright} NK cells was not affected by CMV seropositivity in young donors, whereas it was decreased in middle age CMV-seronegative individuals. The expression of CD161 was lower on both NK cell subsets in middle age and old CMV-seropositive donors compared with young CMV-seronegative individuals (**Figure 1B**).

These results indicate an expansion of CD300a$^+$ on NK cell subsets (CD56^{bright} and CD56^{dim}) from healthy old individuals (all CMV-seropositive), likely associated with the combined effect of CMV infection and age, and a decrease of CD161$^+$ NK cells related to CMV seropositivity (CD56^{dim} NK cells) and age (CD56^{bright} and CD56^{dim} NK cells).

In this analysis, we have also observed an increased percentage of CD56^{dim}CD57$^+$ NK cells associated with CMV seropositivity. Also, CD56^{bright} NK cells do not express or express very low levels of CD57 on their surface (data not shown).

Coexpression of CD300a, CD161, and CD57 on NK Cell Subsets

The coexpression of CD300a, CD161, and CD57 was measured on CD56^{dim} and CD56^{bright} NK cells using FlowJo's Boolean gating options. The majority of CD56^{bright} and CD56^{dim} NK cells expressed CD300a on their surface, either alone or in combination with CD161 or CD57. A minor subset of CD300a$^+$ NK cells coexpressed CD161 and CD57 (**Figure 2A**).

The results on the effect of age and CMV seropositivity on the coexpression of CD300a, CD161, and CD57, revealed that CMV seropositivity, but not age, was associated with an increase of CD56^{dim} NK cells coexpressing CD300a and CD57. We also observed a decrease of CD56^{dim}CD300a$^+$CD161$^+$ NK cells related to both CMV seropositivity and age. No significant differences were found in the percentage of CD56^{dim}CD300a$^+$CD57$^-$CD161$^-$ NK cells among the different groups studied (data not shown). On the other hand, we have also observed an increase of CD56^{bright}CD300a$^+$CD57$^-$CD161$^-$ NK cells and a decrease of CD56^{bright} coexpressing CD300a and CD161 in old individuals (**Figures 2A,B**).

Different Expression Patterns of T-bet and Eomes Transcription Factors in NK Cell Subsets

We analyzed the expression of T-bet and Eomes transcription factors in three subpopulations of NK cells, according to the expression of CD56, CD16, and CD57 markers: CD56^{bright}CD16$^{+/-}$CD57^{-}, CD56^{dim}CD16^{+}CD57^{-}, and CD56^{dim}CD16^{+}CD57^{+} NK cells (Figure S2 in Supplementary Material). Although Eomes was expressed in most NK cells in all groups studied, the analysis of its expression pattern showed higher levels in the most immature NK cells (CD56^{bright}CD16$^{+/-}$CD57^{-}) with a decline in its expression in the CD56^{dim}CD16^{+}CD57^{-} and in the most differentiated NK cells (CD56^{dim}CD16^{+}CD57^{+}) (**Table 1**).

The expression of T-bet in NK cell subsets showed a gradient of expression and two subsets can be distinguished: T-bet^{hi} and

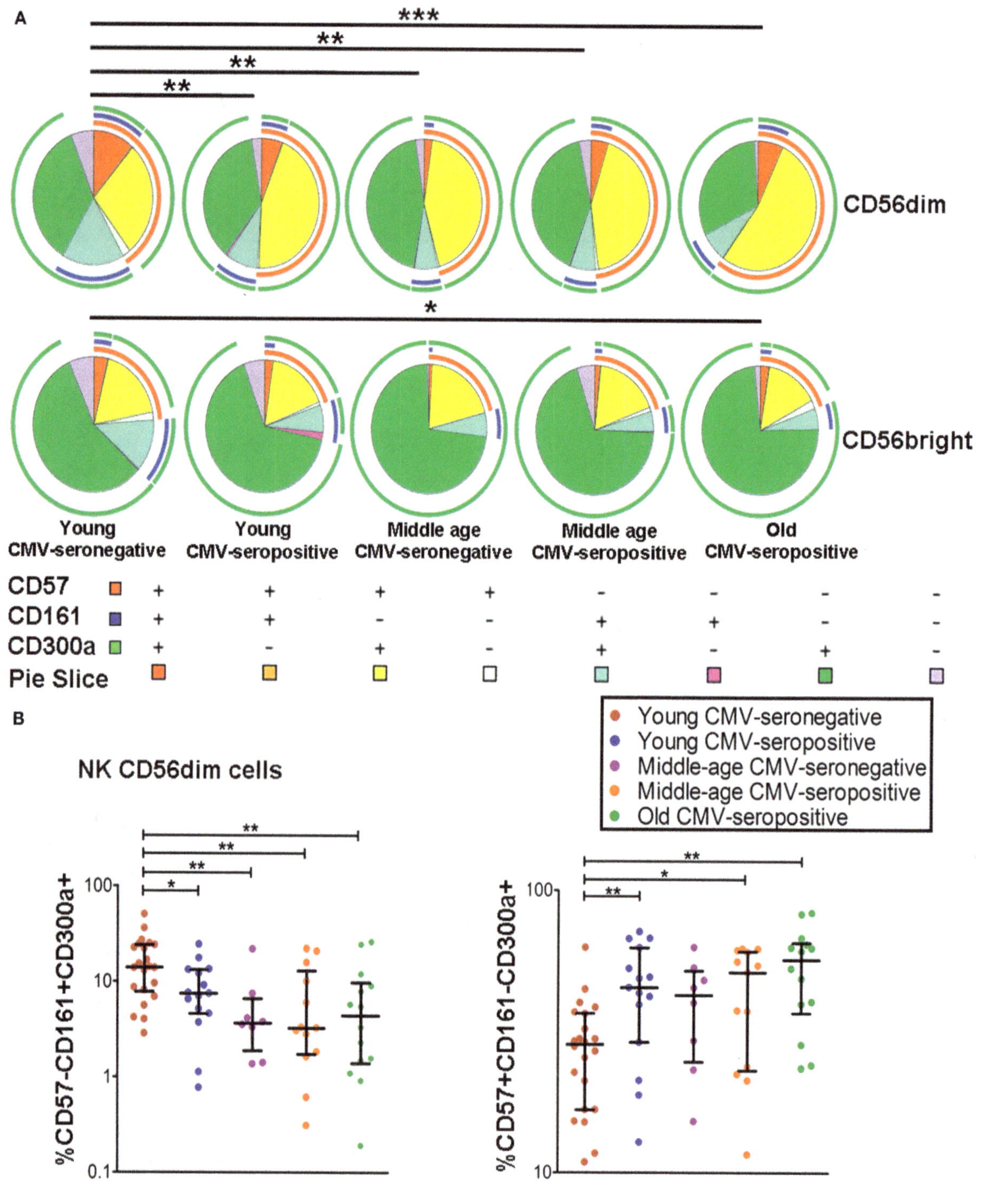

FIGURE 2 | Coexpression of CD57, CD161, and CD300a in NK cell subsets.

(Continued)

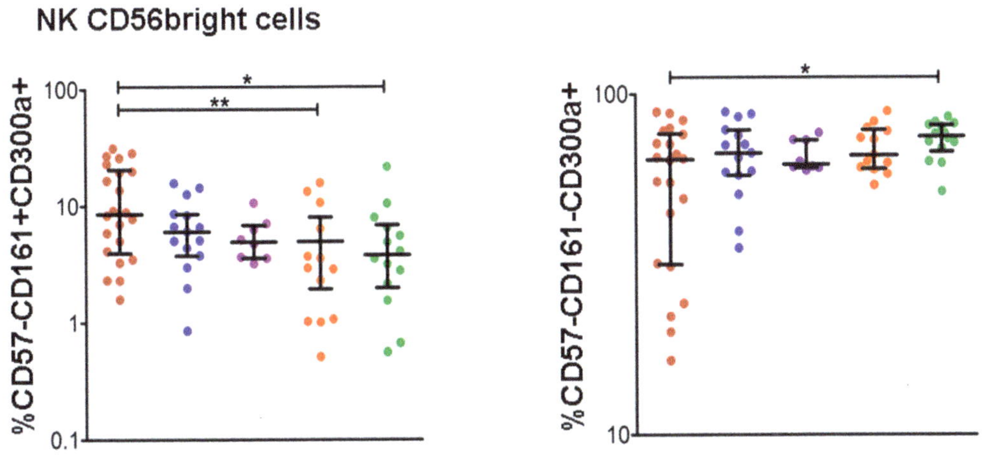

FIGURE 2 | Continued
(A) CD57, CD161, and CD300a expression profile (pie charts) in CD56[dim] and CD56[bright] NK cells from 72 healthy individuals stratified by age and CMV serostatus. Analysis of coexpression of three receptors was performed using FlowJo's Boolean gating options. For analysis of complex multivariate datasets after of flow cytometry, we used the application simplified presentation of incredibly complex evaluations (SPICE). **(B)** Graphs show the coexpression of CD57/CD161/CD300a NK subsets in which we found statistical differences among the four groups studied. Non-parametric Kruskal–Wallis test (for multiple comparisons) and Mann–Whitney test (for paired comparisons) were used. Graphs showed the median with interquartile range. Results were considered significant at $^*p < 0.05$, $^{**}p < 0.01$, and $^{***}p < 0.001$.

TABLE 1 | Eomes expression (MFI) in NK subsets from donors stratified according to age and CMV seropositivity.

	$CD56^{bright}CD16^-CD57^-$ (1)	$CD56^{dim}CD16^+CD57^-$ (2)	$CD56^{dim}CD16^+CD57^+$ (3)	ρ	ρ (1–2)	ρ (2–3)	ρ (1–3)
Young CMV^-	3.47 (4.46–2.58)[a]	2.12 (2.23–1.35)	1.59 (1.78–0.84)	0.000	0.005	0.005	0.005
Young CMV^+	3.75 (4.91–2.35)	2.28 (2.67–1.9)	2.18 (2.34–1.58)	0.007	0.007	0.169	0.007
Middle age CMV^-	3.48 (4.74–2.33)	1.87 (2.41–1.65)	1.45 (2.4–1.28)	0.001	0.018	0.018	0.018
Middle age CMV^+	3.32 (4.54–2.96)	2.15 (2.52–1.67)	1.6 (2.1–1.41)	0.000	0.008	0.008	0.008
Old CMV^+	3.02 (3.79–2.46)	2.01 (2.71–1.59)	1.8 (2.05–1.2)	0.000	0.005	0.005	0.005

[a]*Values expressed as median (interquartile range, 75–25).*
ρ value obtained by the Friedman rank sum test.
ρ values comparing (1–2), (2–3), and (1–3) groups obtained by the Wilcoxon test.

T-bet[int] NK cells. The percentage of NK cells expressing T-bet[hi] was higher than the percentage of NK cells expressing T-bet[int] in the three subsets studied (Figure S2 in Supplementary Material). The analysis of T-bet MFI in the different NK cell subsets showed that T-bet expression was lower in $CD56^{bright}CD16^{+/-}CD57^-$ NK cells from each group studied than in $CD56^{dim}CD16^+CD57^-$ and $CD56^{dim}CD16^+$ $CD57^+$ NK cells into the same group (**Table 2**).

The analysis of Eomes expression in NK cell subsets from CMV-seronegative and CMV-seropositive individuals showed that the percentage of positive cells was significantly increased in $CD56^{bright}CD16^{+/-}CD57^-$ NK cells from CMV-seropositive middle age donors and in $CD56^{dim}CD16^+CD57^+$ and $CD56^{dim}CD16^+CD57^-$ from CMV-seropositive young donors compared with the CMV-seronegative counterparts (**Figures 3A,B**). The analysis of the effect of age on Eomes expression showed a decreased expression in $CD56^{bright}CD16^{+/-}CD57^-$ NK cells from CMV-seronegative middle age donors compared with CMV-seronegative young individuals (**Figure 3A**). Eomes expression in CMV-seropositive donors was significantly higher in the young compared with the middle age and old groups in the three NK cell subsets considered (**Figures 3A,B**).

The effect of age and CMV seropositivity on T-bet expression was also analyzed. The results only showed a significant decrease in the percentage of T-bet[hi] within the $CD56^{dim}CD16^+CD57^+$ NK cell subset from CMV-seropositive compared with CMV-seronegative young donors (**Figures 3C,D**).

DISCUSSION

Cumulative evidences in the last decade support that aging and CMV latent infection combine to influence the immune phenotype and function of immune cells, including NK cells, in different ways that have been often overlooked in studies aiming to analyze the effect of aging on NK cells without considering the CMV serostatus of the individuals studied.

In this work, we have analyzed the effect of CMV seropositivity and aging on the expression of CD300a and CD161 receptors and transcription factors T-bet and Eomes on peripheral blood NK cell subsets with different levels of maturation.

TABLE 2 | T-bet expression (MFI) in NK subsets from donors stratified according to age and CMV seropositivity.

	$CD56^{bright}CD16^{-}CD57^{-}$ (1)	$CD56^{dim}CD16^{+}CD57^{-}$ (2)	$CD56^{dim}CD16^{+}CD57^{+}$ (3)	ρ	ρ (1–2)	ρ (2–3)	ρ (1–3)
Young CMV⁻	6.31 (10.2–2.91)[a]	8.84 (12.6–5.89)	9.31 (12.5–6.24)	0.000	0.008	0.022	0.005
Young CMV⁺	7.17 (10.3–3.47)	10.84 (12.2–7.26)	9.64 (11.4–7.49)	0.000	0.005	0.011	0.008
Middle age CMV⁻	10.10 (13–1.89)	12.8 (13.8–5.77)	13.2 (13.8–5.84)	0.004	0.043	0.042	0.018
Middle age CMV⁺	7.09 (10.8–3.85)	10.3 (13.5–6.97)	10.7 (13.9–6.79)	0.003	0.008	0.285	0.011
Old CMV⁺	4.85 (11.3–2.79)	7.57 (11.6–6.42)	7.22 (10.6–5.95)	0.007	0.017	0.013	0.114

[a]*Values expressed as median (interquartile range, 75–25).*
ρ value obtained by the Friedman rank sum test.
ρ values comparing (1–2), (2–3), and (1–3) groups obtained by the Wilcoxon test.

Our results show that $CD57^{+}CD56^{dim}$ NK cells are expanded in CMV-seropositive individuals and that these cells are not further expanded by aging. In addition, the majority of these cells also coexpress CD300a, but not CD161. The expansion of $CD57^{+}CD56^{dim}$ NK cells support previous data showing that acute and latent CMV infection leads to the expansion of $CD57^{+}CD56^{dim}$ NK cells that also express $NKG2C^{+}$ (28, 49–54). CMV seropositivity is also associated with a decreased expression of other NK receptors, in some cases, as a consequence of the shift of NK cells to the more differentiated $NKG2C^{+}CD57^{+}$ phenotype (14, 27, 30, 54). These cells have some characteristics of adaptive immunity and are considered "memory" or "adaptive" NK cells (55–58). The magnitude of the expansion of $NKG2C^{high}$ NK cells is determined by the magnitude of the proinflammatory cytokine secretion upon NK cell activation (59), and it has been proposed that these cells contribute to the proinflammatory environment based on the relation between the percentage of $NKG2C^{+}$ cells, elevated levels of PCR, and cardiovascular risk determinants of CMV-seropositive individuals (60, 61).

Whereas it has been show that both aging and CMV infection are associated with a decreased expression of several NK activating receptors (with the exception of NKG2C that is increased in CMV-seropositive individuals), their effect on the expression of inhibitory receptors is still controversial. Thus, whereas different studies have shown an age-associated increased expression of KIR (20, 62) and a decreased expression of NKG2A (20), others have not found significant differences in the expression of KIR or NKG2A in the elderly (14, 17, 18, 30) or even a decreased expression of KIR (CD158a) in middle age donors if they are CMV-seropositive (63). The expression of another inhibitory receptor KLRG-1 is significantly reduced with aging (21) and with CMV seropositivity in the young individuals (63).

The inhibitory receptor CD300a is expressed on the majority of NK cells, and its expression increases with aging in CMV-seropositive individuals both in $CD56^{bright}$ and $CD56^{dim}$ NK cell subsets. As stated in Section "Materials and Methods," in our geographic area (Andalusia, Southern Spain), about 99% of individuals over 65 years are CMV-seropositive. Thus, the high prevalence of CMV in our geographic area supposes a limitation of the study as we lack of a group of CMV-seronegative elderly donors. This limitation makes difficult to assess the differential effect of CMV and aging when we observe changes only in the group of old CMV-seropositive individuals compared with the other groups. This is the case of the higher expression of CD300a found in old CMV-seropositive individuals compared with the other groups studied. Since we do not observe an effect of CMV seropositivity between in the young and middle age groups, it could be thought that the higher expression of CD300a found in old CMV-seropositive individuals is likely due to aging. However, due to this limitation, we can only conclude that aging has an effect on CD300a expression on CMV-seropositive donors. CD300a is an inhibitory receptor that can be expressed by NK cells and that deliver inhibitory signals upon binding to PS expressed by tumor cells (64). The binding of CD300a on human or porcine NK cells to the surface of the pseudorabies virus porcine infected cell is increased by the US3 protein kinase of this alpha-herpesvirus in an aminophospholipids and p21-activated kinases dependent way, providing protection of infected cells against NK cell cytotoxicity (65). These results support the possible relevance of CD300a as a possible NK cell evasion strategy by CD300a-modulating viruses and cancer cells.

Although CD300a is highly homologous to CD300c and both receptors are considered as paired receptors with inhibitory and activating roles, respectively, very little is known on the expression and function of CD300c on NK cells. It has been recently shown that NK cells do not express (or express very low levels of) CD300c and that its expression is induced uniquely on $CD56^{bright}$ NK cells after their treatment with IL-2 or IL-15 (66). The analysis of the interaction of CD300a and CD300c with their ligands shows differences on their binding affinity to the lipid ligands. Whereas both CD300a and CD300c show similar binding to PS, it has been shown that CD300a exhibits a stronger binding to dead cells and to PE than CD300c and PE induces a negative response of IL-2 preactivated $CD56^{bright}$ NK cells, supporting that the inhibitory signals triggered by CD300a after binding to its lipid ligands overrides the signals triggered by its activating counterpart CD300c (66). These findings are consistent with these shown for other paired receptors such as KIR2DL1 and CD94/NKG2A that have higher binding affinity to HLA-CLys80 and HLA-E, respectively, than their activating counter parts KIR2DS1 and CD94/NKG2C. The increased expression of CD300a with aging both in $CD56^{bright}$ and $CD56^{dim}$ subsets can contribute together with other age-associated alterations of NK cells to the decreased functional capacity of these cells in the elderly.

The human CD161 receptor, expressed in a subset of NK cells, is an inhibitory receptor that, after interacting with its ligand LLT1, inhibits NK cell cytotoxicity by a mechanism involving the activation of acid sphingomyelinase (37). Our results show that the expression of CD161 on $CD56^{dim}$ NK cells is decreased

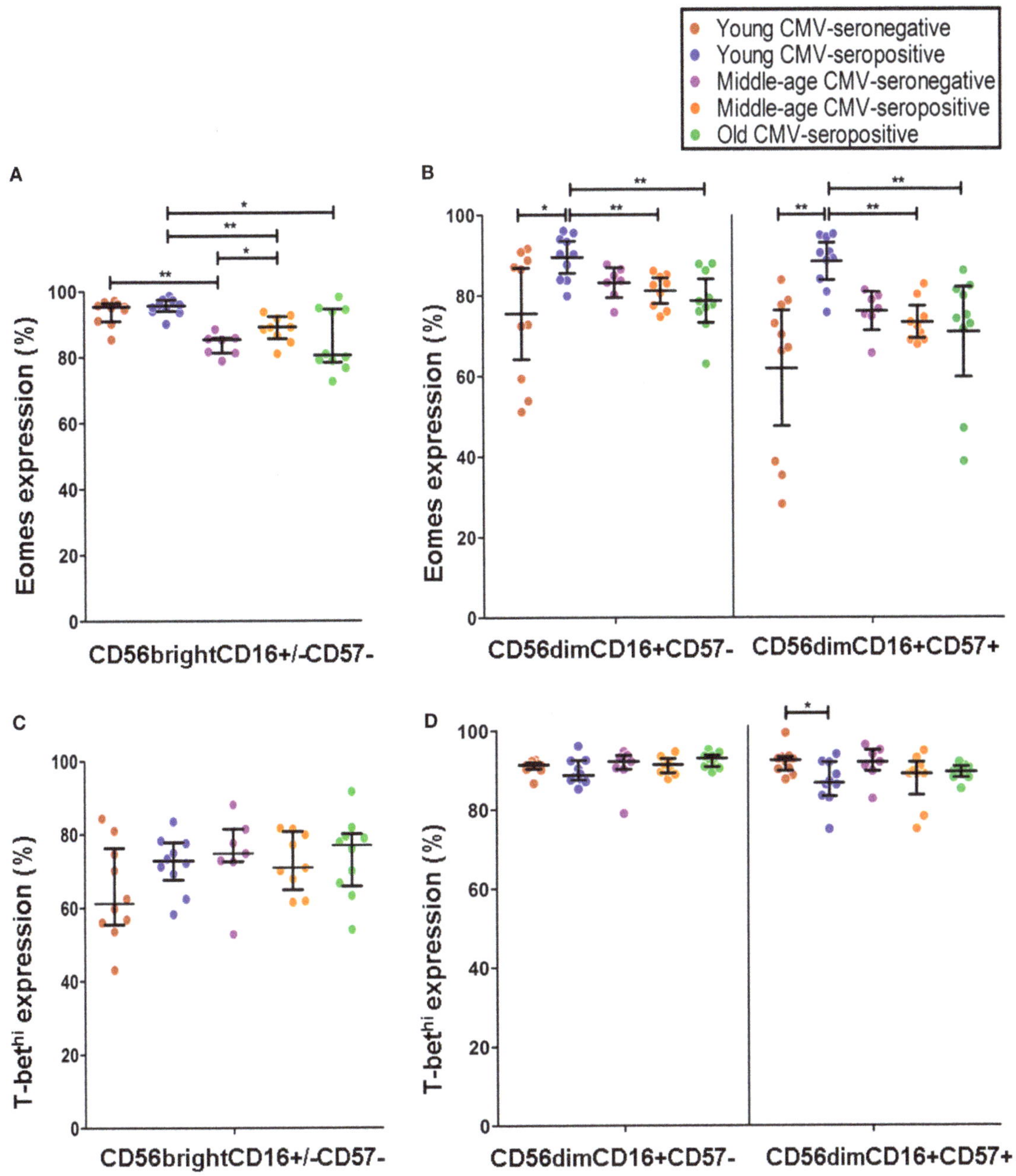

FIGURE 3 | Effect of age and CMV seropositivity on Eomes and T-bet expression. The expression of T-bet and Eomes was measured in CD56^{bright}CD16$^{+/-}$CD57^{-}, CD56^{dim}CD16^{+}CD57^{-}, and CD56^{dim}CD16^{+}CD57^{+} NK cells from 46 healthy individuals (10 young CMV-seronegative, 10 young CMV-seropositive, 7 middle age CMV-seronegative, 9 middle age CMV-seropositive, and 10 old CMV-seropositive donors). **(A)** Analysis of Eomes expression in CD56^{bright}CD16$^{+/-}$CD57^{-} NK cells. **(B)** Analysis of Eomes expression in CD56^{dim}CD16^{+}CD57^{-} and CD56^{dim}CD16^{+}CD57^{+} NK cells **(C)** Analysis of T-bet expression in CD56^{bright}CD16$^{+/-}$CD57^{-} NK cells. **(D)** Analysis of T-bet expression in CD56^{dim}CD16^{+}CD57^{-} and CD56^{dim}CD16^{+}CD57^{+} NK cells. Non-parametric Kruskal–Wallis test (for multiple comparisons) and Mann–Whitney test (for paired comparisons) were used. Graphs showed the median with interquartile range. Results were considered significant at $^{*}p < 0.05$, $^{**}p < 0.01$, and $^{***}p < 0.001$.

in CMV-seropositive donors compared with young CMV-seronegative donors. Little is known regarding the effect of CMV and aging on the expression of CD161 on NK cell subsets. The analysis of telomere length in NK cells has shown that the average telomere length in human NK cells decrease with age. In addition, the telomere length was significantly shorter in CD56^{dim}CD16^{+} NK cells compared to CD56^{bright}CD16^{-} NK cells from the same donor indicating that this subset represents more immature NK cells (67). In this study, it was shown that although CD161 can be expressed on CD56^{bright} and CD56^{dim} subsets, its expression is independent of the level of differentiation estimated by the telomere length (67).

The expression of CD161 in T lymphocytes and NK cells in PBLs from healthy children appears unrelated to CMV serostatus (68). On the contrary, it has been reported that NKG2C$^+$CD56^{dim} NK cells, expanded in CMV-seropositive chronic hepatitis patients, have a significant decrease in the expression of CD161 and a significant increase in the expression of CD57 (69). In agreement with these data, our results show that the majority of NK cells do not coexpress CD57 and CD161, and that CMV seropositivity is associated not only with an increase in the expression of CD57 in CD56^{dim} NK cells but also with a decrease in the expression of CD161 in this subset. The decreased expression of CD161 in NK cells from CMV-seropositive NK cells parallels the finding that CD161 is decreased on CD56$^+$ NKT-like cells in CMV$^+$ subjects compared with CMV-seronegative donors (70). Interestingly, it has also been shown that the expression of CD161 in CMV-specific cytotoxic T lymphocytes is very low (71). Considering the inhibitory capacity of CD161 on NK and T lymphocytes cytokine production, these results support that the CMV-induced downregulation of CD161 receptor together with the expansion of polyfunctional response of CD57$^+$ NK cells and T- and NKT-like lymphocytes (14, 30, 69, 72, 73), contribute to the proinflammatory environment observed in CMV-seropositive healthy individuals.

Recent reports have strengthened the significance of T-bet and Eomes in NK cell biology (47). In mice, T-bet and Eomes are necessary for maintenance of peripheral NK cells, their deletion in mature NK cells results in reversion to an immature phenotype (46). These transcription factors also modulate many NK cell effector functions, including cytotoxicity and cytokine production (74, 75). In human NK cells, T-bet and Eomes are differentially expressed on NK cell subsets (45), supporting that they can regulate different functions in different NK cell subpopulations. Our results confirm and extend these results showing that the levels of T-bet and Eomes are modulated in peripheral blood NK cell subsets representing different maturation stages, independently of aging and CMV serostatus. The expression of T-bet is lower in the CD56^{bright} NK cells than in the CD56^{dim} subset, whereas the expression of Eomes is higher in the CD56^{bright} NK cells. The CD56^{dim}CD57$^+$ subset shows higher levels of T-bet and lower levels of Eomes than the CD56^{bright} and CD56^{dim}CD57$^-$ NK cells. These results agree with the finding that not only the expression of T-bet and Eomes but also the expression of CD57 marker, modulate in parallel with the increase of KIRs on NK cells, but they do not differ in licensed or unlicensed NK cells (76), supporting that NK cell maturation but not NK cell licensing is dependent on T-bet and Eomes modulation. The significance of T-bet and Eomes in the maturation of other NK cell subpopulations is also supported by the demonstration that other subsets of NK cells display distinct patterns of expression of these transcription factors. Thus, a subpopulation of tissue-resident hepatic CD56^{bright} NK-cells, adapted to the tolerogenic liver microenvironment, with reduced proinflammatory potential, and characterized by the expression of CXCR6, express high levels of Eomes and low levels of T-bet, a phenotype virtually absent from peripheral blood (77, 78).

Very little is known on the expression patterns of T-bet and Eomes within human NK cell subpopulations in clinical situations. In this work, we have analyzed the expression of these transcription factors in two circumstances, such as CMV chronic infection and aging, which have a profound impact in NK cell differentiation.

Our results show that CMV seropositivity in young individuals associates with a significant increase in the percentage of CD56^{dim} NK cells expressing Eomes and a decreased percentage of T-bet^{hi} NK cells within the CD56^{dim}CD57$^+$ subset, suggesting that changes in the expression of these transcription factors are involved in CMV-induced remodeling of NK cells, characterized by the expansion of CD56^{dim}CD57$^+$ NK cells coexpressing NKG2C activating receptors (2, 14, 30, 60, 79). It has been reported that the level of expression of Eomes and T-bet is strongly reduced in NK cells from allogeneic hematopoietic stem cell transplantation recipients compared with healthy control subjects and that acute graft-versus-host disease and CMV reactivation are associated with further downregulation of T-bet (80) supporting the importance of these transcription factors in the generation and differentiation of NK cells and in the response against CMV after hematopoietic stem cell transplantation.

The analysis of the effect of age on Eomes expression showed a decreased expression in CD56^{bright} NK cells from CMV-seronegative middle age donors compared with CMV-seronegative young individuals, whereas in CMV-seropositive donors, its expression was significantly higher in the three NK cell subsets, in young donors compared with the middle age and old groups. The percentage of T-bet^{hi} NK cells was not significantly affected by age, independently of CMV serostatus. The effect of aging on the expression of transcription factors in NK cells is not known. It has been recently shown that impaired NK cell maturation in old mice is associated with a decreased expression of T-bet and Eomes in aged bone marrow NK cells. The use of bone marrow chimeras has revealed that the non-hematopoietic environment is responsible for the impaired maturation and function of NK cells, including the defective expression of T-bet and Eomes expression on NK cells (81). Considering that the defect in NK cell generation and in T-bet and Eomes expression in old mice is due to age alterations in non-hematopoietic environment, a better understanding of these non-hematopoietic factors involved in NK cell differentiation is required for the definition of new strategies aiming to improve NK cell function in the elderly.

It has been shown that the expression of T-bet and Eomes is increased in non-naïve CD8$^+$ T cells from aged subjects and that this increase correlated closely with the levels of CD57 and KLRG1. In addition, it was shown that aging is associated with a decreased functionality of influenza virus-specific CD8$^+$ T cells and increased expression of CD57, KLRG1, and T-bet (82), supporting that the increased expression of these transcription factors is related to the expansion of highly differentiated, senescent or exhausted, CD8$^+$ T cells found in elderly individuals.

In conclusion, CMV latent infection has a profound impact on NK cells inducing significant changes in the expression of NK receptors, including the inhibitory receptors CD300a and

CD161. T-bet and Eomes are differentially expressed on NK cell subsets defined by the expression of CD56 and CD57, and its expression is affected by CMV latent infection and aging, which can be involved in the age-associated changes observed in the differentiation and function of NK cells.

AUTHOR CONTRIBUTIONS

RS, CC, and AP designed the study. NL-S collected the data and performed the laboratory experiments. NL-S and FH performed the laboratory analysis. CC and NL-S performed the statistical analysis and wrote the draft. RT, BS-C, CC, and AP made significant conceptual contributions to the manuscript. RS, RT, and CC reviewed the final version of the paper. All the authors provided intellectual content and approved the final version of the paper.

REFERENCES

1. Pera A, Campos C, Lopez N, Hassouneh F, Alonso C, Tarazona R, et al. Immunosenescence: implications for response to infection and vaccination in older people. *Maturitas* (2015) 82:50–5. doi:10.1016/j.maturitas.2015.05.004
2. Solana R, Campos C, Pera A, Tarazona R. Shaping of NK cell subsets by aging. *Curr Opin Immunol* (2014) 29C:56–61. doi:10.1016/j.coi.2014.04.002
3. Solana R, Tarazona R, Gayoso I, Lesur O, Dupuis G, Fulop T. Innate immunosenescence: effect of aging on cells and receptors of the innate immune system in humans. *Semin Immunol* (2012) 24:331–41. doi:10.1016/j.smim.2012.04.008
4. Solana R, Pawelec G, Tarazona R. Aging and innate immunity. *Immunity* (2006) 24:491–4. doi:10.1016/j.immuni.2006.05.003
5. Grubeck-Loebenstein B, Della BS, Iorio AM, Michel JP, Pawelec G, Solana R. Immunosenescence and vaccine failure in the elderly. *Aging Clin Exp Res* (2009) 21:201–9. doi:10.1007/BF03324904
6. Wistuba-Hamprecht K, Haehnel K, Janssen N, Demuth I, Pawelec G. Peripheral blood T-cell signatures from high-resolution immune phenotyping of gammadelta and alphabeta T-cells in younger and older subjects in the Berlin Aging Study II. *Immun Ageing* (2015) 12:25. doi:10.1186/s12979-015-0052-x
7. Pawelec G. T-cell immunity in the aging human. *Haematologica* (2014) 99:795–7. doi:10.3324/haematol.2013.094383
8. Pawelec G, Goldeck D, Derhovanessian E. Inflammation, ageing and chronic disease. *Curr Opin Immunol* (2014) 29:23–8. doi:10.1016/j.coi.2014.03.007
9. Derhovanessian E, Pawelec G. Vaccination in the elderly. *Microb Biotechnol* (2012) 5:226–32. doi:10.1111/j.1751-7915.2011.00283.x
10. Solana R, Tarazona R, Aiello AE, Akbar AN, Appay V, Beswick M, et al. CMV and immunosenescence: from basics to clinics. *Immun Ageing* (2012) 9:23. doi:10.1186/1742-4933-9-23
11. Larbi A, Franceschi C, Mazzatti D, Solana R, Wikby A, Pawelec G. Aging of the immune system as a prognostic factor for human longevity. *Physiology (Bethesda)* (2008) 23:64–74. doi:10.1152/physiol.00040.2007
12. Borrego F, Alonso MC, Galiani MD, Carracedo J, Ramirez R, Ostos B, et al. NK phenotypic markers and IL2 response in NK cells from elderly people. *Exp Gerontol* (1999) 34:253–65. doi:10.1016/S0531-5565(98)00076-X
13. Gayoso I, Sanchez-Correa B, Campos C, Alonso C, Pera A, Casado JG, et al. Immunosenescence of human natural killer cells. *J Innate Immun* (2011) 3:337–43. doi:10.1159/000328005
14. Campos C, Pera A, Sanchez-Correa B, Alonso C, Lopez-Fernandez I, Morgado S, et al. Effect of age and CMV on NK cell subpopulations. *Exp Gerontol* (2014) 54:130–7. doi:10.1016/j.exger.2014.01.008
15. Chidrawar SM, Khan N, Chan YL, Nayak L, Moss PA. Ageing is associated with a decline in peripheral blood CD56bright NK cells. *Immun Ageing* (2006) 3:10. doi:10.1186/1742-4933-3-10
16. Hazeldine J, Lord JM. The impact of ageing on natural killer cell function and potential consequences for health in older adults. *Ageing Res Rev* (2013) 12:1069–78. doi:10.1016/j.arr.2013.04.003
17. Almeida-Oliveira A, Smith-Carvalho M, Porto LC, Cardoso-Oliveira J, Ribeiro AS, Falcao RR, et al. Age-related changes in natural killer cell receptors from childhood through old age. *Hum Immunol* (2011) 72:319–29. doi:10.1016/j.humimm.2011.01.009
18. Le Garff-Tavernier M, Beziat V, Decocq J, Siguret V, Gandjbakhch F, Pautas E, et al. Human NK cells display major phenotypic and functional changes over the life span. *Aging Cell* (2010) 9:527–35. doi:10.1111/j.1474-9726.2010.00584.x
19. Sanchez-Correa B, Morgado S, Gayoso I, Bergua JM, Casado JG, Arcos MJ, et al. Human NK cells in acute myeloid leukaemia patients: analysis of NK cell-activating receptors and their ligands. *Cancer Immunol Immunother* (2011) 60:1195–205. doi:10.1007/s00262-011-1050-2
20. Lutz CT, Moore MB, Bradley S, Shelton BJ, Lutgendorf SK. Reciprocal age related change in natural killer cell receptors for MHC class I. *Mech Ageing Dev* (2005) 126:722–31. doi:10.1016/j.mad.2005.01.004
21. Hayhoe RP, Henson SM, Akbar AN, Palmer DB. Variation of human natural killer cell phenotypes with age: identification of a unique KLRG1-negative subset. *Hum Immunol* (2010) 71:676–81. doi:10.1016/j.humimm.2010.03.014
22. Hazeldine J, Hampson P, Lord JM. Reduced release and binding of perforin at the immunological synapse underlies the age-related decline in natural killer cell cytotoxicity. *Aging Cell* (2012) 11(5):751–9. doi:10.1111/j.1474-9726.2012.00839.x
23. Koch S, Larbi A, Ozcelik D, Solana R, Gouttefangeas C, Attig S, et al. Cytomegalovirus infection: a driving force in human T cell immunosenescence. *Ann N Y Acad Sci* (2007) 1114:23–35. doi:10.1196/annals.1396.043
24. Pawelec G, Derhovanessian E, Larbi A, Strindhall J, Wikby A. Cytomegalovirus and human immunosenescence. *Rev Med Virol* (2009) 19:47–56. doi:10.1002/rmv.598
25. Cannon MJ, Schmid DS, Hyde TB. Review of cytomegalovirus seroprevalence and demographic characteristics associated with infection. *Rev Med Virol* (2010) 20:202–13. doi:10.1002/rmv.655
26. deOry F, Ramirez R, Garcia CL, Leon P, Sagues MJ, Sanz JC. Is there a change in cytomegalovirus seroepidemiology in Spain? *Eur J Epidemiol* (2004) 19:85–9. doi:10.1023/B:EJEP.0000013253.56343.6f
27. Muntasell A, Vilches C, Angulo A, Lopez-Botet M. Adaptive reconfiguration of the human NK-cell compartment in response to cytomegalovirus: a different perspective of the host-pathogen interaction. *Eur J Immunol* (2013) 43:1133–41. doi:10.1002/eji.201243117
28. Lopez-Verges S, Milush JM, Schwartz BS, Pando MJ, Jarjoura J, York VA, et al. Expansion of a unique CD57NKG2Chi natural killer cell subset during acute human cytomegalovirus infection. *Proc Natl Acad Sci U S A* (2011) 108:14725–32. doi:10.1073/pnas.1110900108

29. Della CM, Muccio L, Moretta A. CMV induces rapid NK cell maturation in HSCT recipients. *Immunol Lett* (2013) 155(1-2):11–3. doi:10.1016/j.imlet.2013.09.020
30. Campos C, Lopez N, Pera A, Gordillo JJ, Hassouneh F, Tarazona R, et al. Expression of NKp30, NKp46 and DNAM-1 activating receptors on resting and IL-2 activated NK cells from healthy donors according to CMV-serostatus and age. *Biogerontology* (2015) 16:671–83. doi:10.1007/s10522-015-9581-0
31. Fildes JE, Walker AH, Williams SG, Yonan N, Leonard CT. CMV infection is associated with the depletion but lack of activation of peripheral blood natural killer cells in a lung transplant cohort. *Transpl Immunol* (2008) 19:235–7. doi:10.1016/j.trim.2008.05.007
32. Simhadri VR, Andersen JF, Calvo E, Choi SC, Coligan JE, Borrego F. Human CD300a binds to phosphatidylethanolamine and phosphatidylserine, and modulates the phagocytosis of dead cells. *Blood* (2012) 119:2799–809. doi:10.1182/blood-2011-08-372425
33. Borrego F. The CD300 molecules: an emerging family of regulators of the immune system. *Blood* (2013) 121:1951–60. doi:10.1182/blood-2012-09-435057
34. Lanier LL, Chang C, Phillips JH. Human NKR-P1A. A disulfide-linked homodimer of the C-type lectin superfamily expressed by a subset of NK and T lymphocytes. *J Immunol* (1994) 153:2417–28.
35. Aldemir H, Prod'homme V, Dumaurier MJ, Retiere C, Poupon G, Cazareth J, et al. Cutting edge: lectin-like transcript 1 is a ligand for the CD161 receptor. *J Immunol* (2005) 175:7791–5. doi:10.4049/jimmunol.175.12.7791
36. Rosen DB, Bettadapura J, Alsharifi M, Mathew PA, Warren HS, Lanier LL. Cutting edge: lectin-like transcript-1 is a ligand for the inhibitory human NKR-P1A receptor. *J Immunol* (2005) 175:7796–9. doi:10.4049/jimmunol.175.12.7796
37. Pozo D, Vales-Gomez M, Mavaddat N, Williamson SC, Chisholm SE, Reyburn H. CD161 (human NKR-P1A) signaling in NK cells involves the activation of acid sphingomyelinase. *J Immunol* (2006) 176:2397–406. doi:10.4049/jimmunol.176.4.2397
38. Bartel Y, Bauer B, Steinle A. Modulation of NK cell function by genetically coupled C-type lectin-like receptor/ligand pairs encoded in the human natural killer gene complex. *Front Immunol* (2013) 4:362. doi:10.3389/fimmu.2013.00362
39. Kirkham CL, Carlyle JR. Complexity and diversity of the NKR-P1:Clr (Klrb1:Clec2) recognition systems. *Front Immunol* (2014) 5:214. doi:10.3389/fimmu.2014.00214
40. Spits H, Artis D, Colonna M, Diefenbach A, Di Santo JP, Eberl G, et al. Innate lymphoid cells – a proposal for uniform nomenclature. *Nat Rev Immunol* (2013) 13:145–9. doi:10.1038/nri3365
41. Walker JA, Barlow JL, McKenzie AN. Innate lymphoid cells – how did we miss them? *Nat Rev Immunol* (2013) 13:75–87. doi:10.1038/nri3349
42. Spits H, Bernink JH, Lanier L. NK cells and type 1 innate lymphoid cells: partners in host defense. *Nat Immunol* (2016) 17:758–64. doi:10.1038/ni.3482
43. Intlekofer AM, Takemoto N, Wherry EJ, Longworth SA, Northrup JT, Palanivel VR, et al. Effector and memory CD8+ T cell fate coupled by T-bet and eomesodermin. *Nat Immunol* (2005) 6:1236–44. doi:10.1038/ni1268
44. Luetke-Eversloh M, Cicek BB, Siracusa F, Thom JT, Hamann A, Frischbutter S, et al. NK cells gain higher IFN-gamma competence during terminal differentiation. *Eur J Immunol* (2014) 44:2074–84. doi:10.1002/eji.201344072
45. Knox JJ, Cosma GL, Betts MR, McLane LM. Characterization of T-bet and eomes in peripheral human immune cells. *Front Immunol* (2014) 5:217. doi:10.3389/fimmu.2014.00217
46. Gordon SM, Chaix J, Rupp LJ, Wu J, Madera S, Sun JC, et al. The transcription factors T-bet and eomes control key checkpoints of natural killer cell maturation. *Immunity* (2012) 36:55–67. doi:10.1016/j.immuni.2011.11.016
47. Simonetta F, Pradier A, Roosnek E. T-bet and eomesodermin in NK cell development, maturation, and function. *Front Immunol* (2016) 7:241. doi:10.3389/fimmu.2016.00241
48. Roederer M, Nozzi JL, Nason MC. SPICE: exploration and analysis of post-cytometric complex multivariate datasets. *Cytometry A* (2011) 79:167–74. doi:10.1002/cyto.a.21015
49. Guma M, Budt M, Saez A, Brckalo T, Hengel H, Angulo A, et al. Expansion of CD94/NKG2C+ NK cells in response to human cytomegalovirus-infected fibroblasts. *Blood* (2006) 107:3624–31. doi:10.1182/blood-2005-09-3682
50. Guma M, Angulo A, Vilches C, Gomez-Lozano N, Malats N, Lopez-Botet M. Imprint of human cytomegalovirus infection on the NK cell receptor repertoire. *Blood* (2004) 104:3664–71. doi:10.1182/blood-2004-05-2058
51. Foley B, Cooley S, Verneris MR, Curtsinger J, Luo X, Waller EK, et al. Human cytomegalovirus (CMV)-induced memory-like NKG2C(+) NK cells are transplantable and expand in vivo in response to recipient CMV antigen. *J Immunol* (2012) 189:5082–8. doi:10.4049/jimmunol.1201964
52. Della CM, Falco M, Bertaina A, Muccio L, Alicata C, Frassoni F, et al. Human cytomegalovirus infection promotes rapid maturation of NK cells expressing activating killer Ig-like receptor in patients transplanted with NKG2C-/- umbilical cord blood. *J Immunol* (2014) 192:1471–9. doi:10.4049/jimmunol.1302053
53. Della CM, Falco M, Muccio L, Bertaina A, Locatelli F, Moretta A. Impact of HCMV infection on NK cell development and function after HSCT. *Front Immunol* (2013) 4:458. doi:10.3389/fimmu.2013.00458
54. Beziat V, Liu LL, Malmberg JA, Ivarsson MA, Sohlberg E, Bjorklund AT, et al. NK cell responses to cytomegalovirus infection lead to stable imprints in the human KIR repertoire and involve activating KIRs. *Blood* (2013) 121:2678–88. doi:10.1182/blood-2012-10-459545
55. Lee J, Zhang T, Hwang I, Kim A, Nitschke L, Kim M, et al. Epigenetic modification and antibody-dependent expansion of memory-like NK cells in human cytomegalovirus-infected individuals. *Immunity* (2015) 42:431–42. doi:10.1016/j.immuni.2015.02.013
56. O'Sullivan TE, Johnson LR, Kang HH, Sun JC. BNIP3- and BNIP3L-mediated mitophagy promotes the generation of natural killer cell memory. *Immunity* (2015) 43:331–42. doi:10.1016/j.immuni.2015.07.012
57. Schlums H, Cichocki F, Tesi B, Theorell J, Beziat V, Holmes TD, et al. Cytomegalovirus infection drives adaptive epigenetic diversification of NK cells with altered signaling and effector function. *Immunity* (2015) 42:443–56. doi:10.1016/j.immuni.2015.02.008
58. Holmes TD, Bryceson YT. Natural killer cell memory in context. *Semin Immunol* (2016) 28(4):368–76. doi:10.1016/j.smim.2016.05.008
59. Costa-Garcia M, Vera A, Moraru M, Vilches C, Lopez-Botet M, Muntasell A. Antibody-mediated response of NKG2Cbright NK cells against human cytomegalovirus. *J Immunol* (2015) 194:2715–24. doi:10.4049/jimmunol.1402281
60. Lopez-Botet M, Muntasell A, Martinez-Rodriguez JE, Lopez-Montanes M, Costa-Garcia M, Pupuleku A. Development of the adaptive NK cell response to human cytomegalovirus in the context of aging. *Mech Ageing Dev* (2016) 158:23–6. doi:10.1016/j.mad.2016.06.010
61. Martinez-Rodriguez JE, Munne-Collado J, Rasal R, Cuadrado E, Roig L, Ois A, et al. Expansion of the NKG2C+ natural killer-cell subset is associated with high-risk carotid atherosclerotic plaques in seropositive patients for human cytomegalovirus. *Arterioscler Thromb Vasc Biol* (2013) 33:2653–9. doi:10.1161/ATVBAHA.113.302163
62. Mariani E, Monaco MC, Cattini L, Sinoppi M, Facchini A. Distribution and lytic activity of NK cell subsets in the elderly. *Mech Ageing Dev* (1994) 76:177–87. doi:10.1016/0047-6374(94)91592-X
63. Bigley AB, Spielmann G, Agha N, Simpson RJ. The effects of age and latent cytomegalovirus infection on NK-cell phenotype and exercise responsiveness in man. *Oxid Med Cell Longev* (2015) 2015:979645. doi:10.1155/2015/979645
64. Lankry D, Rovis TL, Jonjic S, Mandelboim O. The interaction between CD300a and phosphatidylserine inhibits tumor cell killing by NK cells. *Eur J Immunol* (2013) 43:2151–61. doi:10.1002/eji.201343433
65. Grauwet K, Vitale M, De PS, Jacob T, Laval K, Moretta L, et al. Pseudorabies virus US3 protein kinase protects infected cells from NK cell-mediated lysis via increased binding of the inhibitory NK cell receptor CD300a. *J Virol* (2016) 90:1522–33. doi:10.1128/JVI.02902-15
66. Dimitrova M, Zenarruzabeitia O, Borrego F, Simhadri VR. CD300c is uniquely expressed on CD56(bright) natural killer cells and differs from CD300a upon ligand recognition. *Sci Rep* (2016) 6:23942. doi:10.1038/srep23942
67. Ouyang Q, Baerlocher G, Vulto I, Lansdorp PM. Telomere length in human natural killer cell subsets. *Ann N Y Acad Sci* (2007) 1106:240–52. doi:10.1196/annals.1392.001
68. Monsivais-Urenda A, Noyola-Cherpitel D, Hernandez-Salinas A, Garcia-Sepulveda C, Romo N, Baranda L, et al. Influence of human cytomegalovirus infection on the NK cell receptor repertoire in children. *Eur J Immunol* (2010) 40:1418–27. doi:10.1002/eji.200939898

69. Beziat V, Dalgard O, Asselah T, Halfon P, Bedossa P, Boudifa A, et al. CMV drives clonal expansion of NKG2C(+) NK cells expressing self-specific KIRs in chronic hepatitis patients. *Eur J Immunol* (2012) 42:447–57. doi:10.1002/eji.201141826
70. Almehmadi M, Flanagan BF, Khan N, Alomar S, Christmas SE. Increased numbers and functional activity of CD56(+) T cells in healthy cytomegalovirus positive subjects. *Immunology* (2014) 142:258–68. doi:10.1111/imm.12250
71. Northfield JW, Kasprowicz V, Lucas M, Kersting N, Bengsch B, Kim A, et al. CD161 expression on hepatitis C virus-specific CD8+ T cells suggests a distinct pathway of T cell differentiation. *Hepatology* (2008) 47:396–406. doi:10.1002/hep.22040
72. Hassouneh F, Campos C, Lopez-Sejas N, Alonso C, Tarazona R, Solana R, et al. Effect of age and latent CMV infection on CD8+ CD56+ T cells (NKT-like) frequency and functionality. *Mech Ageing Dev* (2016) 158:38–45. doi:10.1016/j.mad.2015.12.003
73. Pera A, Campos C, Corona A, Sanchez-Correa B, Tarazona R, Larbi A, et al. CMV latent infection improves CD8+ T response to SEB due to expansion of polyfunctional CD57+ cells in young individuals. *PLoS One* (2014) 9:e88538. doi:10.1371/journal.pone.0088538
74. Daussy C, Faure F, Mayol K, Viel S, Gasteiger G, Charrier E, et al. T-bet and eomes instruct the development of two distinct natural killer cell lineages in the liver and in the bone marrow. *J Exp Med* (2014) 211:563–77. doi:10.1084/jem.20131560
75. Gotthardt D, Prchal-Murphy M, Seillet C, Glasner A, Mandelboim O, Carotta S, et al. NK cell development in bone marrow and liver: site matters. *Genes Immun* (2014) 15:584–7. doi:10.1038/gene.2014.55
76. Pradier A, Simonetta F, Waldvogel S, Bosshard C, Tiercy JM, Roosnek E. Modulation of T-bet and eomes during maturation of peripheral blood NK cells does not depend on licensing/educating KIR. *Front Immunol* (2016) 7:299. doi:10.3389/fimmu.2016.00299
77. Harmon C, Robinson MW, Fahey R, Whelan S, Houlihan DD, Geoghegan J, et al. Tissue-resident Eomes(hi) T-bet(lo) CD56(bright) NK cells with reduced proinflammatory potential are enriched in the adult human liver. *Eur J Immunol* (2016) 46:2111–20. doi:10.1002/eji.201646559
78. Stegmann KA, Robertson F, Hansi N, Gill U, Pallant C, Christophides T, et al. CXCR6 marks a novel subset of T-bet(lo)Eomes(hi) natural killer cells residing in human liver. *Sci Rep* (2016) 6:26157. doi:10.1038/srep26157
79. Lopez-Botet M, Muntasell A, Vilches C. The CD94/NKG2C+ NK-cell subset on the edge of innate and adaptive immunity to human cytomegalovirus infection. *Semin Immunol* (2014) 26:145–51. doi:10.1016/j.smim.2014.03.002
80. Simonetta F, Pradier A, Bosshard C, Masouridi-Levrat S, Chalandon Y, Roosnek E. NK cell functional impairment after allogeneic hematopoietic stem cell transplantation is associated with reduced levels of T-bet and eomesodermin. *J Immunol* (2015) 195:4712–20. doi:10.4049/jimmunol.1501522
81. Shehata HM, Hoebe K, Chougnet CA. The aged nonhematopoietic environment impairs natural killer cell maturation and function. *Aging Cell* (2015) 14:191–9. doi:10.1111/acel.12303
82. Dolfi DV, Mansfield KD, Polley AM, Doyle SA, Freeman GJ, Pircher H, et al. Increased T-bet is associated with senescence of influenza virus-specific CD8 T cells in aged humans. *J Leukoc Biol* (2013) 93:825–36. doi:10.1189/jlb.0912438

23

Antibody-Dependent NK Cell Activation is Associated with Late Kidney Allograft Dysfunction and the Complement-Independent Alloreactive Potential of Donor-Specific Antibodies

*Tristan Legris[1], Christophe Picard[2,3], Dilyana Todorova[4], Luc Lyonnet[5], Cathy Laporte[5], Chloé Dumoulin[5], Corinne Nicolino-Brunet[5], Laurent Daniel[6], Anderson Loundou[7], Sophie Morange[8], Stanislas Bataille[1], Henri Vacher-Coponat[1], Valérie Moal[1], Yvon Berland[1], Francoise Dignat-George[4,5], Stéphane Burtey[1,4] and Pascale Paul[4,5]**

[1] Nephrology Dialysis Renal Transplantation Center, Assistance Publique des Hôpitaux de Marseille, Hospital de la Conception, Marseille, France, [2] Établissement Français du Sang Alpes Méditerranée, Marseille, France, [3] ADES UMR 7268, CNRS, EFS, Aix-Marseille Université, Marseille, France, [4] UMR 1076, Vascular Research Center of Marseille, INSERM, Aix-Marseille University, Marseille, France, [5] Hematology Unit, Assistance Publique des Hôpitaux de Marseille, Hopital de la Conception, Marseille, France, [6] Laboratory for Anatomy, Pathology, Neuropathology, Hôpital de la Timone, Aix-Marseille University, Marseille, France, [7] Unité d'Aide méthodologique à la Recherche Clinique et Epidémiologique, DRRC, Assistance Publique Hôpitaux de Marseille, Marseille, France, [8] Centre d'Investigation Clinique, Hôpital de la Conception, Marseille, France

Edited by:
Emanuela Marcenaro,
University of Genoa, Italy

Reviewed by:
Rafael Solana,
University of Córdoba, Spain
Federico Simonetta,
Hôpitaux Universitaires de Genève, Switzerland

****Correspondence:***
Pascale Paul
pascale.paul@univ-amu.fr,
pascale.paul@ap-hm.fr

Although kidney transplantation remains the best treatment for end-stage renal failure, it is limited by chronic humoral aggression of the graft vasculature by donor-specific antibodies (DSAs). The complement-independent mechanisms that lead to the antibody-mediated rejection (ABMR) of kidney allografts remain poorly understood. Increasing lines of evidence have revealed the relevance of natural killer (NK) cells as innate immune effectors of antibody-dependent cellular cytotoxicity (ADCC), but few studies have investigated their alloreactive potential in the context of solid organ transplantation. Our study aimed to investigate the potential contribution of the antibody-dependent alloreactive function of NK cells to kidney graft dysfunction. We first conducted an observational study to investigate whether the cytotoxic function of NK cells is associated with chronic allograft dysfunction. The NK-Cellular Humoral Activation Test (NK-CHAT) was designed to evaluate the recipient and antibody-dependent reactivity of NK cells against allogeneic target cells. The release of CD107a/Lamp1^{+} cytotoxic granules, resulting from the recognition of rituximab-coated B cells by NK cells, was analyzed in 148 kidney transplant recipients (KTRs, mean graft duration: 6.2 years). Enhanced ADCC responsiveness was associated with reduced graft function and identified as an independent risk factor predicting a decline in the estimated glomerular filtration rate over a 1-year period (hazard ratio: 2.83). In a second approach, we used the NK-CHAT to reveal the cytotoxic potential of circulating alloantibodies *in vitro*. The level of CD16 engagement resulting from the *in vitro* recognition of serum-coated allogeneic B cells or splenic cells was further identified as a specific marker of DSA-induced ADCC. The NK-CHAT scoring of sera obtained from 40 patients at the time of transplant biopsy was associated with ABMR diagnosis. Our findings indicate that despite the administration of immunosuppressive treatments, robust ADCC responsiveness can be maintained in some KTRs. Because it evaluates both the Fab recognition of alloantigens and Fc-driven NK cell activation, the NK-CHAT represents a potentially valuable tool for the non-invasive and individualized evaluation of humoral risk during transplantation.

Keywords: natural killer cells, kidney transplantation, donor-specific antibodies, antibody-dependent cellular cytotoxicity, antibody-mediated rejection

INTRODUCTION

Kidney transplantation is the best treatment for patients with end-stage chronic kidney disease (CKD). However, long-term graft survival is limited by numerous factors, including inadequate control of the immune response to the allograft vasculature (1–3). Antibody-mediated rejection (ABMR) is a major cause of long-term transplant failure (2, 4, 5). *De novo* donor-specific antibodies (*dn*DSA) have been identified as major contributors to chronic ABMR and have been associated with graft microvascular injury (6–9) and arterial intimal fibrosis (10, 11). Despite advances in the development of the multiplex solid-phase single-antigen bead assay (SABA), which detects circulating donor-specific antibodies (DSAs) (12, 13) and their capacity to bind C1q or C3d (14–18), predicting the graft outcome in sensitized transplant recipients remains difficult (16, 19–23). A better understanding of the underlying mechanisms that contribute to ABMR is thus of key importance for improving therapeutic strategies. Recent findings indicate that, in addition to complement-dependent mechanisms, antibody-dependent cellular cytotoxicity (ADCC) involving γδ T cells or natural killer (NK) cells contributes to DSA-mediated graft injury (24–33). The NK cell molecular signatures from transplant biopsies of patients with ABMR (32, 34–37) suggest that NK cell activation is associated with humoral allograft vasculopathy (38, 39). Although NK cells are known cytotoxic effectors of the innate immune responses to antibodies, their potential pathogenic role in transplant rejection remains poorly documented. The antitumor efficiency of therapeutic monoclonal antibodies has been shown to be dependent on the expression of CD16-FcRγIIIA on NK cells (40, 41). FcγR polymorphisms have also been suggested as predisposing factors to infectious complications after liver transplantation (42). Similarly, we hypothesized that antibody-dependent cytotoxicity may be partly conditioned by the intrinsic capacity of the recipient NK cells to form conjugates with antibody-coated donor cells. In the transplant setting, the recipient NK cell alloreactivity may be affected by numerous factors, such as immunosuppressive drugs, infection, uremia, and inflammatory stresses (27, 43–48). Therefore, we aimed to investigate the potential link between NK cell cytotoxicity and allograft function in a cohort of late kidney transplant recipients (KTRs). An *in vitro* NK-Cellular Humoral Activation Test (NK-CHAT) was designed to address the following: (1) the potential link between NK cell activation and transplant dysfunction and (2) the potential toxicity of *dn*DSAs in promoting NK cell activation.

MATERIALS AND METHODS

Patients and Study Design

Study approval was obtained from the Agence Française de Sécurité Sanitaire (Afssaps Ref B805-1860) and from the Comité de Protection des Personnes SUD Méditerranée I. The study was supervised by the Institut National de la Santé et de la Recherche Médicale (INSERM, protocol granted in 2008 under Ref ID RCB 2008-A00604-51, C07-17).

Kidney transplant recipients were prospectively enrolled in the study from November 2008 to November 2011 at the Centre de Nephrologie et Transplantation Rénale, Hôpital de la Conception, Marseille. Patients considered for inclusion underwent transplantation between 2001 and 2005 (>3 years post graft) as part of a follow-up for a previous study (49). Patients with insufficient peripheral blood mononuclear cells (PBMCs) to perform the functional test or patients lacking follow-up data for kidney graft function after 1 year were not considered in the analysis ($n = 42$). The distribution of the variables for this group of 42 patients with missing data were not significantly different from that found for the cohort of 148 KTRs who were included in the final study cohort. The baseline characteristics of the 148 KTRs are summarized in **Table 1**. Renal function decline during follow-up was defined as either a 10% loss in the estimated glomerular filtration rate (eGFR) (mL/min/1.73 m^2, defined with the MDRD simplified equation, $n = 51$) or renal graft failure ($n = 9$). The control group was composed of 86 volunteer blood donors without renal failure.

Cells, Plasma, and Sera

Peripheral blood mononuclear cells for use as effector cells were obtained from healthy volunteer blood donors and KTR patients and were isolated using a Ficoll gradient. The cells used as targets in the NK-CHAT assay included the following: cells from the K562 cell line, cells from a B-EBV immortalized cell line [human leukocyte antigen (HLA)-genotype **Table 2**], NK cell-depleted PBMCs, and the residual donor spleen cells from the pretransplant cross-matching. The splenic or renal tissues were minced and homogenized by incubation with 50 μg/mL collagenase 1A (Sigma C9891) for 30 min at 37°C. Plasmapheresis samples (500 mL), containing HLA-A2 or HLA-DR4 DSA, were collected during the first round of ABMR desensitization therapy from two patients. KTR serum samples were obtained before the initiation of ABMR treatment and blindly analyzed for NK-CHAT activity. NK-CHAT standardization was achieved using a monoclonal

TABLE 1 | Characteristics of the 148 KTRs according to transplant function.

Variable	KTR with eGFR < 60	KTR with eGFR ≥ 60	*p* value KTR with eGFR < 60 versus KTR with eGFR ≥ 60	CTL
Recipient ***n*** **= 148**	92 (62%)	56 (38%)		86
Men (%)	55 (60%)	36 (64%)	ns	47 (55%)
Age at transplant (year, mean ± SD)	45 ± 13	32 ± 13	0.07	NA
Body mass index (kg/m^2) at transplant	23 ± 4**	23 ± 4*	ns	25 ± 5
Donor				
Men (%)	63 (67.7%)	41 (74.5%)	ns	N.A
Age (year, mean ± SD)	45 ± 13	32 ± 13	<0.0001	N.A
No of HLA mismatch (min–max)	3 (0–5)	3 (1–4)	ns	N.A
Antecedent/risk factors				
Time on dialysis, months	33 (18.5–55.5)	25 (14–52)	ns	N.A
Second or third transplantation	12 (13%)	6 (11%)	ns	N.A
Delayed graft function	24 (26%)	8 (14.5%)	0.091ᵗ	N.A
eGFR at 1 year posttransplant (M12)	45 ± 13	63 ± 11.5	<0.0001	N.A
Acute rejection	8 (8.7%)	1 (1.8%)	0.088	N.A
Characteristics at time of inclusion				
Time post graft (year, mean ± SD)	6.3 ± 1.8	6.03 ± 1.7	ns	N.A
Serum creatinine, µmol/L (mean ± SD)	161 ± 58***	89 ± 14***	<0.0001	73 ± 14
eGFR (mL/min/1.73 m^2, mean ± SD)	41.3 ± 11***	75.6 ± 13***	<0.0001	96 ± 18
Obesity (BMI > 30)	13 (14.1%)	4 (7.1%)	0.196	N.A
Current smokers	17 (18.5%)	4 (7.1%)	0.055	N.A
Diabetes	18 (19.6%)	8 (14.3%)	ns	N.A
Hypertension	91 (99%)	51 (91.1%)	0.019	N.A
Cardiovascular history	16 (17.4%)	17 (30.4%)	ns	N.A
DSA at time of inclusion	13 (14%)	2 (4%)	0.039	N.A
Immunosuppressive therapy				
Cyclosporine	42 (46%)	21 (37.5%)	ns	N.A
Mycophenolate mofetil	46 (49.5%)	32 (58.2%)	ns	N.A
Azathioprine	22 (24%)	16 (29%)	ns	N.A
Steroids	86 (92.5%)	51 (92.8%)	ns	N.A
NK cell number and cytotoxic function				
% NK cells (median, 25–75 pct)	9.1 (5.6–15)	10.1 (6.5–15)	ns	10.4 (9–14.5)
Number of NK cells/mm^3 (median, 25–75 pct)	118 (88–180)	149 (93–224)	0.103ᵗ	197 (142–264)
NK cell activation toward target (%CD107a/Lamp1)				
Natural cytotoxicity toward B cells	5.8 (4.3–9.1)	5.4 (3.7–8.6)	ns	6.9 (4.3–11)
ADCC: B cells + rituximab	29.4 (18.7–37)	22.6 (15–39)	0.008	31 (25–40)
eGFR loss ≥10%	33 (36%)	18 (32%)	ns	N.A

*The p values from the comparison of KTRs and healthy individuals (eGFR ≥ 60, CTL) were used to assess the significance of the differences (*p < 0.05, **p < 0.01, ***p < 0.0001); ᵗshows the differences with p values <0.2, and ns indicates the non-significant differences (p > 0.2). The progression of CKD was evaluated as an eGFR decrease of at least 10% from the time of inclusion in the study (time of NK cell evaluation) to 1 year later. According to the distribution of variables, the results are shown as numbers and percentages, means ± SD, or medians (25–75th percentile ranges).*

therapeutic IgG-recognizing CD20, rituximab (obtained from the residual samples that could not be used in the clinics or that were beyond the expiration date). The samples were obtained under institutional approval provided by the Pharmacy and Nephrology Department of Assistance Publique des Hôpitaux de Marseille.

The plasma samples obtained during the first round of plasmapheresis treatment from two ABMR patients were used to validate the performance of the NK-CHAT in detecting anti HLA-A2 and HLA-DR4 DSA reactivity. Serum and plasmapheresis samples (500 mL) were collected and aliquoted during the first round of ABMR desensitization therapy from two KTRs completing 10 plasmapheresis sessions, specifically before the initiation of other treatments using pulse methylprednisolone, intravenous immunoglobulin, and rituximab treatment to ensure that rituximab reactivity does not interfere in the experiments conducted with these plasma samples. The first patient (patient 1) was a 59-year-old man with HLA-A2 *dn*DSAs who experienced an acute C4d-positive ABMR rejection 8 years after his first renal transplantation. The second patient (patient 2) was a 29-year-old woman with HLA-DR4 *dn*DSAs who experienced an acute C4d-positive ABMR rejection episode with cellular borderline changes 8 years after her first transplantation. Antibodies from serum or plasma were purified using Protein A columns and Gentle Ag/A binding and elution buffers (Pierce 20356 and 21030). Before introduction into the NK-CHAT assay, the purified antibodies were dialyzed against PBS buffer (2 × 2 h and overnight) using a 10-kDa dialysis cassette.

TABLE 2 | Univariate and multivariate analysis of factors associated with the progression to CKD at 1 year post-enrollment into the study.

≥10% eGFR loss	Univariate analysis	Estimates: asymptotic Cox regression multivariate					
Variables in the model	***p***	***B***	**SE**	***p***	**HR**	**95% CI for Exp (*B*) (lower–upper)**	
CD107a/Lamp1URI >3	0.033	1.04	0.50	0.038	2.83	1.06	7.57
eGFR 1 year after transplantation	0.001	−0.06	0.02	0.001	0.94	0.91	0.98
eGFR at time of evaluation	0.386	0.03	0.01	0.036	1.03	1.00	1.05
Maintenance treatment: azathioprine	0.044	0.93	0.48	0.052	2.53	0.99	6.47
Smoking of time of enrollment	0.179^t	−0.05	0.44	0.273	0.62	0.26	1.46
Presence of DSA at time of enrollment	0.214	−0.55	0.65	0.401	0.58	0.16	2.07
Recipient age at transplant ≥55 years	0.157^t	−0.32	0.39	0.416	0.73	0.34	1.57
Time posttransplant at time of enrollment	0.333	0.05	0.11	0.640	1.05	0.85	1.30
Number of NK cells at time of enrollment	0.192^t	−0.001	0.002	0.688	1.00	1.00	1.00
Donor age at time of transplant	0.049	0.004	0.01	0.761	1.00	0.98	1.03

The univariate analysis defined variables that were significantly associated ($p < 0.05$) or tended to associate ($^t p < 0.2$) with deterioration of kidney graft function (≥10% eGFR loss 1 year post-enrollment). These variables were used to construct a multivariate Cox regression model to analyze the link between NK cell activation corresponding to the intermediate and high ADCC responses observed in the KTR cohort (CD107a/Lamp1URI ≥ 2.98) and risk of a further 10% decline in allograft function, evaluated in reference to the eGFR values obtained at the time of enrollment and after a median period of 13 months post-enrollment (min–max, 11–41 months post-inclusion). Although found to not be significantly associated with the degradation of kidney graft function using univariate analysis, variables such as the eGFR at the time of NK cell evaluation, the time posttransplant, and the presence of circulating dnDSAs at the time of inclusion were introduced into the multivariate model to control whether the association between NK activation and CKD progression was related to initial graft dysfunction and immunization status.

Identification of Donor-Specific Anti-HLA Alloantibodies

The detection of HLA-specific antibodies in serum samples was performed using standard techniques. The presence of allograft-specific antibodies was screened through CDC and Luminex screening assays (LAScreen® mixed, One Lambda, Canoga Park, CA, USA) using Luminex flow beads (LAScan™ 100, Luminex, Austin, TX, USA). To determine their antibody specificity, all samples with a positive screening result were further evaluated using "Single-Antigen" Gen-Probe Lifecodes reagents (Lifecodes LSA class I and Lifecodes LSA class II kits, Immucor, Norcross, GA, USA) according to the recommendations of the manufacturer and current guidelines (16, 50). The DSAs present in the samples were analyzed at the time of biopsy and were further characterized using single-antigen flow bead assays according to the manufacturer's recommended protocol (LAScreen® Single Antigen class I or LAScreen® Single Antigen class II, One Lambda, Canoga Park, CA, USA). Median fluorescence intensity (MFI) values were obtained using the baseline formula proposed by Fusion 3.2 software. The percentage of PRAs for the single-antigen assays were calculated according to the manufacturer's instructions as the percentage of positive bead reactions among the 99 class I beads and 97 class II beads. Cytotoxic cross-match assays were performed with donor PBMCs or splenic cells according to the protocol recommended by the Eurotransplant Organization using a standard microcytotoxicity assay. The CDC showed positive results when at least 50% of the cells were dead. Autoreaction was detected by incubating the patient's serum with autologous effector lymphocytes. IgM reactivity was excluded through prior treatment of the tested sera with dithiothreitol (Fluka BioChemika).

Flow Cytometry Analysis of Antibody Ligation

Antibody ligation to target cells was analyzed using anti-human Fc antibody. The target cells (10^5) were incubated with 10 µL of the FcR-blocking reagent (Miltenyi 130-059-901) and 90 µL for serum 0 min at 4°C, washed in PBS, incubated with 20 µL of sera or plasma and 80 µL of PBS for 30 min at 4°C, and then washed again in PBS. Binding was evaluated by incubating the serum- or plasma-coated cells with a secondary goat F(ab′)^2 anti-human-Fc antibody conjugated to PE for 20 min at 4°C (Beckman Coulter IM0550). Acquisition and analysis of the MFI of the gated allogeneic target cells were performed using a Beckman Coulter Navios Cytometer.

Absorption of HLA Class I Antibodies

Platelets were obtained from the blood bank service as units of platelet concentrate and were processed as follows. First, the platelets were centrifuged at 200 *g* for 40 min in 50-mL centrifuge tubes. The supernatant was removed, and the platelets were centrifuged again at 2,000 *g* for 15 min. After removal of the supernatant, 20 mL of 0.8% ammonium chloride was added to achieve red blood cell lysis, and the mixture was placed on a rotary mixer for 50 min. The platelets were washed twice with 1% Tris-buffered EDTA/saline and stored in a solution containing 0.1% sodium azide until their use for antibody absorption. Prior to absorption, the platelets were centrifuged at 2,000 *g* for 20 min, the supernatant was removed, and the platelets were washed twice with complement fixing buffer (Ovoid). A 50% volume of complement fixing buffer was added to packed platelets. Then, 1 mL of the above-described mixture was placed in a microcentrifuge tube and centrifuged at 10,000 *g* for 5 min, and the supernatant was removed. A volume of 0.25 mL of each sera sample was mixed, incubated at 22°C for 2 h, and centrifuged at 10,000 *g* for 5 min, and the absorption procedure was repeated with an overnight incubation at 22°C. Non-platelet- and platelet-absorbed sera were stored at 4°C until further use.

Phenotypic Analysis of Antibody-Dependent NK Cell Activation

The NK-CHAT was performed to analyze the antibody-dependent activation potential of NK effector cells resulting from their exposure to rituximab or DSA-coated target cells. Briefly, 500,000 target cells (B-EBV cell lines, NK cell-depleted PBMCs, or spleen cells) were incubated with control (CTL) unsensitized male human AB serum (CTL, Lonza) to block FcRs, rinsed, and incubated for 15 min in the presence of 20% KTR serum or CTL serum either supplemented or not supplemented with 10 μg/mL rituximab or purified IgG. The samples were then rinsed to remove any unbound antibodies. Effector cell PBMCs were incubated with antibody-coated targets for 3 h at 37°C using a 1:1 effector-to-target ratio in the presence of Golgi Stop (Becton Dickinson 554724) and CD107a-PC5 (Becton Dickinson 555802). In several experiments, serum was incubated in the presence of 200 μg/mL of Protein A to block antibody Fc fragment reactivity. The cells were then washed and labeled with CD3-ECD (Beckman Coulter A07748), CD16-PE (Beckman Coulter A07766), and CD56-PC7 (Beckman Coulter A21692) for 15 min at room temperature. Data acquisition and analysis were performed using a Beckman Coulter Navios cytometer. The NK lymphocyte subset within the PBMCs was gated through CD3/CD56-labeling ($CD3^-CD56^+$ population). The CD16 and Lamp1/CD107a expression patterns within the $CD3^-CD56^+$ NK subset were analyzed. ADCC was further analyzed by calculating the rituximab–CD107a/Lamp1 upregulation index (CD107a/Lamp1URI), which is expressed as the ratio between the percentage of CD107a/Lamp1 NK cell activation toward B cells in the presence (ADCC) or absence of rituximab (natural cytotoxicity). The level of CD16 engagement was quantified as the ratio between the MFIs of NK CD16 expression observed after effector PBMCs were incubated with B cell targets exposed to 20% human CTL DSA^- serum in the presence or absence of rituximab and was further defined as the CD16 downregulation index (CD16DRI).

When PBMCs or splenic cells were used as the target cells, the cell preparations were depleted of NK cells prior to being used as targets in the NK-CHAT. NK cell depletion was achieved through CD16 and CD56 magnetic bead selection (Miltenyi 130-045-701 and 130-050-401). Briefly, pelleted cells were resuspended in 60 μL of PBS containing 2 mM EDTA, 0.5% human AB serum (SAB), and 20 μL of each set of beads and then incubated on ice for 30 min. After a wash step, the fixed NK cells were depleted using adapted columns and magnets as recommended by the manufacturer (Miltenyi 130-042-201).

Transplant Histological Assessment

Forty transplant biopsies were retrospectively reassessed and scored using the updated conventional Banff diagnosis criteria (51) by one experimental renal pathologist (LD) blinded to the clinical information. C4d staining of paraffin section was performed through immunochemistry (51, 52). The biopsies were graded (from 0 to 3) according to the following histological parameters: glomerulitis (g), tubulitis (t), interstitial inflammation (i), intimal arteritis (v), peritubular capillaritis (ptc), chronic glomerulopathy (cg), interstitial fibrosis and tubular atrophy (ci and ct), and arterial fibrous intimal thickening (cv). The humoral parameters were integrated into a humoral histological score by adding the following variables: (g + ptc + v + cg + C4d) (36).

Statistical Analysis

The statistical analyses were performed using Graph Pad Prism 5 software (GraphPad Software, La Jolla, CA, USA) and IBM SPSS Statistics for Windows, Version 20.0 (IBM Corp., Armonk, NY, USA). Categorical variables are reported as counts or percentages. The associations between continuous variables were analyzed using Spearman's rank correlation analysis. The Chi-square test (or Fisher's exact test when appropriate) was used for the comparisons of categorical variables. Group comparisons were performed through one-way analysis of variance (ANOVA). A Mann–Whitney *t* test was performed for the comparison of non-parametric data from two groups. A *p* value <0.05 was considered to represent a statistically significant difference. In all figures, one asterisk (*) denotes $p < 0.05$, two asterisks (**) denote $p < 0.01$, and three asterisks denote $p < 0.001$. Clinical, histological, functional, and immunological factors associated with the degradation of kidney graft function were assessed in a separate univariate analysis. The variables identified as significantly associated with outcome variables ($p < 0.05$), that were marginally significant ($p < 0.20$, *t*) in the univariate analysis, or that are considered clinically relevant were selected for inclusion in the Cox regression model, which analyzed parameters associated with a loss in kidney graft function equal to at least 10% of the eGFR over the follow-up period of the 148 patients (median duration of follow-up: 13.3 months, 25–75th percentile: 12.1–15.1 months post-evaluation).

RESULTS

Analysis of Antibody-Dependent NK Cell Responses in Kidney Transplant Recipients

The NK-CHAT assay was designed to evaluate the contribution of the three components of ADCC responses, i.e., NK effector cells, antibodies, and target cells (**Figure 1**). In the first approach, peripheral NK cell cytotoxic activity was evaluated using a standardized combination of targets and antibodies, and variable sources of circulating effector cells isolated from 148 late KTRs were compared to those from 86 healthy controls (**Table 1**). Flow cytometry analysis of $CD107a/Lamp1^+$ NK cell granule exocytosis allowed quantification of natural and antibody-dependent cytotoxic activation against a standardized combination of HLA-negative (K562) target cells or HLA- and CD20-positive B cell target cells evaluated in the presence or absence of rituximab. As observed for the control cells, the NK cell cytotoxic function was highly variable in KTRs (**Figure 2A**). The variations in the levels of NK cell cytotoxic activation toward K562 ($p = 0.206$) or B cell target cells ($p = 0.141$) were not significantly different between the KTR and control groups. Interestingly, although the ADCC levels of KTRs with an eGFR < 60 mL/min/1.73 m^2 were comparable to those observed in the control group, the NK cell ADCC responses were significantly decreased in transplant recipients with preserved graft function (eGFR > 60 mL/min/1.73 m^2). To normalize the interindividual variability of natural NK cell cytotoxicity toward B cells, NK-ADCC was further indexed by calculating the rituximab–CD107a/Lamp1 upregulation index

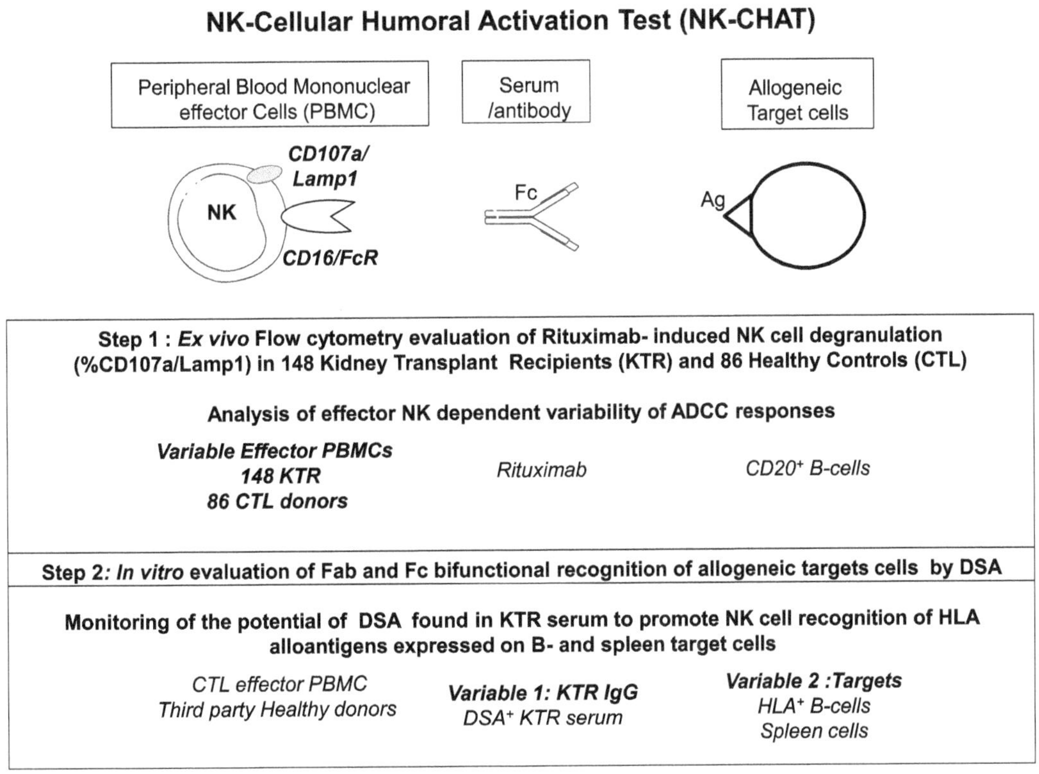

FIGURE 1 | Description of the observational and *in vitro* steps of the study of antibody-dependent NK cell activation using a standardized NK-Cellular Humoral Activation Test (NK-CHAT). Step 1. The standardized *ex vivo* NK-CHAT evaluation of the variability of KTR NK cell activation toward B cell targets coated with anti-CD20 monoclonal antibodies (rituximab). Step 2. *In vitro* scoring of serum DSA activity toward allogeneic target cells using third-party PBMCs isolated from healthy blood donors as immune effector cells.

(CD107a/Lamp1URI), a ratio of the percentage of CD107a/Lamp1$^+$ NK cell activation toward B cells in the presence to that observed in the absence of rituximab (**Figure 2B**). Although highly variable, the level of rituximab-induced CD107a/Lamp1 upregulation was maintained across KTRs (median CD107a/Lamp1URI: 4.1, 25–75th percentile: 2.98–5.79) and was not significantly different from that observed in a cohort of healthy control individuals (median CD107a/Lamp1URI: 4.5, 25–75th percentile: 3.16–6.25, **Figure 2B**). A multivariate Cox regression analysis of the factors associated with CKD progression further indicated that this intermediate level of the ADCC responsiveness observed in KTRs (CD107a/Lamp1URI > 3) was associated with the occurrence of graft function decline (≥10% eGFR loss) or graft failure in 34% of the transplanted recipients during the mean follow-up period of 13.3 months (**Figure 2C** and **Table 2**).

Quantification of DSA-Dependent CD16 Engagement and NK Cell Activation toward Allogeneic Cell Targets

Through the introduction of KTR serum as a variable in the NK-CHAT assay, we further investigated whether this test could assess the variability in DSA-mediated NK cell cytotoxic activity, which relies on the structural features of IgGs found in complexes in the KTR serum (**Figure 1**, step 2). Consistent with studies that utilized C1q binding as a tool to characterize the complement-dependent pathogenicity of DSA, we explored whether indexing the Fc–FcR interactions can reflect the cytotoxic potential of DSAs found in plasmapheresis samples collected during ABMR therapy. Specific binding of the DSAs found in plasma to B cells expressing either HLA-A2 or DR4 antigen was confirmed by flow cytometry cross-match (FCXM, **Figure 3A**). Plasma containing anti-HLA-A2 DSAs induced a significant and reproducible modulation of both CD107a/Lamp1 and CD16 expression on NK cells incubated with distinct HLA-A2$^+$ B cell targets (Figure S1A in Supplementary Material). A phenotypic analysis of NK cells identified CD16 expression as a surrogate marker of NK cell cytotoxic granule exocytosis and shows that it has the capacity to reflect the specific humoral component of NK cell activation (**Figure 3B**). A comparative NK-CHAT evaluation of DSA$^+$ plasma samples and rituximab reactivity was performed using different batches of effector PBMCs from distinct healthy donors ($n = 25$). The NK-CHAT scores resulting from the specific recognition of allogeneic B cells by DSAs were not significantly different from those observed with rituximab (**Figure 3B**). As observed with rituximab, the plasma-driven modulation of CD107a/Lamp1 and CD16 expression was specific, exhibiting HLA alloantigen recognition and effector-dependent variability (**Figure 3B**). The same results were observed using freshly isolated NK cell-depleted human

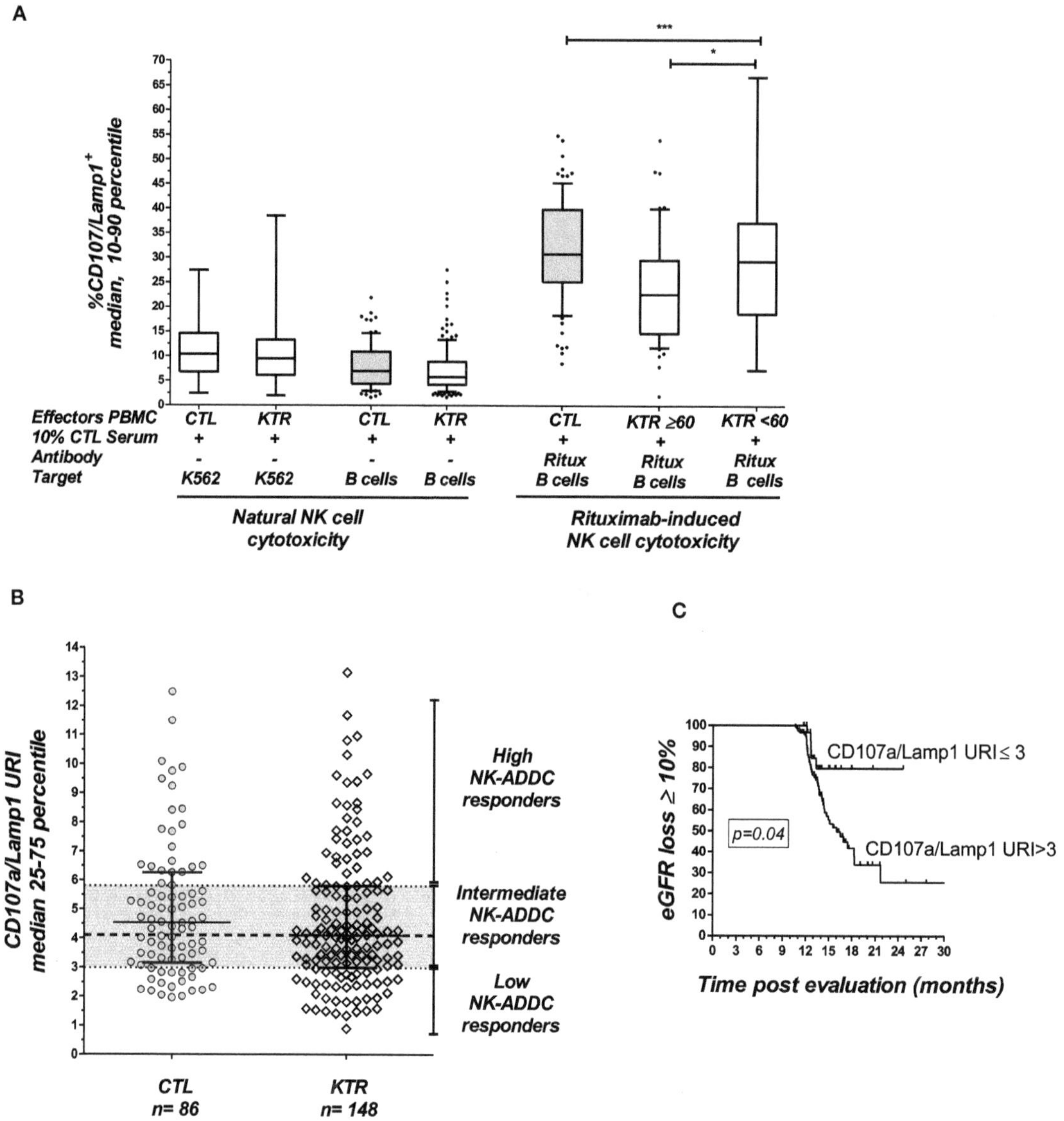

FIGURE 2 | Analysis of NK cell activation in late kidney transplant recipients. (A) The natural cytotoxicity of NK cells toward HLA-negative K562 cells or CD20+ B-lymphocyte target cells was analyzed by multicolor flow cytometry. The expression of CD107a/Lamp1 on the surface of CD3-CD56+ NK cells gated within the PBMCs was analyzed. NK cell ADCC toward the same B cell targets was evaluated in the presence of rituximab (Ritux). NK cell activation was evaluated in 148 KTRs and 86 healthy controls (CTL). The KTR patients were grouped according to graft function as follows: normal graft function (eGFR ≥ 60 mL/min/1.73 m^2, $n = 56$) and moderate-to-severe graft dysfunction (eGFR < 60 mL/min/1.73 m^2, $n = 92$). One-way ANOVA was performed to test the significance of NK-ADCC using PBMC effector cells obtained from CTLs and the KTR subgroups (Kruskal–Wallis, $p = 0.0002$). Asterisks indicate the p values as follows: $^*p \leq 0.05$, $^{**}p \leq 0.01$, and $^{***}p \leq 0.001$. The box plots show the median values, and the whiskers represent the 5–95 percentiles. **(B)** Specific analysis of rituximab-induced NK cell activation. To analyze the specific contribution of antibodies, NK cell activation toward rituximab-coated B cells was normalized in reference to the baseline cytotoxicity of the NK cells toward B cells evaluated in the absence of rituximab by calculating the CD107a/Lamp1 upregulation index (CD107a/Lamp1URI). The CD107a/Lamp1URI values observed in the KTR cohort allowed classification of the KTRs as low ADCC responders if their CD107a/Lamp1URI was below 3 (25th percentile) or as intermediate or high responders if their CD107a/Lamp1URI value was greater than 3. **(C)** Kaplan–Meir curves for an eGFR loss of at least 10% or graft loss ($n = 51$) observed during the follow-up period (months post-inclusion in the study) in KTRs with low NK cell activation (CD107a/Lamp1URI ≤ 3) and intermediate or high NK cell activation (CD107a/Lamp1URI > 3).

PBMCs as the allogeneic target cells expressing HLA alloantigens (Figure S1B in Supplementary Material). To normalize the variability in the baseline expression of CD16 in KTRs, a CD16 downregulation index (CD16DRI) was calculated as the ratio of the CD16 MFI measured in response to B cells coated with DSA-negative control serum (baseline) to the CD16 MFI results from B cells coated with KTR sera. In most cases in which the same sources of effector and target B cells were used in the NK-CHAT assay, the CD16DRI values measured in response to DSA-coated cells were correlated with the CD16DRI values measured in response to rituximab-coated cells (Spearman $r = 0.7, p < 0.0001$).

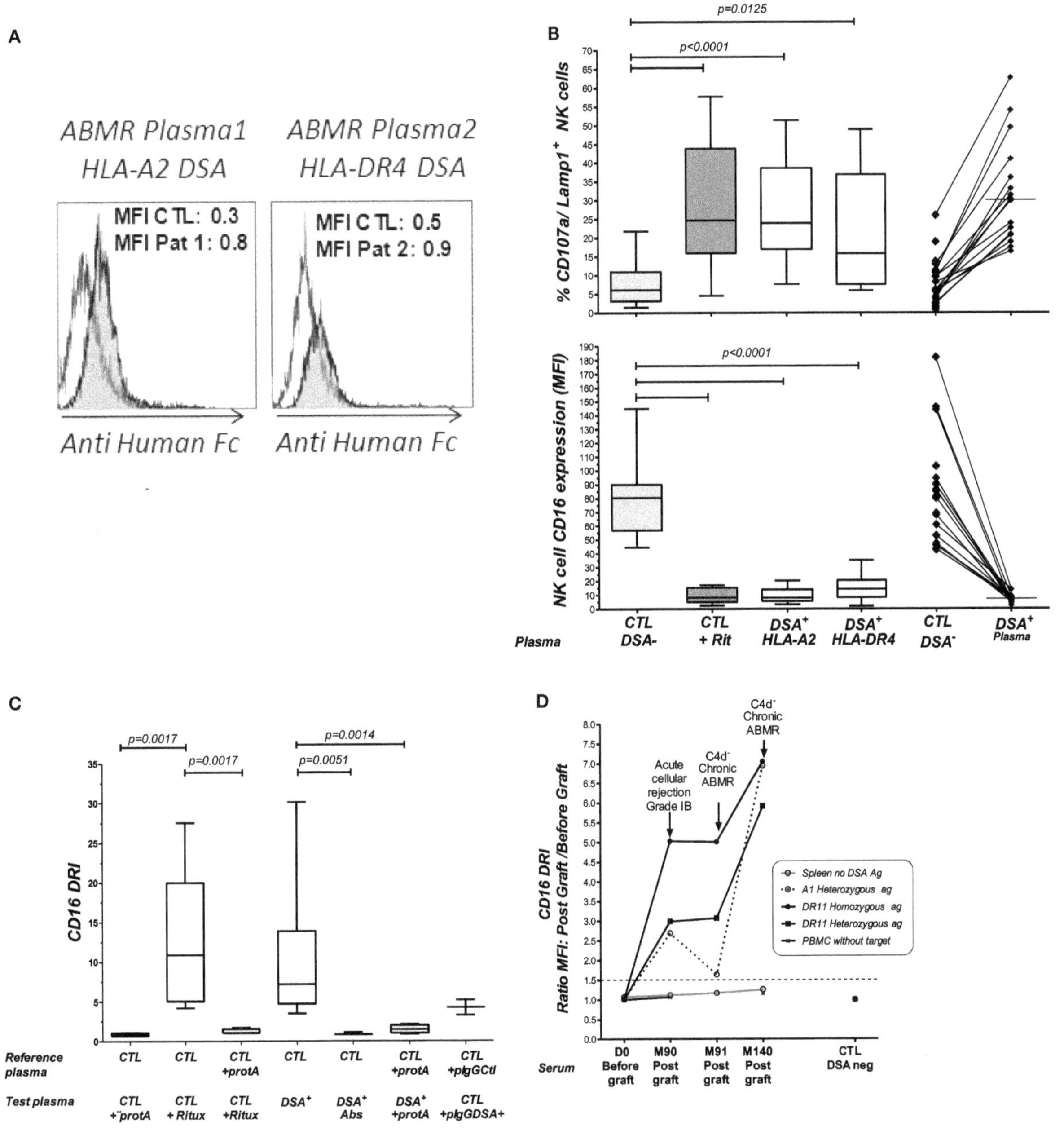

FIGURE 3 | Comparative analysis of rituximab and DSA-specific NK cell activation. **(A)** Plasmapheresis samples were obtained during plasma exchange therapy from two patients with circulating DSAs and a biopsy-proven ABMR. The binding of anti HLA-A2 (patient 1, Pat1) or anti HLA-DR4 DSA (patient 2, Pat2) to B cell targets was revealed by flow cytometry cross-match (FCMX) using a secondary antibody (gray shaded area). FCXM was analyzed in reference to binding and was observed when B cells were coated with the control plasma, showing no humoral reactivity (white area). MFI, median fluorescence intensity. **(B)** Comparative analysis of rituximab and DSA-induced NK cell activation toward target cells expressing HLA alloantigens. Flow cytometric analysis of antibody-mediated CD107a/Lamp1 (upper panel) and CD16 (MFI) (lower panel) expression was performed using NK cells prepared from distinct third-party healthy donors after incubation with HLA-A2 homozygous and DR4 heterozygous B cell targets and HLA-A2 DSA+ plasma (n = 17) or HLA-DR4 DSA+ plasma (n = 11). CTL, control plasma with no detectable anti HLA antibodies. **(C)** NK-CHAT analysis of DSA+ plasma. CD16DRI was evaluated using effector cells exposed to B cell targets in the presence of rituximab, DSA+ plasma, or DSA− CTL plasma. The specificity of antibody-driven CD16 engagement was analyzed by the addition of Protein A (53), by prior platelet absorption (Abs) of the anti-HLA antibodies in plasma, or through the use of IgG purified from ABMR plasma (pIgG DSA+) or control plasmas with no detectable HLA antibodies (pIgG CTL). **(D)** CD16DRI was used to evaluate the reactivity of serum samples obtained during the follow-up period of KTRs toward four different allogeneic splenic cells from the day of transplant (D0) to months (M) 90, 91, and 140 after transplant. The target splenic cells were selected according to the presence or absence of HLA antigens recognized by anti HLA-A1 or HLA-DR11 DSAs: (o) no relevant HLA antigen, (---) HLA-A1 heterozygote, and (-■-) HLA-DR11 homozygous (-◆-) or HLA-DR11 heterozygous antigens. An enhanced CD16DRI of NK cells that were exposed to splenic cells expressing DSA cognate HLA alloantigen was obtained using serum obtained at M90 from this KTR patient at the time of acute cellular rejection diagnosis. Persistent elevated CD16DRI levels were obtained using serum obtained at the times of two consecutive episodes of C4d− diagnosis of chronic ABMR, which occurred at M91 and M140 after transplant. The values of CD16 MFI obtained when effector PBMCs exposed to serum in the absence of splenic cells served as controls for baseline activity (-+-).

Antibody-mediated CD16 engagement was partially inhibited by Fc blockage using Protein A and by prior removal of DSAs from the serum. Specific CD16 engagement *via* the Fc fragment of DSAs was also observed after the exposure of target B cells to IgG purified from DSA⁺ plasma samples (**Figure 3C**).

The analysis of the use of NK cell-depleted splenic cells, a common source of donor cells used in cross-match assays, as cell targets in the NK-CHAT showed that CD16 engagement also reflected the anti-HLA-A2 DSA-specific recognition of donor cells (Figure S2 in Supplementary Material). We used different batches of splenic cells to monitor the NK-CHAT reactivity of the pre- and posttransplant serum samples obtained from a patient who developed HLA-A1 and DR11 *dn*DSAs associated with graft dysfunction and consecutive ABMR episodes (**Figure 3D**). The CD16DRI values evaluated against HLA-DR11 homozygous spleen cells were higher than those against DR11 heterozygous spleen target cells. Interestingly, in this patient, the increased CD16DRI value was associated with the progression of histological lesions consistent with C4d–negative ABMR.

NK-CHAT Evaluation of Serum Sampled at Time of Biopsy-Proven ABMR Diagnosis

To investigate the potential link between CD16DRI and DSA-mediated allograft injury, the NK-CHAT was performed using 46 sera samples obtained from 40 KTRs who had undergone concomitant transplant biopsies for clinical evaluation (**Table 3**). Among these KTRs, 18 patients were not sensitized, whereas the sera collected from 22 patients showed detectable levels of *dn*DSAs (HLA specificities listed in **Table 4**). The serum-driven modulation of the CD16 and CD107a/Lamp1 levels was associated with the detection of circulating DSAs and the histological diagnosis of ABMR (**Figure 4A**). The CD16DRI values were positively correlated with CD107/Lamp1URI (**Figure 4B**) and varied among DSA⁺ sera (**Figure 4C**). Although anti-DQ7 DSAs with similar MFI values sometimes exhibited variable levels of CD16DRI (**Table 4**, illustration in Figure S3 in Supplementary Material), the median CD16DRI values were found to correlate with the MFI intensity of DSAs detected by Luminex assays (Spearman $r = 0.46$, $p = 0.016$). For patients with DSAs and ABMR, histological scores (g + ptc + v + cg + C4d) above 3 were associated with a higher CD16DRI value (**Figure 4C**). In particular, the NK-CHAT was sufficiently sensitive to index the reactivity of an anti-Cw04 DSA in a patient with an ABMR-related microangiopathy (**Table 4**, patient 19). Monitoring of the pre- and posttransplant sera from 18 of these *de novo* sensitized patients revealed that NK activation was only detected posttransplant, specifically at the time of *dn*DSA detection (**Figure 4C**), thus allowing evaluation of the CD16DRI values in reference to the pretransplant sera.

To illustrate the ability of the NK-CHAT to monitor the evolution of DSA reactivity and the potential relevance of NK cell activation during humoral rejection, we evaluated a patient experiencing her first acute ABMR attack at month 44 with subsequent graft loss. The NK-CHAT analysis revealed a significant increase in the CD16DRI value associated with biopsy-proven ABMR (**Figure 5A**). Moreover, the CD16DRI value decreased after ABMR treatment (plasmapheresis and rituximab) but was markedly increased at the time of graft failure. A transplant nephrectomy was performed at month 76 due to graft intolerance syndrome (**Figure 5B**). A comparative phenotypic analysis of circulating and kidney-infiltrating NK cells showed that the NK cells constituted 24% of the intra-graft lymphocytes, whereas only 2.7% of the peripheral blood lymphocytes at the time of transplant nephrectomy were NK cells. When analyzed in reference to circulating NK cells, a sixfold decrease in the CD16 MFI was also observed in the intra-graft NK cells, suggesting that the *dn*DSA-mediated engagement of the CD16 receptor detected by the NK-CHAT may reflect the level of humoral *in situ* NK cell activation within the transplant site (**Figure 5C**).

DISCUSSION

Our study highlights the potential value of NK cell evaluation in monitoring the deleterious effects of alloantibodies in solid organ transplantation. Our observational study shows that the intensity of NK cell activation, evaluated through a standardized assay in response to a monoclonal therapeutic IgG, such as rituximab, may be associated with the progression of late graft dysfunction in KTRs. In addition, our results demonstrate the potential of the NK-CHAT to individualize the intrinsic capacity of an individual KTR to mount deleterious humoral immune cytotoxic responses against the graft, independently of other factors that are expected to impact NK cell and transplant function (step 1, **Figure 1**). The patients with the poorest transplant prognoses 1 year after analysis were predominantly those that exhibit at least a threefold increase in ADCC reactivity in response to rituximab. Our observations indicate that, independently of the detection of circulating DSAs, the individualized NK-CHAT scoring of peripheral NK cell activity may provide relevant information concerning complement-independent mechanisms that favor the development of chronic allograft injury. Because the evaluation of KTRs was only previously performed at one time point years after transplant, we cannot exclude the possibility that subclinical levels of DSAs, which may have developed prior to evaluation, had already promoted chronic NK cell activation *in vivo*, thus restraining the capacity of NK cells from sensitized patients to further respond to rituximab. A recent study demonstrated that pre-ligation of CD16 by anti-CD20 antibodies could impair antibody-mediated NK cell cytotoxicity against tumors (54), thus highlighting the exquisite plasticity of CD16-mediated NK cell responses. If translated into a simplified test, we expect that such individualized appraisal of the NK-ADCC activation potential may warrant closer humoral monitoring and recommend biopsy for a subset of immunized patients with high ADCC responsiveness, whereas less stringent monitoring of DSAs may be considered in patients with lower ADCC reactivity. This variability in KTRs regarding NK cell responses also suggests that the NK-CHAT may also be of value for anticipating or monitoring the efficiency of rituximab treatment in sensitized patients (55, 56).

Consistent with mechanistic evidence showing that NK cells play a critical role in mediating long-term transplant kidney injury (29), such intrinsic variability in the NK-ADCC responsiveness is also expected to condition the level of DSA toxicity *in vivo*

TABLE 3 | Characteristics and histological Banff scores of the 40 patients subjected to evaluation of the serum CD16DRI obtained at the time of histological diagnosis.

Parameters	Patients with DSA directed against B cell targets (*n* = 22)	Patients without DSA (*n* = 18)	*p* value
Recipient age at biopsy (years)	52.1 (15)	51.4 (15.4)	ns
Male	13 (59%)	13 (72%)	ns
Preemptive graft	1 (4%)	1 (6%)	ns
Risk factors for HLA sensitization			
Second transplantation	2 (9%)	0	ns
Blood transfusion before graft	9 (41%)	7 (39%)	ns
HLA A+ mismatch	2.3 (1)	2.5 (0.9)	ns
HLA DR mismatch	1.0 (0.6)	1.0 (0.7)	ns
Deceased donors	19 (86%)	18 (100%)	ns
Donor age (years)	37.2 (17)	46.1 (17.1)	0.11
Expanded criteria donors	4 (18%)	5 (28%)	ns
Delayed graft function	2 (9%)	3 (17%)	ns
Time since transplantation (Mo)	124 (72)	62 (69)	0.008
Indication for biopsy			
Deterioration of graft function	20 (91%)	14 (78%)	ns
Investigate proteinuria	2 (9%)	1 (5%)	ns
BK virus viremia	0	3 (17%)	0.08
Maintenance immunosuppressive regimen at biopsy			
Tacrolimus	10 (45%)	9 (50%)	ns
Cyclosporine	12 (54%)	9 (50%)	ns
MMF	10 (45%)	8 (44%)	ns
Azathioprine	3 (14%)	7 (39%)	0.06
Steroid	17 (77%)	18 (100%)	0.03
Graft function at biopsy			
Serum creatinine (μmol/L)	230 (80)	200 (66)	0.13
eGFR (mL/min/1.73 m^2)	28 (10)	35 (15)	0.15
Proteinuria (g/L)	1.0 (0.20–2)	0.45 (0.10–0.80)	0.04
Panel-reactive antibody at biopsy			
Class I	9% (19)	0%	0.004
Class II	16% (14)	0%	<0.0001
Histological findings			
Sclerotic glomeruli (%)	33%(23)	26%(24)	ns
g score (0–3)	0.9	0	0.0004
ptc score (0–3)	1.8	0.1	<0.0001
Microcirculation inflammation (g + ptc)	2.7	0.1	<0.0001
g + ptc > 0	19	2	<0.0001
V score (0–3)	0.2	0	ns
cg score (0–3)	1.1	0	0.0004
IF/TA (0–3)	1.7	1.5	ns
cv score (0–3)	1.6	1.1	0.01
Humoral histologic score (g + ptc + v + cg + C4d)	6 (4–8)	0	<0.0001
T-cell-mediated rejection	7 (32%)	3 (17%)	ns
Antibody-mediated rejection	20 (91%)	0	<0.0001
NK cell activation			
CD107a/Lamp1 URI median (25–75p)	2.1 (1.9–3.9)	1.1 (0.99–1.3)	<0.0001
CD16 DRI median (25–75p)	5 (3–17)	0.96 (0.8–1.16)	<0.0001

The sera samples were classified into two groups according to the presence or absence of DSAs directed against B cell targets. g, glomerulitis; ptc, peritubular capillaritis, IF/TA, interstitial fibrosis/tubular atrophy. The variables are shown as the means (SD), medians (25–75th percentile), or n (%). χ^2 tests were used for comparisons of the proportions, and unpaired t tests or Mann–Whitney tests were used for the comparisons of continuous variables.

(24–30). Consistently, we observed that NK cells infiltrating the graft of a KTR with ABMR exhibited enhanced *in situ* CD16 engagement compared with peripheral NK cells.

In the second part of our study, we obtained *in vitro* evidence that the NK-CHAT may also act as a powerful tool to discriminate the differential ability of DSAs to stimulate CD16-dependent NK cell activation (step 2, **Figure 1**). One unique feature of the NK-CHAT is its potential to combine the analysis of two parameters that control the strength of the target/effector cell interactions *via* DSAs through the following mechanisms: (i) Fab recognition

TABLE 4 | PRA and MFI of 25 DSA+ serum samples subjected to evaluation of the CD16DRI score at the time of biopsy.

KTR serum	PRA CLI	PRA CLII	Circulating DSA	Specific MFI toward HLA alloantigens target B-cell 2	A2	A2	B44	B56	Cw1	DR1	DR4	DR53	DQ5	DQ7	Median CD16DRI ($n = 2$–6)
Pat. 01 – S01	14	33	A3 A29 B56 DR11 **DQ7**	*23,000*	0	0	0	0	0	0	0	0	0	**23,000**	17.7
Pat. 02 – S02	15	20	B44 **DQ5**	*9,500*	0	0	0	0	0	0	0	**3,500**	**6,000**	0	2.0
Pat. 03 – S03		13	**DQ5**	*15,000*	0	0	0	0	0	0	0	0	**15,000**	0	3.5
Pat. 04 – S04		6	DR12 **DQ7**	*12,000*	0	0	0	0	0	0	0	0	0	**12,000**	4.2
Pat. 05 – S05	35		**DR53**	*12,000*	0	0	0	0	0	0	0	12,000	0	0	5.8
Pat. 06 – S06		21	**DQ7**	*13,000*	0	0	0	0	0	0	0	0	0	**13,000**	33.8
Pat. 07 – S07		9	**DQ7**	*2,000*	0	0	0	0	0	0	0	0	0	**2,000**	1.7
Pat. 08 – S08		21	**DQ7**	*15,000*	0	0	0	0	0	0	0	0	0	**15,000**	6.0
Pat. 09 – S09 Bio 1		18	**DQ7**	*9,000*	0	0	0	0	0	0	0	0	0	**9,000**	24.7
Pat. 09 – S09 Bio 2		22	**DQ7**	*15,000*	0	0	0	0	0	0	0	0	0	**15,000**	37.1
Pat. 10 – S10 Bio 1	37		**A2, B44**	*20,000*	8,000	8,000	4,000	0	0	0	0	0	0	0	3.6
Pat. 10 – S10 Bio 2	34		**A2, B44**	*5,000*	1,500	1,500	2,000	0	0	0	0	0	0	0	1.7
Pat. 10 – S10 Bio 3	90		**A2, B44**	[a]	[a]	[a]	[a]	0	0	0	0	0	0	0	5.2
Pat. 11 – S11	5	5	A1 **DR53**	*10,000*	0	0	0	0	0	0	0	**10,000**	0	0	12.7
Pat. 12 – S12		20	**DQ5**	*10,000*	0	0	0	0	0	0	0	0	**10,000**	0	4.2
Pat. 13 – S13	20	20	**A2 B44 DQ5**	*24,000*	**2,800**	**2,800**	**7,500**	0	0	0	0	0	**14,000**	0	5.1
Pat. 14 – S14		20	**DR53 DQ7**	*19,000*	0	0	0	0	0	0	0	**9,000**	0	**10,000**	27.6
Pat. 15 – S15	15	10	A2 DQ5	*32,000*	14,000	14,000	0	0	0	0	0	0	4,000	0	17.4
Pat. 16 – S16		23	DQ7	*14,000*	0	0	0	0	0	0	0	0	0	14,000	46.9
Pat. 17 – S17	33	5	B56 DR4	*12,500*	0	0	0	10,000	0	0	2,500	0	0	0	16.2
Pat. 18 – S18		1	**DR1**	*500*	0	0	0	0	0	**500**	0	0	0	0	2.9
				Specific MFI toward HLA alloantigens Target B-cell 2	**A3**	**A3**	**B7**	**B35**	**Cw4**	**DR10**	**DR15**	**DR51**	**DQ5**	**DQ6**	
Pat. 19 – S19	15		**Cw4**	9,500	0	0	0	0	**9,500**	0	0	0	0	0	4.8
Pat. 20 – S20	15	3	**A3** A26 C7 **DQ6**	16,000	**5,000**	**5,000**	0	0	0	0	0	0	0	**6,000**	9.6
Pat. 21 – S21		60	**DQ6**	15,000	0	0	0	0	0	0	0	0	0	**15,000**	1.5
Pat. 22 – S22		19	DR17 **DQ6**	3,000	0	0	0	0	0	0	0	0	0	**3,000**	2.3
Pat. 01 – S01	14	33	A3 A29 B56 DR11 **DQ7**	10,000	**5,000**	**5,000**	0	0	0	0	0	0	0	0	4.2
Pat. 02 – S02	15	20	B44 **DQ5**	10,000	0	0	4,000	0	0	0	0	0	**6,000**	0	1.5
Pat. 03 – S03		13	**DQ5**	15,000	0	0	0	0	0	0	0	0	**15,000**	0	4.9

A total of 25 DSA+ sera samples obtained from 22 KTR patients (Pat.) at the time of histological diagnosis were evaluated by CD16DRI scoring. Two distinct B cell targets (HLA genotype-indicated) were used to detect HLA alloantigen recognition. Sera obtained from patients 1–3 exhibited DSA+ reactivity toward both target B cells, and tests were evaluated on the two cell targets. The CD16DRI scores were evaluated through at least two independent experiments (n = 2–6), and the presented CD16DRI values are the medians of these tests.

HLA specificities and specific MFI of circulating donor specific antibodies (DSA) towards target alloantigens expressed on B cell lines 1 and 2 are indicated in bold characters.

[a]The circulating DSAs evaluated in Pat. 10 at the time of graft nephrectomy showed very high titers, which did not enable accurate MFI quantification. HLA class I (CLI) or class II (CLII) panel-reactive antigens (PRA) are indicated as percentages.

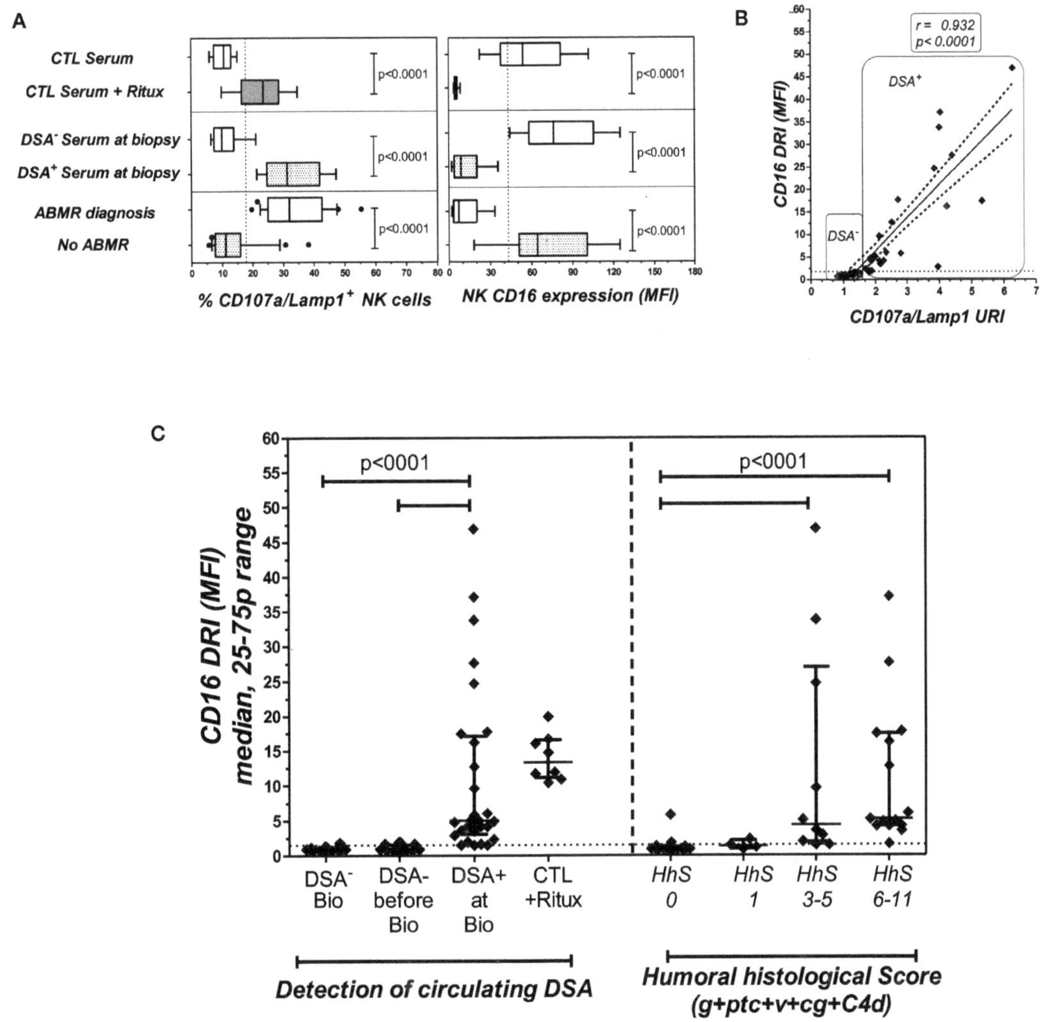

FIGURE 4 | The NK-CHAT evaluation of DSA reactivity toward B cell targets in sera collected at the time of transplant biopsy. (A) The modulation of CD16 (MFI) and CD107a/Lamp1 on NK effector cells exposed to serum-coated B cell targets was analyzed. Forty-three sera samples were collected from 40 KTRs at the time of biopsy and grouped according to the detection (DSA⁺, *n* = 25 sera obtained in 22 patients) or absence (DSA⁻, *n* = 18 patients) of B cell-specific circulating DSAs (the DSA specificities and MFI are detailed in **Table 4**) or to ABMR diagnosis. Rituximab was added to DSA⁻ serum and used as a positive control for the test. **(B)** Correlation of CD16DRI and CD107a/Lamp1URI scores for sera samples collected at the time of biopsy. **(C)** At the time of transplant biopsy (Bio), higher CD16DRI scores were associated with the presence of circulating DSAs (left panel) and humoral histological scores greater than 3 (g + ptc + v + cg + C4d of Banff's classification, right panel). The CD16RI scores of DSA⁻ serum from the same sensitized KTR, obtained prior to the detection of *dn*DSAs (before Bio), were also evaluated in reference to DSA⁺ serum sampled at the time of biopsy.

of *ex vivo* conformational antigens and (ii) Fc structural changes. Because a strong CD16DRI was shown to reflect the humoral component of NK cell cytotoxic activation in response to DSA, we expect that the NK-CHAT evaluation could be limited to the measurement of serum-induced CD16 expression.

The scoring of DSA-mediated NK cell activation was not restricted to B cell targets expressing high levels of HLA antigens but was also performed toward peripheral PBMCs and spleen cell targets that express lower physiological levels of HLA antigens. The NK-CHAT also detected functional immune activation resulting from less commonly evaluated alloantibodies, such as the anti-HLA-Cw and DQ7 antibodies. Our observations suggest that the sensitivity of the NK-CHAT to evaluate the ability of DSAs to bind to alloantigens could be comparable or even superior to that of current methods used to characterize DSAs.

A major feature of the NK-CHAT is its potential to detect Fc-dependent variations in alloantibodies and index their potential to trigger CD16-mediated NK cell activation in an individual KTR. Interestingly, DSAs with a comparable MFI against DQ7 alloantigens were, in some cases, associated with different ranges of CD16DRI values. Thus, our test may also reflect structural features of DSAs, which cannot be fully revealed by a SABA. The efficiency of FcR-mediated NK cell activation has been reported to be dependent on the IgG1/IgG3 subclasses (57) and on the glycosylation/sialylation status of the Fc fragments within these

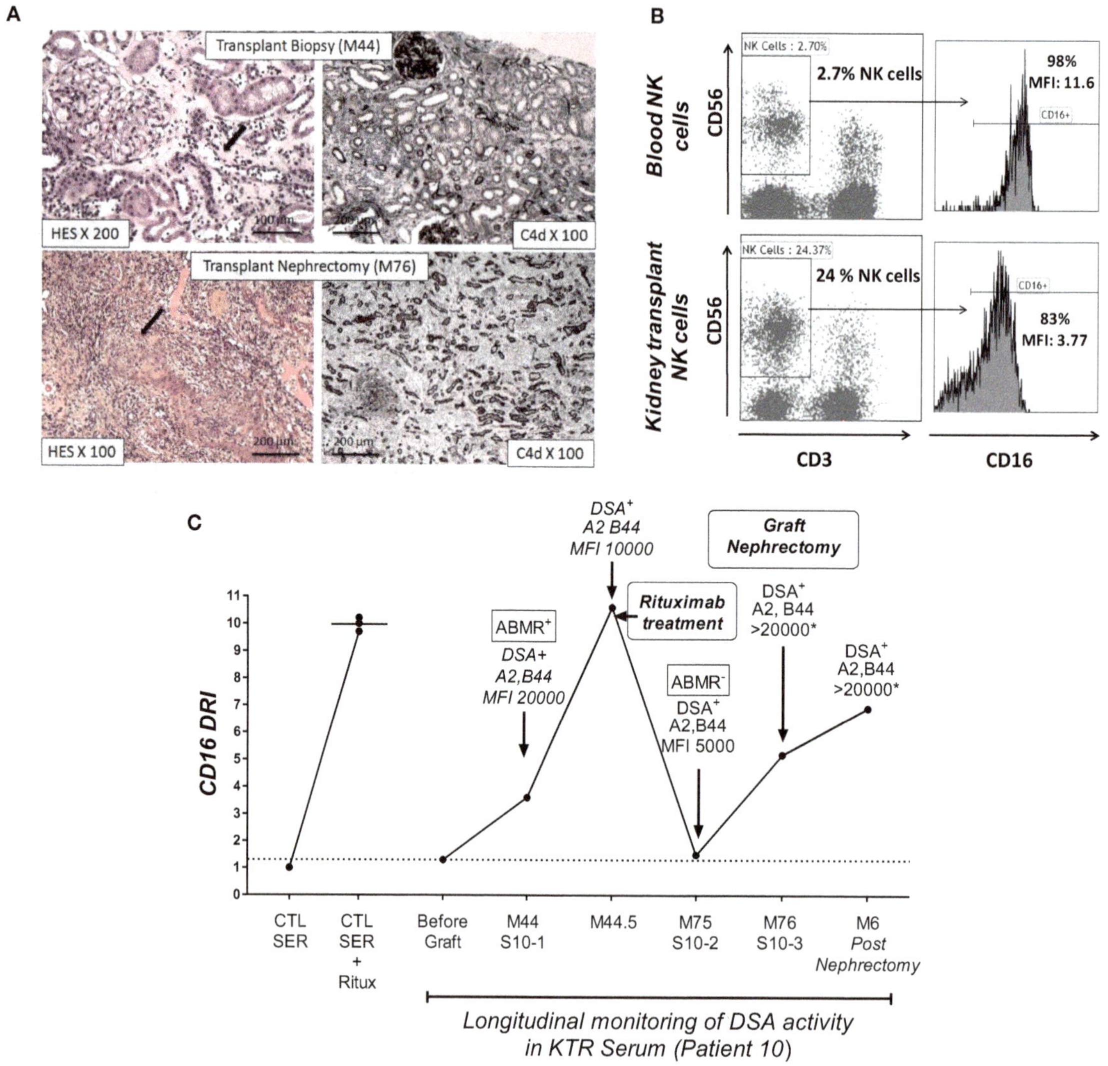

FIGURE 5 | The NK-CHAT monitoring of DSA reactivity and characterization of CD16 engagement of NK cells infiltrating a kidney transplant site. **(A)** Histological assessment of lesions from biopsies of patient 10 at M44 and M76 (transplant nephrectomy). The M44 histological analysis showed peritubular capillaritis (cpt = 2, black arrow). At M76, histology revealed arteritis (v = 2, black arrow) and confirmed severe acute inflammatory lesions (t = 3, i = 3, g = 3, ptc = 3, v = 2, C4d = 3) with diffuse interstitial edema and chronic vascular lesions (cv = 2) as well as tubular atrophy (ct = 2). The left and right panels show the hematoxylin–eosin–saffron (HES) and C4d immunochemical staining of paraffin sections, respectively. **(B)** The NK-CHAT activity of five sera samples obtained during the longitudinal monitoring of patient 10 was analyzed in reference to serum obtained before transplantation (no sensitization). A significant increase in CD16DRI was concomitant with the detection of HLA-A2 and anti-B44 *dn*DSA at M44 as well as histological lesions of ABMR. The CD16DRI level decreased after rituximab treatment. The detection of low levels of circulating DSAs at M75 (before graft nephrectomy) was associated with a lowered CD16DRI and no evidence of humoral activity within the biopsy. An increased CD16DRI was observed at the time of graft nephrectomy and persisted for 6 months. At these two time points (M76 and 6 months after graft nephrectomy), the circulating DSA were at saturating titers, which did not allow accurate MFI quantification. **(C)** Comparative flow cytometric analysis of CD16 expression in blood $CD3^{-}CD56^{+}$ NK cells and infiltrating NK cells isolated from the transplant after nephrectomy.

antibodies (58). Altered fucosylation patterns in the Fc tails of DSAs have been shown to alter Fc binding to FcγRIIIA and C1q (59) and may represent a critical regulatory determinant of both ADCC- and complement-dependent DSA toxicity (60). The CD16 expression levels and polymorphisms have also been shown to influence NK cell activation (40, 57). Although the KTR CD16 genotype was not included in the initial study design, we expect that depending on the FcγRIIIA genetic background of the recipient, CD16-mediated NK-CHAT responses may be differentially triggered by IgG based on the presence of a valine or phenylalanine residue at position 158 of the FcγRIIIA receptor.

Although the NK-CHAT was mostly associated with the results of the C4d histological staining of the biopsy, we demonstrated that the NK-CD16DRI-based monitoring of DSA activity toward splenic cells was associated with the progression of ABMR lesions that occurred in the absence or prior to the detection of C4d histological staining. A longitudinal follow-up of serum reactivity against allogeneic targets revealed that the NK-CHAT scores were lowered after ABMR desensitization therapy and were associated with the progression of biopsy-proven graft lesions.

Altogether, the results of our study provide evidence that independently of the DSA MFI intensity, the NK-CHAT reactivity may serve as a hallmark of clinically relevant features that discriminate the ability of DSAs to target allograft injury.

Our work has several limitations. First, we only analyzed the degradation of eGFR in 148 KTRs, and the ability of the NK-CHAT to predict kidney dysfunction needs to be demonstrated through CD16 monitoring in a larger cohort of KTRs and over a longer follow-up period. We also need to challenge our test with more stringent end-points, such as graft failure and patient mortality. Considering the major role of complement in the pathogenesis of ABMR, the absence of DSA C1q and C3d-binding tests and the use of complement-depleted serum constitute additional limitations. Moreover, the NK-CHAT scores were obtained using sera that were collected at the time of the clinically indicated transplant biopsy, and the value of the test requires control testing of sera that develop *dn*DSAs in the absence of allograft rejection or exhibit signs of vascular lesions in the absence of detectable levels of DSAs. Due to a low number of cases, the establishment of a firm link between the intensity of NK cell activation and the severity of ABMR lesions or clinical outcome was not possible at this stage. Therefore, it may be relevant to investigate whether the NK-CHAT only serves as a powerful detector of DSAs or if it could predict subclinical ABMR in a cohort of KTRs developing *dn*DSAs as well as anticipate the evolution and severity of ABMR histological lesions.

Despite these limitations, we provide novel evidence of the role of NK cells in renal allograft dysfunction and tools to monitor the interindividual variability of humoral alloimmune responses in immunized patients. A full use of the NK-CHAT using a combination of recipient NK cells and serum combined with donor target cells should improve the ability to identify immunized patients who are at higher risk of subsequent graft dysfunction. Our study highlights the importance of designing and translating novel assays that combine the simultaneous evaluation of the recipient IgG and FcR to individualize the evaluation of the humoral risk of a given recipient. In particular, we developed a simple phenotypic assay that integrates both environmental and genetic parameters that condition antibody–antigen interactions combined with FcR engagement (43, 45). Because this non-invasive assay of CD16 engagement was shown to be reproducible, inexpensive, and easy to implement for routine monitoring, prospective studies are warranted to assess the clinical relevance of the NK-CHAT using recipient NK cells and serum reactivity against target cells of donor origin.

ETHICS STATEMENT

Study approval was obtained from the Agence Française de Sécurité Sanitaire (Afssaps Ref B805-1860) and from the Comité de Protection des Personnes SUD Méditerranée I. The study was supervised by the Institut National de la Santé et de la Recherche Médicale (INSERM, protocol granted in 2008 under Ref ID RCB 2008-A00604-51, C07-17).

AUTHOR CONTRIBUTIONS

TL, SBa, SBu, HVC, VM, and YB were responsible for patient care, the selection of patients recruited in this study, the generation of clinical data, and the writing of the clinical sections of the manuscript. LL, DT, CL, and CD performed the flow cytometric analysis and the *in vitro* serum alloreactivity assay using NK-CHAT. CP was responsible for anti-HLA antibody characterization and the collection of HLA-typing and cross-match data and helped with the data interpretation and writing of the manuscript. AL contributed to the methodological design of the study and performed the statistical analyses. CB generated the data of the lymphocyte cell subset (% and cell counts) in KTR and healthy donors. LD was responsible for the anatomopathological characterization of biopsies and ABMR diagnosis. SM was responsible for the quality control of the regulatory aspects of the study, the coordination of ethical committee agreement, and the selection of healthy donors. FDG was responsible of the data generated in the hematology unit and the revision of the manuscript. PP was responsible for the design and coordination of the study, the quality control, analysis and interpretation of the data, and the writing of the manuscript.

ACKNOWLEDGMENTS

We thank the technicians from the Laboratory of Hematology (Hopital de la Conception, Marseille) and Etablissement Français du Sang Alpes Méditerranée (EFS-AM) for their expert assistance and the nurses and clinicians of the "Centre Investigation Clinique" and Department of Nephrology for the patient care and protocol follow-up. We thank Charles Eric Bornet, Pharmacist at Hopital de La Conception, for providing monoclonal therapeutic IgG-recognizing CD20 for research purposes when residual batches of rituximab were beyond the expiration date or not reusable in a clinical setting. We are grateful to Aurélie Larosa, Laurent Dalzon, and Matthieu Pelardy, technicians and nurses at the CIC, for their assistance in the protocol and contribution to data collection, to Louise Paul Delvaux for her contribution to the data analysis, and to Prof. Florence Sabatier for her help in proofreading the manuscript.

REFERENCES

1. von Rossum A, Laher I, Choy JC. Immune-mediated vascular injury and dysfunction in transplant arteriosclerosis. *Front Immunol* (2014) 5:684. doi:10.3389/fimmu.2014.00684
2. Sellares J, de Freitas DG, Mengel M, Reeve J, Einecke G, Sis B, et al. Understanding the causes of kidney transplant failure: the dominant role of antibody-mediated rejection and nonadherence. *Am J Transplant* (2012) 12:388–99. doi:10.1111/j.1600-6143.2011.03840.x
3. Walsh L, Dinavahi R. Current unmet needs in renal transplantation: a review of challenges and therapeutics. *Front Biosci (Elite Ed)* (2016) 8:1–14. doi:10.2741/746
4. Einecke G, Sis B, Reeve J, Mengel M, Campbell PM, Hidalgo LG, et al. Antibody-mediated microcirculation injury is the major cause of late kidney transplant failure. *Am J Transplant* (2009) 9:2520–31. doi:10.1111/j.1600-6143.2009.02799.x
5. Halloran PF, Reeve JP, Pereira AB, Hidalgo LG, Famulski KS. Antibody-mediated rejection, T cell-mediated rejection, and the injury-repair response: new insights from the Genome Canada studies of kidney transplant biopsies. *Kidney Int* (2014) 85:258–64. doi:10.1038/ki.2013.300
6. Nguan C. Antibody-mediated vascular rejection of kidney allografts: a population-based study. *Urology* (2013) 82:503–4. doi:10.1016/j.urology.2013.04.003
7. Nankivell BJ. Antibody-mediated vascular rejection: relation to causation. *Lancet* (2013) 381:275–7. doi:10.1016/S0140-6736(12)61577-3
8. Lefaucheur C, Loupy A, Vernerey D, Duong-Van-Huyen JP, Suberbielle C, Anglicheau D, et al. Antibody-mediated vascular rejection of kidney allografts: a population-based study. *Lancet* (2013) 381:313–9. doi:10.1016/S0140-6736(12)61265-3
9. Hickey MJ, Valenzuela NM, Reed EF. Alloantibody generation and effector function following sensitization to human leukocyte antigen. *Front Immunol* (2016) 7:30. doi:10.3389/fimmu.2016.00030
10. Torres IB, Salcedo M, Moreso F, Sellares J, Castella E, Azancot MA, et al. Comparing transplant glomerulopathy in the absence of C4d deposition and donor-specific antibodies to chronic antibody-mediated rejection. *Clin Transplant* (2014) 28:1148–54. doi:10.1111/ctr.12433
11. Hanf W, Bonder CS, Coates PT. Transplant glomerulopathy: the interaction of HLA antibodies and endothelium. *J Immunol Res* (2014) 2014:549315. doi:10.1155/2014/549315
12. Loupy A, Hill GS, Suberbielle C, Charron D, Anglicheau D, Zuber J, et al. Significance of C4d Banff scores in early protocol biopsies of kidney transplant recipients with preformed donor-specific antibodies (DSA). *Am J Transplant* (2011) 11:56–65. doi:10.1111/j.1600-6143.2010.03364.x
13. Bettinotti MP, Zachary AA, Leffell MS. Clinically relevant interpretation of solid phase assays for HLA antibody. *Curr Opin Organ Transplant* (2016) 21(4):453–8. doi:10.1097/MOT.0000000000000326
14. Sicard A, Ducreux S, Rabeyrin M, Couzi L, McGregor B, Badet L, et al. Detection of C3d-binding donor-specific anti-HLA antibodies at diagnosis of humoral rejection predicts renal graft loss. *J Am Soc Nephrol* (2015) 26(2):457–67. doi:10.1681/ASN.2013101144
15. Loupy A, Lefaucheur C, Vernerey D, Prugger C, Duong van Huyen JP, Mooney N, et al. Complement-binding anti-HLA antibodies and kidney-allograft survival. *N Engl J Med* (2013) 369:1215–26. doi:10.1056/NEJMoa1302506
16. Tait BD, Susal C, Gebel HM, Nickerson PW, Zachary AA, Claas FH, et al. Consensus guidelines on the testing and clinical management issues associated with HLA and non-HLA antibodies in transplantation. *Transplantation* (2013) 95:19–47. doi:10.1097/TP.0b013e31827a19cc
17. Mujtaba MA, Goggins W, Lobashevsky A, Sharfuddin AA, Yaqub MS, Mishler DP, et al. The strength of donor-specific antibody is a more reliable predictor of antibody-mediated rejection than flow cytometry crossmatch analysis in desensitized kidney recipients. *Clin Transplant* (2011) 25:E96–102. doi:10.1111/j.1399-0012.2010.01341.x
18. Gombos P, Opelz G, Scherer S, Morath C, Zeier M, Schemmer P, et al. Superiority of AbCross enzyme-linked immunosorbent assay cross-match over the B-cell complement-dependent lymphocytotoxicity cross-match. *Transplant Proc* (2013) 45:1383–5. doi:10.1016/j.transproceed.2013.01.039
19. Guidicelli G, Guerville F, Lepreux S, Wiebe C, Thaunat O, Dubois V, et al. Non-complement-binding de novo donor-specific anti-HLA antibodies and kidney allograft survival. *J Am Soc Nephrol* (2016) 27(2):615–25. doi:10.1681/ASN.2014040326
20. Mohan S, Palanisamy A, Tsapepas D, Tanriover B, Crew RJ, Dube G, et al. Donor-specific antibodies adversely affect kidney allograft outcomes. *J Am Soc Nephrol* (2012) 23:2061–71. doi:10.1681/ASN.2012070664
21. Bartel G, Regele H, Wahrmann M, Huttary N, Exner M, Horl WH, et al. Posttransplant HLA alloreactivity in stable kidney transplant recipients-incidences and impact on long-term allograft outcomes. *Am J Transplant* (2008) 8:2652–60. doi:10.1111/j.1600-6143.2008.02428.x
22. Hachem RR, Reinsmoen NL. What is the definition of a clinically relevant donor HLA-specific antibody (DSA)? *Am J Transplant* (2015) 15:299–300. doi:10.1111/ajt.13079
23. Frank R, Lal P, Kearns J, Molina MR, Wald JW, Goldberg LR, et al. Correlation of circulating complement-fixing donor-specific antibodies identified by the C1q assay and presence of C4d in endomyocardial biopsy specimens. *Am J Clin Pathol* (2016) 145:62–8. doi:10.1093/ajcp/aqv016
24. Strom TB, Tilney NL, Carpenter CB, Busch GJ. Identity and cytotoxic capacity of cells infiltrating renal allografts. *N Engl J Med* (1975) 292:1257–63. doi:10.1056/NEJM197506122922402
25. Fishman JA. What's new and what's hot? Basic science at the American Transplant Congress 2012. *Am J Transplant* (2013) 13:275–80. doi:10.1111/ajt.12015
26. Bachelet T, Couzi L, Pitard V, Sicard X, Rigothier C, Lepreux S, et al. Cytomegalovirus-responsive gammadelta T cells: novel effector cells in antibody-mediated kidney allograft microcirculation lesions. *J Am Soc Nephrol* (2014) 25(11):2471–82. doi:10.1681/ASN.2013101052
27. Crespo M, Yelamos J, Redondo D, Muntasell A, Perez-Saez MJ, Lopez-Montanes M, et al. Circulating NK-Cell subsets in renal allograft recipients with anti-HLA donor-specific antibodies. *Am J Transplant* (2015) 15:806–14. doi:10.1111/ajt.13010
28. Resch T, Fabritius C, Ebner S, Ritschl P, Kotsch K. The role of natural killer cells in humoral rejection. *Transplantation* (2015) 99:1335–40. doi:10.1097/TP.0000000000000757
29. Zhang ZX, Huang X, Jiang J, Lau A, Yin Z, Liu W, et al. Natural killer cells mediate long-term kidney allograft injury. *Transplantation* (2015) 99:916–24. doi:10.1097/TP.0000000000000665
30. Maier S, Tertilt C, Chambron N, Gerauer K, Huser N, Heidecke CD, et al. Inhibition of natural killer cells results in acceptance of cardiac allografts in CD28-/- mice. *Nat Med* (2001) 7:557–62. doi:10.1038/87880
31. Farkash EA, Colvin RB. Diagnostic challenges in chronic antibody-mediated rejection. *Nat Rev Nephrol* (2012) 8:255–7. doi:10.1038/nrneph.2012.61
32. Hirohashi T, Chase CM, Della Pelle P, Sebastian D, Alessandrini A, Madsen JC, et al. A novel pathway of chronic allograft rejection mediated by NK cells and alloantibody. *Am J Transplant* (2012) 12:313–21. doi:10.1111/j.1600-6143.2011.03836.x
33. Kohei N, Tanaka T, Tanabe K, Masumori N, Dvorina N, Valujskikh A, et al. Natural killer cells play a critical role in mediating inflammation and graft failure during antibody-mediated rejection of kidney allografts. *Kidney Int* (2016) 89:1293–306. doi:10.1016/j.kint.2016.02.030
34. Hidalgo LG, Sis B, Sellares J, Campbell PM, Mengel M, Einecke G, et al. NK cell transcripts and NK cells in kidney biopsies from patients with donor-specific antibodies: evidence for NK cell involvement in antibody-mediated rejection. *Am J Transplant* (2010) 10:1812–22. doi:10.1111/j.1600-6143.2010.03201.x
35. Hidalgo LG, Sellares J, Sis B, Mengel M, Chang J, Halloran PF. Interpreting NK cell transcripts versus T cell transcripts in renal transplant biopsies. *Am J Transplant* (2012) 12:1180–91. doi:10.1111/j.1600-6143.2011.03970.x
36. Loupy A, Lefaucheur C, Vernerey D, Chang J, Hidalgo LG, Beuscart T, et al. Molecular microscope strategy to improve risk stratification in early antibody-mediated kidney allograft rejection. *J Am Soc Nephrol* (2014) 25:2267–77. doi:10.1681/ASN.2013111149
37. Gill RG. NK cells: elusive participants in transplantation immunity and tolerance. *Curr Opin Immunol* (2010) 22:649–54. doi:10.1016/j.coi.2010.09.005
38. Venner JM, Hidalgo LG, Famulski KS, Chang J, Halloran PF. The molecular landscape of antibody-mediated kidney transplant rejection: evidence for NK involvement through CD16a Fc receptors. *Am J Transplant* (2015) 15:1336–48. doi:10.1111/ajt.13115

39. Couzi L, Pitard V, Sicard X, Garrigue I, Hawchar O, Merville P, et al. Antibody-dependent anti-cytomegalovirus activity of human gammadelta T cells expressing CD16 (FcgammaRIIIa). *Blood* (2012) 119:1418–27. doi:10.1182/blood-2011-06-363655
40. Veeramani S, Wang SY, Dahle C, Blackwell S, Jacobus L, Knutson T, et al. Rituximab infusion induces NK activation in lymphoma patients with the high-affinity CD16 polymorphism. *Blood* (2011) 118:3347–9. doi:10.1182/blood-2011-05-351411
41. Hatjiharissi E, Xu L, Santos DD, Hunter ZR, Ciccarelli BT, Verselis S, et al. Increased natural killer cell expression of CD16, augmented binding and ADCC activity to rituximab among individuals expressing the Fc{gamma}RIIIa-158 V/V and V/F polymorphism. *Blood* (2007) 110:2561–4. doi:10.1182/blood-2007-01-070656
42. Shimizu S, Tanaka Y, Tazawa H, Verma S, Onoe T, Ishiyama K, et al. Fc-gamma receptor polymorphisms predispose patients to infectious complications after liver transplantation. *Am J Transplant* (2016) 16:625–33. doi:10.1111/ajt.13492
43. Neudoerfl C, Mueller BJ, Blume C, Daemen K, Stevanovic-Meyer M, Keil J, et al. The peripheral NK cell repertoire after kidney transplantation is modulated by different immunosuppressive drugs. *Front Immunol* (2013) 4:46. doi:10.3389/fimmu.2013.00046
44. Vacher-Coponat H, Brunet C, Moal V, Loundou A, Bonnet E, Lyonnet L, et al. Tacrolimus/mycophenolate mofetil improved natural killer lymphocyte reconstitution one year after kidney transplant by reference to cyclosporine/azathioprine. *Transplantation* (2006) 82:558–66. doi:10.1097/01.tp.0000229390.01369.4a
45. Vacher-Coponat H, Brunet C, Lyonnet L, Bonnet E, Loundou A, Sampol J, et al. Natural killer cell alterations correlate with loss of renal function and dialysis duration in uraemic patients. *Nephrol Dial Transplant* (2008) 23:1406–14. doi:10.1093/ndt/gfm596
46. Morteau O, Blundell S, Chakera A, Bennett S, Christou CM, Mason PD, et al. Renal transplant immunosuppression impairs natural killer cell function in vitro and in vivo. *PLoS One* (2010) 5:e13294. doi:10.1371/journal.pone.0013294
47. Hadaya K, de Rham C, Bandelier C, Ferrari-Lacraz S, Jendly S, Berney T, et al. Natural killer cell receptor repertoire and their ligands, and the risk of CMV infection after kidney transplantation. *Am J Transplant* (2008) 8:2674–83. doi:10.1111/j.1600-6143.2008.02431.x
48. Shin BH, Ge S, Mirocha J, Karasyov A, Vo A, Jordan SC, et al. Regulation of anti-HLA antibody-dependent natural killer cell activation by immunosuppressive agents. *Transplantation* (2014) 97:294–300. doi:10.1097/01.TP.0000438636.52085.50
49. Vacher-Coponat H, Moal V, Indreies M, Purgus R, Loundou A, Burtey S, et al. A randomized trial with steroids and antithymocyte globulins comparing cyclosporine/azathioprine versus tacrolimus/mycophenolate mofetil (CATM2) in renal transplantation. *Transplantation* (2012) 93:437–43. doi:10.1097/TP.0b013e31824215b7
50. Wettstein D, Opelz G, Susal C. HLA antibody screening in kidney transplantation: current guidelines. *Langenbecks Arch Surg* (2014) 399:415–20. doi:10.1007/s00423-013-1138-6
51. Haas M, Sis B, Racusen LC, Solez K, Glotz D, Colvin RB, et al. Banff 2013 meeting report: inclusion of c4d-negative antibody-mediated rejection and antibody-associated arterial lesions. *Am J Transplant* (2014) 14:272–83. doi:10.1111/ajt.12590
52. Sis B, Jhangri GS, Riopel J, Chang J, de Freitas DG, Hidalgo L, et al. A new diagnostic algorithm for antibody-mediated microcirculation inflammation in kidney transplants. *Am J Transplant* (2012) 12:1168–79. doi:10.1111/j.1600-6143.2011.03931.x
53. Banasik M, Boratynska M, Koscielska-Kasprzak K, Kaminska D, Bartoszek D, Zabinska M, et al. The influence of non-HLA antibodies directed against angiotensin II type 1 receptor (AT1R) on early renal transplant outcomes. *Transpl Int* (2014) 27:1029–38. doi:10.1111/tri.12371
54. Capuano C, Romanelli M, Pighi C, Cimino G, Rago A, Molfetta R, et al. Anti-CD20 therapy acts via FcgammaRIIIA to diminish responsiveness of human natural killer cells. *Cancer Res* (2015) 75:4097–108. doi:10.1158/0008-5472.CAN-15-0781
55. Sood P, Hariharan S. Transplantation: rituximab induction for prevention of HLA-antibody rebound. *Nat Rev Nephrol* (2014) 10:682–3. doi:10.1038/nrneph.2014.179
56. Rostaing L, Kamar N. Can we prevent donor-specific antibodies from developing after ABO-incompatible kidney transplantation? *Kidney Int* (2014) 85:245–7. doi:10.1038/ki.2013.425
57. Vidarsson G, Dekkers G, Rispens T. IgG subclasses and allotypes: from structure to effector functions. *Front Immunol* (2014) 5:520. doi:10.3389/fimmu.2014.00520
58. Pincetic A, Bournazos S, DiLillo DJ, Maamary J, Wang TT, Dahan R, et al. Type I and type II Fc receptors regulate innate and adaptive immunity. *Nat Immunol* (2014) 15:707–16. doi:10.1038/ni.2939
59. Kellner C, Derer S, Valerius T, Peipp M. Boosting ADCC and CDC activity by Fc engineering and evaluation of antibody effector functions. *Methods* (2014) 65:105–13. doi:10.1016/j.ymeth.2013.06.036
60. Mishima Y, Terui Y, Kuniyoshi R, Matsusaka S, Mikuniya M, Kojima K, et al. High reproducible ADCC analysis revealed a competitive relation between ADCC and CDC and differences between FcgammaRllla polymorphism. *Int Immunol* (2012) 24:477–83. doi:10.1093/intimm/dxs048

24

Sustained Immune Complex-Mediated Reduction in CD16 Expression after Vaccination Regulates NK Cell Function

*Martin R. Goodier[1], Chiara Lusa[1], Sam Sherratt[1], Ana Rodriguez-Galan[1], Ron Behrens[2] and Eleanor M. Riley[2]**

[1] *Department of Immunology and Infection, London School of Hygiene and Tropical Medicine, London, UK,* [2] *Department of Clinical Research, London School of Hygiene and Tropical Medicine, London, UK*

Edited by:
Emanuela Marcenaro, University of Genoa, Italy

Reviewed by:
Kasper Hoebe, The Cincinnati Children's Hospital Medical Center, USA
Stephen Noel Waggoner, Cincinnati Children's Hospital Medical Center, USA

***Correspondence:**
Eleanor M. Riley
eleanor.riley@lshtm.ac.uk

Cross-linking of FcγRIII (CD16) by immune complexes induces antibody-dependent cellular cytotoxicity (ADCC) by natural killer (NK) cells, contributing to control of intracellular pathogens; this pathway can also be targeted for immunotherapy of cancerous or otherwise diseased cells. However, downregulation of CD16 expression on activated NK cells may limit or regulate this response. Here, we report sustained downregulation of CD16 expression on NK cells *in vivo* after intramuscular (but not intranasal) influenza vaccination. CD16 downregulation persisted for at least 12 weeks after vaccination and was associated with robust enhancement of influenza-specific plasma antibodies after intramuscular (but not intranasal) vaccination. This effect could be emulated *in vitro* by co-culture of NK cells with influenza antigen and immune serum and, consistent with the sustained effects after vaccination, only very limited recovery of CD16 expression was observed during long-term *in vitro* culture of immune complex-treated cells. CD16 downregulation was most marked among normally CD16^{high} CD57^{+} NK cells, irrespective of NKG2C expression, and was strongly positively associated with degranulation (surface CD107a expression). CD16 downregulation was partially reversed by inhibition of ADAM17 matrix metalloprotease, leading to a sustained increase in both CD107a and CD25 (IL-2Rα) expression. Both the degranulation and CD25 responses of CD57+ NK cells were uniquely dependent on trivalent influenza vaccine-specific IgG. These data support a role for CD16 in early activation of NK cells after vaccination and for CD16 downregulation as a means to modulate NK cell responses and maintain immune homeostasis of both antibody and T cell-dependent pathways.

Keywords: NK cells, CD16, CD57, degranulation, CD25, vaccination, influenza

INTRODUCTION

Natural killer (NK) cell effector function can be augmented by vaccination as a result of antibody-dependent cellular cytotoxicity (ADCC) and CD4^{+} T cell/IL-2-driven cytokine secretion (1–4). Moreover, we have recently observed that vaccination induces intrinsic changes in NK cell function, resulting in enhanced responsiveness to innate cytokines that may synergize with adaptive immunity to further potentiate NK cell responses (3).

Abbreviations: ADCC, antibody-dependent cellular cytotoxicity; AEU, arbitrary ELISA units; HCC, high concentration of cytokines; NK, natural killer; TIV, trivalent influenza vaccine.

Natural killer cell-mediated ADCC plays an important role in the control of infections and cancers (5, 6). NK cells express a number of immunoglobulin Fc receptors, with high expression of the activating receptor FcγRIIIA (CD16) and low expression of FCγRII (CD32) being a key feature of $CD56^{dim}$ NK cell populations (7, 8). Cross-linking of CD16 by IgG bound to target cell surface antigens leads to degranulation and release of perforin and granzymes. One particular subset of NK cells, characterized by low expression of CD56 but high levels of expression of CD16, CD57, and CD94/NKG2C, has recently been reported to be particularly efficient at mediating ADCC and may indeed be highly specialized for this particular effector function (9, 10). Furthermore, adaptive expansions of NK cells from HCMV-infected individuals have high frequencies of $Fc\varepsilon R\gamma 1^-$, $PLZF^-$ NK cells with potent ADCC activity against virus-infected target cells (11, 12).

Regulation of NK cell ADCC is achieved in part by the relative strength of signals transduced through activating Fc receptors and NK cell inhibitory receptors but may also rely on matrix-metalloproteinase-9 (MMP9) or ADAM17-mediated cleavage of CD16 from the cell surface after cross-linking by IgG (13–15). For example, rituximab-mediated targeting of $CD20^+$ tumor cells results in potent downregulation of NK cell CD16 that is dependent on MMP9-mediated cleavage (13, 14); exposure of NK cells to HCMV-infected fibroblasts leads to loss of CD16 concurrent with increased degranulation (9); MMP-dependent downregulation of CD16 is a feature of chronic HIV-1 infection (16, 17); and soluble NK cell-derived CD16 is elevated during rheumatoid arthritis, a chronic immune complex-associated inflammatory disease (18). However, similar effects have also been reported among cytokine-activated NK cells (stimulated *in vitro* with IL-2, IL-12, and IL-18) (19–21), suggesting that cross-linking of CD16 may not be essential for its downregulation. Importantly, neither the kinetics of CD16 expression after cross-linking nor the functional consequences of CD16 downregulation have been explored in any depth.

Here, we have investigated CD16 expression by NK cells from healthy subjects and find that CD16 is downregulated for many weeks after influenza vaccination, that $CD56^{dim}$ $CD57^+$ NK cells are particularly prone to losing CD16 after vaccination, and that this is mediated by vaccine antigen–antibody complexes. Furthermore, we show that ADAM-17 inhibitors or blocking antibodies to ADAM-17 prevent shedding of CD16 in response to vaccine antigens and that sustained CD16 signaling potentiates NK cell degranulation and CD25 expression. These data support a role for CD16 downregulation in regulating NK cell responses *in vivo* and maintaining homeostasis of both antibody and T cell-dependent pathways of NK cell activation.

MATERIALS AND METHODS

Subject Recruitment and Sample Collection

Venous blood was taken from a total of 47 healthy volunteers. The precise number of study subjects for each experiment is stated in the respective figure legends. The impact of recent vaccination on NK cells was studied in 37 healthy adult volunteers (median age 37.5 years; range of 21–63 years). None of the subjects had been previously vaccinated against influenza and none had experienced influenza-like symptoms during the previous 6 months. Subjects were randomly assigned to receive a single dose of 2012–2013 seasonal trivalent influenza vaccine (TIV) by either the intramuscular (Split Virion BP, Sanofi Pasteur MSD) or intranasal (Fluenz, AstraZeneca, UK) route. Randomization was structured so that participants in the two arms of the study could be matched according to age and sex. The intramuscular vaccine contains chemically inactivated virus, while the intranasal vaccine contains live attenuated virus. The vaccines were preservative free and were not adjuvanted. Venous blood samples were obtained immediately prior to vaccination and then at 2, 4, 12, and up to 36 weeks after vaccination. The study was approved by the ethical review committee of the London School of Hygiene and Tropical Medicine (Ref 6237). Locally recruited volunteers participating in influenza vaccination studies were provided with a participant information sheet detailing the studies. All participating volunteers provided written consent. The study made use of fully licensed vaccines which are routinely used in clinical practice. The study Clinician (Dr. Behrens) provided medical supervision for all procedures during the baseline visit and was available for emergencies during subsequent visits and was on hand to provide follow-up care for volunteers who experience side effects of the procedures.

Plasma was stored for assay of antibodies to influenza and for use in autologous cell cultures. PBMC were separated by standard Histopaque (Sigma, UK) gradient centrifugation and stimulated within 3 h of blood collection (for immediate culture experiments) or cryopreserved at 1×10^7 cells/ml in RPMI 1640, 40% fetal calf serum (FCS), 10% DMSO (Sigma, UK), within 4 h of blood collection. Cells were stored for 18 h at –80°C in Nalgene™ cryoboxes with isopropanol coolant prior to transfer to liquid nitrogen for longer term storage (22, 23).

Cell Culture Conditions, NK Cell Activation

For each individual, cells collected at baseline and at each post-vaccination time point were tested side-by-side. Cryopreserved PBMC were thawed, washed, and counted in Fastread™ counting slides (Immune Systems, UK), as previously described (22, 23), with a median yield of 56% and viability by trypan blue exclusion of 98%. Cells were rested for 4–6 h, in the absence of exogenous cytokines, prior to stimulation. Briefly, 2×10^5 PBMC were cultured for a total of 6 h, or where indicated for 18 h, in culture medium alone or with inactivated TIV (Split Virion BP, Sanofi Pasteur MSD). Cells were also stimulated with high concentrations of cytokines (HCC): IL-12 (5 ng/ml) plus IL-18 (50 ng/ml). For *in vitro* assays, FITC-conjugated anti-CD107a antibody (clone HP9, Beckton Dickinson) was added at the beginning of the culture, according to established protocols (24). GolgiStop (containing Monensin; 1/1500 concentration; BD Biosciences, Oxford, UK) and GolgiPlug (containing Brefeldin A; 1/1000 final concentration; BD Biosciences, Oxford, UK) were added 3 h before the end of the incubation. Assays were performed in 1% pooled AB serum, batch tested for performance in NK cell assays (Sigma, UK) unless otherwise stated. To determine the role of IgG on NK cell responses, pooled AB plasma was depleted of IgG using a protein G sepharose column (Millipore, UK), as previously described (23). *In vitro* neutralization experiments were performed using a rat-anti human IL-2 antibody (Rat IgG2a, clone MQ1-17H12, NA/LE, BD Biosciences) or a rat IgG2a control reagent (eBioscience, UK).

ADAM 17/MMP-dependent cleavage of CD16 was tested using the inhibitor TAPI-1 at a concentration of 10 μM (Merck Millipore, UK) and the active site-specific anti-ADAM17 monoclonal antibody D1 (A12) at a concentration of 6 μg/ml (Millipore, UK) and responses compared with the DMSO vehicle (Sigma, UK) and isotype-matched mouse IgG1 control (eBioscience, UK) treatments. Optimal concentrations of inhibitors were based on published protocols (25, 26) and confirmed by titration. Inhibitors were added 30 min prior to stimuli unless otherwise stated.

Long-term, *in vitro* culture of NK cells was performed after stimulation of PBMC for 18 h and washing cells (three times) to remove stimuli. Cells were maintained in RPMI 1640 supplemented with 5% AB serum and 0.75 ng/ml of IL-15, replacing the medium every 3 days. CD16 expression was monitored by flow cytometric analysis up to 18 days after initial stimulation, as described below.

Flow Cytometric and ImageStream™ Analysis

Phenotypic and functional analysis of NK cells was performed with the following monoclonal antibodies: anti-CD3-V500 (Clone UCHT1), anti-CD56-PeCy7 (B159), anti-CD107a-FITC (H4A3) (all from BD Biosciences), anti-CD57-e450 (TB01), anti-CD16-APC (CB16), and anti-CD69 PE (all from eBioscience). Dead/apoptotic cells were excluded using APC-efluor780-conjugated fixable viability dye (eBioscience). Cells were acquired on an LSRII flow cytometer (BD Biosciences, Oxford, UK) using FACSDiva® software.

Data analysis was performed using FlowJo V10 (Tree Star). FACS gates set on unstimulated cells (medium alone or isotype controls) were applied across all samples and all conditions. Responses where the gated subset contained <100 events were excluded. $CD57^+$ subsets were gated using an isotype-matched control reagent (mIgG1-eFluor450, eBioscience). Sample gating strategies are shown in **Figure 1**.

Image-stream™ analysis was performed to test for internalization of CD16 after activation. 2×10^6 freshly isolated PBMC were cultured for a total of 5 h in the presence of antigen and human plasma. FCS was substituted as a negative control for human Ig-containing plasma. After 5 h, cells were labeled with the following monoclonal antibodies for detection of surface antigens: anti-CD3 PE, anti-CD56PeCy7, anti-CD16APC (Clone 3G8, Biolegend), and anti-CD57e450 and were blocked for an additional 15 min with unconjugated anti-CD16 (Clone 3G8, BD Biosciences) antibody to ensure occupation of all relevant epitopes prior to intracellular staining. After fixation and permeabilzation [Fix-Perm kit (BD Biosciences)], cells were incubated for a further 30 min with anti-CD16 Pe-Dazzle 594 (Clone 3G8, Biolegend) to detect any internalized CD16. PBMC were then washed in permwash buffer (BD Biosciences) and resuspended in PBS at a minimum concentration of 2×10^7/ml. Events were acquired on an ImageStream® X Mark II Imaging Flow Cytometer (Amnis®) using the associated INSPIRE® software for acquisition and IDEAS® software for analysis. FACS gates were set based on FCS controls and applied across all samples and conditions. $CD56^+$ $CD3^-$ NK cells were inspected visually to determine the presence or absence of internalized CD16.

Statistical Analysis

Statistical analysis was performed using GraphPad Prism version 6.02. Linear trends were evaluated using repeated measures ANOVA. Functional responses between different culture conditions or between vaccination time points were compared using Wilcoxon signed-rank test. Significance levels are assigned as $^*p < 0.05$, $^{**}p < 0.01$, $^{***}p < 0.001$, and $^{****}p < 0.0001$ for all tests.

RESULTS

CD16 Is Downregulated on NK Cells after Intramuscular Influenza Vaccination and Is Antibody Dependent

Intramuscular vaccination with inactivated, trivalent seasonal influenza vaccine (TIV) significantly enhanced plasma concentrations of influenza-specific IgG up to 36 weeks after vaccination (**Figure 1A**) whereas little or no change in systemic IgG concentrations was seen after intranasal vaccination with the live attenuated influenza vaccine (LAIV) (**Figure 1B**).

To see whether vaccination affected the phenotype of peripheral blood NK cells, PBMC collected before vaccination and 2, 4, and 12 weeks after vaccination were stained immediately *ex vivo* for surface expression of CD56, CD16 and, as a marker of NK cell maturity, CD57. Examples of the flow cytometric gating strategy are shown in **Figures 1C–F**. Among those vaccinated intramuscularly with TIV, CD16 expression (MFI) on $CD56^{dim}$ NK cells was significantly reduced 2 weeks after vaccination in comparison to the pre-vaccination (baseline level) (**Figure 1G**); this was sustained for at least 12 weeks post-vaccination and was most pronounced among the $CD57^+$ subset. In complete contrast, there was no significant change in CD16 expression on NK cells of those vaccinated intra-nasally with LAIV (**Figure 1H**).

As downregulation of CD16 correlated with induction of anti-influenza antibodies, we hypothesized that these two observations were causally linked. To determine whether anti-influenza IgG contributed to CD16 downregulation, we cultured PBMC (collected at baseline, 0 weeks) with or without TIV antigens in the presence of autologous plasma collected at baseline (0 weeks) or after vaccination (2 weeks) (**Figures 1I,J**). Significant downregulation of CD16 was observed in cells cultured with autologous week 0 plasma and TIV antigen compared to without antigen, presumably reflecting the presence of pre-existing, influenza-specific antibodies resulting from prior infection (**Figures 1I,J**). However, when cells were cultured with TIV and 2 weeks post-vaccination plasma from TIV-vaccinated donors, there was a pronounced further reduction in CD16 expression on both $CD57^-$ and $CD57^+$ NK cells compared to cells cultured without antigen (**Figure 1I**)

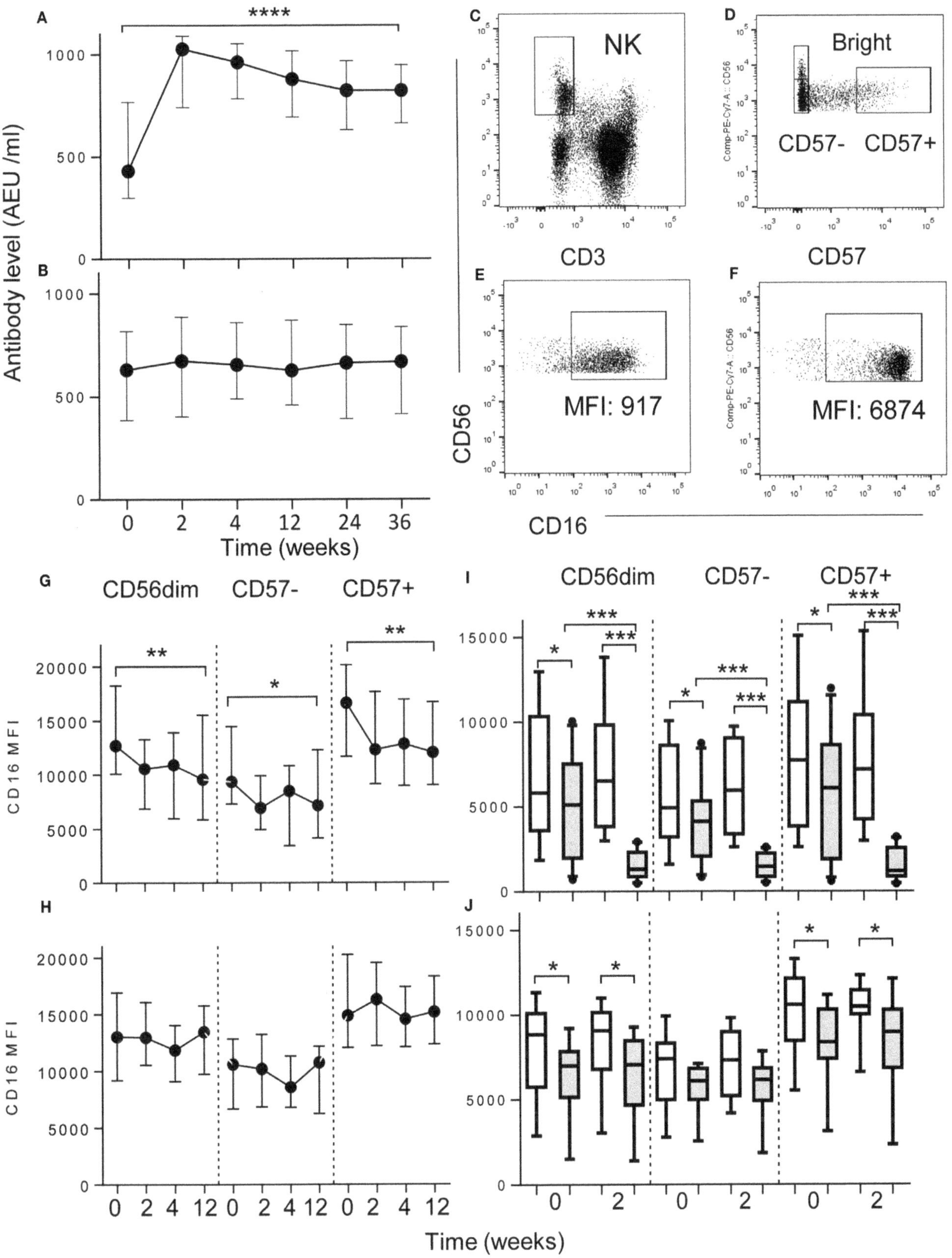

FIGURE 1 | **Downregulation of NK cell CD16 expression after vaccination.**

(Continued)

FIGURE 1 | Continued

(A,B) Increased plasma concentrations of anti-influenza IgG after **(A)** intramuscular (I.M, $n = 17$) but not after **(B)** intranasal vaccination (I. N, $n = 18$) up to 36 weeks after vaccination. Data represent median values with interquartile ranges. **(C–F)** Flow cytometric gating to assess the impact of vaccination on CD16 expression in CD56^{dim}, CD57$^-$, or CD57$^+$ NK cell subsets, *ex vivo*. NK cells were gated as CD56$^+$ CD3$^-$ **(C)** and then as CD56^{bright}, CD56^{dim}CD57$^-$ or CD56^{dim}CD57$^+$ **(D)**. **(E,F)** For determination of antigen-driven effects CD16$^+$ CD56$^+$ NK cells were gated among PBMC cultured with TIV and immune plasma **(E)** or immune plasma without TIV **(F)**. *Ex vivo* analysis of CD16 expression at baseline (0), 2, 4, and 12 weeks after intramuscular (I. M) vaccination with TIV **(G)** or intranasal (I.N.) vaccination with LAIV **(H)**. Impact on NK cell CD16 expression of culturing baseline PBMC with TIV (shaded bars) or without TIV (open bars) and either pre- (0) and post- (2) vaccination I.M. **(I)** or I.N. **(J)** plasma; data are shown for CD56^{dim}, CD57$^-$, and CD57$^+$ NK subsets after 6 h of *in vitro* culture with TIV. Trend analysis was performed using a one-way repeated measures ANOVA. Paired comparisons between pre- and post-vaccination plasma were made using Mann–Whitney *U* test. $^*p < 0.05$, $^{**}p < 0.01$, $^{***}p < 0.001$, $^{****}p < 0.0001$.

but no such effect was observed with post-vaccination plasma from LAIV-vaccinated donors (**Figure 1J**). Furthermore, for TIV-vaccinated donors, the extent of CD16 downregulation was dependent on plasma concentration, and was seen in both CD57$^-$ and CD57$^+$ NK cells and in both NKG2C$^+$ and NKG2C$^-$ NK cell subsets (Figures S1A,B Supplementary Material). Together, these data indicate that downregulation of CD16 is a sensitive indicator of vaccine-induced IgG concentration.

Downregulation of CD16 on NK Cells Is Due to CD16 Shedding and Is Slow to Recover

IgG-dependent downregulation of CD16 may be due to internalization of CD16–IgG–Ag complexes or to cleavage of CD16 at the cell surface and shedding into the extracellular milieu. Downregulation of CD16 did not occur in cells cultured in FCS or in either IgG-depleted or IgG-replete human plasma in the absence of TIV antigen. CD16 downregulation was seen only when cells were cultured in IgG-replete plasma with TIV antigen but not when TIV antigen was added to cultures containing FCS or IgG-depleted plasma (**Figure 2A**). To determine whether CD16 was internalized after IgG–Ag cross-linking, NK cells were incubated with TIV plus IgG-replete (immune) human plasma or FCS for 5 h and analyzed by Imagestream™ for extracellular and intracellular CD16 (**Figure 2B**). Surface staining for CD16 was clearly visible on CD56^{dim} CD57$^+$ NK cells cultured with TIV + FCS but not on cells cultured with TIV and immune human plasma. In neither case was CD16 detected intracellularly, indicating that loss of CD16 from the cell surface was not accompanied by internalization of intact CD16.

The sustained downregulation of CD16 observed *ex vivo* after TIV vaccination suggested that recovery of CD16 expression may be a slow process. To test this, downregulation of CD16 was induced by culturing PBMC with TIV and autologous plasma for 18 h; after washing to remove unbound antigen and antibody, the cells were maintained in a low concentration of IL-15 for 18 days (**Figure 2C**). CD16 expression was totally ablated on CD56^{dim} NK cells 24 h after activation and remained low until day 7 when there was a partial but sustained recovery; a further slight increase in CD16 expression was seen among CD57$^+$ NK cells between 15 and 18 days. However, CD16 expression remained significantly below baseline levels in both CD57$^+$ and CD57$^-$ NK cells until at least 18 days after activation. These data, taken together with our *ex vivo* observations, suggest that CD16 downregulation persists for several weeks after exposure to antigen in the presence of specific antibody.

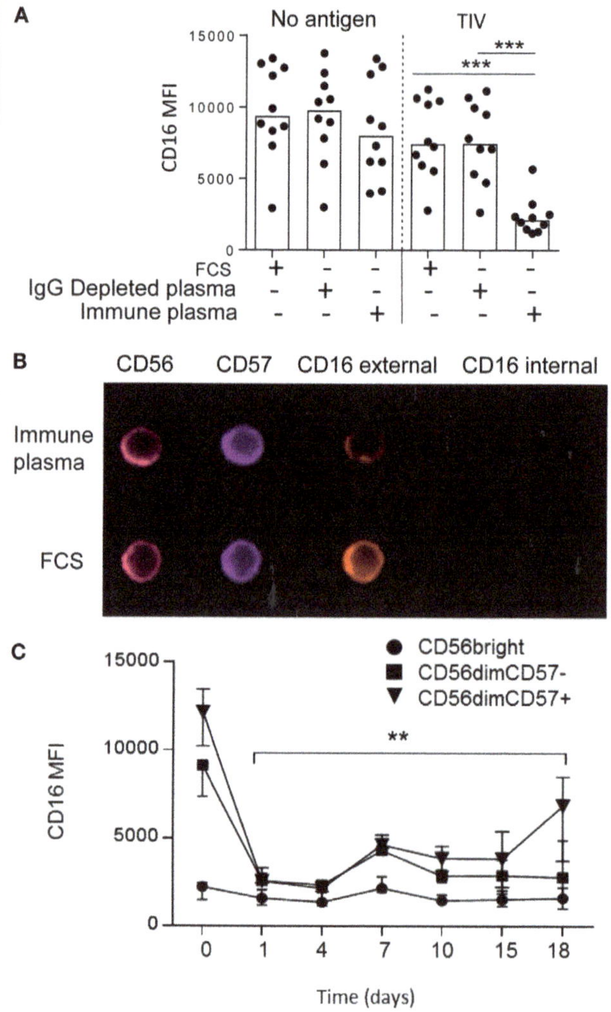

FIGURE 2 | Immune complex-induced shedding and sustained downregulation of NK cell CD16 expression. **(A)** Downregulation of CD16 requires IgG-replete human plasma and TIV antigen. PBMC were cultured for 5 h in FCS, IgG-depleted human plasma or IgG-replete plasma in the presence or absence of TIV. **(B)** ImageStream™ analysis of NK cells after 5 h culture with TIV plus human plasma (containing anti-influenza IgG) or FCS. Examples are shown of CD56^{dim} CD57$^+$ NK cells stained for surface (external) and intracellular (internal) CD16. **(C)** Time course for recovery of CD16 expression *in vitro*. Expression of CD16 was tracked by flow cytometry up to 18 days after culture of PBMC from five separate donors with TIV for 18 h in the presence of 1% human plasma (containing anti-influenza IgG). CD16 expression is shown for CD56^{bright}, CD56^{dim}CD57$^-$, and CD56^{dim} CD57$^+$ NK cell subsets. Trend analysis was performed using a one-way repeated measures ANOVA ($^*p < 0.05$).

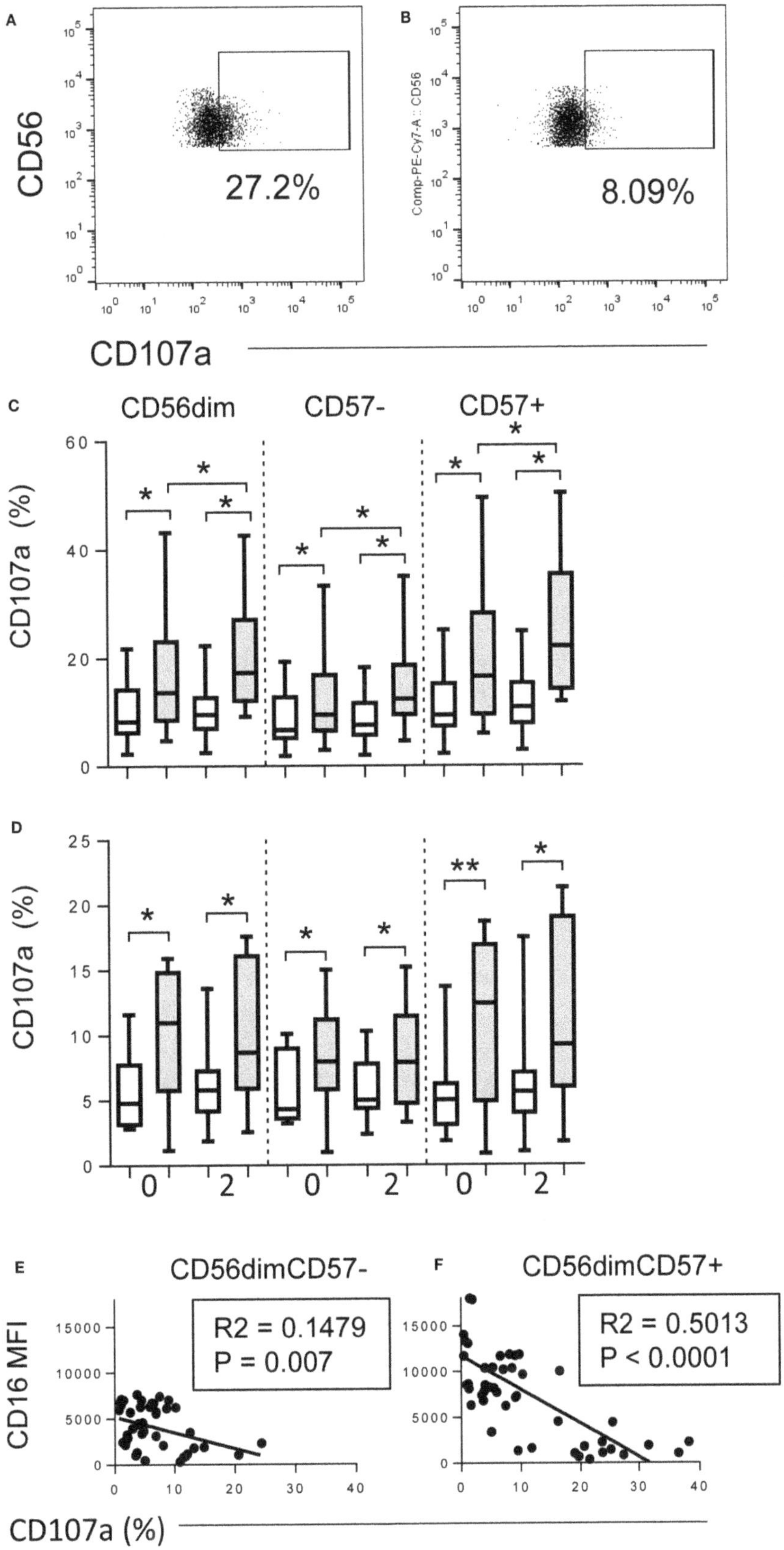

FIGURE 3 | Downregulation of CD16 correlates with degranulation.

(Continued)

FIGURE 3 | Continued

(A,B) Flow cytometry gating strategy for CD107a expression on CD56+ NK cells cultured with immune plasma **(A)** with or **(B)** without TIV. **(C,D)** Proportions of CD56^{dim}, CD56^{dim}CD57$^-$ and CD56^{dim} CD57$^+$ NK cells expressing CD107a after culture of baseline PBMC for 6 h with TIV (shaded bars) or without TIV (open bars) in the presence of baseline (0) or 2 weeks post-vaccination (2) plasma. Individuals (n = 10 per group) were vaccinated I.M. with TIV (C) or I.N. with LAIV (D). **(E,F)** Correlation between CD16 expression (MFI) and frequency of CD107a-expressing NK cells within CD57$^-$ **(E)** and CD57$^+$ **(F)** NK cells after culture for 6 h with TIV and immune plasma. Comparisons between conditions were made using Mann–Whitney U test. Correlations were performed using linear regression. $^*p < 0.05$, $^{**}p < 0.01$, $^{***}p < 0.001$.

Downregulation of CD16 Correlates with Degranulation

As CD16 downregulation is IgG–Ag-mediated and, thus, likely linked to ADCC, we explored the relationship between influenza vaccination, NK cell CD16 expression and degranulation (**Figure 3**). PBMCs were collected at baseline from TIV- and LAIV-vaccinated donors and cultured for 6 h with or without TIV and autologous plasma collected at baseline or 2 weeks after vaccination. Sample plots for analysis of CD107a expression are shown for NK cells cultured in autologous plasma with TIV or, as a negative control, without TIV (**Figures 3A,B**). Modest, but statistically significant, degranulation was seen in cells cultured with baseline (week 0) plasma for both TIV-vaccinated (**Figure 3C**) and LAIV-vaccinated (**Figure 3D**) individuals, again likely reflecting the presence of anti-influenza antibodies due to environmental exposure. Degranulation was, however, further enhanced in the presence of post-vaccination plasma (2 weeks) from TIV-vaccinated individuals (**Figure 3C**) but not from individuals receiving LAIV (**Figure 3D**). As expected, degranulation responses were stronger among CD57$^+$ NK cells than among CD57$^-$ NK cells and CD16 expression was strongly inversely associated with CD107a expression (**Figures 3E,F**). As expected, for TIV-vaccinated donors, the extent of CD107a expression was dependent upon plasma concentration, and was greater among CD57$^+$ NK cells than among CD57$^-$ NK cells, although there was no evidence that NKG2C$^+$ cells responded more strongly than NKG2C$^-$ cells (Supplementary Figure S1 C,D).

Shedding of CD16 Is Mediated by ADAM17

CD16 shedding was significantly reduced on both CD57$^-$ and CD57$^+$ NK cells when they were incubated with TIV and immune plasma in the presence of the ADAM17 inhibitor, TAPI-1 (**Figure 4A**) but not when incubated with the DMSO vehicle control. Notably, TAPI-1-mediated maintenance of CD16 expression was associated with enhanced degranulation as indicated by a significant increase in the frequency of CD107a$^+$ cells, particularly within the highly cytotoxic CD56^{dim}CD57$^+$ subset (**Figure 4B**). While there is a modest trend toward an increase in CD16 MFI when NK cells are cultured with FCS in the presence of TAPI-1, this is not statistically significant and there is no effect on degranulation. Moreover, blocking the active site of ADAM17 with a specific monoclonal antibody also prevented IgG–TIV-mediated shedding of CD16 (**Figure 4C**) and enhanced CD107a expression, although to a somewhat lesser extent than the TAPI-1 inhibitor (**Figure 4D**); these effects were not observed when cells were cultured in FCS rather than immune human serum. However, although specific monoclonal antibody blockade of ADAM17 was sufficient to prevent CD16 shedding, the more effective enhancement of degranulation by the TAPI-1 inhibitor suggests that this may additionally target other molecules involved in the discharge or recycling of cytotoxic granules.

Early Immune Complex-Mediated Responses Condition Later Events

Our previous studies have suggested that early, antibody-dependent mechanisms may synergize with antigen-specific T cell responses to enhance NK cell responses to vaccines (3, 23). Since sustained expression of CD16 in the presence of ADAM17/MMP inhibitors led to enhanced degranulation (**Figure 5**), we investigated whether prevention of CD16 shedding would affect other NK cell responses. ADAM17/MMP blockade had no impact on IgG–TIV-induced CD69 expression, nor was there any impact on CD25 expression after 6 h of culture (Figure S2 in Supplementary Material). However in 18 h cultures, addition of TAPI-1 (**Figures 5A–C**) or blocking antibody to ADAM17 (**Figures 5D–F**) 30 min before the start of the culture (time − 0.5 h) not only sustained CD16 expression (**Figures 5A,D**) and enhanced CD107a expression (**Figure 5B**) but also significantly enhanced CD25 expression compared to control cultures (**Figures 5C,F**). However, when addition of TAPI-1 or anti-ADAM17 was delayed for 6 h after the initiation of cultures (+ 6 h) CD16 expression could not be rescued (**Figures 5A,D**) and there was no enhancement of degranulation or CD25 expression (**Figures 5B,C,E,F**).

Bi-directional Cross-talk between CD16-Mediated and Cytokine-Mediated Pathways of NK Cell Activation

The data presented in **Figure 5** are consistent with a role for antigen–antibody complex signaling via CD16 in induction of CD25 as well as in degranulation/cytotoxicity. To explore this further, PBMCs were cultured with TIV for 18 h in the presence of intact immune plasma or immune plasma that had been depleted of IgG by passage over a protein G column (**Figure 6**). The concentration of anti-TIV–IgG in intact plasma was 413.4 arbitrary ELISA units (AEU)/ml and this was reduced to 7.4 AEU/ml after protein G depletion.

As expected, there was complete loss of CD16 from the surface of NK cells cultured for 18 h with TIV and immune plasma. CD16 shedding was seen in both CD57$^-$ and CD57$^+$ NK cells and was completely prevented by depleting the immune plasma of IgG (**Figure 6A**). Also as expected from our previous data, degranulation was inversely correlated with CD16 expression: CD107a expression was induced by TIV plus immune plasma

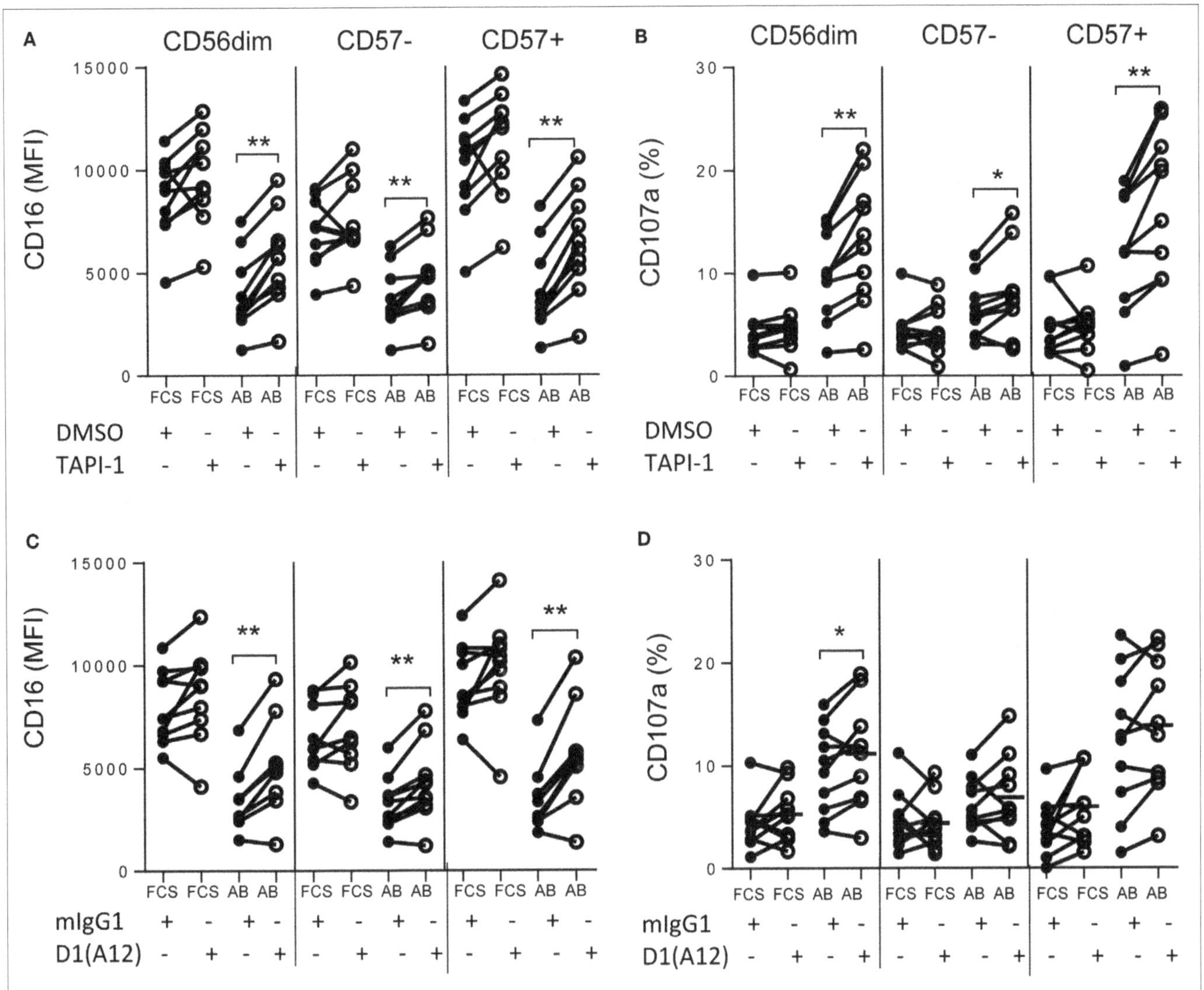

FIGURE 4 | Shedding of CD16 is mediated by ADAM17. PBMCs were cultured for 6 h with TIV and FCS or immune AB plasma (AB) in the presence or absence of the MMP inhibitor TAP1-1 **(A,B)** or the D1(A12) blocking antibody to ADAM17 **(C,D)** and the relevant negative controls (DMSO and mIgG1, respectively). CD16 (MFI) **(A,C)** and CD107a (%) **(B,D)** were assessed by flow cytometry on CD56^{dim}, CD56^{dim}CD57$^-$ and CD56^{dim}CD57$^+$ NK cells. Data are presented for 10 different individuals. Paired comparisons between conditions were made using Mann–Whitney *U* test. $^*p < 0.05$, $^{**}p < 0.01$, $^{***}p < 0.001$.

in both CD57$^+$ and CD57$^-$ subsets (**Figure 6B**). However, this effect was reversed by culture with IgG-depleted only for CD57$^+$ NK cells, consistent with this subset being more dependent on direct activation by TIV–IgG complexes (**Figure 6B**). In line with observations in **Figure 5**, TIV plus immune plasma also significantly upregulated CD25 expression on all NK cell subsets (**Figure 6C**). CD25 upregulation was dependent upon IgG–Ag immune complexes, particularly within CD57+ NK cells, since no induction of CD25 was observed when cells were cultured with TIV and IgG-depleted plasma (**Figure 6C**).

These observations suggest that there is cross-talk between the CD16 immune complex-mediated pathway and the cytokine-mediated pathway of NK cell activation. To determine whether this cross-talk was bi-directional, CD16, CD107a, and CD25 expression were characterized on PBMCs incubated for 18 h with recombinant IL-12 and IL-18. As shown previously (3, 23, 27), IL-12+IL-18 induces degranulation and strongly upregulates CD25 expression on both CD57$^-$ and CD57$^+$ NK cells (**Figures 6E,F**). Importantly, however, IL-12+IL-18 also induced significant shedding of CD16 from the surface of NK cells (**Figure 6D**), indicating that cytokine-mediated activation of NK cells may restrict their subsequent antibody-dependent responses. Also, as high concentrations of IL-2 have previously been reported to downregulate CD16 (19), and as IL-2 might be being produced in these cultures by TIV-specific T cells, we added neutralizing anti-IL-2 antibodies to our cultures to test whether IL-2 might be contributing to CD16 downregulation. However, neutralization of IL-2 did not affect downregulation of CD16 by TIV/IgG nor the associated induction of CD107a and CD25 (**Figures 6A–C**).

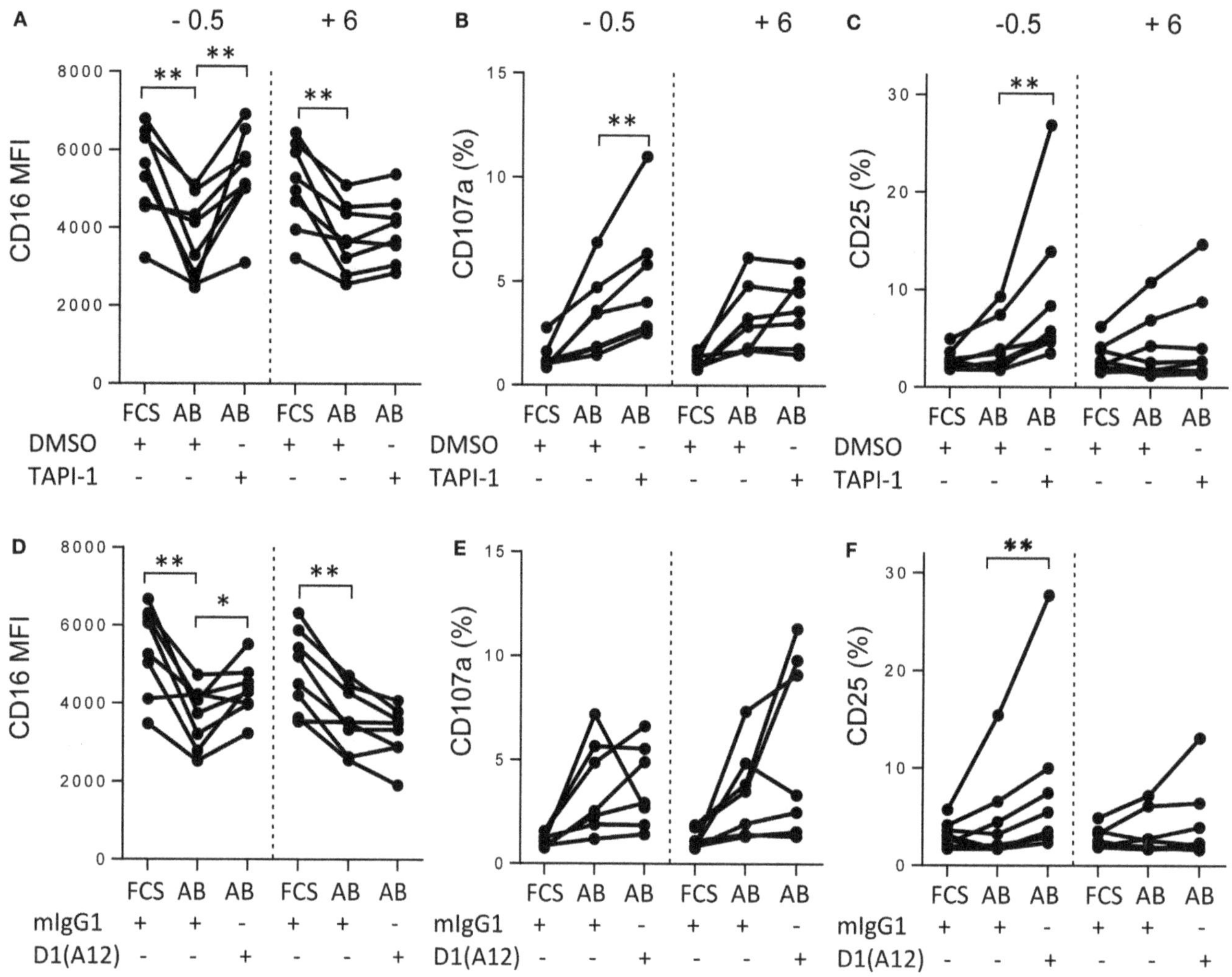

FIGURE 5 | Early immune complex-mediated responses condition later events. PBMCs were cultured for a total of 18 h with TIV and FCS or immune AB plasma (AB) in the presence or absence of the MMP inhibitor TAP1-1 **(A–C)** or the D1(A12) blocking antibody to ADAM17 **(D–F)** and the relevant negative controls (DMSO and mIgG1, respectively). CD16 (MFI) **(A,D)**, CD107a (%) **(B,E)**, and CD25 **(C,F)** expression were assessed by flow cytometry. TAPI-1, DMSO, D1(A12), or mIgG1 were added either 30 min prior to (−0.5), or 6 h after (+6) TIV antigen. Data are presented for eight different individuals. Paired comparisons between conditions were made using the Mann–Whitney *U* test. *$p < 0.05$, **$p < 0.01$, ***$p < 0.001$.

DISCUSSION

It is increasingly being recognized that human NK cells are functionally as well as phenotypically heterogeneous, with markers of NK cell differentiation and maturation correlating with effector function. More differentiated NK cells, bearing high levels of CD57 and also expressing CD16, NKG2C, KIR, and the FcεRγ adaptor protein, appear specialized for ADCC, whereas $CD57^-$ NK cells (both $CD56^{bright}$ and $CD56^{dim}$ subsets) are highly sensitive to exogenous cytokines. However, NK cell function is not binary: $CD57^-$ NK cells express CD16 and can mediate ADCC whereas $CD57^+$ NK cells can be induced to secrete IFN-γ under certain circumstances. Moreover, as shown here, immune complex-mediated pathways of NK cell activation influence cytokine-induced pathways, and vice versa.

Natural killer cells are essential for resistance to infection and cancer but – if unregulated – have the potential to cause significant immunopathology (28); NK cell activation is well known, therefore, to be very tightly controlled by the balance of activating and inhibitory signals. However, rather less attention has been paid to regulating the consequences of NK cell activation and restoring immune homeostasis after an infection or other threat has been controlled or eliminated. It is in this context that downregulation of CD16 after cross-linking by immune complexes is likely to be of functional significance.

Internalization, degradation, or shedding of cell surface receptors after ligation by their cognate ligand is a characteristic of the immune system and appears designed to release active soluble mediators or to ensure that immune responses are self-limiting (29). Here, we have demonstrated that downregulation of CD16

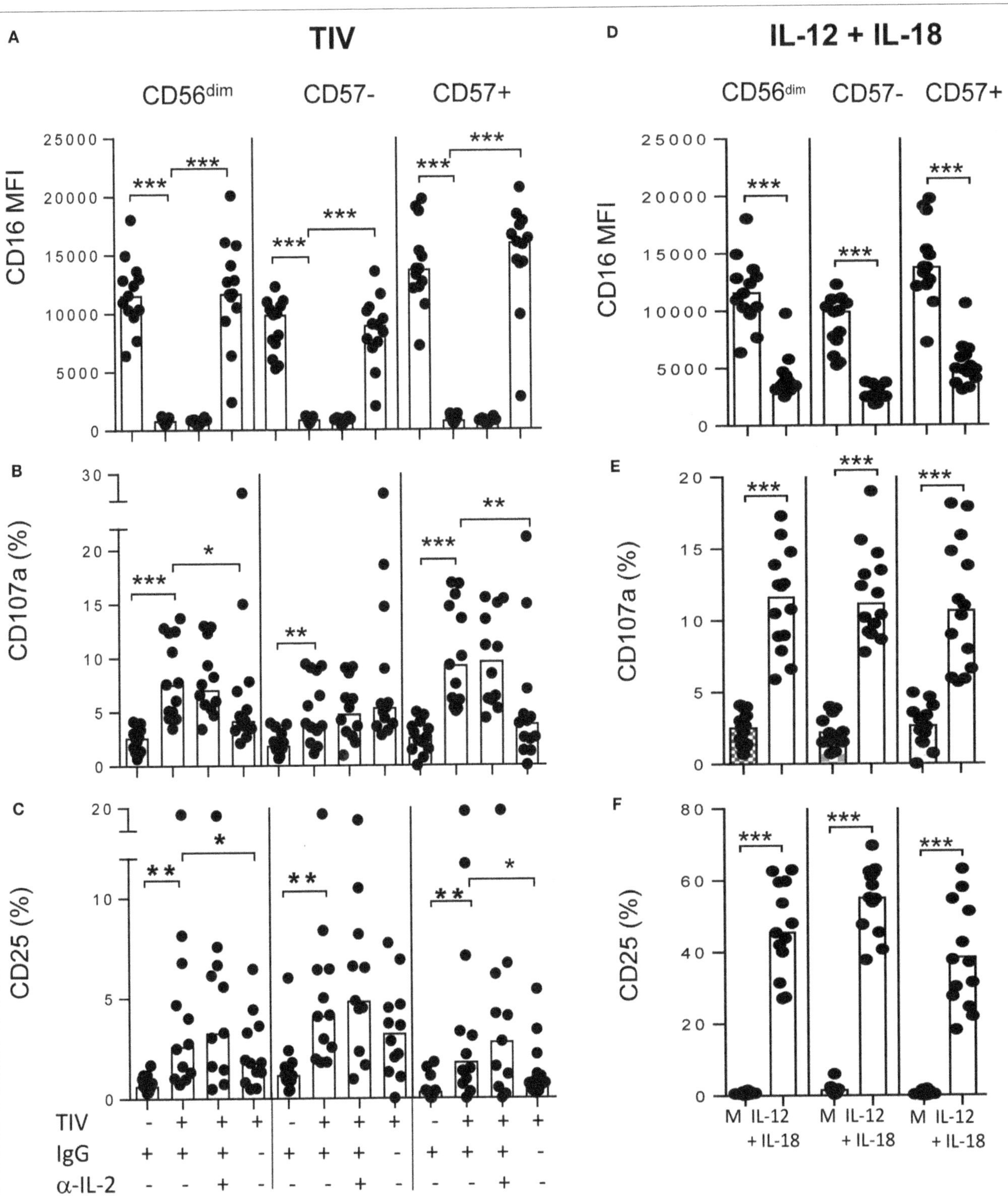

FIGURE 6 | **Bi-directional cross-talk between CD16-mediated and cytokine-mediated pathways of NK cell activation.** **(A–C)** PBMCs were cultured for 18 h with or without TIV in the presence of immune plasma (IgG+) or IgG-depleted immune plasma (IgG−) or after treatment with anti-IL-2 neutralizing antibody or isotype-matched control and expression of CD16 MFI **(A)**, CD107a (%) **(B)**, or CD25 (%) **(C)** on CD56^{dim}, CD56^{dim}CD57^{-}, and CD56^{dim}CD57^{+} NK cells was assessed by flow cytometry. **(D–F)** PBMCs were cultured for 18 h with or without IL-12 (5 ng/ml) combined with IL-18 (50 ng/ml) and expression of CD16 MFI **(D)**, CD107a (%) **(E)**, or CD25 (%) **(F)** on CD56^{dim}, CD56^{dim}CD57^{-}, and CD56^{dim}CD57^{+} NK cells was assessed by flow cytometry. Paired comparisons were made using the Mann–Whitney *U* test. $^{*}p < 0.05$, $^{**}p < 0.01$, $^{***}p < 0.001$.

on NK cells after cross-linking by IgG–Ag immune complexes is an example of such self-regulation, curtailing NK cell degranulation and limiting their ability to respond to exogenous cytokines by constraining expression of the high-affinity IL-2 receptor. Moreover, we have shown that CD16 downregulation is due to ADAM17-mediated shedding of CD16 from the NK cell surface, is induced *in vivo* after systemic vaccination, and is sustained for at least 12 weeks *in vivo* and at least 18 days *in vitro*. Taken together, these data suggest that tightly controlled surface expression of CD16 represents an important mechanism for regulating NK cell function *in vivo*.

It has previously been shown that broad spectrum matrix metalloprotease inhibitors allow recovery of CD16 expression in cytokine-maintained NK cells from HIV-1 infected individuals (16) and prevent downregulation of CD16 *in vitro* in response to cytokines or PMA (19, 20). We have confirmed these observations, revealing a very specific role for the ADAM17 class of proteases in this process, and extended them to show that inhibition of CD16 shedding can potentiate NK cell cytotoxic function, opening up potential therapeutic applications. However, TAPI-1 treatment enhanced degranulation more consistently than blockade with the ADAM-17-specific monoclonal antibody, suggesting that although reported as an ADAM-17 specific, this inhibitor may target additional MMPs involved in cytolytic granule processing (30, 31).

The prolonged period of CD16 downregulation *in vivo/ex vivo* after immunization (at least 3 months) was a surprise and raises questions about the ability of NK cells to mediate ADCC reactions in the immediate aftermath of vaccination or infection. Given that this is a global, generic effect rather than an antigen-specific effect, the consequences could be wide-ranging and might conceivably contribute to the increased risk of secondary infection after a primary viral infection, for example. Indeed, cross-linking of CD16 with rituximab (anti-CD20) has been shown to induce SHP-1-dependent hypo-responsiveness of NK cells to a diverse array of activating signals, including third-party tumor cell lines and cross-linking of NKp46, 2B4, NKG2D, and DNAM-1 (32). CD16 cross-linking and downregulation may, therefore, represent a dominant early pathway for controlling NK cell activation, affecting diverse NK cell signaling pathways. In line with this, we observed that CD16 downregulation also affects cytokine-mediated pathways of NK cell activation, and vice versa, raising further questions about the consequences of this homeostatic mechanism. It suggests, for example, that CD4+ T cell/IL-2-mediated activation of NK cells (3, 23, 27) may be unable to fully compensate for the inhibition of the ADDC pathway. On the other hand, CD25 expression by $CD57^-$ NK cells – which are the major cytokine-producing subset of NK cells (27, 33) – was less affected by loss of CD16 (by comparison with $CD57^+$ cells), indicating that cytokine responsiveness may be retained by a subset of the NK cell population despite widespread downregulation of CD16.

Persistence of $CD16^{low}$ NK cells has been reported in chronically HIV-1 infected individuals despite effective suppression of viral replication by anti-retroviral therapy (16), but persistence of antigen – and, thus, of immune complexes – cannot be ruled out in this case. While long-term persistence of antigen cannot be entirely ruled out in our TIV model, the vaccine is inactivated. On the other hand, our data do imply that individual NK cells may have very limited capacity to re-express CD16 once it has been lost: NK cells that had been induced to shed CD16 showed very little propensity to re-express CD16 over a period of more than 2 weeks in culture, despite the absence of IgG or antigen. Recovery of the $CD16^+$ NK cell population *in vivo* may, thus, rely on repopulation of the periphery from NK cell precursors.

Finally, our study raises questions about the nature of protection induced by influenza vaccination. The very low levels of circulating IgG induced by the live attenuated intra-nasal vaccine, and the correspondingly low levels of NK cell degranulation/ADCC/CD16 shedding, suggest that NK cell ADCC to influenza is not potentiated by LAIV. Anti-influenza IgG antibodies not only mediate NK cell ADCC but are sufficient to control virulent (H1N1 pandemic) infection in rhesus macaques in the absence of neutralizing antibodies (34, 35). It is interesting to speculate, therefore, that the lack of this ADCC response may in part explain the apparently much lower efficacy of LAIV compared to TIV that is now being recognized (CDC Advisory Committee on Immunization Practices, http://www.cdc.gov/media/releases/2016/s0622-laiv-flu.html).

AUTHOR CONTRIBUTIONS

MG directed research, designed the study, designed and performed experiments, analysed and interpreted data, and wrote the manuscript. CL designed and performed antibody determinations and antibody-dependent NK cell assays, and analyzed data. SS designed and performed the imagestream experiments and anayzed the data. AR-G designed and performed *ex vivo* and *in vitro* vaccine experiments and analyzed data. RB designed vaccination protocol, performed seasonal influenza vaccinations, and reviewed data. ER directed research and wrote the manuscript.

ACKNOWLEDGMENTS

We would like to thank Carolynne Stanley for the recruitment and consenting of study subjects and for blood sample collection. We would also like to thank Dr. Christian Bottomley for statistical advice.

SUPPLEMENTARY MATERIAL

The Supplementary Material for this article can be found online at http://journal.frontiersin.org/article/10.3389/fimmu.2016.00384

FIGURE S1 | Sensitivity of CD57-defined NK cell subsets to antigen–antibody-mediated downregulation of CD16. Effect of TIV plus varying concentrations of immune plasma on CD16 (MFI) **(A,B)** and CD107a (%)

(C,D) expression on CD56^{bright}, CD56^{dim}CD57^{-}, CD56^{dim}CD57^{+} subsets **(A,C)** and on CD57^{+}NKG2C^{-}, and CD57^{+}NKG2C^{+} NK cells **(B,D)**. Paired comparisons were made between CD56dimCD57^{-} and CD56dimCD57 + NK cell subsets using Mann–Whitney *U* test. $^{*}p < 0.05$, $^{**}p < 0.01$, $^{***}p < 0.001$.

FIGURE S2 | The impact of MMP inhibition and ADAM17 blockade on early CD69 and CD25 expression. PBMCs were cultured for 6 h with TIV and FCS or immune AB plasma (AB) in the presence or absence of the MMP inhibitor TAP1-1 **(A,C)** or the D1(A12) blocking antibody to ADAM17 **(B,D)** and the relevant negative controls (DMSO and mIgG1, respectively). CD69 (%) **(A,B)** and CD25 (%) **(C,D)** expression was assessed by flow cytometry on CD56^{dim}, CD56^{dim}CD57^{-}, and CD56^{dim}CD57^{+} NK cells. Data are presented for 10 different individuals. Paired comparisons between conditions were made using Mann–Whitney *U* test. $^{*}p < 0.05$, $^{**}p < 0.01$, $^{***}p < 0.001$.

REFERENCES

1. Horowitz A, Behrens RH, Okell L, Fooks AR, Riley EM. NK cells as effectors of acquired immune responses: effector CD4+ T cell-dependent activation of NK cells following vaccination. *J Immunol* (2010) 185(5):2808–18. doi:10.4049/jimmunol.1000844
2. Jost S, Tomezsko PJ, Rands K, Toth I, Lichterfeld M, Gandhi RT, et al. CD4+ T-cell help enhances NK cell function following therapeutic HIV-1 vaccination. *J Virol* (2014) 88(15):8349–54. doi:10.1128/JVI.00924-14
3. Goodier MR, Rodriguez-Galan A, Lusa C, Nielsen CM, Darboe A, Moldoveanu AL, et al. Influenza vaccination generates cytokine-induced memory-like NK cells: impact of human cytomegalovirus infection. *J Immunol* (2016) 197(1):313–25. doi:10.4049/jimmunol.1502049
4. Suliman S, Geldenhuys H, Johnson JL, Hughes JE, Smit E, Murphy M, et al. Bacillus Calmette-Guerin (BCG) revaccination of adults with latent *Mycobacterium tuberculosis* infection induces long-lived BCG-reactive NK cell responses. *J Immunol* (2016) 197(4):1100–10. doi:10.4049/jimmunol.1501996
5. Biron CA, Nguyen KB, Pien GC, Cousens LP, Salazar-Mather TP. Natural killer cells in antiviral defense: function and regulation by innate cytokines. *Annu Rev Immunol* (1999) 17:189–220. doi:10.1146/annurev.immunol.17.1.189
6. Cerwenka A, Lanier LL. Natural killer cell memory in infection, inflammation and cancer. *Nat Rev Immunol* (2016) 16(2):112–23. doi:10.1038/nri.2015.9
7. Metes D, Galatiuc C, Moldovan I, Morel PA, Chambers WH, DeLeo AB, et al. Expression and function of Fc gamma RII on human natural killer cells. *Nat Immun* (1994) 13(6):289–300.
8. Stewart-Akers AM, Cunningham A, Wasko MC, Morel PA. Fc gamma R expression on NK cells influences disease severity in rheumatoid arthritis. *Genes Immun* (2004) 5(7):521–9. doi:10.1038/sj.gene.6364121
9. Wu Z, Sinzger C, Frascaroli G, Reichel J, Bayer C, Wang L, et al. Human cytomegalovirus-induced NKG2C(hi) CD57(hi) natural killer cells are effectors dependent on humoral antiviral immunity. *J Virol* (2013) 87(13):7717–25. doi:10.1128/JVI.01096-13
10. Costa-Garcia M, Vera A, Moraru M, Vilches C, Lopez-Botet M, Muntasell A. Antibody-mediated response of NKG2Cbright NK cells against human cytomegalovirus. *J Immunol* (2015) 194(6):2715–24. doi:10.4049/jimmunol.1402281
11. Zhang T, Scott JM, Hwang I, Kim S. Cutting edge: antibody-dependent memory-like NK cells distinguished by FcRgamma deficiency. *J Immunol* (2013) 190(4):1402–6. doi:10.4049/jimmunol.1203034
12. Lee J, Zhang T, Hwang I, Kim A, Nitschke L, Kim M, et al. Epigenetic modification and antibody-dependent expansion of memory-like NK cells in human cytomegalovirus-infected individuals. *Immunity* (2015) 42(3):431–42. doi:10.1016/j.immuni.2015.02.013
13. Dall'Ozzo S, Tartas S, Paintaud G, Cartron G, Colombat P, Bardos P, et al. Rituximab-dependent cytotoxicity by natural killer cells: influence of FCGR3A polymorphism on the concentration-effect relationship. *Cancer Res* (2004) 64(13):4664–9. doi:10.1158/0008-5472.CAN-03-2862
14. Bowles JA, Wang SY, Link BK, Allan B, Beuerlein G, Campbell MA, et al. Anti-CD20 monoclonal antibody with enhanced affinity for CD16 activates NK cells at lower concentrations and more effectively than rituximab. *Blood* (2006) 108(8):2648–54. doi:10.1182/blood-2006-04-020057
15. Konjevic G, Vuletic A, Mirjacic Martinovic K, Colovic N, Colovic M, Jurisic V. Decreased CD161 activating and increased CD158a inhibitory receptor expression on NK cells underlies impaired NK cell cytotoxicity in patients with multiple myeloma. *J Clin Pathol* (2016) 15:203614. doi:10.1136/jclinpath-2016-203614
16. Liu Q, Sun Y, Rihn S, Nolting A, Tsoukas PN, Jost S, et al. Matrix metalloprotease inhibitors restore impaired NK cell-mediated antibody-dependent cellular cytotoxicity in human immunodeficiency virus type 1 infection. *J Virol* (2009) 83(17):8705–12. doi:10.1128/JVI.02666-08
17. Tang CC, Isitman G, Bruneau J, Tremblay C, Bernard NF, Kent SJ, et al. Phenotypical and functional profiles of natural killer cells exhibiting matrix metalloproteinase-mediated CD16 cleavage after anti-HIV antibody-dependent activation. *Clin Exp Immunol* (2015) 181(2):275–85. doi:10.1111/cei.12593
18. Masuda M, Morimoto T, De Haas M, Nishimura N, Nakamoto K, Okuda K, et al. Increase of soluble FcgRIIIa derived from natural killer cells and macrophages in plasma from patients with rheumatoid arthritis. *J Rheumatol* (2003) 30(9):1911–17.
19. Peruzzi G, Femnou L, Gil-Krzewska A, Borrego F, Weck J, Krzewski K, et al. Membrane-type 6 matrix metalloproteinase regulates the activation-induced downmodulation of CD16 in human primary NK cells. *J Immunol* (2013) 191(4):1883–94. doi:10.4049/jimmunol.1300313
20. Romee R, Foley B, Lenvik T, Wang Y, Zhang B, Ankarlo D, et al. NK cell CD16 surface expression and function is regulated by a disintegrin and metalloprotease-17 (ADAM17). *Blood* (2013) 121(18):3599–608. doi:10.1182/blood-2012-04-425397
21. Zhou Q, Gil-Krzewska A, Peruzzi G, Borrego F. Matrix metalloproteinases inhibition promotes the polyfunctionality of human natural killer cells in therapeutic antibody-based anti-tumour immunotherapy. *Clin Exp Immunol* (2013) 173(1):131–9. doi:10.1111/cei.12095
22. Goodier MR, White MJ, Darboe A, Nielsen CM, Goncalves A, Bottomley C, et al. Rapid NK cell differentiation in a population with near-universal human cytomegalovirus infection is attenuated by NKG2C deletions. *Blood* (2014) 124(14):2213–22. doi:10.1182/blood-2014-05-576124
23. Nielsen CM, White MJ, Bottomley C, Lusa C, Rodriguez-Galan A, Turner SE, et al. Impaired NK cell responses to pertussis and H1N1 influenza vaccine antigens in human cytomegalovirus-infected individuals. *J Immunol* (2015) 194(10):4657–67. doi:10.4049/jimmunol.1403080
24. Alter G, Malenfant JM, Altfeld M. CD107a as a functional marker for the identification of natural killer cell activity. *J Immunol Methods* (2004) 294(1–2):15–22. doi:10.1016/j.jim.2004.08.008
25. Mullberg J, Durie FH, Otten-Evans C, Alderson MR, Rose-John S, Cosman D, et al. A metalloprotease inhibitor blocks shedding of the IL-6 receptor and the p60 TNF receptor. *J Immunol* (1995) 155(11):5198–205.
26. Romee R, Schneider SE, Leong JW, Chase JM, Keppel CR, Sullivan RP, et al. Cytokine activation induces human memory-like NK cells. *Blood* (2012) 120(24):4751–60. doi:10.1182/blood-2012-04-419283
27. White MJ, Nielsen CM, McGregor RH, Riley EH, Goodier MR. Differential activation of CD57-defined natural killer cell subsets during recall responses to vaccine antigens. *Immunology* (2014) 142(1):140–50. doi:10.1111/imm.12239
28. Popko K, Gorska E. The role of natural killer cells in pathogenesis of autoimmune diseases. *Cent Eur J Immunol* (2015) 40(4):470–6. doi:10.5114/ceji.2015.56971
29. Marco M, Fortin C, Fulop T. Membrane-type matrix metalloproteinases: key mediators of leukocyte function. *J Leukoc Biol* (2013) 94(2):237–46. doi:10.1189/jlb.0612267
30. Crowe PD, Walter BN, Mohler KM, Otten-Evans C, Black RA, Ware CF. A metalloprotease inhibitor blocks shedding of the 80-kD TNF receptor and TNF processing in T lymphocytes. *J Exp Med* (1995) 181(3):1205–10. doi:10.1084/jem.181.3.1205

31. DasGupta S, Murumkar PR, Giridhar R, Yadav MR. Studies on novel 2-imidazolidinones and tetrahydropyrimidin-2(1H)-ones as potential TACE inhibitors: design, synthesis, molecular modeling, and preliminary biological evaluation. *Bioorg Med Chem* (2009) 17(10):3604–17. doi:10.1016/j.bmc.2009.04.003
32. Capuano C, Romanelli M, Pighi C, Cimino G, Rago A, Molfetta R, et al. Anti-CD20 therapy acts via FcgammaRIIIA to diminish responsiveness of human natural killer cells. *Cancer Res* (2015) 75(19):4097–108. doi:10.1158/0008-5472.CAN-15-0781
33. Lopez-Verges S, Milush JM, Pandey S, York VA, Arakawa-Hoyt J, Pircher H, et al. CD57 defines a functionally distinct population of mature NK cells in the human CD56dimCD16+ NK-cell subset. *Blood* (2010) 116(19):3865–74. doi:10.1182/blood-2010-04-282301
34. Jegaskanda S, Job ER, Kramski M, Laurie K, Isitman G, de Rose R, et al. Cross-reactive influenza-specific antibody-dependent cellular cytotoxicity antibodies in the absence of neutralizing antibodies. *J Immunol* (2013) 190(4):1837–48. doi:10.4049/jimmunol.1201574
35. Jegaskanda S, Weinfurter JT, Friedrich TC, Kent SJ. Antibody-dependent cellular cytotoxicity is associated with control of pandemic H1N1 influenza virus infection of macaques. *J Virol* (2013) 87(10):5512–22. doi:10.1128/JVI.03030-12

25

NK Cell Influence on the Outcome of Primary Epstein–Barr Virus Infection

Obinna Chijioke[1,2], *Vanessa Landtwing*[2] *and Christian Münz*[2]*

[1] *Institute of Surgical Pathology, University Hospital Zürich, Zürich, Switzerland,* [2] *Viral Immunobiology, Institute of Experimental Immunology, University of Zürich, Zürich, Switzerland*

Edited by:
Eric Vivier,
Centre d'Immunologie de Marseille-Luminy, France

Reviewed by:
Bojan Polić,
University of Rijeka, Croatia
Nicholas D. Huntington,
Walter and Eliza Hall Institute of Medical Research, Australia

***Correspondence:**
Christian Münz
christian.muenz@uzh.ch

The herpesvirus Epstein–Barr virus (EBV) was discovered as the first human candidate tumor virus in Burkitt's lymphoma more than 50 years ago. Despite its strong growth transforming capacity, more than 90% of the human adult population carries this virus asymptomatically under near perfect immune control. The mode of primary EBV infection is in part responsible for EBV-associated diseases, including Hodgkin's lymphoma. It is, therefore, important to understand which circumstances lead to symptomatic primary EBV infection, called infectious mononucleosis (IM). Innate immune control of lytic viral replication by early-differentiated natural killer (NK) cells was found to attenuate IM symptoms and continuous loss of the respective NK cell subset during the first decade of life might predispose for IM during adolescence. In this review, we discuss the evidence that NK cells are involved in the immune control of EBV, mechanisms by which they might detect and control lytic EBV replication, and compare NK cell subpopulations that expand during different human herpesvirus infections.

Keywords: lytic EBV infection, NKG2D, DNAM-1, infectious mononucleosis, humanized mice

EPSTEIN–BARR VIRUS INFECTION AND PREDISPOSING FACTORS FOR EBV DISEASE

The herpesvirus Epstein–Barr virus (EBV) was discovered in 1964 by electron microscopy in Burkitt's lymphoma, the most common childhood tumor in sub-Saharan Africa (1). It is arguably the most potent human tumor virus, because it readily transforms primary human B cells into immortalized lymphoblastoid cell lines (LCLs) in culture (2). This strong growth transforming capacity is due to the latent EBV proteins, six nuclear antigens (EBNAs) and two latent membrane proteins (LMPs), which are expressed as the default infection program in B cells (3). Lytic EBV replication occurs in LCLs only at low levels and triggers the expression of around 80 gene products under the guidance of the immediate early lytic transactivator BZLF-1 for the production of infectious DNA virus particles (4). In addition to Burkitt's lymphoma, EBV is associated with numerous malignancies, mostly of B and epithelial cell origin, such as Hodgkin's lymphoma and nasopharyngeal carcinoma (3). Despite this strong growth transforming capacity, EBV is carried by more than 90% of the human adult population as an asymptomatic persistent infection.

Epstein–Barr virus infection remains asymptomatic in most persistently infected individuals despite transforming latent EBV protein expression (5). In healthy EBV carriers, the expression of all six EBNAs and the two LMPs can be found in naïve B cells of secondary lymphoid organs like the tonsils (6) (**Figure 1**). In germinal center B cells, only the subset of viral proteins that is also present in Hodgkin's lymphoma is expressed (EBNA1, LMP1 and 2). Finally, in homeostatically proliferating memory B cells, the latency pattern of Burkitt's lymphoma is present with EBNA1 as the only expressed protein (7). Reactivation from this persistent EBV reservoir of memory B cells

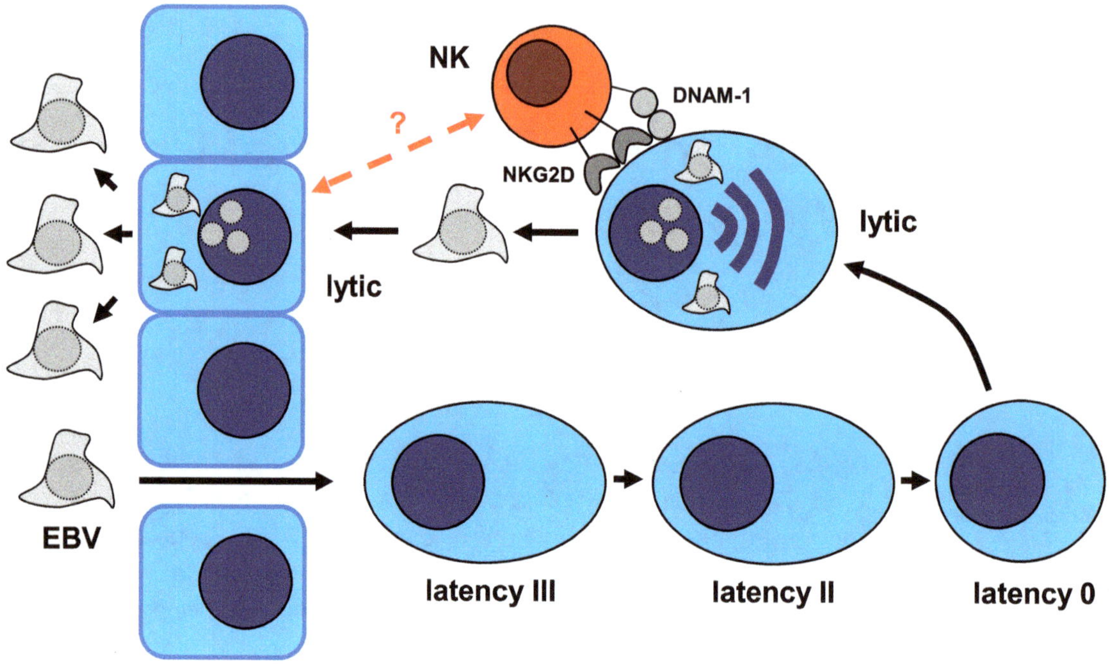

FIGURE 1 | Role of NK cells in the immune control of the EBV life cycle. Epstein–Barr virus (EBV) is transmitted via saliva exchange and infects submucosal B cells. In infected naïve B cells, the latency III EBV program can be found (EBNA1, 2, 3A–C, LP, and LMP1 and 2). Activation via EBV infection drives infected B cells into differentiation. In resulting germinal center B cells, latency II EBV infection can be found (EBNA1, LMP1 and 2). These EBV proteins allow EBV-infected B cells to survive and enter the memory B cell pool. In memory B cells, all EBV proteins expression is switched off (latency 0). Upon B cell receptor cross-linking, the lytic EBV cycle is activated due to plasma cell differentiation, which allows epithelial cell infection for further amplification of infectious virus before shedding into saliva. NK cells target lytically EBV replicating cells via their activating NKG2D and DNAM-1 receptors. It remains unclear if also lytically EBV replicating epithelial cells can be recognized by NK cells.

into lytic replication seems to occur after B cell activation and plasma cell differentiation (8). These findings, however, indicated that healthy EBV carriers are continuously challenged with transforming latent EBV expression programs, which could result in tumor formation without immune control.

Indeed, immune suppression after organ transplantation or due to human immunodeficiency virus (HIV) co-infection causes the occurrence of EBV-associated malignancies, such as post-transplant lymphoproliferative disease (PTLD) and immunoblastic lymphoma (3). Some of these lymphoproliferations can be treated by the adoptive transfer of EBV-specific T cell lines (9). Moreover, some individuals remain EBV seronegative despite carrying the virus, and seem to control persistent EBV infection entirely by cell-mediated immunity (10). Thus, cell-mediated immunity by T cells seems to be sufficient to control persistent EBV infection (11). In addition to direct immune suppression of cell-mediated immunity, the conditions under which this immune response is primed, seems to be decisive for an asymptomatic immune control of EBV infection. Indeed, if primary infection with EBV is delayed into adulthood, the virus is more frequently acquired with symptomatic primary infection, called infectious mononucleosis (IM) (12). This immunopathology by massive $CD8^+$ T cell expansion and activation seems to result from an uncontrolled lytic EBV replication, because most of the expanding $CD8^+$ T cells are directed against lytic EBV antigens (13). This massive lymphocytosis seems to transiently compromise EBV-specific immune control with an increased susceptibility to some EBV-associated malignancies, such as Hodgkin's lymphoma up to 5–10 years after IM (14). In this review, we discuss the innate arm of cell-mediated immune control of EBV, which could explain the different outcomes of primary infection with this tumor virus and might be required to limit initial viral titers so that long-lasting adaptive cell-mediated immune control can be efficiently primed.

PRIMARY IMMUNODEFICIENCIES THAT AFFECT NK CELL FUNCTION AND PREDISPOSE FOR EBV DISEASE

Evidence that loss of cytotoxic cell-mediated immune control predisposes for EBV-associated diseases comes from primary immunodeficiencies that sensitize for EBV-associated malignancies (15, 16). A subset of these affect, in addition to T cell responses, natural killer (NK) cell responses and hint toward an important function of cell-mediated innate immunity in EBV-specific immune control. The underlying genetic lesions affect gene products that are involved in NK cell differentiation, stimulation, and cytotoxic effector function.

Natural killer cell differentiation is disrupted by mutations in GATA-binding protein 2 (GATA2) and minichromosome maintenance complex component 4 (MCM4). Accordingly, a GATA2 mutation was identified later in the first indicator patient with susceptibility to herpesvirus induced diseases (17, 18). GATA2 is a hematopoietic transcription factor that is required for the development of several immune cell lineages, including B cells, CD4$^+$ T cells, dendritic cells, neutrophils, and monocytes in addition to NK cells (19). With respect to EBV-associated diseases, patients with GATA2 mutations have been diagnosed with chronic active EBV infection (CAEBV) and virus-positive smooth muscle tumors (20, 21). In contrast to this multilineage deficiency in patients with GATA2 mutations, partial deficiency of the DNA helicase MCM4 blocked differentiation of the human CD56^{dim} NK cell subset, while other hematopoietic lineages seemed to be unaffected (22). One of the afflicted patients suffered from an EBV-associated lymphoma (23). Thus, compromised NK cell differentiation is associated with uncontrolled EBV infection.

In addition to NK cell differentiation, some mutations that impact NK cell stimulation are associated with EBV disease. These include deficiencies in SLAM-associated protein (SAP) of X-linked lymphoproliferative disease type 1 (XLP1), in magnesium transporter 1 (MAGT1) of X-linked immunodeficiency with magnesium defect, EBV infection, and neoplasia (XMEN), in CD27, in phosphatidylinositol 3 kinase (PI3K) 110δ and in FcγR3A (CD16). XLP1, also known as Duncan's disease, primarily manifests in boys (24). Primary infection with EBV often leads to fatal IM in the affected patients, if they cannot be identified early enough and treated with bone marrow transplantation (25). The underlying mutations in SAP were identified in 1998 (26–28) and affect the adaptor protein of SLAM receptors that mediate their co-activating function in T and NK cells. Two of these SLAM receptors, 2B4 and NTB-A, increase NK cell cytotoxicity (29, 30), but XLP1-associated SAP mutations might primarily compromise EBV-specific CD8$^+$ T cell immune control (31–33). Furthermore, deficiency in the magnesium transporter MAGT1 results in diminished free magnesium levels within cells, which is associated with downregulation of the activating NKG2D receptor on cytotoxic lymphocytes, T, and NK cells (34). Supplementation of magnesium results in decreased EBV loads in the affected XMEN patients. Another activating co-receptor on T and NK cells is CD27. Mutations in this CD70 engaging co-receptor predispose for EBV-associated lymphoproliferations (35, 36). Also, loss-of-function mutations in the signaling molecule PI3K 110δ of activating receptors are associated with persistent EBV viremia (37). Finally, the activating FcγR on NK cells, CD16, seems to be required for EBV-specific immune control. Mutations in CD16 were reported to be associated with persisting IM symptoms (38, 39). These primary immunodeficiencies identify 2B4, NKG2D, CD27, and FcγR as important receptors in EBV-specific cell-mediated immune control.

Apart from these activating receptors, the cytotoxic effector machinery also seems to be important in EBV-specific immune control. Accordingly, mutations in perforin, Munc13-4, and Munc18-2 have been identified in patients with EBV-associated diseases. Mutations in perforin are responsible for type 2 familial hemophagocytic lymphohistiocytosis (FHL2). Persistent IM has been described in one FHL2 patient (40). Munc18-2 and 13-4 mediate docking and activation of syntaxin 11 for cytotoxic granule fusion with the cell membrane, respectively. Mutations in these two components of the cytotoxic machinery were found in patients with CAEBV (41). These genetic lesions point toward a role of cytotoxic lymphocytes in EBV-specific immune control. Primarily, prolonged IM resulting in CAEBV seems to be associated with primary immunodeficiencies that affect NK cell function.

NK CELL EXPANSION DURING PRIMARY EBV INFECTION

Natural killer cell expansion during primary EBV infection has first been reported in a study by Tomkinson et al. (42), in which peripheral NK cells (identified as CD16$^+$ lymphocytes) were described to be significantly increased in both frequency (1.5-fold) and absolute number (4-fold) in – by these measures – a similar manner to CD8$^+$ T cells in a cohort of IM patients. However, since the authors had to use a strategy for gating lymphocytes that included activated and, thus, blasted cells, CD16$^+$ monocytes could not be excluded from the analysis and might account for some of the quantitative changes ascribed to the NK cell compartment. Still, threefold to sixfold increases in the number of bulk NK cells in IM patients were found by other groups as well (43, 44) and these increases were found to be inversely correlated with viral load in blood (43). Likewise, higher NK cell counts tended to be associated with less severe disease (43). On the contrary, a large and, notably, prospective study of primary EBV infection (45), while also reporting expansions of NK cells during the acute phase, positively correlated NK cell numbers with blood viral load and also positively correlated blood viral load with disease severity (45). Similarly, the increase in NK cells in IM patients was related to greater disease severity by another group, although the small number of subjects in that study precluded statistical significance (44). A study by Azzi et al. (46) detailed the phenotype of NK cells during IM and convalescence in pediatric patients and demonstrated the lack of influence of primary EBV infection on the expression of killer cell immunoglobulin-like receptors (KIRs) but instead noted an up to fivefold expansion of an early-differentiated NK cell subset (**Figure 2**). This accumulated NKG2A$^+$KIR$^-$CD57$^-$ NK cell subset was the only identifiable subset within the NK cell compartment that proliferated in the acute phase and importantly, this proliferating early-differentiated NK cell subset also correlated with viral load in PBMCs (46). Although overall NK cell numbers and frequencies contract early after the onset of symptoms (43–46), these early-differentiated NK cells remain elevated in frequency up to 6 months after the acute symptomatic phase (46–48), but over time accumulate signs of differentiation (46, 47). Asymptomatic primary EBV infection is mostly found in young children (49, 50) compared to a symptomatic outcome, i.e., IM, in up to 75% of cases of primary EBV infection in adolescents (45). While asymptomatic infection was associated with high viral load, phenotype, and frequencies of antigen-specific CD8$^+$ T cells similar to IM, the massive expansion of CD8$^+$ T cell numbers typically seen in IM was absent (49). It might be speculated that the confinement of CD8$^+$ T cell expansion is exerted by the EBV-responsive early-differentiated NKG2A$^+$KIR$^-$ NK cell subset, especially since this subset is highest in both frequency and numbers in newborns and

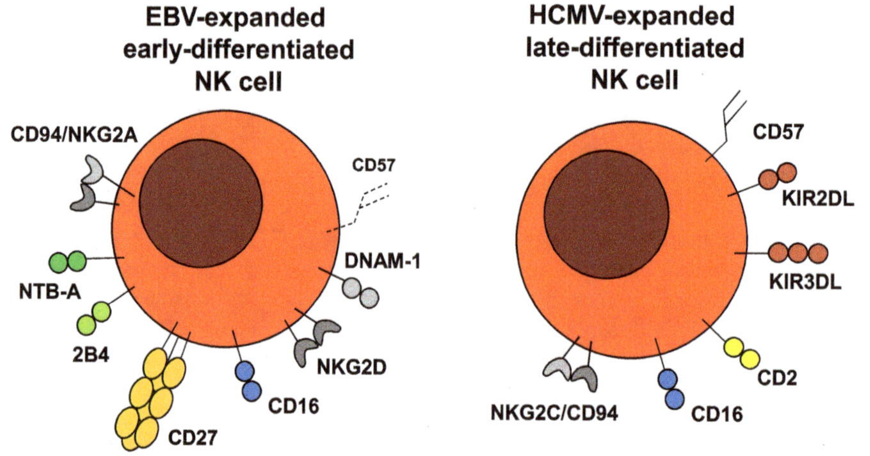

FIGURE 2 | EBV and HCMV expand different NK cell populations. Epstein–Barr virus (EBV) expands early-differentiated NKG2A/CD94, NTB-A, 2B4, CD27, CD16, NKG2D, and DNAM-1-positive NK cells, which after expansion upregulate the senescence marker CD57. Human cytomegalovirus (HCMV) in contrast expands late-differentiated CD94/NKG2C, CD16, CD2, KIR, and CD57-positive NK cells.

young children but decreases with age (46). Whether the loss of early-differentiated NK cells during adolescence is associated with a specific molecular imprint that affects NK cell homeostasis, e.g., the result of changes in the expression of transcription factors, has not yet been explored in the current literature. One explanation for such an age-dependent effect, however, is an increased burden and accumulation of various infectious challenges with advancing years that can likely be expected to have an impact on the differentiation of NK cells. One of these challenges, namely infection with the human cytomegalovirus (HCMV), that seems to drive NK cell differentiation via IL-12 and IL-15 production, is discussed below. Thus, dynamics within the NK cell compartment over time might in part explain the age-dependent occurrence of symptomatic primary EBV infection.

NK CELL REACTIVITY AGAINST LYTIC EBV INFECTION

Indeed, the trigger of peripheral NK cell accumulation in primary EBV infection does not seem to be caused by the inflammatory status of IM itself, e.g., increased levels of pro-inflammatory cytokines, since patients with equally inflammatory conditions but lacking evidence of EBV seroconversion do not show any expansions in their NK cell compartment (46). Instead, there is evidence that the state of the infectious cycle of EBV, either latent or lytic, drives the expansion of NK cells during infection, specifically changes that are inherent to lytic replication. In mice with reconstituted human immune system components (HIS mice), NK cell expansion only occurs during infection with wild-type EBV, but not with recombinant EBV engineered to only establish latent infection (EBV BZLF-1 knockout or BZ1KO EBV) (51). Furthermore, proliferation of NKG2A$^+$KIR$^-$ NK cells was only seen after *in vitro* infection of cord blood with wild-type EBV but not with BZ1KO EBV (46). It is, therefore, conceivable that the expansion of the cytotoxic lymphocyte populations, namely NK and CD8$^+$ T cells, during EBV infection is driven by the amount of available antigen (45, 46, 51), since the expansion of total CD8$^+$ T cell and NK cell numbers as well as viral load correlate (45). Actually, lytic replication might not only be responsible for the expansion of the early-differentiated NK cell subset, but seems to also be a target of NK cells itself (**Figure 1**). In EBV-infected HIS mice, NK cells protect from high viral load, elevated cytokine levels, splenomegaly, weight loss, and occurrence of lymphoproliferative tumors, as well as limit the expansion of CD8$^+$ T cells (51). Most of the protective effects of NK cells are lost in HIS mice only latently infected with EBV, but regained when these mice are infected with a recombinant virus reverted to allow for lytic replication (51). Also, in EBV-infected HIS mice depleted of NK cells, there is an increased abundance of lytic proteins and cell-free viral DNA indicative of ongoing uncontrolled lytic replication (51). *In vitro*, NK cells respond to and kill an EBV-positive B cell line more efficiently when these cells are in the lytic as compared to the latent phase of infection (51–53), in particular NK cells with the NKG2A$^+$KIR$^-$ phenotype (46). The preferential killing of lytic cells was sensitive to blocking of CD112 and ULBP-1, ligands of the activating NK cell receptors DNAM-1 and NKG2D, respectively (52) as well as directly blocking DNAM-1 (53). Therefore, the identification of activating receptors or combinations thereof crucial in NK cell-mediated protection *in vivo* holds promise to further our understanding of the intricate interplay between EBV with the host's immune system and HIS mice might constitute a feasible model to answer such questions (51).

DIFFERENCES BETWEEN EBV-DRIVEN NK CELL EXPANSION AND OTHER HUMAN HERPESVIRUS INFECTIONS

In contrast to EBV infection, other herpesviruses either do not change the NK cell composition, such as recurrent α-herpesvirus infection by herpes simplex virus 2 (HSV2) (54), or expand terminally differentiated NKG2C$^+$KIR$^+$CD16$^+$ NK cells, such as the β-herpesvirus HCMV (55–57) or the γ-herpesvirus Kaposi sarcoma-associated herpesvirus (KSHV) in HIV-infected individuals (58). Accumulation of terminally differentiated NK cells is primarily connected to HCMV infection (**Figure 2**) and it has been argued that in other viral infections, for which such terminal NK cell differentiation can be observed, such as with Hantavirus (59), Chikungunya virus (60), HIV (61), and hepatitis virus (62), mainly HCMV-positive individuals are affected by this alteration in NK cell repertoire composition (63–65).

This HCMV-driven terminal NK cell expansion has been linked to NK cell stimulation by cells that produce the NKG2C ligand HLA-E plus the NK cell proliferation stimulating cytokine IL-15 on their surface (59, 66). Expansion of NKG2C-positive NK cells could be obtained with HCMV infected fibroblasts plus IL-15 (67) and bystander monocytes were able to provide NK cell stimulating cytokines, including IL-12 (68). However, HCMV-infected individuals with NKG2C deficiency also develop NK cell populations that more vigorously secrete IFN-γ upon stimulation, the so-called adaptive NK cell populations (69), and the NKG2C genotype does not affect the outcome of congenital HCMV infection (57). Therefore, HLA-E-mediated NK cell stimulation might not be essential for the expansion and anti-viral function of NKG2C-positive NK cell populations, but IL-15 and IL-12 might be more important (65). Accordingly, one patient with IL-12Rβ1 deficiency did not carry adaptive NK cell populations (70). Therefore, cytokines might be one of the main drivers of adaptive NK cell expansion, as originally proposed in mice (71). These adaptive NK cells are terminally differentiated NKG2C-positive NK cells during HCMV infection, while for the early-differentiated NK cells that expand and persist for 6 months during acute EBV infection adaptive features have still to be investigated.

In contrast to direct immune control of lytic EBV replication by early-differentiated NK cells (51), the role of terminally differentiated NK cells is less clear during HCMV infection. Only for decidual NKG2C-positive NK cells it was shown that they directly kill HCMV-infected autologous decidual fibroblasts in an HLA-E dependent fashion (72). Most studies, however, implicate the NKG2C-positive NK cell subset that expands during HCMV infection in mediating superior antibody-dependent cellular cytotoxicity (ADCC) against antibody opsonized HCMV-infected macrophages or fibroblasts (73, 74). In these studies, both IFN-γ production and degranulation of NKG2C-positive NK cells of HCMV-infected donors were superior upon opsonized target recognition compared to NKG2C-negative NK cell populations. These superior effector functions most likely result from epigenetic modifications, as has been shown for the IFN-γ gene locus in NKG2C-positive NK cells of HCMV infected individuals (75–78). Thus, early-differentiated NK cells that expand during EBV infection might directly recognize lytically EBV replicating targets, while the terminally differentiated NK cells in HCMV-infected individuals mainly promote ADCC.

AUTHOR CONTRIBUTIONS

All authors listed have made substantial, direct, and intellectual contribution to the work and approved it for publication.

REFERENCES

1. Epstein MA, Henle G, Achong BG, Barr YM. Morphological and biological studies on a virus in cultured lymphoblasts from Burkitt's lymphoma. *J Exp Med* (1964) 121:761–70. doi:10.1084/jem.121.5.761
2. Nilsson K, Klein G, Henle W, Henle G. The establishment of lymphoblastoid lines from adult and fetal human lymphoid tissue and its dependence on EBV. *Int J Cancer* (1971) 8:443–50. doi:10.1002/ijc.2910080312
3. Cesarman E. Gammaherpesviruses and lymphoproliferative disorders. *Annu Rev Pathol* (2014) 9:349–72. doi:10.1146/annurev-pathol-012513-104656
4. Mckenzie J, El-Guindy A. Epstein-Barr virus lytic cycle reactivation. *Curr Top Microbiol Immunol* (2015) 391:237–61. doi:10.1007/978-3-319-22834-1_8
5. Thorley-Lawson DA. EBV persistence – introducing the virus. *Curr Top Microbiol Immunol* (2015) 390:151–209. doi:10.1007/978-3-319-22822-8_8
6. Babcock JG, Hochberg D, Thorley-Lawson AD. The expression pattern of Epstein-Barr virus latent genes in vivo is dependent upon the differentiation stage of the infected B cell. *Immunity* (2000) 13:497–506. doi:10.1016/S1074-7613(00)00049-2
7. Hochberg D, Middeldorp JM, Catalina M, Sullivan JL, Luzuriaga K, Thorley-Lawson DA. Demonstration of the Burkitt's lymphoma Epstein-Barr virus phenotype in dividing latently infected memory cells in vivo. *Proc Natl Acad Sci U S A* (2004) 101:239–44. doi:10.1073/pnas.2237267100
8. Laichalk LL, Thorley-Lawson DA. Terminal differentiation into plasma cells initiates the replicative cycle of Epstein-Barr virus in vivo. *J Virol* (2005) 79:1296–307. doi:10.1128/JVI.79.2.1296-1307.2005
9. Gottschalk S, Rooney CM, Heslop HE. Post-transplant lymphoproliferative disorders. *Annu Rev Med* (2005) 56:29–44. doi:10.1146/annurev.med.56.082103.104727
10. Savoldo B, Cubbage ML, Durett AG, Goss J, Huls MH, Liu Z, et al. Generation of EBV-specific CD4$^+$ cytotoxic T cells from virus naive individuals. *J Immunol* (2002) 168:909–18. doi:10.4049/jimmunol.168.2.909
11. Taylor GS, Long HM, Brooks JM, Rickinson AB, Hislop AD. The immunology of Epstein-Barr virus-induced disease. *Annu Rev Immunol* (2015) 33:787–821. doi:10.1146/annurev-immunol-032414-112326
12. Luzuriaga K, Sullivan JL. Infectious mononucleosis. *N Engl J Med* (2010) 362:1993–2000. doi:10.1056/NEJMcp1001116
13. Callan MF, Tan L, Annels N, Ogg GS, Wilson JD, O'Callaghan CA, et al. Direct visualization of antigen-specific CD8$^+$ T cells during the primary immune response to Epstein-Barr virus in vivo. *J Exp Med* (1998) 187:1395–402. doi:10.1084/jem.187.9.1395
14. Hjalgrim H, Askling J, Rostgaard K, Hamilton-Dutoit S, Frisch M, Zhang JS, et al. Characteristics of Hodgkin's lymphoma after infectious mononucleosis. *N Engl J Med* (2003) 349:1324–32. doi:10.1056/NEJMoa023141
15. Jouanguy E, Gineau L, Cottineau J, Beziat V, Vivier E, Casanova JL. Inborn errors of the development of human natural killer cells. *Curr Opin Allergy Clin Immunol* (2013) 13:589–95. doi:10.1097/ACI.0000000000000011

16. Cohen JI. Primary immunodeficiencies associated with EBV disease. *Curr Top Microbiol Immunol* (2015) 390:241–65. doi:10.1007/978-3-319-22822-8_10
17. Biron CA, Byron KS, Sullivan JL. Severe herpesvirus infections in an adolescent without natural killer cells. *N Engl J Med* (1989) 320:1731–5. doi:10.1056/NEJM198906293202605
18. Mace EM, Hsu AP, Monaco-Shawver L, Makedonas G, Rosen JB, Dropulic L, et al. Mutations in GATA2 cause human NK cell deficiency with specific loss of the $CD56^{bright}$ subset. *Blood* (2013) 121:2669–77. doi:10.1182/blood-2012-09-453969
19. Collin M, Dickinson R, Bigley V. Haematopoietic and immune defects associated with GATA2 mutation. *Br J Haematol* (2015) 169:173–87. doi:10.1111/bjh.13317
20. Hsu AP, Sampaio EP, Khan J, Calvo KR, Lemieux JE, Patel SY, et al. Mutations in GATA2 are associated with the autosomal dominant and sporadic monocytopenia and mycobacterial infection (MonoMAC) syndrome. *Blood* (2011) 118:2653–5. doi:10.1182/blood-2011-05-356352
21. Spinner MA, Sanchez LA, Hsu AP, Shaw PA, Zerbe CS, Calvo KR, et al. GATA2 deficiency: a protean disorder of hematopoiesis, lymphatics, and immunity. *Blood* (2014) 123:809–21. doi:10.1182/blood-2013-07-515528
22. Gineau L, Cognet C, Kara N, Lach FP, Dunne J, Veturi U, et al. Partial MCM4 deficiency in patients with growth retardation, adrenal insufficiency, and natural killer cell deficiency. *J Clin Invest* (2012) 122:821–32. doi:10.1172/JCI61014
23. Eidenschenk C, Dunne J, Jouanguy E, Fourlinnie C, Gineau L, Bacq D, et al. A novel primary immunodeficiency with specific natural-killer cell deficiency maps to the centromeric region of chromosome 8. *Am J Hum Genet* (2006) 78:721–7. doi:10.1086/503269
24. Purtilo DT, Cassel C, Yang JP. Letter: fatal infectious mononucleosis in familial lymphohistiocytosis. *N Engl J Med* (1974) 291:736. doi:10.1056/NEJM197410032911415
25. Marsh RA, Bleesing JJ, Chandrakasan S, Jordan MB, Davies SM, Filipovich AH. Reduced-intensity conditioning hematopoietic cell transplantation is an effective treatment for patients with SLAM-associated protein deficiency/X-linked lymphoproliferative disease type 1. *Biol Blood Marrow Transplant* (2014) 20:1641–5. doi:10.1016/j.bbmt.2014.06.003
26. Coffey AJ, Brooksbank RA, Brandau O, Oohashi T, Howell GR, Bye JM, et al. Host response to EBV infection in X-linked lymphoproliferative disease results from mutations in an SH2-domain encoding gene. *Nat Genet* (1998) 20:129–35. doi:10.1038/2424
27. Nichols KE, Harkin DP, Levitz S, Krainer M, Kolquist KA, Genovese C, et al. Inactivating mutations in an SH2 domain-encoding gene in X-linked lymphoproliferative syndrome. *Proc Natl Acad Sci U S A* (1998) 95:13765–70. doi:10.1073/pnas.95.23.13765
28. Sayos J, Wu C, Morra M, Wang N, Zhang X, Allen D, et al. The X-linked lymphoproliferative-disease gene product SAP regulates signals induced through the co-receptor SLAM. *Nature* (1998) 395:462–9. doi:10.1038/26683
29. Parolini S, Bottino C, Falco M, Augugliaro R, Giliani S, Franceschini R, et al. X-linked lymphoproliferative disease. 2B4 molecules displaying inhibitory rather than activating function are responsible for the inability of natural killer cells to kill Epstein-Barr virus-infected cells. *J Exp Med* (2000) 192:337–46. doi:10.1084/jem.192.3.337
30. Bottino C, Falco M, Parolini S, Marcenaro E, Augugliaro R, Sivori S, et al. NTB-A, a novel SH2D1A-associated surface molecule contributing to the inability of natural killer cells to kill Epstein-Barr virus-infected B cells in X-linked lymphoproliferative disease. *J Exp Med* (2001) 194:235–46. doi:10.1084/jem.194.3.235
31. Palendira U, Low C, Chan A, Hislop AD, Ho E, Phan TG, et al. Molecular pathogenesis of EBV susceptibility in XLP as revealed by analysis of female carriers with heterozygous expression of SAP. *PLoS Biol* (2011) 9:e1001187. doi:10.1371/journal.pbio.1001187
32. Palendira U, Low C, Bell AI, Ma CS, Abbott RJM, Phan TG, et al. Expansion of somatically reverted memory $CD8^{+}$ T cells in patients with X-linked lymphoproliferative disease caused by selective pressure from Epstein-Barr virus. *J Exp Med* (2012) 209:911–22. doi:10.1084/jem.20112391
33. Chijioke O, Marcenaro E, Moretta A, Capaul R, Münz C. The SAP-dependent 2B4 receptor mediates $CD8^{+}$ T cell dependent immune control of Epstein Barr virus infection in mice with reconstituted human immune system components. *J Infect Dis* (2015) 212:803–7. doi:10.1093/infdis/jiv114
34. Chaigne-Delalande B, Li FY, O'Connor GM, Lukacs MJ, Jiang P, Zheng L, et al. $Mg2^{+}$ regulates cytotoxic functions of NK and CD8 T cells in chronic EBV infection through NKG2D. *Science* (2013) 341:186–91. doi:10.1126/science.1240094
35. Salzer E, Daschkey S, Choo S, Gombert M, Santos-Valente E, Ginzel S, et al. Combined immunodeficiency with life-threatening EBV-associatedlymphoproliferative disorder in patients lacking functional CD27. *Haematologica* (2012) 98:473–8. doi:10.3324/haematol.2012.068791
36. Van Montfrans JM, Hoepelman AI, Otto S, Van Gijn M, Van De Corput L, De Weger RA, et al. CD27 deficiency is associated with combined immunodeficiency and persistent symptomatic EBV viremia. *J Allergy Clin Immunol* (2012) 129(3):787–93.e6. doi:10.1016/j.jaci.2011.11.013
37. Kuehn HS, Niemela JE, Rangel-Santos A, Zhang M, Pittaluga S, Stoddard JL, et al. Loss-of-function of the protein kinase C delta (PKCdelta) causes a B-cell lymphoproliferative syndrome in humans. *Blood* (2013) 121:3117–25. doi:10.1182/blood-2012-12-469544
38. De Vries E, Koene HR, Vossen JM, Gratama JW, Von Dem Borne AE, Waaijer JL, et al. Identification of an unusual Fc gamma receptor IIIa (CD16) on natural killer cells in a patient with recurrent infections. *Blood* (1996) 88:3022–7.
39. Grier JT, Forbes LR, Monaco-Shawver L, Oshinsky J, Atkinson TP, Moody C, et al. Human immunodeficiency-causing mutation defines CD16 in spontaneous NK cell cytotoxicity. *J Clin Invest* (2012) 122:3769–80. doi:10.1172/JCI64837
40. Katano H, Ali MA, Patera AC, Catalfamo M, Jaffe ES, Kimura H, et al. Chronic active Epstein-Barr virus infection associated with mutations in perforin that impair its maturation. *Blood* (2004) 103:1244–52. doi:10.1182/blood-2003-06-2171
41. Rohr J, Beutel K, Maul-Pavicic A, Vraetz T, Thiel J, Warnatz K, et al. Atypical familial hemophagocytic lymphohistiocytosis due to mutations in UNC13D and STXBP2 overlaps with primary immunodeficiency diseases. *Haematologica* (2010) 95:2080–7. doi:10.3324/haematol.2010.029389
42. Tomkinson BE, Wagner DK, Nelson DL, Sullivan JL. Activated lymphocytes during acute Epstein-Barr virus infection. *J Immunol* (1987) 139:3802–7.
43. Williams H, Mcaulay K, Macsween KF, Gallacher NJ, Higgins CD, Harrison N, et al. The immune response to primary EBV infection: a role for natural killer cells. *Br J Haematol* (2005) 129:266–74. doi:10.1111/j.1365-2141.2005.05452.x
44. Zhang Y, Wallace DL, De Lara CM, Ghattas H, Asquith B, Worth A, et al. In vivo kinetics of human natural killer cells: the effects of ageing and acute and chronic viral infection. *Immunology* (2007) 121:258–65. doi:10.1111/j.1365-2567.2007.02573.x
45. Balfour HH Jr, Odumade OA, Schmeling DO, Mullan BD, Ed JA, Knight JA, et al. Behavioral, virologic, and immunologic factors associated with acquisition and severity of primary Epstein-Barr virus infection in university students. *J Infect Dis* (2013) 207:80–8. doi:10.1093/infdis/jis646
46. Azzi T, Lünemann A, Murer A, Ueda S, Beziat V, Malmberg KJ, et al. Role for early-differentiated natural killer cells in infectious mononucleosis. *Blood* (2014) 124:2533–43. doi:10.1182/blood-2014-01-553024
47. Hendricks DW, Balfour HH Jr, Dunmire SK, Schmeling DO, Hogquist KA, Lanier LL. Cutting edge: $NKG2C^{hi}CD57^{+}$ NK cells respond specifically to acute infection with cytomegalovirus and not Epstein-Barr virus. *J Immunol* (2014) 192:4492–6. doi:10.4049/jimmunol.1303211
48. Dunmire SK, Grimm JM, Schmeling DO, Balfour HH Jr, Hogquist KA. The incubation period of primary Epstein-Barr virus infection: viral dynamics and immunologic events. *PLoS Pathog* (2015) 11:e1005286. doi:10.1371/journal.ppat.1005286
49. Jayasooriya S, De Silva TI, Njie-Jobe J, Sanyang C, Leese AM, Bell AI, et al. Early virological and immunological events in asymptomatic Epstein-Barr virus infection in African children. *PLoS Pathog* (2015) 11:e1004746. doi:10.1371/journal.ppat.1004746
50. Gantt S, Orem J, Krantz EM, Morrow RA, Selke S, Huang ML, et al. Prospective characterization of the risk factors for transmission and symptoms of primary human herpesvirus infections among Ugandan infants. *J Infect Dis* (2016) 214:36–44. doi:10.1093/infdis/jiw076
51. Chijioke O, Muller A, Feederle R, Barros MH, Krieg C, Emmel V, et al. Human natural killer cells prevent infectious mononucleosis features by targeting

lytic Epstein-Barr virus infection. *Cell Rep* (2013) 5:1489–98. doi:10.1016/j.celrep.2013.11.041

52. Pappworth IY, Wang EC, Rowe M. The switch from latent to productive infection in Epstein-Barr virus-infected B cells is associated with sensitization to NK cell killing. *J Virol* (2007) 81:474–82. doi:10.1128/JVI.01777-06
53. Williams LR, Quinn LL, Rowe M, Zuo JM. Induction of the lytic cycle sensitizes Epstein-Barr virus-infected B cells to NK cell killing that is counteracted by virus-mediated NK cell evasion mechanisms in the late lytic cycle. *J Virol* (2016) 90:947–58. doi:10.1128/JVI.01932-15
54. Bjorkstrom NK, Svensson A, Malmberg KJ, Eriksson K, Ljunggren HG. Characterization of natural killer cell phenotype and function during recurrent human HSV-2 infection. *PLoS One* (2011) 6:e27664. doi:10.1371/journal.pone.0027664
55. Guma M, Angulo A, Vilches C, Gomez-Lozano N, Malats N, Lopez-Botet M. Imprint of human cytomegalovirus infection on the NK cell receptor repertoire. *Blood* (2004) 104:3664–71. doi:10.1182/blood-2004-05-2058
56. Monsivais-Urenda A, Noyola-Cherpitel D, Hernandez-Salinas A, Garcia-Sepulveda C, Romo N, Baranda L, et al. Influence of human cytomegalovirus infection on the NK cell receptor repertoire in children. *Eur J Immunol* (2010) 40:1418–27. doi:10.1002/eji.200939898
57. Noyola DE, Alarcon A, Noguera-Julian A, Muntasell A, Munoz-Almagro C, Garcia J, et al. Dynamics of the NK-cell subset redistribution induced by cytomegalovirus infection in preterm infants. *Hum Immunol* (2015) 76:118–23. doi:10.1016/j.humimm.2015.01.017
58. Goodier MR, Mela CM, Steel A, Gazzard B, Bower M, Gotch F. NKG2C^{+} NK cells are enriched in AIDS patients with advanced-stage Kaposi's sarcoma. *J Virol* (2007) 81:430–3. doi:10.1128/JVI.01567-06
59. Bjorkstrom NK, Lindgren T, Stoltz M, Fauriat C, Braun M, Evander M, et al. Rapid expansion and long-term persistence of elevated NK cell numbers in humans infected with hantavirus. *J Exp Med* (2011) 208:13–21. doi:10.1084/jem.20100762
60. Petitdemange C, Becquart P, Wauquier N, Beziat V, Debre P, Leroy EM, et al. Unconventional repertoire profile is imprinted during acute chikungunya infection for natural killer cells polarization toward cytotoxicity. *PLoS Pathog* (2011) 7:e1002268. doi:10.1371/journal.ppat.1002268
61. Mavilio D, Lombardo G, Benjamin J, Kim D, Follman D, Marcenaro E, et al. Characterization of CD56^{-}/CD16^{+} natural killer (NK) cells: a highly dysfunctional NK subset expanded in HIV-infected viremic individuals. *Proc Natl Acad Sci U S A* (2005) 102:2886–91. doi:10.1073/pnas.0409872102
62. Oliviero B, Varchetta S, Paudice E, Michelone G, Zaramella M, Mavilio D, et al. Natural killer cell functional dichotomy in chronic hepatitis B and chronic hepatitis C virus infections. *Gastroenterology* (2009) 137:1151–60, 1160.e1–7. doi:10.1053/j.gastro.2009.05.047
63. Guma M, Cabrera C, Erkizia I, Bofill M, Clotet B, Ruiz L, et al. Human cytomegalovirus infection is associated with increased proportions of NK cells that express the CD94/NKG2C receptor in aviremic HIV-1-positive patients. *J Infect Dis* (2006) 194:38–41. doi:10.1086/504719
64. Beziat V, Dalgard O, Asselah T, Halfon P, Bedossa P, Boudifa A, et al. CMV drives clonal expansion of NKG2C+ NK cells expressing self-specific KIRs in chronic hepatitis patients. *Eur J Immunol* (2012) 42:447–57. doi:10.1002/eji.201141826
65. Rolle A, Brodin P. Immune adaptation to environmental influence: the case of NK cells and HCMV. *Trends Immunol* (2016) 37:233–43. doi:10.1016/j.it.2016.01.005
66. Beziat V, Liu LL, Malmberg JA, Ivarsson MA, Sohlberg E, Bjorklund AT, et al. NK cell responses to cytomegalovirus infection lead to stable imprints in the human KIR repertoire and involve activating KIRs. *Blood* (2013) 121:2678–88. doi:10.1182/blood-2012-10-459545
67. Guma M, Budt M, Saez A, Brckalo T, Hengel H, Angulo A, et al. Expansion of CD94/NKG2C^{+} NK cells in response to human cytomegalovirus-infected fibroblasts. *Blood* (2006) 107:3624–31. doi:10.1182/blood-2005-09-3682
68. Rolle A, Pollmann J, Ewen EM, Le VT, Halenius A, Hengel H, et al. IL-12-producing monocytes and HLA-E control HCMV-driven NKG2C^{+} NK cell expansion. *J Clin Invest* (2014) 124:5305–16. doi:10.1172/JCI77440
69. Liu LL, Landskron J, Ask EH, Enqvist M, Sohlberg E, Traherne JA, et al. Critical role of CD2 co-stimulation in adaptive natural killer cell responses revealed in NKG2C-deficient humans. *Cell Rep* (2016) 15:1088–99. doi:10.1016/j.celrep.2016.04.005
70. Simhadri VR, Mariano JL, Zenarruzabeitia O, Seroogy CM, Holland SM, Kuehn HS, et al. Intact IL-12 signaling is necessary for the generation of human natural killer cells with enhanced effector function after restimulation. *J Allergy Clin Immunol* (2014) 134(5):1190–3.e1. doi:10.1016/j.jaci.2014.06.006
71. Cooper MA, Elliott JM, Keyel PA, Yang L, Carrero JA, Yokoyama WM. Cytokine-induced memory-like natural killer cells. *Proc Natl Acad Sci U S A* (2009) 106:1915–9. doi:10.1073/pnas.0813192106
72. Siewiera J, El Costa H, Tabiasco J, Berrebi A, Cartron G, Le Bouteiller P, et al. Human cytomegalovirus infection elicits new decidual natural killer cell effector functions. *PLoS Pathog* (2013) 9:e1003257. doi:10.1371/journal.ppat.1003257
73. Wu Z, Sinzger C, Frascaroli G, Reichel J, Bayer C, Wang L, et al. Human cytomegalovirus-induced NKG2C^{hi} CD57^{hi} natural killer cells are effectors dependent on humoral antiviral immunity. *J Virol* (2013) 87:7717–25. doi:10.1128/JVI.01096-13
74. Costa-Garcia M, Vera A, Moraru M, Vilches C, Lopez-Botet M, Muntasell A. Antibody-mediated response of NKG2Cbright NK cells against human cytomegalovirus. *J Immunol* (2015) 194:2715–24. doi:10.4049/jimmunol.1402281
75. Zhang T, Scott JM, Hwang I, Kim S. Cutting edge: antibody-dependent memory-like NK cells distinguished by FcRgamma deficiency. *J Immunol* (2013) 190:1402–6. doi:10.4049/jimmunol.1203034
76. Luetke-Eversloh M, Hammer Q, Durek P, Nordstrom K, Gasparoni G, Pink M, et al. Human cytomegalovirus drives epigenetic imprinting of the IFNG locus in NKG2C^{hi} natural killer cells. *PLoS Pathog* (2014) 10:e1004441. doi:10.1371/journal.ppat.1004441
77. Lee J, Zhang T, Hwang I, Kim A, Nitschke L, Kim M, et al. Epigenetic modification and antibody-dependent expansion of memory-like NK cells in human cytomegalovirus-infected individuals. *Immunity* (2015) 42:431–42. doi:10.1016/j.immuni.2015.02.013
78. Schlums H, Cichocki F, Tesi B, Theorell J, Beziat V, Holmes TD, et al. Cytomegalovirus infection drives adaptive epigenetic diversification of NK cells with altered signaling and effector function. *Immunity* (2015) 42:443–56. doi:10.1016/j.immuni.2015.02.008

26

Regulatory Functions of Natural Killer Cells in Multiple Sclerosis

*Catharina C. Gross[1], Andreas Schulte-Mecklenbeck[1], Heinz Wiendl[1], Emanuela Marcenaro[2,3], Nicole Kerlero de Rosbo[4], Antonio Uccelli[2,4,5] and Alice Laroni[2,4,5]**

[1] Department of Neurology, University Hospital Münster, Münster, Germany, [2] Centre of Excellence for Biomedical Research, University of Genova, Genova, Italy, [3] Department of Experimental Medicine, University of Genova, Genova, Italy, [4] Department of Neuroscience, Rehabilitation, Ophthalmology, Genetics, Maternal and Child Health, University of Genova, Genova, Italy, [5] IRCCS San Martino-IST, Genova, Italy

Edited by:
Miguel López-Botet,
IMIM (Hospital del Mar Research Institute), Spain

Reviewed by:
Debby Burshtyn,
University of Alberta, Canada
Jacques Zimmer,
Centre de Recherche Public de la Santé, Luxembourg

***Correspondence:**
Alice Laroni
alice.laroni@unige.it

There is increasing evidence that natural killer (NK) cells exhibit regulatory features. Among them, CD56^{bright} NK cells have been suggested to play a major role in controlling T cell responses and maintaining homeostasis. Dysfunction in NK cell-mediated regulatory features has been recently described in untreated multiple sclerosis (MS), suggesting a contribution to MS pathogenesis. Moreover, biological disease-modifying treatments effective in MS apparently enhance the frequencies and/or regulatory function of NK cells, further pointing toward an immunoprotective role of NK cells in MS. Here, we summarize the current knowledge on the regulatory functions of NK cells, based on their interactions with other cells belonging to the innate compartment, as well as with adaptive effector cells. We review the more recent data reporting disruption of NK cell/T cell interactions in MS and discuss how disease-modifying treatments for MS affect NK cells.

Keywords: natural killer cells, CD56^{bright} NK cells, CD56^{dim} NK cells, regulatory immune cells, innate immune system, multiple sclerosis

NATURAL KILLER (NK) CELLS AS CONTROLLERS OF (AUTO)IMMUNE RESPONSES

Regulatory Features of NK Cells

Autoimmune reactivity occurs in every subject, but only 5–10% of humans develop an autoimmune disease (1). Keeping autoreactive cells under control and, thus, preventing them to cause disease is the task of specialized immune cell subsets, called "regulatory" cells. The best characterized regulatory immune cell populations belong to the adaptive immune system (IS) and include regulatory T cells (T_{regs}), type-1 regulatory T cells, and regulatory B cells. However, there is increasing evidence that the innate IS also plays an important role in controlling autoreactive cells.

An important population of regulatory immune cells belongs to the natural killer (NK) cells. These, so-called CD56^{bright} NK cells, owe their name to high surface expression of CD56 (also known as neural cell adhesion molecule), are CD16$^{-/dim}$, express the inhibitory receptor NKG2A, and do not express killer cell immunoglobulin-like receptors (KIR). CD56^{bright} NK cells were first considered "immunoregulatory" by Cooper et al., due to increased production of cytokines and reduced cytotoxicity compared to CD56^{dim} NK cells (2).

It is now established that CD56^{bright} NK cells regulate other immune cells belonging to both the innate and adaptive IS. Although most studies on CD56^{bright} NK cell function have been conducted *ex vivo* with cells purified from peripheral blood, lymph nodes (LNs) are likely a key place where CD56^{bright} NK cells exert their regulatory function (3), since they preferentially home to parafollicular

T cell areas (4) where immune responses develop. In addition to $CD56^{bright}$ NK cells, the major NK cell subset in peripheral blood, $CD56^{dim}$ NK cells, which derive from $CD56^{bright}$ NK cells and are more differentiated, also exert regulatory functions as discussed below.

Interactions between Regulatory NK Cells and Innate Immune Cells

$CD56^{bright}$ NK cells express receptors for cytokines such as interleukin (IL)-12, IL-15, and IL-18 (5–7), which are produced by activated antigen-presenting cells (APCs). These cytokines can trigger proliferation of $CD56^{bright}$ NK cells and their production of molecules such as IFN-γ, IL-10 and IL-13, TNF-β, and GM-CSF (2). In this context, Ferlazzo et al. demonstrated that dendritic cells (DCs) are a key source of IL-12 and IL-15 for activation of $CD56^{bright}$ NK cells (8), and we have shown that DC-derived IL-27 can modulate proliferation and function of these cells (9). Thus, APCs modulate NK cell functions and phenotype (10–13). Infections most likely modulate the function of $CD56^{bright}$ NK cells indirectly through APCs, because co-culturing $CD56^{bright}$ with APCs activated via TLR4 (macrophages, DC) or TLR9 (plasmacytoid DCs) stimulates their proliferation and cytokine production (2, 8, 14, 15). Conversely, activated NK cells modulate the function of APCs: they stimulate monocytes to produce TNF-α (16) and kill immature DCs in a process called DC editing (17, 18).

Interactions between Regulatory NK Cells and Adaptive Immune Cells

Natural killer cells also interact with adaptive effector cells. IFN-γ secreted by $CD56^{bright}$ NK cells in response to T cell-derived IL-2 has been demonstrated to stimulate T cells in LNs (4). Along this line, increased local bioavailability of IL-2 by blocking the IL-2Rα chain (CD25) on recently activated T cells upon treatment with daclizumab is associated with expansion and activation of $CD56^{bright}$ NK cells in multiple sclerosis (MS) patients (19–21). Indeed, while T cells express the high-affinity form of the IL-2 receptor, which comprises CD25, $CD56^{bright}$ NK cells express both high-affinity and intermediate-affinity (not comprising CD25) forms of the IL-2 receptor (20, 22). Thus, upon daclizumab treatment, NK cells are stimulated through binding of IL-2 to their intermediate-affinity receptor. This results in control of T cell activation through direct killing (19, 21), which, for the $CD56^{bright}$ subset, involves release of cytotoxic granzyme K (23). Furthermore, IL-27-stimulated $CD56^{bright}$ NK cells have been shown to suppress the proliferation of autologous $CD4^+$ T cells in a contact-dependent manner associated with increased perforin content (9). $CD56^{bright}$ NK cells, stimulated with the pro-inflammatory cytokines IL-12 and IL-15, prevent autologous $CD4^+$ T cell proliferation through a cytotoxic mechanism involving the engagement of the natural cytotoxicity receptors (NCRs), such as NKp30 and NKp46 (24), on NK cells and the release of granzyme B (25). $CD56^{bright}$ NK cells were also shown to inhibit proliferation of autologous $CD4^+$ T cells by secreting the immunosuppressive molecule adenosine. Inhibition of CD38 ("ADP ribosyl-cyclase"), an enzyme involved in the production of adenosine, restored proliferation of T cells in the presence of $CD56^{bright}$ NK cells (26). While these studies described the effects of $CD56^{bright}$ NK cells on T cells undergoing activation, others reported direct cytotoxicity of $CD56^{bright}$ NK cells on previously activated T cells. Nielsen and coauthors found that killing of pre-activated T cells by $CD56^{bright}$ NK cells involves the activating receptors NKG2D, LFA-1, and TRAIL and is enhanced when blocking NKG2A (27). Another study demonstrated that both $CD56^{bright}$ and $CD56^{dim}$ NK cells kill autologous antigen-activated $CD4^+$ T cells through engagement of DNAM-1 and 2B4 and their cognate receptors CD155 and CD48, respectively (21). These and other studies reveal that different stimuli activate NK cells toward cytotoxicity and/or suppression of T cell proliferation (**Figure 1**).

POSSIBLE ROLE OF NK CELLS IN MS AND ITS ANIMAL MODEL

Multiple sclerosis is an autoimmune disease of the central nervous system (CNS) characterized by an attack of the myelin sheath that surrounds and protects CNS axons by autoreactive T cells. Its murine model, experimental autoimmune encephalomyelitis (EAE), is triggered by active immunization with myelin antigens or transfer of activated autoreactive myelin-specific T cells to naïve recipients. Until recently, defects in regulatory mechanisms had only been described in MS in cells of the adaptive compartment (28).

Conflicting data on a beneficial vs. detrimental role of NK cells in EAE have been published (29–31), but studies on regulatory NK cells in mice are difficult to translate into humans, because murine NK cells do not express CD56 and the murine counterparts of $CD56^{bright}$ and $CD56^{dim}$ subsets have not been identified with certainty.

Enhancing regulatory features of NK cells ameliorates the disease course of EAE. In particular, NK cells expressing NKG2A (which is expressed by all $CD56^{bright}$ NK cells in humans) were shown to decrease CNS inflammation by killing T cells and microglial cells, when the interaction between NKG2A and its ligand Qa-1 (the murine equivalent to the human HLA-E) expressed on the target cells was blocked by antibodies specific for either antigen (32, 33). Decreased expression of Qa-1 on microglial cells upon CNS inflammation rendered them more sensitive to $NKG2A^+$ NK cell-mediated lysis. Importantly, enrichment of NK cells through treatment with IL-2 coupled with IL-2 mAb ("IL-2 complexes") (34) ameliorated EAE (35). A recent work from the same group has shed light on the differential effects of NK cells in early vs. late stages of EAE and possibly MS (36), which may depend on their interactions with neural stem cells (NSCs). Indeed, Liu et al. found that, in MS and EAE brains, NK cells are in contact with NSCs and that, in EAE NSCs released IL-15 upon contact with NK cells, thereby supporting NK cell proliferation and survival; in turn, NK cells killed NSCs, particularly during the late stages of EAE, as a result of reduced expression of Qa-1 on NSCs. Accordingly, removal of NK cells during the late phase of EAE reduced disease severity (36). These observations suggest that cytotoxicity of NK cells may be a double-edged sword in EAE, with NK cells attenuating inflammation in the acute phase of disease by killing immune cells (T cells, microglial cells), but impairing potential repair during the late stage by killing NSCs.

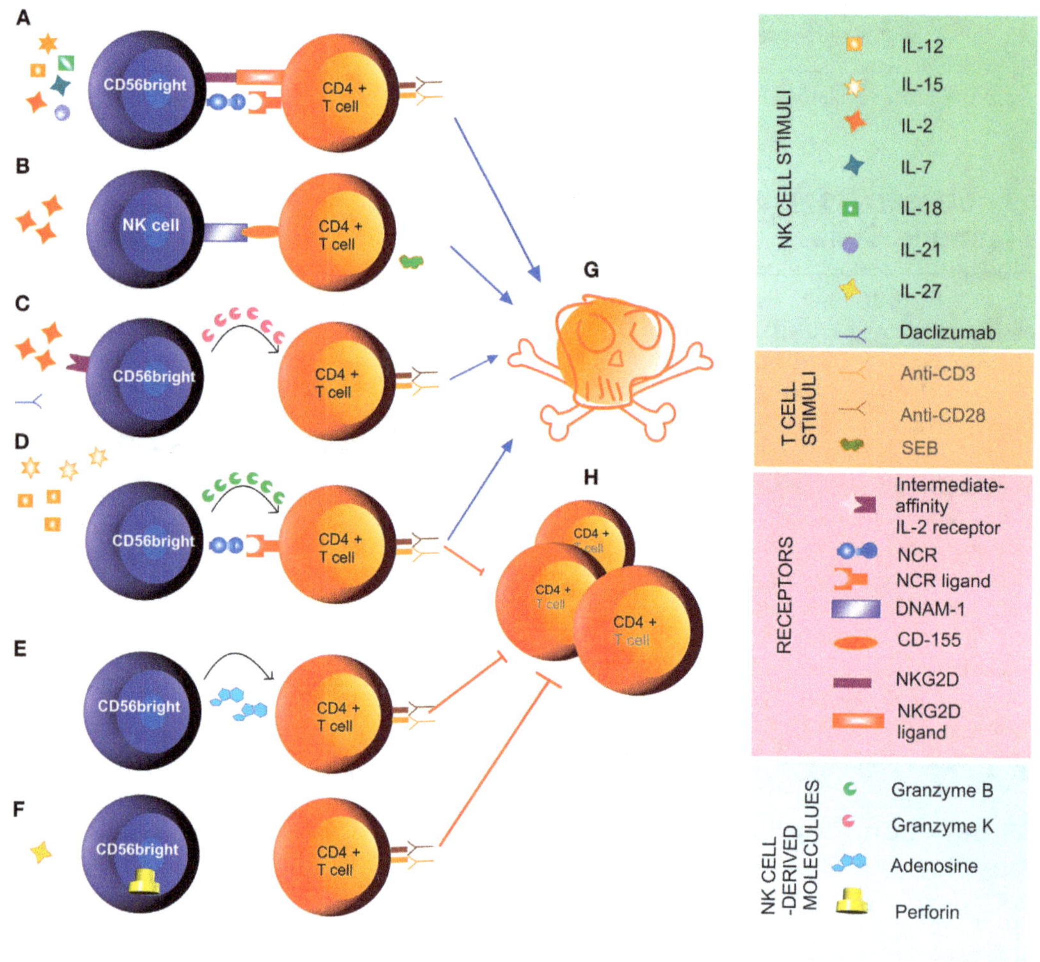

FIGURE 1 | Natural killer (NK) cell-mediated control of T cell responses. (A) Stimulus of CD56^{bright} with IL-2, IL-7, IL-12, IL-15, IL-18, and IL-21 induces cytotoxicity **(G)** toward previously activated autologous CD4^{+} T cells through the engagement of NKG2D and the natural cytotoxicity receptors (NCRs) (27). **(B)** Pre-activation of NK cells with IL-2 and of autologous CD4^{+} T cells with staphylococcal enterotoxin B (SEB) induces cytotoxicity of NK cells toward autologous T cells **(G)** through engagement of the activating receptor DNAM-1 on NK cells and its ligand CD155 on T cells (21). **(C)** In the presence of the anti-IL-2Rα monoclonal antibody daclizumab, IL-2 signal through the intermediate-affinity receptor induces cytotoxicity of CD56^{bright} NK cells toward autologous activated CD4^{+} T cells **(G)** involving the transfer of granzyme K to target cells (23). **(D)** The pro-inflammatory cytokines IL-12 and IL-15 induce anti-proliferative **(H)** and cytotoxic function of CD56^{bright} NK cells toward CD4^{+} T cells undergoing activation through engagement of the NCRs NKp30 and NKp46 and release of granzyme B (25). **(E)** CD56^{bright} suppress proliferation of autologous CD4^{+} T cells **(H)** by releasing adenosine (26). **(F)** IL-27 induces suppressor function of CD56^{bright} NK cells toward autologous CD4^{+} T cells **(H)**, which is associated with increased perforin content (9).

While Liu et al. did not ascertain whether NK cells in contact with NSCs belonged to the CD56^{bright} or CD56^{dim} NK subset (36), others analyzed the phenotype of NK cells in the cerebrospinal fluid (CSF). The majority of intrathecal NK cells in healthy individuals, MS patients and patients with other neurological diseases are CD56^{bright} NK cells (21, 37, 38), suggesting that CSF enrichment in CD56^{bright} NK cells is not MS specific, but rather CNS specific (37). This may reflect organ-specific, rather than blood-specific, tropism of CD56^{bright} NK cells (39). The recently discovered lymphatic vessels in the brain (40, 41) may be the route of entry for CD56^{bright} NK cells, which were shown to circulate in the lymph (42). Of note, a higher migratory capacity of CD56^{bright} compared to CD56^{dim} NK cells was observed in a model of the human blood–brain barrier (BBB) (21).

Given the evidence that CD56^{bright} NK cells are a regulatory population of the IS in healthy individuals, we have explored their function in untreated MS patients or patients with clinically isolated syndrome suggestive of MS (25). The number of CD56^{bright} NK cells was similar in MS patients and healthy subjects (HS). However, upon stimulus with pro-inflammatory cytokines, CD56^{bright} NK cells from MS patients suppressed much less efficiently the proliferation of autologous CD4^{+} T cells compared to those from HS. This was associated with an increased expression of HLA-E on CD4^{+} T cells in MS and was reverted by blocking HLA class I on T cells, suggesting that the cytotoxic function of CD56^{bright} NK cells on their targets is inhibited through binding of HLA-E on T cells to the NK cell inhibitory ligand NKG2A. HLA-E is a non-classic major HLA class I molecule expressed by immune cells and, outside the immune compartment, by endothelial cells, which release its soluble form upon inflammation (43). HLA-E upregulation was found in MS CNS within white matter lesions, in endothelial cells and astrocytes (44, 45). Similarly, Morandi et al. detected increased levels of soluble HLA-E in the CSF and expression of HLA-E within immune cells and neural cells in MS plaques, which correlated with decreased NK cell cytotoxicity (46). The causes of such upregulation are, as yet, unknown. Thus, an impairment of CD56^{bright} NK cell immunoregulatory function in MS may occur not only in the periphery but also within the CNS. Recently, we also described decreased cytolytic activity of NK cells in MS as a consequence of reduced upregulation of CD155 on T cells after activation, concomitantly with a reduced NK cell surface expression of DNAM-1 (21). These studies point toward a resistance of T cells to NK cell suppressive functions rather than an intrinsic defect in the NK cells in MS.

The relevance of the immunoregulatory function of NK cells in MS is emphasized by studies from the group of Takashi Yamamura, who described a particular peripheral NK phenotype to be characteristic of MS patients in remission, with increased production of the anti-inflammatory cytokine, IL-5 ("NK2" cells), and high expression of CD95 (47), which inhibited the production of IFN-γ by Th1 clones (48). Among these patients, a further division of NK2 cells as CD11c-high (not producing IL-5) and CD11c-low (producing IL-5) subsets identified CD11c-high patients at risk of relapse (49), suggesting that CD11c + NK cells are pro-inflammatory. In another study, a subpopulation of NK cells characterized by low expression of CD8, a phenotype associated with CD56^{bright} NK cells, was observed to be reduced in untreated patients with relapsing-remitting MS (50).

Previous infections may influence the development of MS (51). Interestingly, infections not only activate NK cells but also shape NK cell functions (52–54). Thus, cytomegalovirus (CMV) induces expansion of NK cells that produce IL-10 in mice, to prevent excess of activation of CD8 T cells (55, 56). In humans, CMV infection is associated with expansion of terminally differentiated NK cells bearing the NKG2C receptor and has been implicated both in MS etiology and/or "protection" (57). In this context, Martinez-Rodriguez et al. explored the expression of NKG2C on NK cells from MS patients and controls, in relation to their CMV^{+} serostatus and to the NKG2C genotype, finding that the expansion of NKG2C^{+} NK cells in CMV^{+} patients was associated to lower risk of disease progression, suggesting that CMV may exert a beneficial influence on MS, either through expansion of NKG2C^{+} NK cells or through other mechanisms (58). Differently to CMV, infection with Epstein–Barr virus, which has also been associated with an increased risk of MS, expands early differentiated NKG2A + CD56^{dim}NK cells (59, 60), but whether such cells have any role in the pathogenesis of MS is unknown.

THE IMPACT OF MS THERAPIES ON NK CELLS

In addition to first-line MS therapies, interferon beta (IFN-β) and glatiramer acetate, novel immune-modulating therapies such as the anti-inflammatory dimethyl fumarate, the T cell proliferation inhibitor teriflunomide, the migration inhibitors natalizumab and fingolimod (FTY720), the IL-2 receptor-modulating daclizumab, and the immune cell-depleting alemtuzumab are now available for treatment of MS (61, 62). Many of these immune-modulating biologicals alter the NK cell compartment (**Figures 2A,B**) by increasing NK cell frequencies (**Figure 2A**) and/or NK-mediated immune regulatory functions (**Figure 2B**) (61, 63), which points to an immune-protective role of NK cells in MS. Furthermore, antibody-dependent cell-mediated cytotoxicity (ADCC) by CD56^{dim} NK cells has been proposed as an essential therapeutic mechanism in alemtuzumab-mediated T and B cell depletion as well as rituximab-mediated B cell depletion (**Figure 2C**) (64, 65).

One therapeutic approach in MS is the reduction of inflammatory lesions by inhibiting infiltration of autoreactive lymphocytes into the CNS. While the humanized monoclonal antibody (hMA) anti-CD49d (alpha 4 integrin) natalizumab prevents transmigration of circulating lymphocytes across the BBB (67), the sphingosine 1-phosphate receptor (S1P) modulator FTY720 reduces CNS inflammation in MS (68) indirectly by preventing lymphocyte egress from LNs (69). A natalizumab-induced increase of total NK cells and CD56^{bright} NK cells in blood concomitantly with reduced NK cell numbers in CSF (**Figure 2A**) (21, 70, 71) suggests CD49d-dependent transmigration of NK cells into the CNS. Trafficking of NK cells in steady state and inflammatory conditions requires S1P (72) and decreased numbers of circulating CD56^{bright} NK cells have been observed 6 h after FTY720 treatment (73) (**Figure 2A**). Long-term treatment also resulted in reduced numbers of both NK cell subsets in the CSF (own observations). Despite reduced NK cell numbers, a relative increase in peripheral and intrathecal NK cell subsets within the lymphocyte compartment (74) indicates that FTY720 inhibits NK cell emigration less than that of other lymphocytes (**Figure 2A**). This might be due to the fact that egress of NK cells from LNs is regulated through both $S1P_1$ and $S1P_5$ (72, 75), whereas other lymphocytes use only $S1P_1$ (76). Since $S1P_1$ and $S1P_5$ seem to trigger the activation of distinct intracellular signal transduction pathways, it has been suggested that $S1P_5$ might be less susceptible to FTY720 than $S1P_1$ (72). Along this line, a higher expression of $S1P_5$ on CD56^{dim} NK cells than on CD56^{bright} ones (72) might explain the relative increase of the CD56^{dim} NK cell subset (77). While treatment of MS patients with FTY720 alters NK cell trafficking, it has no impact on cytokine secretion or cytolytic function of NK cells (77).

In contrast, treatment with daclizumab affects both peripheral (20, 78) and intrathecal CD56^{bright} NK cell numbers (79), as well as their immunoregulatory function (19, 21) (**Figures 1** and **2B**). Daclizumab is a recently approved hMA directed against IL-2Rα, which showed enhanced efficacy in MS compared to IFN-β [DECIDE trial (80)]. Daclizumab enhances endogenous mechanisms of immune tolerance by reducing early T cell activation (81), expanding CD56^{bright} NK cells (20, 78), while reducing lymphoid tissue-inducer cells (82), and restoring defective NK cell-mediated control of T cell activity in MS (19, 21). The mechanism of the effect of daclizumab on CD56^{bright} NK cells has been discussed in chapter (**Figure 2C**). Daclizumab both boosts NK cell cytolytic function in a DC-dependent manner and renders antigen-activated T cells more sensitive toward NK-mediated

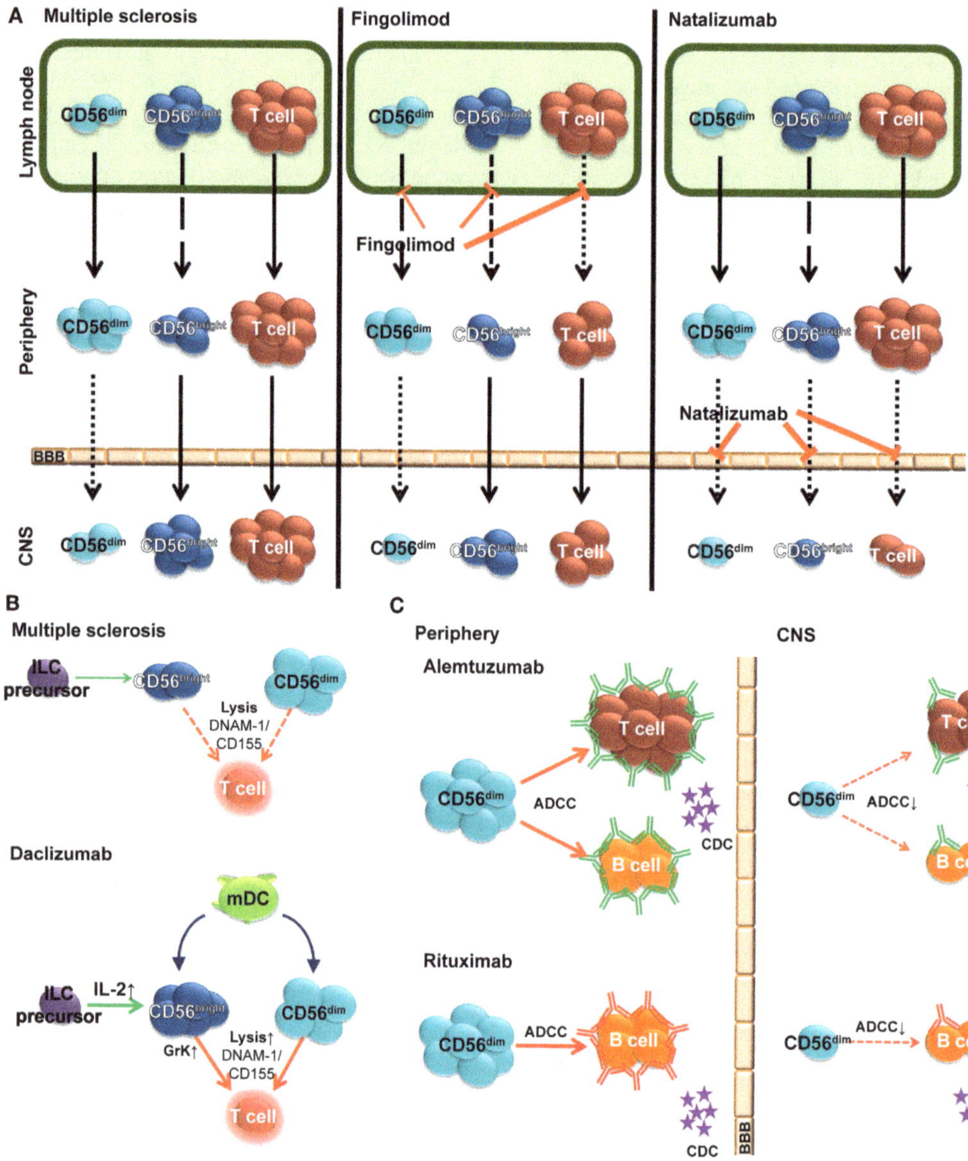

FIGURE 2 | Impact of multiple sclerosis (MS) therapies on the natural killer (NK) cell compartment. (A) Fingolimod (FTY720) inhibits egress of $CD56^{bright}$, and to a lower degree, $CD56^{dim}$ NK cells from the lymph node (LN), resulting in a relative increase of the latter subset in the periphery (middle). Natalizumab inhibits transmigration of lymphocytes including NK cells across the blood–brain barrier (BBB). **(B)** Elevated levels of IL-2 in daclizumab-treated patients promote differentiation and expansion of $CD56^{bright}$ NK cells. Daclizumab both boosts NK cell cytolytic function in a DC-dependent manner and renders antigen-activated T cells more sensitive toward NK-mediated lysis, thus restoring defective NK-mediated control of T cell activity in MS. **(C)** In addition to complement-dependent cytotoxicity (CDC), alemtuzumab (top) and rituximab (bottom) use $CD56^{dim}$ NK cell-mediated antibody-dependent cellular cytotoxicity (ADCC) to deplete T and/or B cells in peripheral blood. However, in the CSF, sparseness of $CD56^{dim}$ NK cells, reduced levels of complement proteins, and lack of antibodies crossing the BBB limit local immune-modulating efficacy [**(C)** adapted from Ref. (66)].

lysis, thus restoring defective NK cell-mediated control of T cell activity in MS (19, 21) (**Figure 2B**).

Alemtuzumab is a hMA directed against the cell surface molecule CD52, which demonstrated a high clinical efficacy in MS (83, 84). CD52 is highly expressed on T and B cells and to a lower degree on NK cell subsets (85, 86). Accordingly, a relative increase of circulating NK cells with increased numbers of $CD56^{bright}$ NK cells was observed 6 months after alemtuzumab therapy, whereas the $CD56^{dim}$ subset remained unaltered. However, there was no change in NK cell cytolytic function (87). ADCC is mediated via the FcγRIII (CD16)-expressing $CD56^{dim}$ NK cell subset (88). Since intrathecal $CD56^{dim}$ NK cells are sparse (21, 89), the therapeutic efficacy of alemtuzumab within the CNS might be limited (**Figure 2C**). Along this line, insufficient disease inhibition in progressive MS by intrathecal application of rituximab was proposed to be due to low numbers of $CD56^{dim}$ NK cells and reduced levels of complement proteins within the CNS (66). Further studies are required to shed more light on NK cell-mediated ADCC as a mechanism of action of human monoclonal antibody-mediated depleting therapies in MS.

SUMMARY/OUTLOOK

NK cells are important players in controlling T cell activation in CNS autoimmunity, and impaired immune regulatory function of NK cells might be one of the driving forces in the pathogenesis of MS. Thus, a better understanding of the underlying mechanisms of NK cell-mediated regulation of T cell activation might help to improve treatment strategies in MS.

AUTHOR CONTRIBUTIONS

CG wrote chapter, conceptualized **Figure 2**, revised the manuscript, and approved the final version. AS-M drew **Figure 2**, revised the manuscript, and approved the final version. HW wrote chapter, revised the manuscript, and approved the final version. EM critically reviewed the manuscript for important intellectual content and approved the final version. NKdeR conceptualized **Figure 1**, critically reviewed the manuscript fo-r important intellectual content, and approved the final version. AU critically reviewed the manuscript for important intellectual content and approved the final version. AL organized the sections of the manuscript, wrote chapters 1 and 2, conceptualized and drew **Figure 1**, revised the manuscript, and approved the final version.

REFERENCES

1. Cooper GS, Bynum MLK, Somers EC. Recent insights in the epidemiology of autoimmune diseases: improved prevalence estimates and understanding of clustering of diseases. *J Autoimmun* (2009) 33:197–207. doi:10.1016/j.jaut.2009.09.008
2. Cooper MA, Fehniger TA, Turner SC, Chen KS, Ghaheri BA, Ghayur T, et al. Human natural killer cells: a unique innate immunoregulatory role for the CD56bright subset. *Blood* (2001) 97:3146–51. doi:10.1182/blood.V97.10.3146
3. Ferlazzo G, Morandi B. Cross-talks between natural killer cells and distinct subsets of dendritic cells. *Front Immunol* (2014) 5:159. doi:10.3389/fimmu.2014.00159
4. Fehniger TA, Cooper MA, Nuovo GJ, Cella M, Facchetti F, Colonna M, et al. CD56bright natural killer cells are present in human lymph nodes and are activated by T cell-derived IL-2: a potential new link between adaptive and innate immunity. *Blood* (2003) 101:3052–7. doi:10.1182/blood-2002-09-2876
5. Carson WE, Giri JG, Lindemann MJ, Linett ML, Ahdieh M, Paxton R, et al. Interleukin (IL) 15 is a novel cytokine that activates human natural killer cells via components of the IL-2 receptor. *J Exp Med* (1994) 180:1395–403. doi:10.1084/jem.180.4.1395
6. Poli A, Michel T, Thérésine M, Andrès E, Hentges F, Zimmer J. CD56(bright) natural killer (NK) cells: an important NK cell subset. *Immunology* (2009) 126:458–65. doi:10.1111/j.1365-2567.2008.03027.x
7. Cichocki F, Schlums H, Theorell J, Tesi B, Miller JS, Ljunggren HG, et al. Diversification and functional specialization of human NK cell subsets. *Curr Top Microbiol Immunol* (2016) 395:63–94. doi:10.1007/82_2015_487
8. Ferlazzo G, Pack M, Thomas D, Paludan C, Schmid D, Strowig T, et al. Distinct roles of IL-12 and IL-15 in human natural killer cell activation by dendritic cells from secondary lymphoid organs. *Proc Natl Acad Sci U S A* (2004) 101:16606–11. doi:10.1073/pnas.0407522101
9. Laroni A, Gandhi R, Beynon V, Weiner HL. IL-27 imparts immunoregulatory function to human NK cell subsets. *PLoS One* (2011) 6:e26173. doi:10.1371/journal.pone.0026173
10. Piccioli D, Sbrana S, Melandri E, Valiante NM. Contact-dependent stimulation and inhibition of dendritic cells by natural killer cells. *J Exp Med* (2002) 195:335–41. doi:10.1084/jem.20010934
11. Mailliard RB, Alber SM, Shen H, Watkins SC, Kirkwood JM, Herberman RB, et al. IL-18-induced CD83+CCR7+ NK helper cells. *J Exp Med* (2005) 202:941–53. doi:10.1084/jem.20050128
12. Marcenaro E, Ferranti B, Moretta A. NK-DC interaction: on the usefulness of auto-aggression. *Autoimmun Rev* (2005) 4:520–5. doi:10.1016/j.autrev.2005.04.015
13. Marcenaro E, Cantoni C, Pesce S, Prato C, Pende D, Agaugue S, et al. Uptake of CCR7 and acquisition of migratory properties by human KIR+ NK cells interacting with monocyte-derived DC or EBV cell lines: regulation by KIR/HLA-class I interaction. *Blood* (2009) 114:4108–16. doi:10.1182/blood-2009-05-222265
14. Vitale M, Chiesa MD, Carlomagno S, Romagnani C, Thiel A, Moretta L, et al. The small subset of CD56brightCD16 – natural killer cells is selectively responsible for both cell proliferation and interferon-γ production upon interaction with dendritic cells. *Eur J Immunol* (2004) 34:1715–22. doi:10.1002/eji.200425100
15. Romagnani C, Della Chiesa M, Kohler S, Moewes B, Radbruch A, Moretta L, et al. Activation of human NK cells by plasmacytoid dendritic cells and its modulation by CD4+ T helper cells and CD4+ CD25hi T regulatory cells. *Eur J Immunol* (2005) 35:2452–8. doi:10.1002/eji.200526069
16. Dalbeth N, Gundle R, Davies RJ, Lee YC, Mcmichael AJ, Callan MF. CD56bright NK cells are enriched at inflammatory sites and can engage with monocytes in a reciprocal program of activation. *J Immunol* (2004) 173:6418–26. doi:10.4049/jimmunol.173.10.6418
17. Della Chiesa M, Vitale M, Carlomagno S, Ferlazzo G, Moretta L, Moretta A. The natural killer cell-mediated killing of autologous dendritic cells is confined to a cell subset expressing CD94/NKG2A, but lacking inhibitory killer Ig-like receptors. *Eur J Immunol* (2003) 33:1657–66. doi:10.1002/eji.200323986
18. Morandi B, Mortara L, Chiossone L, Accolla RS, Mingari MC, Moretta L, et al. Dendritic cell editing by activated natural killer cells results in a more protective cancer-specific immune response. *PLoS One* (2012) 7:e39170. doi:10.1371/journal.pone.0039170
19. Bielekova B, Catalfamo M, Reichert-Scrivner S, Packer A, Cerna M, Waldmann TA, et al. Regulatory CD56(bright) natural killer cells mediate immunomodulatory effects of IL-2Ralpha-targeted therapy (daclizumab) in multiple sclerosis. *Proc Natl Acad Sci U S A* (2006) 103:5941–6. doi:10.1073/pnas.0601335103
20. Martin JF, Perry JS, Jakhete NR, Wang X, Bielekova B. An IL-2 paradox: blocking CD25 on T cells induces IL-2-driven activation of CD56(bright) NK cells. *J Immunol* (2010) 185:1311–20. doi:10.4049/jimmunol.0902238
21. Gross CC, Schulte-Mecklenbeck A, Runzi A, Kuhlmann T, Posevitz-Fejfar A, Schwab N, et al. Impaired NK-mediated regulation of T-cell activity in multiple sclerosis is reconstituted by IL-2 receptor modulation. *Proc Natl Acad Sci U S A* (2016) 113:E2973–82. doi:10.1073/pnas.1524924113
22. Caligiuri MA, Zmuidzinas A, Manley TJ, Levine H, Smith KA, Ritz J. Functional consequences of interleukin 2 receptor expression on resting human lymphocytes. Identification of a novel natural killer cell subset with high affinity receptors. *J Exp Med* (1990) 171:1509–26. doi:10.1084/jem.171.5.1509
23. Jiang W, Chai NR, Maric D, Bielekova B. Unexpected role for granzyme K in CD56bright NK cell-mediated immunoregulation of multiple sclerosis. *J Immunol* (2011) 187:781–90. doi:10.4049/jimmunol.1100789
24. Moretta A, Bottino C, Vitale M, Pende D, Cantoni C, Mingari MC, et al. Activating receptors and coreceptors involved in human natural killer cell-mediated cytolysis. *Annu Rev Immunol* (2001) 19:197–223. doi:10.1146/annurev.immunol.19.1.197
25. Laroni A, Armentani E, Kerlero De Rosbo N, Ivaldi F, Marcenaro E, Sivori S, et al. Dysregulation of regulatory CD56(bright) NK cells/T cells interactions in multiple sclerosis. *J Autoimmun* (2016) 72:8–18. doi:10.1016/j.jaut.2016.04.003
26. Morandi F, Horenstein AL, Chillemi A, Quarona V, Chiesa S, Imperatori A, et al. CD56brightCD16- NK cells produce adenosine through a CD38-mediated pathway and act as regulatory cells inhibiting autologous CD4+ T cell proliferation. *J Immunol* (2015) 195(3):965–72. doi:10.4049/jimmunol.1500591

27. Nielsen N, Odum N, Urso B, Lanier LL, Spee P. Cytotoxicity of CD56(bright) NK cells towards autologous activated CD4+ T cells is mediated through NKG2D, LFA-1 and TRAIL and dampened via CD94/NKG2A. *PLoS One* (2012) 7:e31959. doi:10.1371/journal.pone.0031959
28. Kleinewietfeld M, Hafler DA. Regulatory T cells in autoimmune neuroinflammation. *Immunol Rev* (2014) 259:231–44. doi:10.1111/imr.12169
29. Gandhi R, Laroni A, Weiner HL. Role of the innate immune system in the pathogenesis of multiple sclerosis. *J Neuroimmunol* (2010) 221:7–14. doi:10.1016/j.jneuroim.2009.10.015
30. Lunemann JD, Munz C. Do natural killer cells accelerate or prevent autoimmunity in multiple sclerosis? *Brain* (2008) 131:1681–3. doi:10.1093/brain/awn132
31. Morandi B, Bramanti P, Bonaccorsi I, Montalto E, Oliveri D, Pezzino G, et al. Role of natural killer cells in the pathogenesis and progression of multiple sclerosis. *Pharmacol Res* (2008) 57:1–5. doi:10.1016/j.phrs.2007.11.003
32. Lu L, Ikizawa K, Hu D, Werneck MB, Wucherpfennig KW, Cantor H. Regulation of activated CD4+ T cells by NK cells via the Qa-1-NKG2A inhibitory pathway. *Immunity* (2007) 26:593–604. doi:10.1016/j.immuni.2007.03.017
33. Leavenworth JW, Schellack C, Kim HJ, Lu L, Spee P, Cantor H. Analysis of the cellular mechanism underlying inhibition of EAE after treatment with anti-NKG2A F(ab')2. *Proc Natl Acad Sci U S A* (2010) 107:2562–7. doi:10.1073/pnas.0914732107
34. Boyman O, Kovar M, Rubinstein MP, Surh CD, Sprent J. Selective stimulation of T cell subsets with antibody-cytokine immune complexes. *Science* (2006) 311:1924–7. doi:10.1126/science.1122927
35. Hao J, Liu R, Piao W, Zhou Q, Vollmer TL, Campagnolo DI, et al. Central nervous system (CNS)-resident natural killer cells suppress Th17 responses and CNS autoimmune pathology. *J Exp Med* (2010) 207:1907–21. doi:10.1084/jem.20092749
36. Liu Q, Sanai N, Jin WN, La Cava A, Van Kaer L, Shi FD. Neural stem cells sustain natural killer cells that dictate recovery from brain inflammation. *Nat Neurosci* (2016) 19:243–52. doi:10.1038/nn.4211
37. Han S, Lin YC, Wu T, Salgado AD, Mexhitaj I, Wuest SC, et al. Comprehensive immunophenotyping of cerebrospinal fluid cells in patients with neuroimmunological diseases. *J Immunol* (2014) 192:2551–63. doi:10.4049/jimmunol.1302884
38. Rodriguez-Martin E, Picon C, Costa-Frossard L, Alenda R, Sainz De La Maza S, Roldan E, et al. Natural killer cell subsets in cerebrospinal fluid of patients with multiple sclerosis. *Clin Exp Immunol* (2015) 180:243–9. doi:10.1111/cei.12580
39. Carrega P, Bonaccorsi I, Di Carlo E, Morandi B, Paul P, Rizzello V, et al. CD56(bright)perforin(low) noncytotoxic human NK cells are abundant in both healthy and neoplastic solid tissues and recirculate to secondary lymphoid organs via afferent lymph. *J Immunol* (2014) 192:3805–15. doi:10.4049/jimmunol.1301889
40. Louveau A, Smirnov I, Keyes TJ, Eccles JD, Rouhani SJ, Peske JD, et al. Structural and functional features of central nervous system lymphatic vessels. *Nature* (2015) 523(7560):337–41. doi:10.1038/nature14432
41. Aspelund A, Antila S, Proulx ST, Karlsen TV, Karaman S, Detmar M, et al. A dural lymphatic vascular system that drains brain interstitial fluid and macromolecules. *J Exp Med* (2015) 212:991–9. doi:10.1084/jem.20142290
42. Montalto E, Mangraviti S, Costa G, Carrega P, Morandi B, Pezzino G, et al. Seroma fluid subsequent to axillary lymph node dissection for breast cancer derives from an accumulation of afferent lymph. *Immunol Lett* (2010) 131:67–72. doi:10.1016/j.imlet.2010.03.002
43. Coupel S, Moreau A, Hamidou M, Horejsi V, Soulillou JP, Charreau B. Expression and release of soluble HLA-E is an immunoregulatory feature of endothelial cell activation. *Blood* (2007) 109:2806–14. doi:10.1182/blood-2006-06-030213
44. Durrenberger PF, Webb LV, Sim MJ, Nicholas RS, Altmann DM, Boyton RJ. Increased HLA-E expression in white matter lesions in multiple sclerosis. *Immunology* (2012) 137:317–25. doi:10.1111/imm.12012
45. Pannemans K, Broux B, Goris A, Dubois B, Broekmans T, Van Wijmeersch B, et al. HLA-E restricted CD8+ T cell subsets are phenotypically altered in multiple sclerosis patients. *Mult Scler* (2014) 20:790–801. doi:10.1177/1352458513509703
46. Morandi F, Venturi C, Rizzo R, Castellazzi M, Baldi E, Caniatti ML, et al. Intrathecal soluble HLA-E correlates with disease activity in patients with multiple sclerosis and may cooperate with soluble HLA-G in the resolution of neuroinflammation. *J Neuroimmune Pharmacol* (2013) 8:944–55. doi:10.1007/s11481-013-9459-3
47. Takahashi K, Miyake S, Kondo T, Terao K, Hatakenaka M, Hashimoto S, et al. Natural killer type 2 bias in remission of multiple sclerosis. *J Clin Invest* (2001) 107:R23–9. doi:10.1172/JCI11819
48. Takahashi K, Aranami T, Endoh M, Miyake S, Yamamura T. The regulatory role of natural killer cells in multiple sclerosis. *Brain* (2004) 127:1917–27. doi:10.1093/brain/awh219
49. Aranami T, Miyake S, Yamamura T. Differential expression of CD11c by peripheral blood NK cells reflects temporal activity of multiple sclerosis. *J Immunol* (2006) 177:5659–67. doi:10.4049/jimmunol.177.8.5659
50. De Jager PL, Rossin E, Pyne S, Tamayo P, Ottoboni L, Viglietta V, et al. Cytometric profiling in multiple sclerosis uncovers patient population structure and a reduction of CD8low cells. *Brain* (2008) 131:1701–11. doi:10.1093/brain/awn118
51. Sospedra M, Martin R. Immunology of multiple sclerosis. *Annu Rev Immunol* (2005) 23:683–747. doi:10.1146/annurev.immunol.23.021704.115707
52. Saghafian-Hedengren S, Sohlberg E, Theorell J, Carvalho-Queiroz C, Nagy N, Persson JO, et al. Epstein-Barr virus coinfection in children boosts cytomegalovirus-induced differentiation of natural killer cells. *J Virol* (2013) 87:13446–55. doi:10.1128/JVI.02382-13
53. Schlums H, Cichocki F, Tesi B, Theorell J, Beziat V, Holmes TD, et al. Cytomegalovirus infection drives adaptive epigenetic diversification of NK cells with altered signaling and effector function. *Immunity* (2015) 42:443–56. doi:10.1016/j.immuni.2015.02.008
54. Muntasell A, Pupuleku A, Cisneros E, Vera A, Moraru M, Vilches C, et al. Relationship of NKG2C copy number with the distribution of distinct cytomegalovirus-induced adaptive NK cell subsets. *J Immunol* (2016) 196:3818–27. doi:10.4049/jimmunol.1502438
55. Lee SH, Kim KS, Fodil-Cornu N, Vidal SM, Biron CA. Activating receptors promote NK cell expansion for maintenance, IL-10 production, and CD8 T cell regulation during viral infection. *J Exp Med* (2009) 206:2235–51. doi:10.1084/jem.20082387
56. Tarrio ML, Lee SH, Fragoso MF, Sun HW, Kanno Y, O'shea JJ, et al. Proliferation conditions promote intrinsic changes in NK cells for an IL-10 response. *J Immunol* (2014) 193:354–63. doi:10.4049/jimmunol.1302999
57. Vanheusden M, Stinissen P, 't Hart BA, Hellings N. Cytomegalovirus: a culprit or protector in multiple sclerosis? *Trends Mol Med* (2015) 21:16–23. doi:10.1016/j.molmed.2014.11.002
58. Martinez-Rodriguez JE, Cobo-Calvo A, Villar LM, Munteis E, Blanco Y, Rasal R, et al. Adaptive natural killer cell response to cytomegalovirus and disability progression in multiple sclerosis. *Mult Scler* (2016) 22:741–52. doi:10.1177/1352458515601215
59. Azzi T, Lünemann A, Murer A, Ueda S, Béziat V, Malmberg K-J, et al. Role for early-differentiated natural killer cells in infectious mononucleosis. *Blood* (2014) 124:2533. doi:10.1182/blood-2014-01-553024
60. Chijioke O, Landtwing V, Munz C. NK cell influence on the outcome of primary Epstein-Barr virus infection. *Front Immunol* (2016) 7:323. doi:10.3389/fimmu.2016.00323
61. Wiendl H, Gross CC. Modulation of IL-2Ralpha with daclizumab for treatment of multiple sclerosis. *Nat Rev Neurol* (2013) 9:394–404. doi:10.1038/nrneurol.2013.95
62. Ziemssen T, De Stefano N, Pia Sormani M, Van Wijmeersch B, Wiendl H, Kieseier BC. Optimizing therapy early in multiple sclerosis: an evidence-based view. *Mult Scler Relat Disord* (2015) 4:460–9. doi:10.1016/j.msard.2015.07.007
63. Chanvillard C, Jacolik RF, Infante-Duarte C, Nayak RC. The role of natural killer cells in multiple sclerosis and their therapeutic implications. *Front Immunol* (2013) 4:63. doi:10.3389/fimmu.2013.00063
64. Anolik JH, Campbell D, Felgar RE, Young F, Sanz I, Rosenblatt J, et al. The relationship of FcgammaRIIIa genotype to degree of B cell depletion by rituximab in the treatment of systemic lupus erythematosus. *Arthritis Rheum* (2003) 48:455–9. doi:10.1002/art.10764
65. Hu Y, Turner MJ, Shields J, Gale MS, Hutto E, Roberts BL, et al. Investigation of the mechanism of action of alemtuzumab in a human CD52 transgenic mouse model. *Immunology* (2009) 128:260–70. doi:10.1111/j.1365-2567.2009.03115.x
66. Komori M, Lin YC, Cortese I, Blake A, Ohayon J, Cherup J, et al. Insufficient disease inhibition by intrathecal rituximab in progressive multiple sclerosis. *Ann Clin Transl Neurol* (2016) 3:166–79. doi:10.1002/acn3.293

67. Coyle PK. The role of natalizumab in the treatment of multiple sclerosis. *Am J Manag Care* (2010) 16:S164–70.
68. Kappos L, Antel J, Comi G, Montalban X, O'Connor P, Polman CH, et al. Oral fingolimod (FTY720) for relapsing multiple sclerosis. *N Engl J Med* (2006) 355:1124–40. doi:10.1056/NEJMoa052643
69. Chun J, Hartung HP. Mechanism of action of oral fingolimod (FTY720) in multiple sclerosis. *Clin Neuropharmacol* (2010) 33:91–101. doi:10.1097/WNF.0b013e3181cbf825
70. Putzki N, Baranwal MK, Tettenborn B, Limmroth V, Kreuzfelder E. Effects of natalizumab on circulating B cells, T regulatory cells and natural killer cells. *Eur Neurol* (2010) 63:311–7. doi:10.1159/000302687
71. Skarica M, Eckstein C, Whartenby KA, Calabresi PA. Novel mechanisms of immune modulation of natalizumab in multiple sclerosis patients. *J Neuroimmunol* (2011) 235:70–6. doi:10.1016/j.jneuroim.2011.02.010
72. Walzer T, Chiossone L, Chaix J, Calver A, Carozzo C, Garrigue-Antar L, et al. Natural killer cell trafficking in vivo requires a dedicated sphingosine 1-phosphate receptor. *Nat Immunol* (2007) 8:1337–44. doi:10.1038/ni1523
73. Mehling M, Burgener AV, Brinkmann V, Bantug GR, Dimeloe S, Hoenger G, et al. Tissue distribution dynamics of human NK cells inferred from peripheral blood depletion kinetics after sphingosine-1-phosphate receptor blockade. *Scand J Immunol* (2015) 82:460–6. doi:10.1111/sji.12347
74. Kowarik MC, Pellkofer HL, Cepok S, Korn T, Kumpfel T, Buck D, et al. Differential effects of fingolimod (FTY720) on immune cells in the CSF and blood of patients with MS. *Neurology* (2011) 76:1214–21. doi:10.1212/WNL.0b013e3182143564
75. Jenne CN, Enders A, Rivera R, Watson SR, Bankovich AJ, Pereira JP, et al. T-bet-dependent S1P5 expression in NK cells promotes egress from lymph nodes and bone marrow. *J Exp Med* (2009) 206:2469–81. doi:10.1084/jem.20090525
76. Schwab SR, Cyster JG. Finding a way out: lymphocyte egress from lymphoid organs. *Nat Immunol* (2007) 8:1295–301. doi:10.1038/ni1545
77. Johnson TA, Evans BL, Durafourt BA, Blain M, Lapierre Y, Bar-Or A, et al. Reduction of the peripheral blood CD56(bright) NK lymphocyte subset in FTY720-treated multiple sclerosis patients. *J Immunol* (2011) 187:570–9. doi:10.4049/jimmunol.1003823
78. Bielekova B, Howard T, Packer AN, Richert N, Blevins G, Ohayon J, et al. Effect of anti-CD25 antibody daclizumab in the inhibition of inflammation and stabilization of disease progression in multiple sclerosis. *Arch Neurol* (2009) 66:483–9. doi:10.1001/archneurol.2009.50
79. Bielekova B, Richert N, Herman ML, Ohayon J, Waldmann TA, Mcfarland H, et al. Intrathecal effects of daclizumab treatment of multiple sclerosis. *Neurology* (2011) 77:1877–86. doi:10.1212/WNL.0b013e318239f7ef
80. Kappos L, Wiendl H, Selmaj K, Arnold DL, Havrdova E, Boyko A, et al. Daclizumab HYP versus interferon beta-1a in relapsing multiple sclerosis. *N Engl J Med* (2015) 373:1418–28. doi:10.1056/NEJMoa1501481
81. Wuest SC, Edwan JH, Martin JF, Han S, Perry JS, Cartagena CM, et al. A role for interleukin-2 trans-presentation in dendritic cell-mediated T cell activation in humans, as revealed by daclizumab therapy. *Nat Med* (2011) 17:604–9. doi:10.1038/nm.2365
82. Perry JS, Han S, Xu Q, Herman ML, Kennedy LB, Csako G, et al. Inhibition of LTi cell development by CD25 blockade is associated with decreased intrathecal inflammation in multiple sclerosis. *Sci Transl Med* (2012) 4:145ra106. doi:10.1126/scitranslmed.3004140
83. Cohen JA, Coles AJ, Arnold DL, Confavreux C, Fox EJ, Hartung HP, et al. Alemtuzumab versus interferon beta 1a as first-line treatment for patients with relapsing-remitting multiple sclerosis: a randomised controlled phase 3 trial. *Lancet* (2012) 380:1819–28. doi:10.1016/S0140-6736(12)61769-3
84. Coles AJ, Twyman CL, Arnold DL, Cohen JA, Confavreux C, Fox EJ, et al. Alemtuzumab for patients with relapsing multiple sclerosis after disease-modifying therapy: a randomised controlled phase 3 trial. *Lancet* (2012) 380:1829–39. doi:10.1016/S0140-6736(12)61768-1
85. Freedman MS, Kaplan JM, Markovic-Plese S. Insights into the mechanisms of the therapeutic efficacy of alemtuzumab in multiple sclerosis. *J Clin Cell Immunol* (2013) 4:1000152. doi:10.4172/2155-9899.1000152
86. Ruck T, Bittner S, Wiendl H, Meuth SG. Alemtuzumab in multiple sclerosis: mechanism of action and beyond. *Int J Mol Sci* (2015) 16:16414–39. doi:10.3390/ijms160716414
87. Gross CC, Ahmetspahic D, Ruck T, Schulte-Mecklenbeck A, Schwarte K, Jörgens S, et al. Alemtuzumab treatment alters circulating innate immune cells in multiple sclerosis. *Neurol Neuroimmunol Neuroinflamm* (2016) 3:e289. doi:10.1212/NXI.0000000000000183
88. Perussia B, Trinchieri G, Jackson A, Warner NL, Faust J, Rumpold H, et al. The Fc receptor for IgG on human natural killer cells: phenotypic, functional, and comparative studies with monoclonal antibodies. *J Immunol* (1984) 133:180–9.
89. Hamann I, Dorr J, Glumm R, Chanvillard C, Janssen A, Millward JM, et al. Characterization of natural killer cells in paired CSF and blood samples during neuroinflammation. *J Neuroimmunol* (2013) 254:165–9. doi:10.1016/j.jneuroim.2012.08.009

Permissions

All chapters in this book were first published in NKCSHDND, by Frontiers; hereby published with permission under the Creative Commons Attribution License or equivalent. Every chapter published in this book has been scrutinized by our experts. Their significance has been extensively debated. The topics covered herein carry significant findings which will fuel the growth of the discipline. They may even be implemented as practical applications or may be referred to as a beginning point for another development.

The contributors of this book come from diverse backgrounds, making this book a truly international effort. This book will bring forth new frontiers with its revolutionizing research information and detailed analysis of the nascent developments around the world.

We would like to thank all the contributing authors for lending their expertise to make the book truly unique. They have played a crucial role in the development of this book. Without their invaluable contributions this book wouldn't have been possible. They have made vital efforts to compile up to date information on the varied aspects of this subject to make this book a valuable addition to the collection of many professionals and students.

This book was conceptualized with the vision of imparting up-to-date information and advanced data in this field. To ensure the same, a matchless editorial board was set up. Every individual on the board went through rigorous rounds of assessment to prove their worth. After which they invested a large part of their time researching and compiling the most relevant data for our readers.

The editorial board has been involved in producing this book since its inception. They have spent rigorous hours researching and exploring the diverse topics which have resulted in the successful publishing of this book. They have passed on their knowledge of decades through this book. To expedite this challenging task, the publisher supported the team at every step. A small team of assistant editors was also appointed to further simplify the editing procedure and attain best results for the readers.

Apart from the editorial board, the designing team has also invested a significant amount of their time in understanding the subject and creating the most relevant covers. They scrutinized every image to scout for the most suitable representation of the subject and create an appropriate cover for the book.

The publishing team has been an ardent support to the editorial, designing and production team. Their endless efforts to recruit the best for this project, has resulted in the accomplishment of this book. They are a veteran in the field of academics and their pool of knowledge is as vast as their experience in printing. Their expertise and guidance has proved useful at every step. Their uncompromising quality standards have made this book an exceptional effort. Their encouragement from time to time has been an inspiration for everyone.

The publisher and the editorial board hope that this book will prove to be a valuable piece of knowledge for researchers, students, practitioners and scholars across the globe.

List of Contributors

Aude Charbonnier and Evelyne D'Incan
Département d'Hématologie, Institut Paoli-Calmettes, Marseille, France

Jérôme Rey
Département d'Hématologie, Institut Paoli-Calmettes, Marseille, France
Plateforme d'Immunomonitoring en Cancérologie de Marseille, Institut Paoli-Calmettes, Marseille, France

Cyril Fauriat, Eloïse Kochbati, Florence Orlanducci, Bernadette Barbarat and Daniel Olive
Plateforme d'Immunomonitoring en Cancérologie de Marseille, Institut Paoli-Calmettes, Marseille, France
Centre de Recherche en Cancérologie de Marseille, INSERM U1068, Institut Paoli-Calmettes, Aix-Marseille Université, UM105, CNRS, UMR7258, Marseille, France

Norbert Vey
Département d'Hématologie, Institut Paoli-Calmettes, Marseille, France
Centre de Recherche en Cancérologie de Marseille, INSERM U1068, Institut Paoli-Calmettes, Aix-Marseille Université, UM105, CNRS, UMR7258, Marseille, France

Pascale Andre
Innate-Pharma, Marseille, France

François Romagne
Innate-Pharma, Marseille, France
Mi-mAbs, Aix-Marseille Université, Marseille, France

Janine E. Melsen, Gertjan Lugthart, Arjan C. Lankester and Marco W. Schilham
Department of Pediatrics, Leiden University Medical Center, Leiden, Netherlands

Emanuela Marcenaro
Department of Experimental Medicine (DIMES), Center of Excellence for Biomedical Research (CEBR), University of Genova, Genova, Italy

Luigi D. Notarangelo
Immune Deficiency Genetics Section, Laboratory of Clinical Immunology and Microbiology, National Institute of Allergy and Infectious Diseases, National Institutes of Health, Bethesda, MD, United States

Jordan S. Orange
Department of Pediatrics, Baylor College of Medicine, Center for Human Immunobiology, Texas Children's Hospital, Houston, TX, United States

Eric Vivier
Centre d'Immunologie de Marseille-Luminy, Aix Marseille Université, INSERM, CNRS, Marseille, France
Service d'Immunologie, Hôpital de la Timone, Assistance Publique-Hôpitaux de Marseille, Marseille, France

Nicholas D. Huntington, Gabrielle T. Belz and Cyril Seillet
Molecular Immunology Division, Walter and Eliza Hall Institute of Medical Research, Melbourne, VIC, Australia
Department of Medical Biology, University of Melbourne, Melbourne, VIC, Australia

Yuhao Jiao
Molecular Immunology Division, Walter and Eliza Hall Institute of Medical Research, Melbourne, VIC, Australia
Department of Medical Biology, University of Melbourne, Melbourne, VIC, Australia
School of Medicine, Tsinghua University, Beijing, China

Costanza Maria Cristiani, Eleonora Palella, Rossana Tallerico and Cinzia Garofalo
Department of Experimental and Clinical Medicine, University Magna Graecia of Catanzaro, Catanzaro, Italy

Rosa Sottile and Ennio Carbone
Department of Experimental and Clinical Medicine, University Magna Graecia of Catanzaro, Catanzaro, Italy
Department of Microbiology, Cell and Tumor Biology, Karolinska Institutet, Stockholm, Sweden

Federica Bozzano
Department of Experimental Medicine (DIMES), University of Genova, Genova, Italy
Center of Excellence for Biomedical Research (CEBR), University of Genova, Genova, Italy

Francesco Marras
Istituto Giannina Gaslini, Genova, Italy

Andrea De Maria
Center of Excellence for Biomedical Research (CEBR), University of Genova, Genova, Italy
Clinica Malattie Infettive, IRCCS AOU San Martino-IST Genova, Istituto Nazionale per la Ricerca sul Cancro, Genova, Italy
Department of Health Sciences, DISSAL, University of Genova, Genova, Italy

Amandine Pradier, Federico Simonetta, Sophie Waldvogel, Carine Bosshard and Eddy Roosnek
Division of Hematology, Department of Medical Specialties, Geneva University Hospitals, University of Geneva, Geneva, Switzerland

Jean-Marie Tiercy
Transplantation Immunology Unit, National Reference Laboratory for Histocompatibility, Department of Genetics and Laboratory Medicine, University Hospital Geneva, Geneva, Switzerland

Mariella Della Chiesa, Silvia Pesce, Letizia Muccio and Simona Carlomagno
Dipartimento di Medicina Sperimentale, Università degli Studi di Genova, Genova, Italy

Simona Sivori, Alessandro Moretta and Emanuela Marcenaro
Dipartimento di Medicina Sperimentale, Università degli Studi di Genova, Genova, Italy
CEBR, Università degli Studi di Genova, Genova, Italy

Katharina Pallmer and Annette Oxenius
Institute of Microbiology, ETH Zürich, Zürich, Switzerland

Fabrizio Antonangeli and Valentina Bonanni
Department of Molecular Medicine, Sapienza University, Rome, Italy

Giovanni Bernardini
Department of Molecular Medicine, Sapienza University, Rome, Italy
IRCCS NEUROMED - Mediterranean Neurological Institute, Isernia, Italy

Angela Santoni
IRCCS NEUROMED - Mediterranean Neurological Institute, Isernia, Italy
Department of Molecular Medicine, Istituto Pasteur-Fondazione Cenci Bolognetti, Rome, Italy

Bo Li, Xifeng Xiao, Shuqiang Chen, Jun Wang, Fang Yang and Xiaohong Wang
Department of Gynecology and Obstetrics, Tangdu Hospital, The Fourth Military Medical University, Xi'an, China

Wenwei Guo, Liang Fang and Lihua Chen
Department of Gynecology and Obstetrics, Tangdu Hospital, The Fourth Military Medical University, Xi'an, China
Department of Immunology, The Fourth Military Medical University, Xi'an, China

Clara Di Vito and Elisa Zaghi
Unit of Clinical and Experimental Immunology, Humanitas Clinical and Research Center, Rozzano, Italy

Joanna Mikulak
Unit of Clinical and Experimental Immunology, Humanitas Clinical and Research Center, Rozzano, Italy
Istituto di Ricerca Genetica e Biomedica, UOS di Milano, Consiglio Nazionale delle Ricerche (UOS/IRGB/CNR), Rozzano, Italy

Domenico Mavilio
Unit of Clinical and Experimental Immunology, Humanitas Clinical and Research Center, Rozzano, Italy
Department of Medical Biotechnologies and Translational Medicine (BioMeTra), University of Milan, Milan, Italy

Silvia Pesce
Dipartimento di Medicina Sperimentale, Università degli Studi di Genova, Genova, Italy

Lorenzo Moretta
Dipartimento di Immunologia, IRCCS Bambino Gesù Ospedale Pediatrico, Rome, Italy

Alessandro Moretta and Emanuela Marcenaro
Dipartimento di Medicina Sperimentale, Università degli Studi di Genova, Genova, Italy
CEBR, Università degli Studi di Genova, Genova, Italy

Dara M. Strauss-Albee
Stanford Immunology, Stanford University School of Medicine, Stanford, CA, USA

Catherine A. Blish
Stanford Immunology, Stanford University School of Medicine, Stanford, CA, USA
Department of Medicine, Stanford University School of Medicine, Stanford, CA, USA

Paolo Carrega
Laboratory of Immunology and Biotherapy, Department of Human Pathology, University of Messina, Messina, Italy
Cell Factory Center, University of Messina, Messina, Italy

Guido Ferlazzo
Laboratory of Immunology and Biotherapy, Department of Human Pathology, University of Messina, Messina, Italy
Cell Factory Center, University of Messina, Messina, Italy
Cell Therapy Program, University Hospital Policlinico G.Martino, Messina, Italy
Division of Clinical Pathology, University Hospital Policlinico G.Martino, Messina, Italy

Emily M. Mace and Jordan S. Orange
Center for Human Immunobiology, Baylor College of Medicine and Texas Children's Hospital, Houston, TX, USA

Steven D. Scoville
Biomedical Sciences Graduate Program, Medical Scientist Training Program, The James Cancer Hospital and Solove Research Institute, The Ohio State University, Columbus, OH, USA
Comprehensive Cancer Center, The James Cancer Hospital and Solove Research Institute, The Ohio State University, Columbus, OH, USA

Aharon G. Freud
Comprehensive Cancer Center, The James Cancer Hospital and Solove Research Institute, The Ohio State University, Columbus, OH, USA
Department of Pathology, The James Cancer Hospital and Solove Research Institute, The Ohio State University, Columbus, OH, USA

Michael A. Caligiuri
Comprehensive Cancer Center, The James Cancer Hospital and Solove Research Institute, The Ohio State University, Columbus, OH, USA
Division of Hematology and Oncology, Department of Internal Medicine, The James Cancer Hospital and Solove Research Institute, The Ohio State University, Columbus, OH, USA

Aura Muntasell
Hospital del Mar Medical Research Institute (IMIM), Barcelona, Spain

Miguel López-Botet
Hospital del Mar Medical Research Institute (IMIM), Barcelona, Spain
Department of Immunology, Hospital del Mar, Barcelona, Spain
Univ. Pompeu Fabra, Barcelona, Spain

José Yélamos
Hospital del Mar Medical Research Institute (IMIM), Barcelona, Spain
Department of Immunology, Hospital del Mar, Barcelona, Spain

Aldi Pupuleku
Univ. Pompeu Fabra, Barcelona, Spain

Carlos Vilches
Immunogenetics-Histocompatibility, Instituto de Investigación Sanitaria Puerta de Hierro, Majadahonda, Spain

Julio Pascual, Marta Crespo1 and Dolores Redondo-Pachón
Hospital del Mar Medical Research Institute (IMIM), Barcelona, Spain
Department of Nephrology, Hospital del Mar, Barcelona, Spain

Christian Chabannon, Bechara Mfarrej and Boris Calmels
CBT-1409: INSERM, Aix Marseille Univ, Institut Paoli-Calmettes, AP-HM, Marseille, France
CRCM: INSERM, CNRS, Aix Marseille Univ, Institut Paoli-Calmettes, CRCM, Marseille, France

Raynier Devillier and Didier Blaise
CRCM: INSERM, CNRS, Aix Marseille Univ, Institut Paoli-Calmettes, CRCM, Marseille, France

Sophie Guia and Sophie Ugolini
UM2, INSERM, Centre d'Immunologie de Marseille-Luminy, U1104, CNRS UMR7280, Aix-Marseille University, Marseille, France

Eric Vivier
UM2, INSERM, Centre d'Immunologie de Marseille-Luminy, U1104, CNRS UMR7280, Aix-Marseille University, Marseille, France
Laboratoire d'Immunologie, Hôpital de la Conception, Assistance Publique – Hôpitaux de Marseille, Aix-Marseille University, Marseille, France

Paola Vacca and Daniele Croxatto
Department of Experimental Medicine, University of Genova, Genova, Italy

Elisa Montaldo
Division of Regenerative Medicine, Stem Cells and Gene Therapy, San Raffaele Telethon Institute for Gene Therapy, IRCCS San Raffaele Scientific Institute, Milano, Italy

Francesca Moretta
Department of Internal Medicine, University of Verona, Verona, Italy
Ospedale Sacro Cuore Negrar, Verona, Italy

Alice Bertaina
Department of Pediatric Hematology and Oncology, IRCCS Bambino Gesù Children's Hospital, Rome, Italy

Chiara Vitale
Department of Experimental Medicine, University of Genova, Genova, Italy
U.O. Immunology IRCCS AOU San Martino-IST, Genova, Italy

Franco Locatelli
Department of Pediatric Hematology and Oncology, IRCCS Bambino Gesù Children's Hospital, Rome, Italy
Department of Pediatrics, University of Pavia, Pavia, Italy

Maria Cristina Mingari
Department of Experimental Medicine, University of Genova, Genova, Italy
U.O. Immunology IRCCS AOU San Martino-IST, Genova, Italy
Centro di Eccellenza per la Ricerca Biomedica - CEBR, Genova, Italy

Lorenzo Moretta
Department of Immunology, IRCCS Bambino Gesù Children's Hospital, Rome, Italy

Federico Simonetta, Amandine Pradier and Eddy Roosnek
Department of Medical Specialties, Division of Hematology, Geneva University Hospitals, University of Geneva, Geneva, Switzerland

Nelson Lopez-Sejas, Carmen Campos, Fakhri Hassouneh, Alejandra Peral and Rafael Solana
Maimonides Biomedicine Institute of Cordoba (IMIBIC), Reina Sofia Hospital, University of Cordoba, Cordoba, Spain

Beatriz Sanchez-Correa and Raquel Tarazona
Immunology Unit, Department of Physiology, University of Extremadura, Cáceres, Spain

Tristan Legris, Stanislas Bataille, Henri Vacher-Coponat, Valérie Moal and Yvon Berland
Nephrology Dialysis Renal Transplantation Center, Assistance Publique des Hôpitaux de Marseille, Hospital de la Conception, Marseille, France

Christophe Picard
Établissement Français du Sang Alpes Méditerranée, Marseille, France
ADES UMR 7268, CNRS, EFS, Aix-Marseille Université, Marseille, France

Dilyana Todorova and Stéphane Burtey
Nephrology Dialysis Renal Transplantation Center, Assistance Publique des Hôpitaux de Marseille, Hospital de la Conception, Marseille, France
UMR 1076, Vascular Research Center of Marseille, INSERM, Aix-Marseille University, Marseille, France

Luc Lyonnet, Cathy Laporte, Chloé Dumoulin, Corinne Nicolino-Brunet
Hematology Unit, Assistance Publique des Hôpitaux de Marseille, Hopital de la Conception, Marseille, France

Francoise Dignat-George and Pascale Paul
UMR 1076, Vascular Research Center of Marseille, INSERM, Aix-Marseille University, Marseille, France
Hematology Unit, Assistance Publique des Hôpitaux de Marseille, Hopital de la Conception, Marseille, France

Laurent Daniel
Laboratory for Anatomy, Pathology, Neuropathology, Hôpital de la Timone, Aix-Marseille University, Marseille, France

Anderson Loundou
Unité d'Aide méthodologique à la Recherche Clinique et Epidémiologique, DRRC, Assistance Publique Hôpitaux de Marseille, Marseille, France

Sophie Morange
Centre d'Investigation Clinique, Hôpital de la Conception, Marseille, France

Martin R. Goodier, Chiara Lusa, Sam Sherratt and Ana Rodriguez-Galan
Department of Immunology and Infection, London School of Hygiene and Tropical Medicine, London, UK

Ron Behrens and Eleanor M. Riley
Department of Clinical Research, London School of Hygiene and Tropical Medicine, London, UK

Obinna Chijioke
Institute of Surgical Pathology, University Hospital Zürich, Zürich, Switzerland
Viral Immunobiology, Institute of Experimental Immunology, University of Zürich, Zürich, Switzerland

Vanessa Landtwing and Christian Münz
Viral Immunobiology, Institute of Experimental Immunology, University of Zürich, Zürich, Switzerland

Catharina C. Gross, Andreas Schulte-Mecklenbeck and Heinz Wiendl
Department of Neurology, University Hospital Münster, Münster, Germany

Emanuela Marcenaro
Centre of Excellence for Biomedical Research, University of Genova, Genova, Italy
Department of Experimental Medicine, University of Genova, Genova, Italy

Nicole Kerlero de Rosbos
Department of Neuroscience, Rehabilitation, Ophthalmology, Genetics, Maternal and Child Health, University of Genova, Genova, Italy

Antonio Uccelli and Alice Laroni
Centre of Excellence for Biomedical Research, University of Genova, Genova, Italy
Department of Neuroscience, Rehabilitation, Ophthalmology, Genetics, Maternal and Child Health, University of Genova, Genova, Italy
IRCCS San Martino-IST, Genova, Italy

Index

www.ingramcontent.com/pod-product-compliance
Lightning Source LLC
LaVergne TN
LVHW080949210926
848485LV00004B/810
* 9 7 8 1 6 3 9 8 7 3 8 2 1 *